STATISTICAL MECHANICS

LUDWIG BOLTZMANN, 1844–1906, *whose H theorem opened the door to an understanding of the macroscopic world on the basis of molecular dynamics.*

STATISTICAL MECHANICS

Kerson Huang

Associate Professor of Physics
Massachusetts Institute of Technology

John Wiley & Sons, Inc., New York · London · Sydney

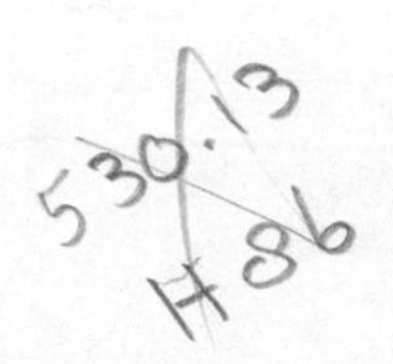

FOURTH PRINTING, JUNE, 1967

Library of Congress Catalog Card Number: 63-11437

Printed in the United States of America

TO CHEN NING YANG

PREFACE

This book is an outgrowth of a year course in statistical mechanics that I have been giving at the Massachusetts Institute of Technology. It is directed mainly to graduate students in physics.

The purpose of the book is to teach statistical mechanics as an integral part of theoretical physics, a discipline that aims to describe all natural phenomena on the basis of a single unifying theory. This theory, at present, is quantum mechanics.

This does not mean that the sole concern of this book is the derivation of statistical mechanics from quantum mechanics, because such a preoccupation would not serve the purpose of teaching. Furthermore, such a derivation does not at present exist.

In this book the starting point of statistical mechanics is taken to be certain phenomenological postulates, whose relation to quantum mechanics I try to state as clearly as I can, and whose physical consequences I try to derive as simply and directly as I can.

Before the subject of statistical mechanics proper is presented, a brief but self-contained discussion of thermodynamics and the classical kinetic theory of gases is given. The order of this development is imperative, from a pedagogical point of view, for two reasons. First, thermodynamics has successfully described a large part of macroscopic experience, which is the concern of statistical mechanics. It has done so not on the basis of molecular dynamics but on the basis of a few simple and intuitive postulates stated in everyday terms. If we first familiarize ourselves with thermodynamics, the task of statistical mechanics reduces to the explanation of thermodynamics. Second, the classical kinetic theory of gases is the only known special case in which thermodynamics can be derived nearly from first principles, i.e., molecular dynamics. A study of this special case will help us understand why statistical mechanics works.

A large part of this book is devoted to selected applications of statistical mechanics. The selection is guided by the interest of the topic to physicists, its value as an illustration of calculating techniques, and my personal taste.

To read the first half of the book the reader needs a good knowledge of classical mechanics and some intuitive feeling for thermodynamics and kinetic theory. To read the second half of the book he needs to have a working knowledge of quantum mechanics. The mathematical knowledge required of the reader does not exceed what he should have acquired in his study of classical mechanics and quantum mechanics.

Certain passages in the book set in reduced type may be omitted on first reading. At the end of most chapters a set of problems is included. They are designed to illustrate or to extend the discussion given in the text. The serious reader should consider them to be an integral part of the book.

The material in this book probably cannot be completely covered in a year's study. It might be helpful, therefore, to give a list of chapters that form the "hard core" of the book. They are the following: Chapters 3, 4, 7, 8 (possibly excluding Sections 8.5, 8.6, and 8.7), 9, 11, and 12.

KERSON HUANG

Cambridge, Massachusetts
February 1963

ACKNOWLEDGMENT

I am indebted to my wife for preparing the index, and to Professor M. J. Klein of the Case Institute of Technology for several useful comments on the manuscript of this book.

CONTENTS

D APPENDICES

A

THERMODYNAMICS AND KINETIC THEORY

chapter 1

THE LAWS OF THERMODYNAMICS

1.1 PRELIMINARIES

Thermodynamics is a phenomenological theory of matter. As such, it draws its concepts directly from experiments. The following is a list of some working concepts which the physicist, through experience, has found it convenient to introduce. We shall be extremely brief, as the reader is assumed to be familiar with these concepts.

(*a*) A *thermodynamic system* is any macroscopic system.

(*b*) *Thermodynamic parameters* are measurable macroscopic quantitics associated with the system, such as the pressure P, the volume V, the temperature T, and the magnetic field B. They are defined experimentally.

(*c*) A *thermodynamic state* is specified by a set of values of all the thermodynamic parameters necessary for the description of the system.

(*d*) *Thermodynamic equilibrium* prevails when the thermodynamic state of the system does not change with time.

(*e*) The *equation of state* is a functional relationship among the thermodynamic parameters for a system in equilibrium. If P, V, and T are the thermodynamic parameters of the system, the equation of state takes the form

$$f(P, V, T) = 0$$

which reduces the number of independent variables of the system from three to two. The function f is assumed to be given as part of the specification

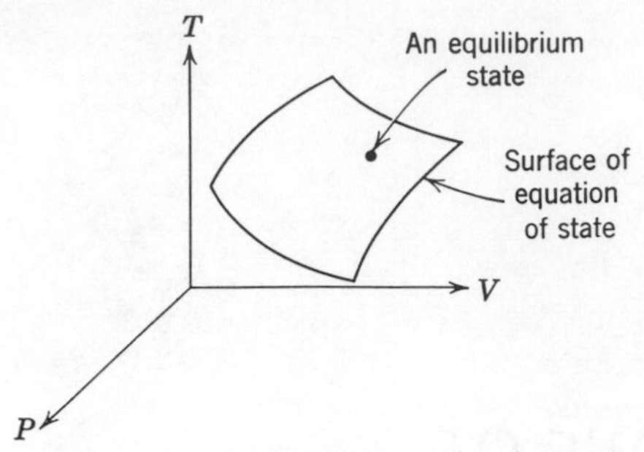

Fig. 1.1. Geometrical representation of the equation of state.

of the system. It is customary to represent the state of such a system by a point in the three-dimensional P-V-T space. The equation of state then defines a surface in this space, as shown in Fig. 1.1. Any point lying on this surface represents a state in equilibrium. In thermodynamics a state automatically means a state in equilibrium unless otherwise specified.

(f) A *thermodynamic transformation* is a change of state. If the initial state is an equilibrium state, the transformation can be brought about only by changes in the external condition of the system. The transformation is *quasi-static* if the external condition changes so slowly that at any moment the system is approximately in equilibrium. It is *reversible* if the transformation retraces its history in time when the external condition retraces its history in time. A reversible transformation is quasi-static, but the converse is not necessarily true. For example, a gas that freely expands into successive infinitesimal volume elements undergoes a quasi-static transformation but not a reversible one.

(g) The *P-V diagram* of a system is the projection of the surface of the equation of state onto the P-V plane. Every point on the P-V diagram therefore represents an equilibrium state. A reversible transformation is a continuous path on the P-V diagram. Reversible transformations of specific types give rise to paths with specific names, such as *isotherms*, *adiabatics*, etc. A transformation that is not reversible cannot be so represented.

(h) The concept of *work* is taken over from mechanics. For example, for a system whose parameters are P, V, and T, the work dW done by a system in an infinitesimal transformation in which the volume increases by dV is given by

$$dW = P\,dV$$

Generalization to other cases is obvious.

(i) *Heat* is what is absorbed by a homogeneous system if its temperature increases while no work is done. If ΔQ is a small amount of the heat absorbed, and ΔT is the small change in temperature accompanying the absorption of heat, the *heat capacity* C is defined by

$$\Delta Q = C\,\Delta T$$

The heat capacity depends on the detailed nature of the system and is given as a part of the specification of the system. It is an experimental fact that, for the same ΔT, ΔQ is different for different ways of heating up the system. Correspondingly, the heat capacity depends on the manner of heating. Commonly considered heat capacities are C_V and C_P, which respectively correspond to heating at constant V and P. Heat capacities per unit mass or per mole of a substance are called *specific heats*.

(*j*) A *heat reservoir*, or simply *reservoir*, is a system so large that the gain or loss of any finite amount of heat does not change its temperature.

(*k*) A system is *thermally isolated* if no heat exchange can take place between it and the external world. Thermal isolation may be achieved by surrounding a system with an *adiabatic wall*. Any transformation the system can undergo in thermal isolation is said to take place adiabatically.

(*l*) A thermodynamic quantity is said to be *extensive* if it is proportional to the amount of substance in the system under consideration and is said to be *intensive* if it is independent of the amount of substance in the system under consideration. It is an important empirical fact that to a good approximation thermodynamic quantities are either extensive or intensive.

(*m*) The *ideal gas* is an important idealized thermodynamic system. Experimentally all gases behave in a universal way when they are sufficiently dilute. The ideal gas is an idealization of this limiting behavior. The parameters for an ideal gas are pressure P, volume V, temperature T, and number of molecules N. The equation of state is given by Boyle's law:

$$\frac{PV}{N} = \text{constant} \qquad \text{(for constant temperature)}$$

The value of this constant depends on the experimental scale of temperature used.

(*n*) The equation of state of an ideal gas in fact defines a temperature scale, the *ideal-gas temperature* T:

$$PV = NkT$$

where

$$k = 1.38 \times 10^{-16} \text{ erg/deg}$$

which is called Boltzmann's constant. Its value is determined by the conventional choice of temperature intervals, namely, the Centigrade degree. This scale has a universal character because the ideal gas has a universal character. The origin $T = 0$ is here arbitrarily chosen. Later we see that it actually has an absolute meaning according to the second law of thermodynamics.

To construct the ideal-gas temperature scale we may proceed as follows. Measure PV/Nk of an ideal gas at the temperature at which water boils and at which water freezes. Plot these two points and draw a straight line through them, as shown in Fig. 1.2. The intercept of this line with the abscissa is chosen to be the origin of the scale. The intervals of the temperature scale are so chosen that there are 100 equal divisions between the boiling and the freezing points of water. The resulting scale is the Kelvin scale (°K). To use the scale, bring anything whose temperature is to be measured into thermal contact with an ideal gas (e.g., helium gas at sufficiently low density), measure PV/Nk of the ideal gas, and read off the temperature from Fig. 1.2. An equivalent form of the equation of state of an ideal gas is

$$PV = nRT$$

where n is the number of moles of the gas and R is the gas constant:

$$\begin{aligned} R &= 8.315 \text{ joule/deg} \\ &= 1.986 \text{ cal/deg} \\ &= 0.0821 \text{ liter-atm/deg} \end{aligned}$$

Its value follows from the value of Boltzmann's constant and Avogadro's number:

$$\text{Avogadro's number} = 6.205 \times 10^{23} \text{ atoms/mole}$$

Most of these concepts are properly understood only in molecular terms. Here we have to be satisfied with empirical definitions.

In the following we introduce thermodynamic laws, which may be regarded as mathematical axioms defining a mathematical model. It is possible to deduce rigorous consequences of these axioms, but it is

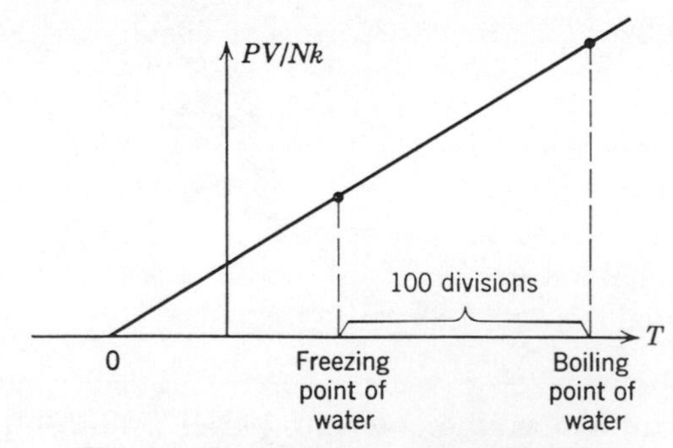

Fig. 1.2. The ideal-gas temperature scale.

important to remember that this model may not rigorously correspond to the physical world; the thermodynamic laws may not be rigorous consequences of the molecular laws, which we take to be the fundamental laws of the physical world. The thermodynamic laws, then, are introduced only as phenomenological statements that conveniently summarize macroscopic experience. As such, they must be at least approximately true for the physical world. The relation between thermodynamic laws and molecular laws is discussed later, in kinetic theory.

Thermodynamics, as a mathematical model, can be axiomatized, in the best tradition of mathematics. Nevertheless, in view of the foregoing discussion, such a formulation contributes little to the understanding of physics.

1.2 THE FIRST LAW OF THERMODYNAMICS

In an *arbitrary* thermodynamic transformation let ΔQ denote the net amount of heat absorbed by the system and ΔW the net amount of work done by the system. The first law of thermodynamics states that the quantity ΔU, defined by

$$\Delta U = \Delta Q - \Delta W \tag{1.1}$$

is the same for all transformations leading from a given initial state to a given final state.

This immediately defines a state function U, called the internal energy. Its value for any state may be found as follows. Choose an arbitrary fixed state as reference. Then the internal energy of any state is $\Delta Q - \Delta W$ in *any* transformation which leads from the reference state to the state in question. It is defined only up to an arbitrary additive constant. Empirically U is an extensive quantity. This follows from the saturation property of molecular forces, namely, that the energy of a substance is doubled if its mass is doubled.

The experimental foundation of the first law is Joule's demonstration of the equivalence between heat and mechanical energy—the feasibility of converting mechanical work completely into heat. The inclusion of heat as a form of energy leads naturally to the inclusion of heat in the statement of the conservation of energy. The first law is precisely such a statement.

In an infinitesimal transformation, the first law reduces to the statement that the differential

$$dU = dQ - dW \tag{1.2}$$

is exact. That is, there exists a function U whose differential is dU; or, the integral $\int dU$ is independent of the path of the integration and depends only on the limits of integration. This property is obviously not shared by dQ or dW.

Given a differential of the form $df = g(A, B)\, dA + h(A, B)\, dB$, the condition that df be exact is $\partial g/\partial B = \partial h/\partial A$. Let us explore some of the consequences of the exactness of dU. Consider a system whose parameters are P, V, T. Any pair of these three parameters may be chosen to be the independent variables that completely specify the state of the system. The other parameter is then determined by the equation of state. We may, for example, consider $U = U(P, V)$. Then*

$$dU = \left(\frac{\partial U}{\partial P}\right)_V dP + \left(\frac{\partial U}{\partial V}\right)_P dV \tag{1.3}$$

The requirement that dU be exact immediately leads to the result

$$\frac{\partial}{\partial V}\left[\left(\frac{\partial U}{\partial P}\right)_V\right]_P = \frac{\partial}{\partial P}\left[\left(\frac{\partial U}{\partial V}\right)_P\right]_V \tag{1.4}$$

The following equations, expressing the heat absorbed by a system during an infinitesimal reversible transformation (in which $dW = P\, dV$), are easily obtained by successively choosing as independent variables the pairs (P, V), (P, T), and (V, T):

$$dQ = \left(\frac{\partial U}{\partial P}\right)_V dP + \left[\left(\frac{\partial U}{\partial V}\right)_P + P\right] dV \tag{1.5}$$

$$dQ = \left[\left(\frac{\partial U}{\partial T}\right)_P + P\left(\frac{\partial V}{\partial T}\right)_P\right] dT + \left[\left(\frac{\partial U}{\partial P}\right)_T + P\left(\frac{\partial V}{\partial P}\right)_T\right] dP \tag{1.6}$$

$$dQ = \left(\frac{\partial U}{\partial T}\right)_V dT + \left[\left(\frac{\partial U}{\partial V}\right)_T + P\right] dV \tag{1.7}$$

Called dQ equations, these are of little practical use in their present form, because the partial derivatives that appear are usually unknown and inaccessible to direct measurement. They will be transformed to more useful forms when we come to the second law of thermodynamics.

It can be immediately deduced from the dQ equations that

$$C_V \equiv \left(\frac{\Delta Q}{\Delta T}\right)_V = \left(\frac{\partial U}{\partial T}\right)_V \tag{1.8}$$

$$C_P \equiv \left(\frac{\Delta Q}{\Delta T}\right)_P = \left(\frac{\partial H}{\partial T}\right)_P \tag{1.9}$$

where $H = U + PV$ is called the enthalpy of the system.

* The symbol $(\partial U/\partial P)_V$ denotes the partial derivative of U with respect to P, with V held constant.

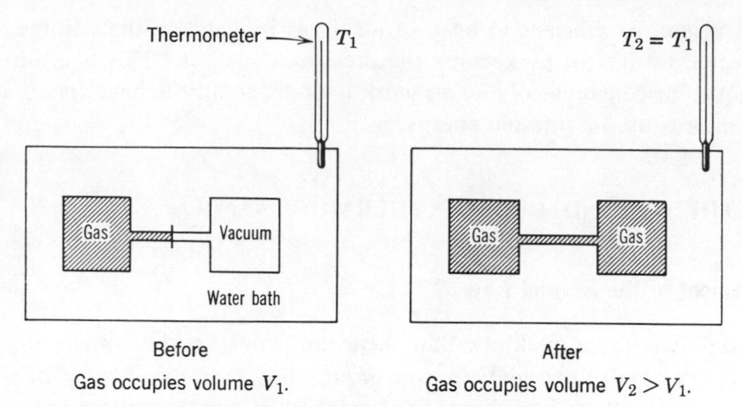

Fig. 1.3. Joule's free-expansion experiment.

We consider the following examples of the application of the first law.

(*a*) *Analysis of Joule's Free-Expansion Experiment.* The experiment in question concerns the free expansion of an ideal gas into a vacuum. The initial and final situations are illustrated in Fig. 1.3.

Experimental Finding. $T_1 = T_2$.

Deductions. $\Delta W = 0$, since the gas performs no work on its external surrounding. $\Delta Q = 0$, since $\Delta T = 0$. Therefore, $\Delta U = 0$ by the first law. Thus two states with the same temperature but different volumes have the same internal energy. Since temperature and volume may be taken to be the independent parameters, and since U is a state function, we conclude that for an ideal gas U is a function of the temperature alone. This conclusion can also be reached theoretically, without reference to a specific experiment, with the help of the second law of thermodynamics.

(*b*) *Internal Energy of Ideal Gas.* Since U depends only on T, (1.8) yields

$$C_V = \left(\frac{\partial U}{\partial T}\right)_V = \frac{dU}{dT}$$

Assuming C_V to be independent of the temperature, we obtain

$$U = C_V T + \text{constant}$$

The additive constant may be arbitrarily set equal to zero.

(*c*) *The Quantity $C_P - C_V$ for an Ideal Gas.* The enthalpy of an ideal gas is a function of T only:

$$H = U + PV = (C_V + Nk)T$$

Hence, from (1.9),

$$C_P = \left(\frac{\partial H}{\partial T}\right)_P = \frac{dH}{dT} = C_V + Nk$$

or

$$C_P - C_V = Nk$$

Thus it is more efficient to heat an ideal gas by keeping the volume constant than to heat it by keeping the pressure constant. This is intuitively obvious; at constant volume no work is done, so all the heat energy goes into increasing the internal energy.

1.3 THE SECOND LAW OF THERMODYNAMICS

Statement of the Second Law

From experience we know that there are processes that satisfy the law of conservation of energy yet never occur. For example, a piece of stone resting on the floor is never seen to cool itself spontaneously and jump up to the ceiling, thereby converting the heat energy given off into potential energy. The purpose of the second law of thermodynamics is to incorporate such experimental facts into thermodynamics. Its experimental foundation is common sense, as the following equivalent statements of the second law will testify.

Kelvin Statement. There exists no thermodynamic transformation whose *sole* effect is to extract a quantity of heat from a given heat reservoir and to convert it entirely into work.

Clausius Statement. There exists no thermodynamic transformation whose *sole* effect is to extract a quantity of heat from a colder reservoir and to deliver it to a hotter reservoir.

In both statements the key word is "sole." An example suffices to illustrate the point. If an ideal gas is expanded reversibly and isothermally, work is done by the gas. Since $\Delta U = 0$ in this process, the work done is equal to the heat absorbed by the gas during the expansion. Hence a certain quantity of heat is converted entirely into work. This is not the sole effect of the transformation, however, because the gas occupies a larger volume in the final state. This process is allowed by the second law.

The Kelvin statement K and the Clausius statement C are equivalent. To prove this we prove that if the Kelvin statement is false, the Clausius statement is false, and vice versa.

Proof that K False $\Rightarrow$ C False. Suppose K is false. Then we can extract heat from a reservoir at temperature T_1 and convert it entirely into work, with no other effect. Now we can convert this work into heat and deliver it to a reservoir at temperature $T_2 > T_1$ with no other effect. (A practical way of carrying out this particular step is illustrated by Joule's experiment on the equivalence of heat and energy.) The net result of this two-step process is the transfer of an amount of heat from a colder reservoir to a hotter one with no other effect. Hence C is false.

Proof that C False ⇒ K False. First define an *engine* to be a thermodynamic system that can undergo a cyclic transformation (i.e., a transformation whose final state is identical with the initial state), in which the system does the following things, and only the following things:

(*a*) absorbs an amount of heat $Q_2 > 0$ from reservoir T_2;
(*b*) rejects an amount of heat $Q_1 > 0$ to reservoir T_1, with $T_1 < T_2$;
(*c*) performs an amount of work $W > 0$.

Suppose C is false. Extract Q_2 from reservoir T_1 and deliver it to reservoir T_2, with $T_2 > T_1$. Operate an engine between T_2 and T_1 for one cycle, and arrange the engine so that the amount of heat extracted by the engine from T_2 is exactly Q_2. The net result is that an amount of heat is extracted from T_1 and entirely converted into work, with no other effect. Hence K is false. This completes the proof of equivalence.

The Carnot Engine

An engine that does all the things required by the definition in a reversible way is called a *Carnot engine*. A Carnot engine consists of any substance that is made to go through the reversible cyclic transformation illustrated in the P-V diagram of Fig. 1.4, where ab is isothermal at temperature T_2, during which the system absorbs heat Q_2; bc is adiabatic;

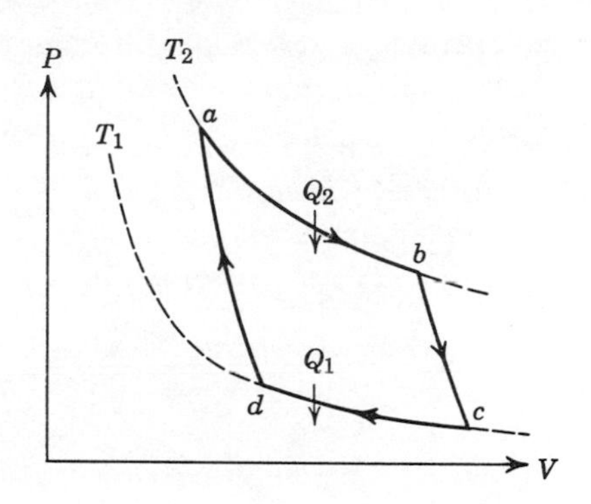

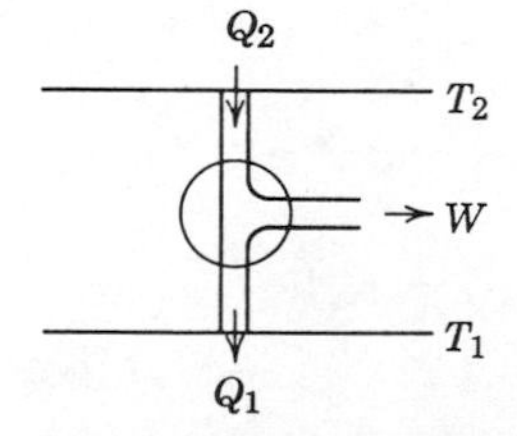

Fig. 1.4. The Carnot engine.

cd is isothermal at temperature T_1, with $T_1 < T_2$, during which time the system rejects heat Q_1; and da is adiabatic. It may also be represented schematically as in the lower part of Fig. 1.4. The work done by the system in one cycle is, according to the first law,

$$W = Q_2 - Q_1$$

since $\Delta U = 0$ in any cyclic transformation. The efficiency of the engine is defined to be

$$\eta = \frac{W}{Q_2} = 1 - \frac{Q_1}{Q_2}$$

We show that if $W > 0$, then $Q_1 > 0$ and $Q_2 > 0$. The proof is as follows. It is obvious that $Q_1 \neq 0$, for otherwise we have an immediate violation of the Kelvin statement. Suppose $Q_1 < 0$. This means that the engine absorbs the amount of heat Q_2 from T_2 and the amount of heat $-Q_1$ from T_1 and converts the net amount of heat $Q_2 - Q_1$ into work. Now we may convert this amount of work, which by assumption is positive, into heat and deliver it to the reservoir at T_2, with no other effect. The net result is the transfer of the positive amount of heat $-Q_1$ from T_1 to T_2 with no other effect. Since $T_2 > T_1$ by assumption, this is impossible by the Clausius statement. Therefore $Q_1 > 0$. From $W = Q_2 - Q_1$ and $W > 0$ it follows immediately that $Q_2 > 0$. (QED)

In the same way we can show that if $W < 0$ and $Q_1 < 0$, then $Q_2 < 0$. In this case the engine operates in reverse and becomes a "refrigerator."

The importance of the Carnot engine lies in the following theorem.

CARNOT'S THEOREM. No engine operating between two given temperatures is more efficient than a Carnot engine.

Proof. Operate a Carnot engine C and an arbitrary engine X between the reservoirs T_2 and T_1 ($T_2 > T_1$), as shown in Fig. 1.5. We have, by the first law,

$$W = Q_2 - Q_1$$
$$W' = Q_2' - Q_1'$$

Let

$$\frac{Q_2}{Q_2'} = \frac{N'}{N}$$

where N' and N are two integers. This equality can be satisfied to any degree of accuracy by making N', N sufficiently large. Now operate the C engine N cycles in reverse, and the X engine N' cycles. At the end of this operation we have

$$W_{\text{total}} = N'W' - NW$$
$$(Q_2)_{\text{total}} = N'Q_2' - NQ_2 = 0$$
$$(Q_1)_{\text{total}} = N'Q_1' - NQ_1$$

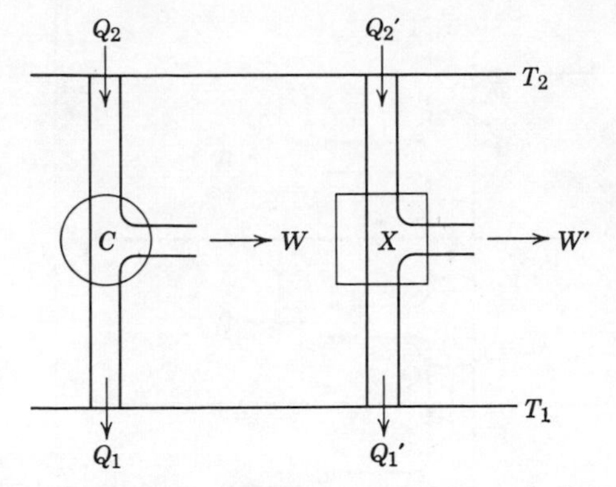

Fig. 1.5. Construction for the proof of Carnot's theorem.

On the other hand we can also write

$$W_{\text{total}} = (Q_2)_{\text{total}} - (Q_1)_{\text{total}} = -(Q_1)_{\text{total}}$$

The net result of the operation is thus a violation of the Kelvin statement, unless

$$W_{\text{total}} \leq 0$$

which implies that

$$(Q_1)_{\text{total}} \geq 0$$

In other words we must have

$$N'Q_1' - NQ_1 \geq 0$$

$$Q_2Q_1' - Q_2'Q_1 \geq 0$$

$$\frac{Q_1}{Q_2} \leq \frac{Q_1'}{Q_2'}$$

Therefore

$$\left(1 - \frac{Q_1}{Q_2}\right) \geq \left(1 - \frac{Q_1'}{Q_2'}\right) \qquad \text{(QED)}$$

Since X is arbitrary, it can be a Carnot engine. Thus we have the trivial **corollary**: All Carnot engines operating between two given temperatures have the same efficiency.

Absolute Scale of Temperature

The corollary to Carnot's theorem furnishes a definition of a scale of temperature, the absolute scale. It is defined as follows. If the efficiency of a Carnot engine operating between two reservoirs of respective absolute temperatures θ_1 and θ_2 ($\theta_2 > \theta_1$) is η, then

$$\frac{\theta_1}{\theta_2} = 1 - \eta$$

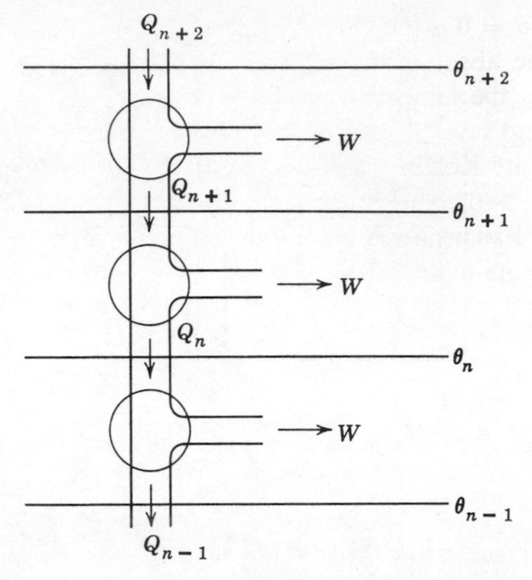

Fig. 1.6. A series of Carnot engines used to define the absolute scale of temperature.

Since $0 \leq \eta \leq 1$, the absolute temperature of any reservoir is always greater than zero. To obtain a uniform temperature scale arrange a series of ordered Carnot engines, all performing the same amount of work W, so that the heat rejected by any Carnot engine is absorbed by the next one, as shown in Fig. 1.6. Obviously, for all n, we have

$$Q_{n+1} - Q_n = W$$

$$\frac{Q_n}{Q_{n+1}} = \frac{\theta_n}{\theta_{n+1}}$$

The last equation can be rewritten in the form

$$\frac{\theta_n}{Q_n} = \frac{\theta_{n+1}}{Q_{n+1}} = x$$

Hence $x \equiv \theta_n/Q_n$ is independent of n. It is easily seen that

$$\theta_{n+1} - \theta_n = xW$$

which is independent of n. Hence a temperature scale of equal intervals results. Choosing $xW = 1°\text{K}$ results in the absolute Kelvin scale of temperature. It is to be noted that

(*a*) The definition of the absolute scale of temperature is independent of the specific properties of any substance. It depends only on a property that is common to all substances, the second law of thermodynamics.

(*b*) The limit $\theta = 0$ is the greatest lower bound of the temperature scale and is called the absolute zero. Actually no Carnot engine exists with absolute zero as the temperature of the lower reservoir, for that would violate the second law. The absolute zero exists only in a limiting sense.

(*c*) The absolute Kelvin scale θ is identical with the ideal-gas temperature scale T, if $T > 0$. This is easily proved by using an ideal gas to form a Carnot engine. From now on we do not distinguish between the two and denote the absolute temperature by T.

1.4 ENTROPY

The second law of thermodynamics enables us to define a state function S, the entropy, which we find useful. We owe this possibility to the following theorem.

CLAUSIUS' THEOREM. In *any* cyclic transformation throughout which the temperature is defined, the following inequality holds:

$$\oint \frac{dQ}{T} \leq 0$$

where the integral extends over one cycle of the transformation. The equality holds if the cyclic transformation is reversible.

Proof. Let the cyclic transformation in question be denoted by $\mathcal{O}$. Divide the cycle into n infinitesimal steps for which the temperature may be considered to be constant in each step. The system is imagined to be brought successively into contact with heat reservoirs at temperatures $T_1, T_2, \ldots, T_n$. Let Q_i be the amount of heat absorbed by the system during the ith step from the heat reservoir of temperature T_i. We shall prove that

$$\sum_{i=1}^{n} \left(\frac{Q_i}{T_i}\right) \leq 0$$

The theorem is obtained as we let $n \to \infty$. Construct a set of n Carnot engines $\{C_1, C_2, \ldots, C_n\}$ such that C_i

(*a*) operates between T_i and T_o ($T_o \geq T_i$, all i),

(*b*) absorbs amount of heat $Q_i^{(o)}$ from T_o,

(*c*) rejects amount of heat Q_i to T_i.

We have, by definition of the temperature scale.

$$\frac{Q_i^{(o)}}{Q_i} = \frac{T_o}{T_i}$$

Consider one cycle of the combined operation $\mathcal{O} + \{C_1 + \cdots + C_n\}$. The net result of this cycle is that an amount of heat

$$Q_o = \sum_{i=1}^{n} Q_i^{(o)} = T_o \sum_{i=1}^{n} \left(\frac{Q_i}{T_i}\right)$$

is absorbed from the reservoir T_o and converted entirely into work, with no other effect. According to the second law this is impossible unless $Q_o \leq 0$. Therefore

$$\sum_{i=1}^{n} \left(\frac{Q_i}{T_i}\right) \leq 0$$

This proves the first part of the theorem.

If $\mathcal{O}$ is reversible, we reverse it. Going through the same arguments, we arrive at the same inequality except that the signs of Q_i are reversed:

$$-\sum_{i=1}^{n} \left(\frac{Q_i}{T_i}\right) \leq 0$$

Combining this with the previous inequality (which of course still holds for a reversible $\mathcal{O}$) we obtain

$$\sum_{i=1}^{n} \left(\frac{Q_i}{T_i}\right) = 0$$

This completes the proof.

Corollary. For a reversible transformation, the integral

$$\int \frac{dQ}{T}$$

is independent of the path and depends only on the initial and final states of the transformation.

Proof. Let the initial state be A and the final state be B. Let I, II denote two arbitrary reversible paths joining A to B, and let II′ be the reverse of II. Clausius' theorem implies that

$$\int_{\mathrm{I}} \frac{dQ}{T} + \int_{\mathrm{II}'} \frac{dQ}{T} = 0$$

But

$$\int_{\mathrm{II}'} \frac{dQ}{T} = -\int_{\mathrm{II}} \frac{dQ}{T}$$

Hence,

$$\int_{\mathrm{I}} \frac{dQ}{T} = \int_{\mathrm{II}} \frac{dQ}{T} \qquad \text{(QED)}$$

This corollary enables us to define a state function, the entropy S. It is defined as follows. Choose an arbitrary fixed state O as reference state. The entropy $S(A)$ of any state A is defined by

$$S(A) \equiv \int_O^A \frac{dQ}{T}$$

where the path of integration is any reversible path joining O to A. Thus the entropy is defined only up to an arbitrary additive constant.* The difference in the entropy of two states, however, is completely defined:

$$S(A) - S(B) = \int_B^A \frac{dQ}{T}$$

where the path of integration is any reversible path joining B to A. It follows from this formula that in any infinitesimal *reversible* transformation the change in S is given by

$$dS = \frac{dQ}{T}$$

which is an exact differential.

We note the following properties of the entropy:

(*a*) For an arbitrary transformation,

$$\int_A^B \frac{dQ}{T} \leq S(B) - S(A)$$

The equality holds if the transformation is reversible.

Proof. Let R and I denote respectively any reversible and any irreversible path joining A to B, as shown in Fig. 1.7. For path R the assertion holds by definition of S. Now consider the cyclic transformation made up of I plus the reverse of R. From Clausius' theorem we have

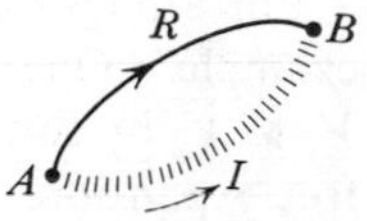

$$\int_{\mathrm{I}} \frac{dQ}{T} - \int_R \frac{dQ}{T} \leq 0$$

or

$$\int_{\mathrm{I}} \frac{dQ}{T} \leq \int_R \frac{dQ}{T} \equiv S(B) - S(A) \qquad \text{(QED)}$$

Fig. 1.7. Reversible path R and irreversible path I connecting states A and B.

(*b*) The entropy of a thermally isolated system never decreases.

* This definition depends on the assumption that every equilibrium state A is accessible from any state O through a reversible transformation. In other words, the surface of the equation of state consists of one continuous sheet. If the surface of the equation of state consists of two disjoint sheets, our definition only defines S for each sheet up to an additive constant. The absolute value of this constant, which becomes relevant when we go from one sheet to another, is the subject of the third law of thermodynamics, which states that at absolute zero $S = O$ for all sheets.

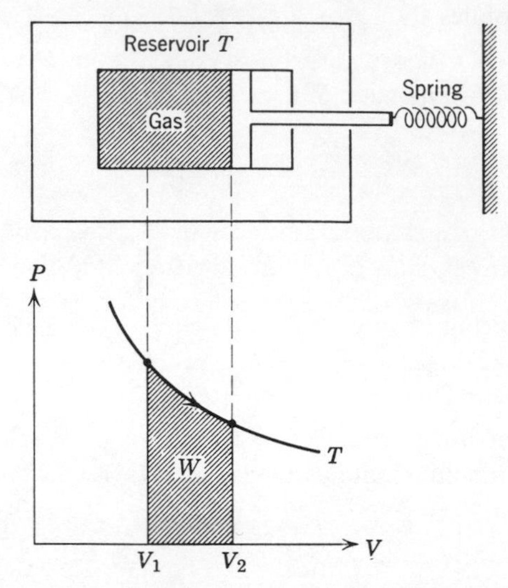

Fig. 1.8. Reversible isothermal expansion of an ideal gas.

Proof. A thermally isolated system cannot exchange heat with the external world. Therefore $dQ = 0$ for any transformation. By the previous property we immediately have

$$S(B) - S(A) \geq 0$$

The equality holds if the transformation is reversible. (QED)

An immediate consequence of this is that for a thermally isolated system the state of equilibrium is the state of maximum entropy consistent with external constraints.

For a physical interpretation of the entropy, we consider the following example. One mole of ideal gas expands isothermally from volume V_1 to V_2 by two routes: Reversible isothermal expansion and irreversible free expansion. Let us calculate the change of entropy of the gas and of the external surroundings.

Reversible Isothermal Expansion. The arrangement is illustrated in Fig. 1.8. In the *P-V* diagram the states of the gas (and not its surroundings) are represented. Since the gas is ideal, $U = U(T)$. Hence $\Delta U = 0$. The amount of heat absorbed is equal to the work done, which is the shaded area in the *P-V* diagram:

$$\Delta Q = RT \log \frac{V_2}{V_1}$$

Hence,

$$(\Delta S)_{\text{gas}} = \int \frac{dQ}{T} = \frac{\Delta Q}{T} = R \log \frac{V_2}{V_1}$$

The reservoir supplies the amount of heat $-\Delta Q$. Hence,

$$(\Delta S)_{\text{reservoir}} = -\frac{\Delta Q}{T} = -R \log \frac{V_2}{V_1}$$

The change in entropy of the whole arrangement is zero. An amount of work

$$W = \Delta Q = RT \log \frac{V_2}{V_1}$$

is stored in the spring connected to the piston. This can be used to compress the gas, reversing the transformation.

Free Expansion. This process is illustrated in Fig. 1.3. The initial and final states are the same as in the reversible isothermal expansion. Therefore $(\Delta S)_{\text{gas}}$ is the same as in that case because S is a state function. Thus

$$(\Delta S)_{\text{gas}} = R \log \frac{V_2}{V_1}$$

Since no heat is supplied by the reservoir we have

$$(\Delta S)_{\text{reservoir}} = 0$$

which leads to an increase of entropy of the entire system of gas plus reservoir:

$$(\Delta S)_{\text{total}} = R \log \frac{V_2}{V_1}$$

In comparison with the previous case, an amount of useful energy

$$W = T(\Delta S)_{\text{total}}$$

is "wasted," for it could have been extracted by expanding the gas reversibly. This example illustrates the fact that irreversibility is generally "wasteful," and is marked by an increase of entropy of the total system under consideration. For this reason the entropy of a state may be viewed as a measure of the unavailability of useful energy in that state.

It may be noted in passing that heat conduction is an irreversible process and thus increases the total entropy. Suppose a metal bar conducts heat from reservoir T_2 to reservoir T_1 at the rate of Q per second. The net increase in entropy per second of the entire system under consideration is

$$Q\left(\frac{1}{T_1} - \frac{1}{T_2}\right) > 0$$

The only reversible way to transfer heat is to operate a Carnot engine between the two reservoirs.

We might indulge in the following thought. The entropy of the entire universe, which is as isolated a system as exists, can never decrease. Furthermore, we have ample evidence, by just looking around us, that the universe is not unchanging, and that most changes are irreversible. It follows that the entropy of the universe constantly increases, and will lead relentlessly to a "heat death" of the universe—a state of maximum entropy. Is this the fate of the universe? In a universe in which the second law of thermodynamics is rigorously correct, the affirmative answer is inescapable. In fact, however, ours is not such a universe, although this conclusion cannot be arrived at within thermodynamics.

Our universe is governed by molecular laws, whose invariance under time reversal denies the existence of any natural phenomenon that absolutely distinguishes between the past and the future. The proper answer to the question we posed is no. The reason is that the second law of thermodynamics cannot be a rigorous law of nature.

This leads to the new question, "In what sense, and to what extent, is the second law of thermodynamics correct?" We examine this question in our discussion of kinetic theory (see Sec. 4.4) where we see that the second law of thermodynamics is correct "on the average," and that in macroscopic phenomena deviations from this law are so rare that for all practical purposes they never occur.

1.5 SOME IMMEDIATE CONSEQUENCES OF THE SECOND LAW

The consequences of the second law that we discuss here are based on the fact that dS is an exact differential. Let us first recall one of the dQ equations, equation (1.7):

$$dQ = C_V\, dT + \left[\left(\frac{\partial U}{\partial V}\right)_T + P\right] dV$$

Putting $dQ = T\, dS$, we obtain

$$dS = \left(\frac{C_V}{T}\right) dT + \frac{1}{T}\left[\left(\frac{\partial U}{\partial V}\right)_T + P\right] dV \tag{1.10}$$

Since dS is an exact differential, we must have

$$\left(\frac{\partial}{\partial V}\right)_T \left(\frac{C_V}{T}\right) = \left(\frac{\partial}{\partial T}\right)_V \left[\frac{1}{T}\left(\frac{\partial U}{\partial V}\right)_T + \frac{P}{T}\right] \tag{1.11}$$

Putting $C_V = (\partial U/\partial T)_V$ and carrying out the differentiations on the right-hand side, we obtain, after some algebra,

$$\left(\frac{\partial U}{\partial V}\right)_T = T\left(\frac{\partial P}{\partial T}\right)_V - P \tag{1.12}$$

It is now possible to calculate $(\partial U/\partial V)_T$ for an ideal gas. Using the equation of state $P = NkT/V$, we obtain

$$\left(\frac{\partial U}{\partial V}\right)_T = \frac{NkT}{V} - P = 0$$

Hence U is a function of T alone. This was earlier deduced from Joule's free expansion experiment with the help of the first law. We now see that it is a logical consequence of the second law.

Substitution of (1.12) into (1.10) yields

$$T\,dS = C_V\,dT + T\left(\frac{\partial P}{\partial T}\right)_V dV \tag{1.13}$$

Going through similar steps for another form of the dQ equation, equation (1.6), we obtain

$$T\,dS = C_P\,dT - T\left(\frac{\partial V}{\partial T}\right)_P dP \tag{1.14}$$

It is possible to re-express (1.13) and (1.14) in such a way that only quantities that are conveniently measurable appear in these equations. To do this we need the following mathematical lemma.

Lemma. Let x, y, z be quantities satisfying a functional relation $f(x, y, z) = 0$. Let w be a function of any two of x, y, z. Then

(*a*)
$$\left(\frac{\partial x}{\partial y}\right)_w \left(\frac{\partial y}{\partial z}\right)_w = \left(\frac{\partial x}{\partial z}\right)_w$$

(*b*)
$$\left(\frac{\partial x}{\partial y}\right)_z = \frac{1}{\left(\frac{\partial y}{\partial x}\right)_z}$$

(*c*)
$$\left(\frac{\partial x}{\partial y}\right)_z \left(\frac{\partial y}{\partial z}\right)_x \left(\frac{\partial z}{\partial x}\right)_y = -1 \qquad \text{(chain relation)}$$

The proof is straightforward.

Let us define the following quantities, which are experimentally measurable.

$$\alpha \equiv \frac{1}{V}\left(\frac{\partial V}{\partial T}\right)_P \qquad \text{(coefficient of thermal expansion)} \tag{1.15}$$

$$\kappa_T \equiv -\frac{1}{V}\left(\frac{\partial V}{\partial P}\right)_T \qquad \text{(isothermal compressibility)} \tag{1.16}$$

$$\kappa_S \equiv -\frac{1}{V}\left(\frac{\partial V}{\partial P}\right)_S \qquad \text{(adiabatic compressibility)} \tag{1.17}$$

Using the lemma we have

$$\left(\frac{\partial P}{\partial T}\right)_V = -\frac{1}{(\partial T/\partial V)_P(\partial V/\partial P)_T} = \frac{(\partial V/\partial T)_P}{-(\partial V/\partial P)_T} = \frac{\alpha}{\kappa_T} \tag{1.18}$$

The equations (1.13) and (1.14) can now be written in the desired forms:

$$T\,dS = C_V\,dT + \frac{\alpha T}{\kappa_T}\,dV \tag{1.19}$$

$$T\,dS = C_P\,dT - \alpha T V\,dP \tag{1.20}$$

These are known as the $T\,dS$ equations.

Next we try to express $C_P - C_V$ for any substance in terms of other experimental quantities. Equating the right-hand side of (1.13) to that of (1.14), we have

$$C_V\,dT + T\left(\frac{\partial P}{\partial T}\right)_V dV = C_P\,dT - T\left(\frac{\partial V}{\partial T}\right)_P dP$$

Choosing P and V to be the independent variables, we write

$$dT = \left(\frac{\partial T}{\partial V}\right)_P dV + \left(\frac{\partial T}{\partial P}\right)_V dP$$

Substitution of this into the previous equation yields

$$\left[(C_P - C_V)\left(\frac{\partial T}{\partial V}\right)_P - T\left(\frac{\partial P}{\partial T}\right)_V\right] dV + \left[(C_P - C_V)\left(\frac{\partial T}{\partial P}\right)_V - T\left(\frac{\partial V}{\partial T}\right)_P\right] dP = 0$$

Since dV and dP are independent of each other, the coefficients of dV and dP must separately vanish. From the first of these we obtain

$$C_P - C_V = \frac{T(\partial P/\partial T)_V}{(\partial T/\partial V)_P} = -T\left[\left(\frac{\partial V}{\partial T}\right)_P\right]^2 \left(\frac{\partial P}{\partial V}\right)_T$$

where the chain relation was used. Therefore

$$C_P - C_V = \frac{TV\alpha^2}{\kappa_T} \tag{1.21}$$

This shows that $(C_P - C_V) > 0$ if $\kappa_T \geq 0$. From experience we know that $\kappa_T \geq 0$ for most substances, but this fact is implied by neither the first law nor the second law. It can be proven in statistical mechanics, where use is made of the nature of intermolecular forces, and where it is known as Van Hove's theorem.

Finally we consider $\gamma \equiv C_P/C_V$. Equations (1.13) and (1.14) remain valid for adiabatic transformations for which $dS = 0$. The following expressions for C_V and C_P are therefore true:

$$C_V = -T\left(\frac{\partial P}{\partial T}\right)_V\left(\frac{\partial V}{\partial T}\right)_S$$

$$C_P = T\left(\frac{\partial V}{\partial T}\right)_P\left(\frac{\partial P}{\partial T}\right)_S$$

Dividing one by the other, we find

$$\frac{C_P}{C_V} = -\frac{(\partial V/\partial T)_P(\partial P/\partial T)_S}{(\partial P/\partial T)_V(\partial V/\partial T)_S} = -\frac{(\partial V/\partial T)_P}{(\partial P/\partial T)_V}\left(\frac{\partial P}{\partial V}\right)_S = \frac{(\partial V/\partial P)_T}{(\partial V/\partial P)_S} \tag{1.22}$$

Combining (1.21) and (1.22), we can solve for C_V and C_P separately:

$$C_V = \frac{TV\alpha^2\kappa_S}{(\kappa_T - \kappa_S)\kappa_T} \tag{1.23}$$

$$C_P = \frac{TV\alpha^2}{\kappa_T - \kappa_S} \tag{1.24}$$

1.6 THERMODYNAMIC POTENTIALS

We introduce two auxiliary state functions, the Helmholtz free energy A (or simply free energy) and the Gibbs thermodynamic potential G (or simply Gibbs potential). They are defined as follows:

$$A = U - TS \tag{1.25}$$

$$G = A + PV \tag{1.26}$$

They are useful in determining the equilibrium state of a system that is not isolated. We discuss their significances separately.

The physical meaning of the free energy A is furnished by the fact that in an isothermal transformation the change of the free energy is the negative of the maximum possible work done by the system. To see this, let a system undergo an arbitrary isothermal transformation from state A to state B. We have from the second law that

$$\int_A^B \frac{dQ}{T} \leq S(B) - S(A)$$

or, since T is constant,

$$\frac{\Delta Q}{T} \leq \Delta S$$

where ΔQ is the amount of heat absorbed during the transformation and $\Delta S = S(B) - S(A)$. By use of the first law we can rewrite the inequality just given as follows:

$$W \leq -\Delta U + T\,\Delta S \tag{1.27}$$

where W is the work done by the system. Since the right-hand side is none other than $-\Delta A$, we have

$$W \leq -\Delta A \tag{1.28}$$

The equality holds if the transformation is reversible. (QED)

Suppose $W = 0$; then (1.28) reduces to a useful theorem.

THEOREM. For a mechanically isolated system kept at constant temperature the Helmholtz free energy never increases.

Corollary. For a mechanically isolated system kept at constant temperature the state of equilibrium is the state of minimum Helmholtz free energy.

In an infinitesimal reversible transformation it is easily verified that

$$dA = -P\,dV - S\,dT \tag{1.29}$$

From this follow the relations

$$P = -\left(\frac{\partial A}{\partial V}\right)_T \tag{1.30}$$

$$S = -\left(\frac{\partial A}{\partial T}\right)_V \tag{1.31}$$

which are members of a class of relations known as Maxwell relations. If the function $A(V, T)$ is known, then P and S are calculable by (1.30) and (1.31).

As an example of the principle of minimization of free energy, consider a gas in a cylinder kept at constant temperature. A sliding piston divides the total volume V into two parts V_1 and V_2, in which the pressures are respectively P_1 and P_2. If the piston is released and allowed to slide freely, what is its equilibrium position? By the principle just stated the position of the piston must minimize the free energy of the total system. Suppose equilibrium has been established. Then a slight change in the position of the piston should not change the free energy, since it is at a minimum. That is, $\delta A = 0$. Now A is a function of V_1, V_2, and T. Hence

$$0 = \delta A = \left(\frac{\partial A}{\partial V_1}\right)_T \delta V_1 + \left(\frac{\partial A}{\partial V_2}\right)_T \delta V_2$$

Since $V_1 + V_2 = V$ remains constant, we must have $\delta V_1 = -\delta V_2$. Therefore

$$0 = \left[\left(\frac{\partial A}{\partial V_1}\right)_T - \left(\frac{\partial A}{\partial V_2}\right)_T\right] \delta V_1$$

As δV_1 is arbitrary, its coefficient must vanish. We thus obtain the equilibrium condition

$$\left(\frac{\partial A}{\partial V_1}\right)_T = \left(\frac{\partial A}{\partial V_2}\right)_T$$

which becomes, through use of (1.30),

$$P_1 = P_2$$

a result that is intuitively obvious.

We consider now the Gibbs potential. Its importance lies with the following theorem.

THEOREM. For a system kept at constant temperature and pressure the Gibbs potential never increases.

Corollary. For a system kept at constant temperature and pressure the state of equilibrium is the state of minimum Gibbs potential.

Proof. If T is kept constant, then in any transformation

$$W \leq -\Delta A$$

Now specialize the situation further by keeping the pressure constant, thereby making $W = P\,\Delta V$. We then have

$$P\,\Delta V + \Delta A \leq 0$$

$$\Delta G \leq 0 \qquad \text{(QED)}$$

In an infinitesimal reversible transformation

$$dG = -S\,dT + V\,dP \tag{1.32}$$

From this we immediately obtain more Maxwell relations:

$$S = -\left(\frac{\partial G}{\partial T}\right)_P \tag{1.33}$$

$$V = \left(\frac{\partial G}{\partial P}\right)_T \tag{1.34}$$

Still two more Maxwell relations may be obtained by considering the differential changes of the enthalpy:

$$H = U + PV \tag{1.35}$$

$$dH = T\,dS + V\,dP$$

from which follow

$$V = \left(\frac{\partial H}{\partial P}\right)_S \tag{1.36}$$

$$T = \left(\frac{\partial H}{\partial S}\right)_P \tag{1.37}$$

Further Maxwell relations are

$$T = \left(\frac{\partial U}{\partial S}\right)_V \tag{1.38}$$

and

$$P = -\left(\frac{\partial U}{\partial V}\right)_S \tag{1.39}$$

which follow from the first law, $dU = -P\,dV + T\,dS$. The eight Maxwell relations* so far derived are convenient for thermodynamic calculations.

1.7 THE THIRD LAW OF THERMODYNAMICS

The second law of thermodynamics enables us to define the entropy of a substance up to an arbitrary additive constant. The definition of the entropy depends on the existence of a reversible transformation connecting an arbitrarily chosen reference state O to the state A under consideration. Such a reversible transformation always exists if both O and A lie on one sheet of the equation of state surface. If we consider two different substances, or meta-stable phases of the same substance, the equation of state surface may consist of more than one disjoint sheets. In such cases the kind of reversible path we have mentioned may not exist. Therefore the second law does not uniquely determine the difference in entropy of two states A and B, if A refers to one substance and B to another. In 1905 Nernst supplied a rule for such a determination. This rule has since been called the *third law of thermodynamics*. It states:

The entropy of a system at absolute zero is a universal constant, which may be taken to be zero.

The generality of this statement rests in the facts that (*a*) it refers to any system, and that (*b*) it states that $S = 0$ at $T = 0$, regardless of the values of any other parameter of which S may be a function. It is obvious that the third law renders the entropy of any state of any system unique.

The third law immediately implies that any heat capacity of a system must vanish at absolute zero. Let R be any reversible path connecting a state of the system at absolute zero to the state A, whose entropy is to be

* The eight Maxwell relations are conveniently summarized by the diagram

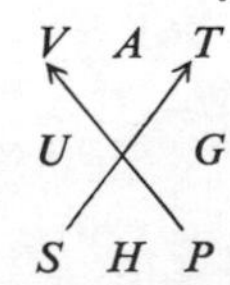

The reader can easily figure out the rules of using it to generate all the Maxwell relations.

calculated. Let $C_R(T)$ be the heat capacity of the system along the path R. Then, by the second law,

$$S(A) = \int_0^{T_A} C_R(T) \frac{dT}{T} \tag{1.40}$$

But according to the third law

$$S(A) \underset{T_A \to 0}{\longrightarrow} 0 \tag{1.41}$$

Hence we must have

$$C_R(T) \underset{T \to 0}{\longrightarrow} 0 \tag{1.42}$$

In particular, C_R may be C_V or C_P. The statement (1.42) is experimentally verified for all substances so far examined.

A less obvious consequence of the third law is that at absolute zero the coefficient of thermal expansion of any substance vanishes. This may be shown as follows. From the $T\,dS$ equations we can deduce the equalities

$$\left(\frac{\partial S}{\partial T}\right)_P = \frac{C_P}{T} \tag{1.43}$$

$$\left(\frac{\partial S}{\partial P}\right)_T = -\left(\frac{\partial V}{\partial T}\right)_P \tag{1.44}$$

Combining these we arrive at

$$\left(\frac{\partial C_P}{\partial P}\right)_T = -T\left(\frac{\partial^2 V}{\partial T^2}\right)_P \tag{1.45}$$

From (1.44) and (1.40) we have, for the coefficient of thermal expansion α, the expression

$$V\alpha \equiv \left(\frac{\partial V}{\partial T}\right)_P = -\left(\frac{\partial S}{\partial P}\right)_T = -\frac{\partial}{\partial P}\int_0^T C_P \frac{dT}{T} = -\int_0^T \left(\frac{\partial C_P}{\partial P}\right)_T \frac{dT}{T} \tag{1.46}$$

where the integrations proceed along a path of constant P. Using (1.45), we rewrite this as

$$V\alpha = \int_0^T \left(\frac{\partial^2 V}{\partial T^2}\right)_P dT = \left(\frac{\partial V}{\partial T}\right)_P - \left[\left(\frac{\partial V}{\partial T}\right)_P\right]_{T=0} \tag{1.47}$$

Therefore

$$\alpha \underset{T \to 0}{\longrightarrow} 0 \tag{1.48}$$

In a similar fashion we can show that

$$\left(\frac{\partial P}{\partial T}\right)_V \underset{T \to 0}{\longrightarrow} 0 \tag{1.49}$$

Combined with (1.48), this implies that on the P-T diagram the melting curve has zero tangent at $T = 0$.

It is experimentally found that C_P can be represented by the following series expansion at low temperatures:

$$C_P = T^x(a + bT + cT^2 + \cdots) \tag{1.50}$$

where x is a positive constant, and $a, b, c, \ldots$ are functions of P. Differentiating (1.50) with respect to P, we find that

$$\left(\frac{\partial C_P}{\partial P}\right)_T = T^x(a' + b'T + c'T^2 + \cdots) \tag{1.51}$$

Substituting this into (1.46), we have

$$V\alpha = -\int_0^T dT(a'T^{x-1} + b'T^x + \cdots) = -T^x\left(\frac{a'}{x} + \frac{b'T}{x+1} + \frac{c'T^2}{x+2} + \cdots\right)$$

Hence

$$\frac{V\alpha}{C_P} \underset{T\to 0}{\longrightarrow} \text{finite constant} \tag{1.52}$$

This has the consequence that a system cannot be cooled to absolute zero by a finite change of the thermodynamic parameters. For example, from one of the $T\,dS$ equations we find that through an adiabatic change dP of the pressure, the temperature changes by

$$dT = \left(\frac{V\alpha}{C_P}\right) T\,dP \tag{1.53}$$

By virtue of (1.52), the change of P required to produce a finite change in the temperature is unbounded as $T \to 0$.

The unattainability of absolute zero is sometimes stated as an alternative formulation of the third law. This statement is independent of the second law, for the latter only implies that there exists no Carnot engine whose lower reservoir is at absolute zero.* Whether it is possible to make a system approach absolute zero from a higher temperature is an independent question. According to (1.53), it depends on the behavior of the specific heat, of which the second law says nothing.

Before experimental techniques were well developed for low temperatures, it was generally believed that heat capacities of substances remain

* The second law requires that in any reversible transformation we must have $dQ = T\,dS$, where dS is an exact differential. Hence $dQ = 0$ when $T = 0$. That is, all transformations at absolute zero are adiabatic. Even if we had a system at absolute zero, there would still be no reversible way to heat it to a higher temperature. Thus we cannot construct a Carnot engine whose lower reservoir is at absolute zero.

constant down to absolute zero, as classical kinetic theory predicts [i.e., $x = 0$ in (1.50)]. If this were so, we see directly from (1.53) that the unattainability of absolute zero would be automatic. This is why the question did not receive attention until the turn of the century, when it was discovered that heat capacities tend to vanish at low temperatures. We now see that even if the heat capacities vanish at absolute zero, the absolute zero is still unattainable.

We see, when we come to quantum statistical mechanics, that the third law of thermodynamics is a macroscopic manifestation of quantum effects (see Sec. 9.4). The foregoing discussions, which are somewhat abstract, become concrete and physical when they are presented in the context of quantum statistical mechanics. The importance of the third law of thermodynamics, therefore, does not lie in these abstract considerations, but in its practical usefulness. We end our discussion of the third law with one of its applications.

The free energy of a system is defined as

$$A = U - TS \tag{1.54}$$

which, according to the third law, can be written in the form

$$A = U - T\int_0^T \frac{C_V}{T'}\,dT' \tag{1.55}$$

where the integral in the second term extends over a path of constant volume. There is no arbitrary additive constant except the one already contained in U. This formula, together with

$$U = \int_0^T C_V\,dT' + \text{constant} \tag{1.56}$$

enables us to determine both U and A up to the same arbitrary additive constant from measurements of C_V.

To illustrate the practical use of these formulas we consider the melting point of solid quartz. The stable phase of quartz at low temperatures is a crystalline solid. The liquid phase (glass), however, can be supercooled and can exist in meta-stable equilibrium far below the melting point. Hence a direct measurement of the melting point is difficult. It can, however, be determined indirectly through the use of (1.55). Let the specific heat c_V of both solid and liquid quartz be measured through a range of temperatures at a fixed volume V. Let Δc_V denote their difference, which is a function of temperature. Then the difference in internal energy per unit mass of the two phases is obtained by numerically integrating Δc_V at constant V:

$$\Delta u = \int \Delta c_V\,dT'$$

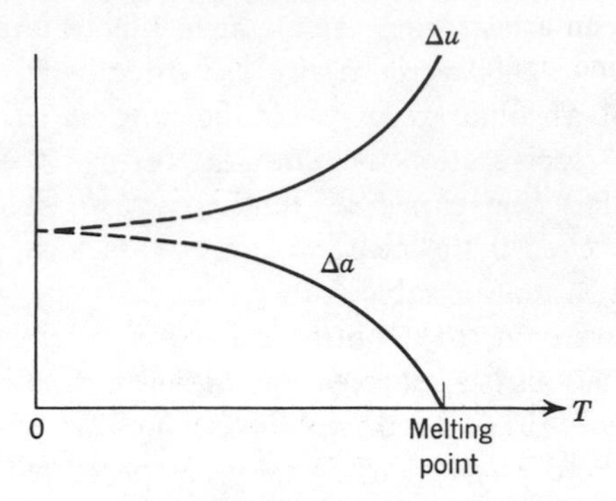

Fig. 1.9. Determination of the melting point through use of the third law.

Using (1.56) we have, for the difference in free energy per unit mass of the two phases,

$$\Delta a = \Delta u - T \int_0^T \frac{\Delta c_V}{T'} \, dT'$$

Plotting Δu and Δa as a function of T at a fixed V, we should obtain a graph that looks qualitatively like that shown in Fig. 1.9. The melting point is the temperature at which $\Delta a = 0$, since the condition of phase equilibrium at fixed T and V is the equality of the free energies per unit mass. In practice the point at which $\Delta a = 0$ may be obtained either by direct integration up to that point, or by extrapolation.

PROBLEMS

1.1. Find the equations governing an adiabatic transformation of an ideal gas.

1.2. (*a*) An engine is represented by the cyclic transformation shown in the accompanying *T-S* diagram, where *A* denotes the area of the shaded region and *B* the area of the region below it. Show that this engine is not as efficient as a Carnot engine operating between the highest and the lowest available temperatures.

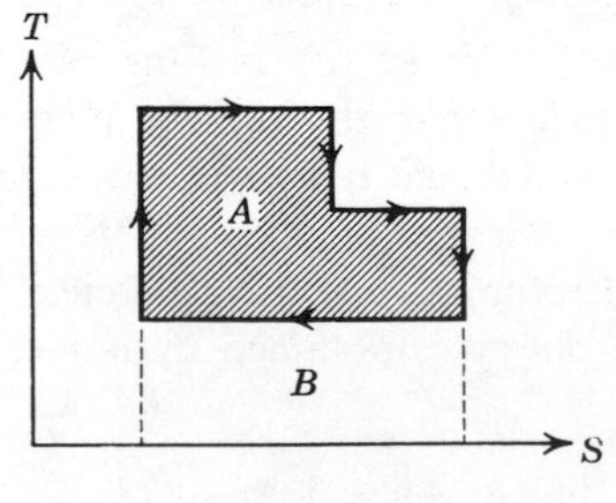

Problem 1.2

(*b*) Show that an arbitrary reversible engine cannot be more efficient than a Carnot engine operating between the highest and the lowest available temperatures.

1.3. Let a real gas undergo the free-expansion process. Supply all relevant arguments to show that the change in temperature ΔT is related to the change in the volume ΔV by the formula

$$\Delta T = \left(\frac{\partial T}{\partial V}\right)_U \Delta V$$

where ΔT, ΔV are small quantities. In particular, explain why this formula looks the same as that for a reversible process, although the process under consideration is irreversible.

1.4. Two isotherms of 1 mole of a substance that can undergo a gas-liquid

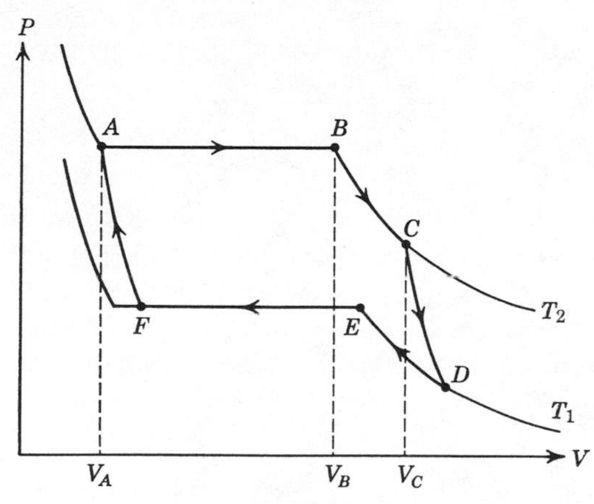

Problem 1.4

transition are shown in the accompanying P-V diagram. The absolute temperatures are T_2 and T_1 respectively. The substance is made to go through one cycle of a cyclic reversible transformation *ABCDEF*, as indicated in the diagram. The following information is given.

(*a*) *ABC* and *DEF* are isothermal transformations.

(*b*) *FA* and *CD* are adiabatic transformations.

(*c*) In the gas phase (*BCDE*) the substance is an ideal gas. At *A* the substance is pure liquid.

(*d*) Latent heat along *AB*: $L = 200$ cal/mole

$T_2 = 300°\text{K}$

$T_1 = 150°\text{K}$

$V_A = 0.5$ liter

$V_B = 1$ liter

$V_C = 2.71828$ liter

Calculate the net amount of work done by the substance.

1.5. A substance has the following properties:

(i) At a constant temperature T_0 the work done by it on expansion from V_0 to V is

$$W = RT_0 \log \frac{V}{V_0}$$

(ii) The entropy is given by

$$S = R\frac{V_0}{V}\left(\frac{T}{T_0}\right)^a$$

where V_0, T_0, and a are fixed constants.

(*a*) Calculate the Helmholtz free energy.

(*b*) Find the equation of state.

(*c*) Find the work done at an arbitrary constant temperature T.

chapter 2

SOME APPLICATIONS OF THERMODYNAMICS

2.1 THERMODYNAMIC DESCRIPTION OF PHASE TRANSITIONS

The surface of the equation of state of a typical substance is shown in Fig. 2.1, where the shaded areas are cylindrical surfaces, representing regions of phase transition. The P-V and P-T diagrams are shown in Fig. 2.2. We study here the implications of the second law for these phase transitions.

Let us consider the transition between the gas phase and the liquid phase. The transition takes place at a constant temperature and pressure,

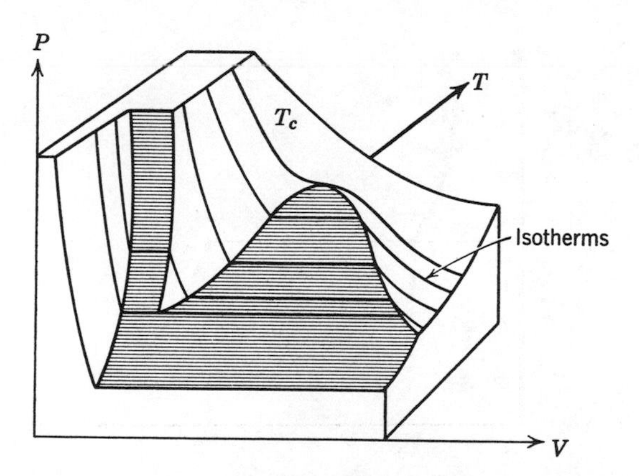

Fig. 2.1. Surface of equation of state of a typical substance (not to scale).

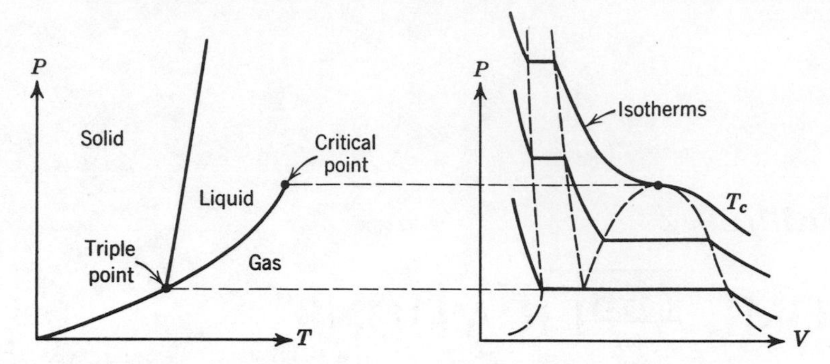

Fig. 2.2. *P-V* and *P-T* diagrams of a typical substance (not to scale).

as shown in Fig. 2.3. This pressure $P(T)$ is called the vapor pressure at the temperature T. Let the system be initially in state 1, where it is all liquid. When heat is added to the system, some of the liquid will be converted into gas, and so on until we reach state 2, where the system is all gas, as schematically shown in Fig. 2.4. The important facts are that

(*a*) during the phase transition both P and T remain constant;

(*b*) in the gas-liquid mixture the liquid exists in the same state as at 1 and the gas exists in the same state as at 2.

As a result, knowing the properties of the states 1 and 2 suffices for a complete description of the phase transition. The isotherm in the *P-V* diagram is horizontal during the phase transition, because the gas phase has a smaller density than the liquid phase. Consequently, when a certain mass of liquid is converted into gas, the total volume of the system expands, although P and T remain unchanged. Such a transition is known as a "first-order transition."

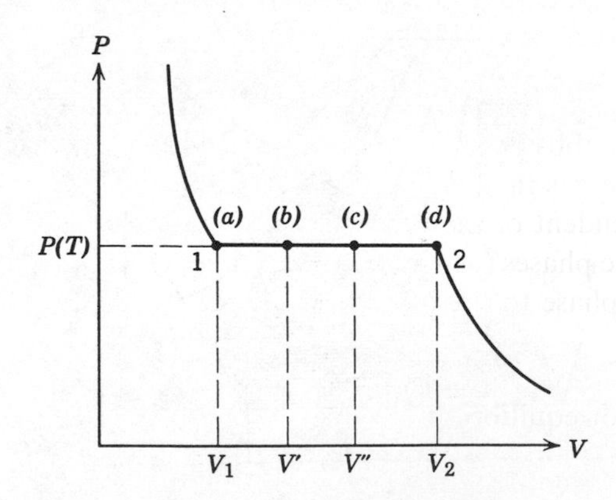

Fig. 2.3. An isotherm exhibiting a phase transition.

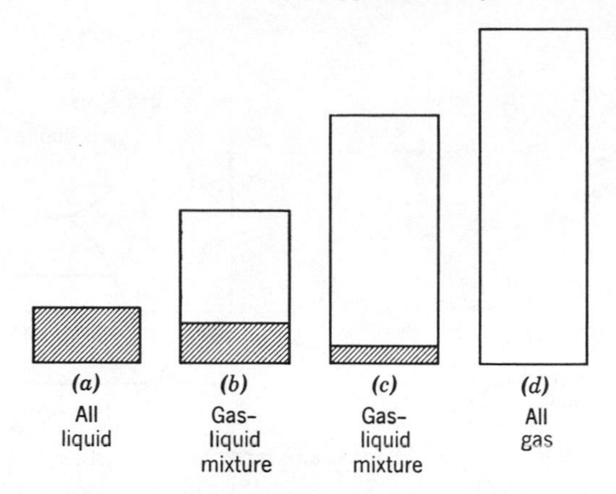

Fig. 2.4. Schematic illustration of a first-order phase transition. The temperature and the pressure of the system remain constant throughout the transition. The total volume of the system changes as the relative amount of the substance in the two phases changes, because the two phases have different densities.

The dependence of the vapor pressure $P(T)$ on the temperature may be found by applying the second law. Consider a gas-liquid mixture in equilibrium at temperature T and vapor pressure $P(T)$. Let the mass of the liquid be m_1 and the mass of the gas be m_2. If the system is in equilibrium with the given T and $P(T)$, the Gibbs potential of this state must be at a minimum. That is, if any parameters other than T and P are varied slightly, we must have $\delta G = 0$. Let us vary the composition of the mixture by converting an amount δm of liquid to gas, so that

$$-\delta m_1 = \delta m_2 = \delta m \tag{2.1}$$

The total Gibbs potential of the gas-liquid mixture may be represented, with neglect of surface effects, as

$$G = m_1 g_1 + m_2 g_2 \tag{2.2}$$

where g_1 is the Gibbs potential per unit mass of the liquid in state 1 and g_2 is that for the gas in state 2. They are also called *chemical potentials.* They are independent of the total mass of the phases but may depend on the density of the phases (which, however, are not altered when we transfer mass from one phase to the other). Thus

$$\delta G = 0 = -(g_1 - g_2)\,\delta m$$

The condition for equilibrium is then

$$g_1 = g_2 \tag{2.3}$$

This condition determines the vapor pressure, as we shall see.

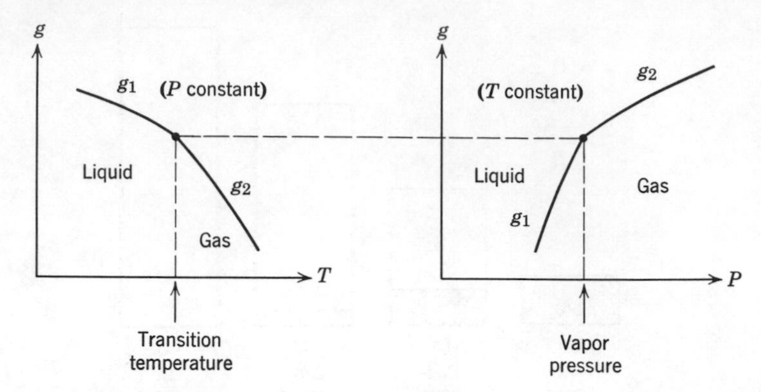

Fig. 2.5. Chemical potentials g_1, g_2 for the two phases in a first-order phase transition.

The chemical potentials $g_1(P, T)$ and $g_2(P, T)$ are two state functions of the liquid and gas respectively. Recall that in each phase we have

$$\left(\frac{\partial g}{\partial T}\right)_P = -s \qquad \text{(entropy per unit mass)} \tag{2.4}$$

$$\left(\frac{\partial g}{\partial P}\right)_T = v \qquad \text{(volume per unit mass)} \tag{2.5}$$

We see that the first derivative of g_1 is different from that of g_2 at the transition temperature and pressure:

$$\left[\frac{\partial(g_2 - g_1)}{\partial T}\right]_P = -(s_2 - s_1) < 0 \tag{2.6}$$

$$\left[\frac{\partial(g_2 - g_1)}{\partial P}\right]_T = v_2 - v_1 > 0 \tag{2.7}$$

This is why the transition is called "first-order." The behavior of $g_1(P, T)$ and $g_2(P, T)$ are qualitatively sketched in Fig. 2.5.

To determine the vapor pressure we proceed as follows. Let

$$\begin{aligned} \Delta g &= g_2 - g_1 \\ \Delta s &= s_2 - s_1 \\ \Delta v &= v_2 - v_1 \end{aligned} \tag{2.8}$$

where all quantities are evaluated at the transition temperature T and vapor pressure P. The condition for equilibrium is that T and P be such as to make $\Delta g = 0$. Dividing (2.6) by (2.7), we have

$$\frac{(\partial\, \Delta g/\partial T)_P}{(\partial\, \Delta g/\partial P)_T} = -\frac{\Delta s}{\Delta v} \tag{2.9}$$

By the chain relation,

$$\left(\frac{\partial\,\Delta g}{\partial T}\right)_P\left(\frac{\partial T}{\partial P}\right)_{\Delta g}\left(\frac{\partial P}{\partial\,\Delta g}\right)_T = -1$$

or

$$\frac{(\partial\,\Delta g/\partial T)_P}{(\partial\,\Delta g/\partial P)_T} = -\left(\frac{\partial P}{\partial T}\right)_{\Delta g} \tag{2.10}$$

The reason the chain relation is valid here is that Δg is a function of T and P, and hence there must exist a relation of the form $f(T, P, \Delta g) = 0$. The derivative

$$\frac{dP(T)}{dT} = \left(\frac{\partial P}{\partial T}\right)_{\Delta g=0} \tag{2.11}$$

is precisely the derivative of the vapor pressure with respect to temperature under equilibrium conditions, for Δg is held fixed at the value zero. Combining (2.11), (2.10), and (2.9), we obtain

$$\frac{dP(T)}{dT} = \frac{\Delta s}{\Delta v} \tag{2.12}$$

The quantity

$$l = T\,\Delta s \tag{2.13}$$

is called the *latent heat of transition*. Thus

$$\frac{dP(T)}{dT} = \frac{l}{T\Delta v} \tag{2.14}$$

This is known as the *Clapeyron equation*. It governs the vapor pressure in any first-order transition.

It may happen in a phase transition that $s_2 - s_1 = 0$ and $v_2 - v_1 = 0$. When this is so the first derivatives of the chemical potentials are continuous across the transition point. Such a transition is not of the first order and would not be governed by the Clapeyron equation, and its isotherm would not have a horizontal part in the P-V diagram. Ehrenfest defines a phase transition to be an nth-order transition if, at the transition point,

$$\frac{\partial^n g_1}{\partial T^n} \neq \frac{\partial^n g_2}{\partial T^n} \qquad \text{and} \qquad \frac{\partial^n g_1}{\partial P^n} \neq \frac{\partial^n g_2}{\partial T^n}$$

whereas all lower derivatives are equal. Apart from the well-known gas-liquid transition, there is one known example of a phase transition that fits into the scheme of Ehrenfest—the second-order transition in superconductivity. On the other hand many examples of phase transitions cannot be described by this scheme. Notable among these are the Curie point transition in ferromagnets, the order-disorder transition in binary alloys, and the λ-transition in liquid helium. In these cases the specific heat diverges logarithmically at the transition point. Since the specific heat is related to the second derivative of g these examples cannot be characterized by the behaviors of the higher derivatives of g, because they do not exist. Thus the Ehrenfest scheme is not the most general classification of phase transitions.

2.2 SURFACE EFFECTS IN CONDENSATION

If we compress a gas isothermally, it is supposed to start condensing at a point O, as shown in the P-V diagram in Fig. 2.6. If we compress the system further, the pressure is supposed to remain constant. Actually the pressure will sometimes follow the dotted line shown in Fig. 2.6; along this dotted line the system is not in stable equilibrium, however, because the slightest jar will abruptly reduce the pressure to the correct vapor pressure. Similarly, if a liquid is expanded beyond the point O', it will sometimes follow the dotted curve shown, but this too would not be a situation of stable equilibrium. These phenomena are respectively known as supersaturation and supercooling. They owe their existence to surface effects, which we have previously ignored. We give a qualitative discussion of the surface effects responsible for supersaturation.

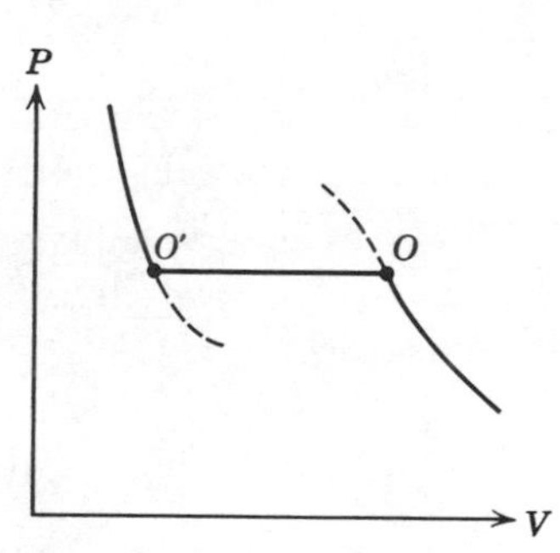

Fig. 2.6. Supersaturation and supercooling.

Vapor pressure as we have defined it is the pressure at which a gas can coexist in equilibrium with an infinitely large body of its own liquid. We now denote it by $P_\infty(T)$. On the other hand, the pressure at which a gas can coexist in equilibrium with a finite droplet (of radius r) of its own liquid is not the vapor pressure $P_\infty(T)$ but a higher pressure $P_r(T)$. The difference between $P_\infty(T)$ and $P_r(T)$ is due to the surface tension of the droplet. Before we try to give a qualitative description of the mechanism of condensation, we calculate $P_r(T)$.

Suppose a droplet of liquid is placed in an external medium that exerts a pressure P on the droplet. Then the work done by the droplet on expansion is empirically given by

$$dW = P\,dV - \gamma\,da \tag{2.15}$$

where da is the increase in the surface area of the droplet and γ the coefficient of surface tension. The first law now takes the form

$$dU = dQ - P\,dV + \gamma\,da \tag{2.16}$$

Integrating this, we obtain for the internal energy of a droplet of radius r the expression

$$U = \tfrac{4}{3}\pi r^3 u_\infty + 4\pi\gamma r^2 \tag{2.17}$$

where u_∞ is the internal energy per unit volume of an infinite droplet. Correspondingly the Gibbs potential takes the form

$$G = \tfrac{4}{3}\pi r^3 g_\infty + 4\pi\gamma r^2$$

Consider a droplet of radius r in equilibrium with a gas of temperature T and pressure P. For given T and P, r must be such that the total Gibbs potential of the entire system is at a minimum. This condition determines a relation between P and r for a given T. Let the mass of the droplet be M_1 and the mass of the gas be M_2. The total Gibbs potential is of the form

$$G_{\text{total}} = M_2 g_2 + M_1 g_1 + 4\pi\gamma r^2 \tag{2.18}$$

Where g_2 and g_1 are respectively the chemical potential of an infinite body of gas and liquid. We now imagine the radius of the drop changed slightly by evaporation, so that $\delta M_1 = -\delta M_2$. The equilibrium condition is

$$\delta G_{\text{total}} = 0 = \delta M_1\left(-g_2 + g_1 + 8\pi\gamma r\,\frac{\partial r}{\partial M_1}\right) \tag{2.19}$$

Since

$$\frac{\partial r}{\partial M_1} = \frac{1}{4\pi\rho r^2} \tag{2.20}$$

where ρ is the mass density of the droplet, we have

$$g_2 - g_1 = \frac{2\gamma}{\rho r} \tag{2.21}$$

as the condition for equilibrium. Differentiating both sides with respect to the pressure P of the gas at constant temperature, and remembering the Maxwell relation $(\partial g/\partial P)_T = 1/\rho$, we obtain

$$\frac{1}{\rho'} - \frac{1}{\rho} = -\frac{2\gamma}{\rho r^2}\left(\frac{\partial r}{\partial P}\right)_T - \frac{2\gamma}{\rho^2 r}\left(\frac{\partial \rho}{\partial P}\right)_T \tag{2.22}$$

where ρ' is the mass density of the gas. We assume that the gas is sufficiently dilute to be considered an ideal gas. Hence

$$\rho' = \left(\frac{m}{kT}\right)P \tag{2.23}$$

where m is the mass of a gas atom. Further, $1/\rho$ may be neglected compared to $1/\rho'$, and $(\partial\rho/\partial P) \approx 0$. Therefore

$$\left(\frac{\partial r}{\partial P}\right)_T = -\left(\frac{kT}{m}\right)\frac{\rho r^2}{2\gamma P} \tag{2.24}$$

Integrating both sides of this equation we find P as a function of r for a given temperature:

$$P_r(T) = P_\infty(T) \exp\left(\frac{2\gamma m}{\rho kT} \frac{1}{r}\right) \tag{2.25}$$

which is the expression we seek. A graph of $P_r(T)$ is shown in Fig. 2.7.

Now we can give a qualitative description of what happens when a gas starts to condense. According to (2.25) only liquid droplets of a given radius r can exist in equilibrium with the gas at a given T and P. The droplets that are too large find the external pressure too high. They attempt to reduce the external pressure by gathering vapor, but this makes them grow still larger. The droplets that are too small, on the other hand, find the pressure too low, and tend to evaporate, but this makes them still smaller, and they eventually disappear. Thus, unless all droplets initially have exactly the same radius r_0 (which is unlikely), the average size will shift towards a larger value. This can only be achieved through a net condensation of vapor onto the droplets, thereby lowering pressure of the vapor. The whole process then repeats. Thus there is a self-sustaining process favoring the formation of larger and larger droplets until all the liquid forms a single body essentially infinite in size. At this point the pressure of the vapor drops to P_∞, the equilibrium vapor pressure.

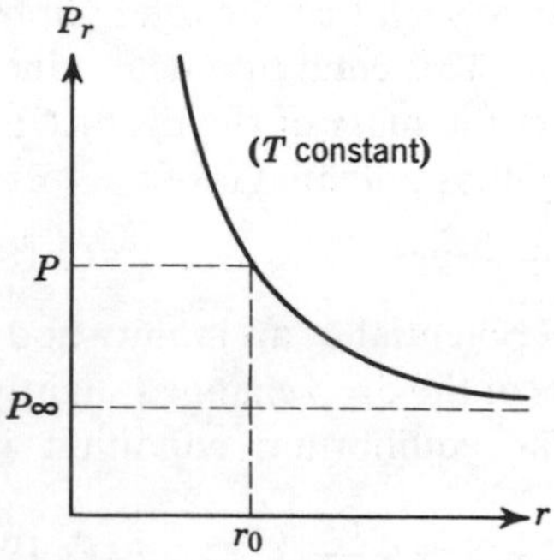

Fig. 2.7. The pressure at which a liquid droplet of radius r can exist in equilibrium with its own vapor.

The instability, as we noted, is triggered by the presence of droplets larger than a critical size. This critical size decreases as the degree of supersaturation increases (since P_r increases with the degree of supersaturation), making the presence of too-large droplets more and more likely. Finally, when P_r becomes so large that the critical size is of the order of the molecular radius, the presence of large droplets becomes a certainty, through the momentary formation of bound states of a few molecules in random collision. Supersaturation cannot be pushed beyond this point.

It is clear that supercooling can be discussed in a similar way by considering bubbles instead of droplets.

2.3 VAN DER WAALS EQUATION OF STATE

Van der Waals attempted to find a simple qualitative way to improve the equation of state of a dilute gas by incorporating the effects of molecular interaction. The result is the Van der Waals equation of state.

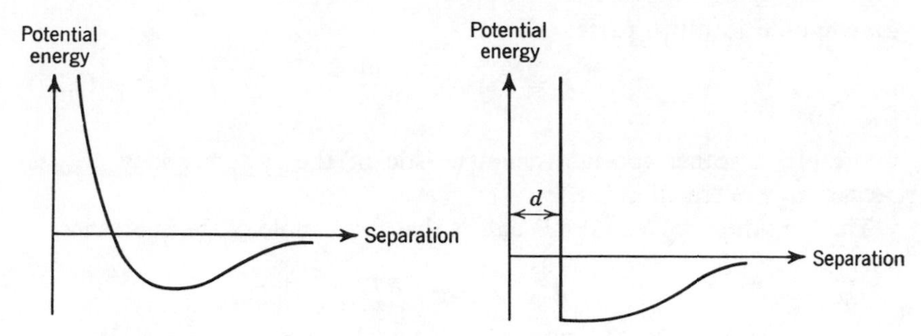

Fig. 2.8. Typical intermolecular potential.

Fig. 2.9. Idealized intermolecular potential.

In most substances the potential energy between two molecules as a function of the intermolecular separation has the qualitative shape shown in Fig. 2.8. The attractive part of the potential energy originates from the mutual electric polarization of the two molecules and the repulsive part from the Coulomb repulsion of the overlapping electronic clouds of the molecules. Van der Waals idealized the situation by approximating the repulsive part by an infinite hard-sphere repulsion, so that the potential energy looks like that illustrated in Fig. 2.9. Thus each molecule is imagined to be a hard elastic sphere of diameter d, surrounded by an attractive force field. The effects of the repulsive and attractive parts are then discussed separately.

The main effect of the hard core would be to forbid the presence of any other molecule in a certain volume centered about a molecule. If V is the total volume occupied by a substance, the effective volume available to one of its molecules would be smaller than V by the totality of such excluded volumes, which is a constant depending on the molecular diameter and the number of molecules present:

$$V_{\text{eff}} = V - b \tag{2.26}$$

where b is a constant characteristic of the substance under discussion.

The qualitative effect of the attractive part of the potential energy is a tendency for the system to form a bound state. If the attraction is sufficiently strong, the system will exist in an N-body bound state, which requires no external wall to contain it. Thus we may assume that the attraction produces a decrease in the pressure that the system exerts on an external wall. The amount of decrease is proportional to the number of pairs of molecules, within interaction range, in a layer near the wall. This in turn is roughly proportional to N^2/V^2. Since N and the range of interaction are constants, the true pressure P of the system may be decomposed into two parts:

$$P = P_{\text{kinetic}} - \frac{a}{V^2} \tag{2.27}$$

where a is another constant characteristic of the system and P_{kinetic} is defined by the equation itself.

The hypothesis of Van der Waals is that for 1 mole of the substance

$$V_{\text{eff}} P_{\text{kinetic}} = RT$$

where R is the gas constant. Therefore the equation of state is

$$(V - b)\left(P + \frac{a}{V^2}\right) = RT \tag{2.28}$$

This is the Van der Waals equation of state. Some isotherms corresponding to this equation of state are shown in Fig. 2.10. There exists a temperature T_c, called the critical temperature, at which the "kink" in the isotherm disappears. The point of inflection c is called the critical point. Its pressure P_c, volume V_c, and temperature T_c can be expressed in terms of a and b as follows. For a given T and P, (2.28) generally has three roots in V (e.g., the values V_1, V_2, V_3 shown in Fig. 2.10). As T increases these roots move together, and at $T = T_c$ they merge into V_c. Thus in the neighborhood of the critical point the equation of state must read

$$(V - V_c)^3 = 0$$

or

$$V^3 - 3V_cV^2 + 3V_c^2V - V_c^3 = 0 \tag{2.29}$$

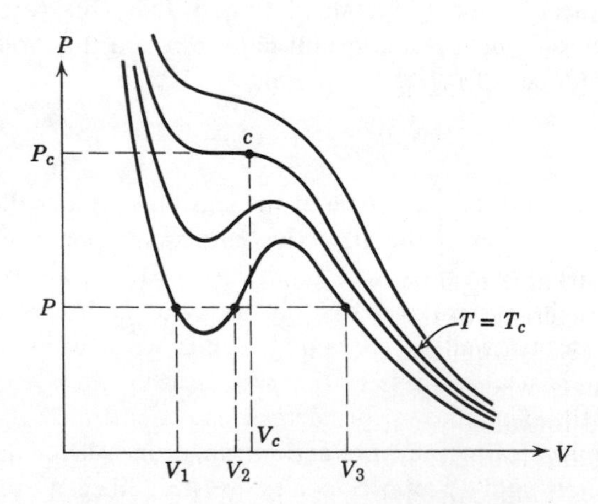

Fig. 2.10. *P-V* diagram of the Van der Waals equation of state.

This is to be compared with (2.28) when we put $T = T_c$ and $P = P_c$:

$$(V - b)\left(P_c + \frac{a}{V^2}\right) = RT_c$$

or

$$V^3 - \left(b + \frac{RT_c}{P_c}\right)V^2 + \frac{a}{P_c}V - \frac{ab}{P_c} = 0 \tag{2.30}$$

We thus obtain the simultaneous equations

$$3V_c = b + \frac{RT_c}{P_c}$$

$$3V_c^2 = \frac{a}{P_c}$$

$$V_c^3 = \frac{ab}{P_c}$$

which may be solved to yield

$$RT_c = \frac{8a}{27b}$$

$$P_c = \frac{a}{27b^2} \tag{2.31}$$

$$V_c = 3b$$

Therefore the Van der Waals constants a and b may be fitted to experiments by measuring any two of T_c, P_c, and V_c.

Let us measure P in units of P_c, T in units of T_c, and V in units of V_c:

$$\bar{P} = \frac{P}{P_c}, \qquad \bar{T} = \frac{T}{T_c}, \qquad \bar{V} = \frac{V}{V_c} \tag{2.32}$$

Then the Van der Waals equation of state becomes

$$\left(\bar{P} + \frac{3}{\bar{V}^2}\right)(\bar{V} - \tfrac{1}{3}) = \tfrac{8}{3}\bar{T} \tag{2.33}$$

This is a remarkable equation because it does not explicitly contain any constant characteristic of the substance. If the Van der Waals hypothesis were correct, (2.33) would hold for all substances. The assertion that the equation of state when expressed in terms of $\bar{P}$, $\bar{T}$, and $\bar{V}$ is a universal equation valid for all substances is called the *law of corresponding states.*

Looking at the isotherms of the Van der Waals equation of state we notice that the "kink" in a typical isotherm is unphysical, for it implies a negative compressibility. Its occurrence may be attributed to the implicit assumption that the system is homogeneous, with no allowance made for the possible coexistence of two phases. The situation may be improved by making what is called a *Maxwell construction* in the following manner. We ask whether it is possible to have two different states of the Van der Waals system coexisting in equilibrium. It is immediately obvious that for this to be possible the two states must have the same P and T. Therefore only states like those at volumes V_1, V_2, V_3 in Fig. 2.10 need be considered as candidates. The further principle we apply is the minimization of free energy. Let the temperature and the total volume of the system be fixed. Then we assume that the system is either in one homogeneous phase, or is composed of more than one phase. The situation that has the lower free energy is the equilibrium situation.

The free energy may be calculated by integrating $-P\,dV$ along an isotherm:

$$A(T, V) = -\int_{\text{isotherm}} P\,dV \tag{2.34}$$

This may be done graphically, as shown in Fig. 2.11. It is seen that the states 1 and 2 can coexist because they have the same T and P. Further, the point b, which lies between 1 and 2 on the common tangent passing through 1 and 2, represents a state in which part of the system is in state 1 and part in state 2, because the free energy of this state is obviously a linear combination of those of 1 and 2. We note that point b lies lower than point a, which represents the free energy of a homogeneous system at the same T and V. Hence b, the phase separation case, is the equilibrium situation. Thus between the points 1 and 2 on the isotherm the system breaks up into two phases, with the pressure remaining constant. In other words the system undergoes a first-order phase transition. In the P-V diagram the points 1 and 2 are so located that the areas A and B are equal. To show this, let us write down all the conditions determining 1 and 2:

$$-\frac{\partial A}{\partial V_1} = -\frac{\partial A}{\partial V_2} \qquad \text{(equal pressure)}$$

$$\frac{A_2 - A_1}{V_2 - V_1} = \frac{\partial A}{\partial V_1} \qquad \text{(common tangent)}$$

Combining these we can write

$$\left(-\frac{\partial A}{\partial V_1}\right)(V_2 - V_1) = -(A_2 - A_1)$$

or

$$P_1(V_2 - V_1) = \int_{V_1}^{V_2} P\,dV$$

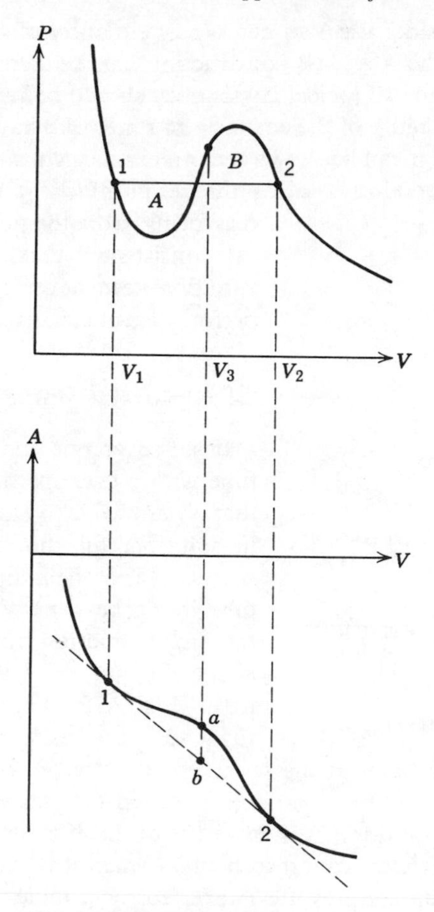

Fig. 2.11. The Maxwell construction.

whose geometrical meaning is precisely $A = B$. This geometrical construction is known as the Maxwell construction.

The Van der Waals equation of state and the Maxwell construction are instructive examples, but they have no foundation other than a heuristic one. The hypothesis that led to the Van der Waals equation is clearly *ad hoc*. The same may be said of the assumption underlying the Maxwell construction.

In particular, in making the Maxwell construction, it is necessary to use the portion of the isotherm that we declare to be unphysical and eventually discard. To criticize the Maxwell construction for its logical inconsistency would be out of place here; if we adhere to a strict logical standard, the Van der Waals equation of state should not even be contemplated. If however, we consider the Van der Waals equation of state purely for its heuristic value, the Maxwell construction can be admitted in the same spirit. The question of logical consistency should be raised only when we have a complete theory of the equation of state. We have this in statistical mechanics, where it can be shown that an exact calculation of the equation of state of matter always yields the result $\partial P/\partial V \leq 0$. It is therefore reassuring, from the point of view of logical consistency, that the Maxwell construction need never be used, if we can perform exact calculations.

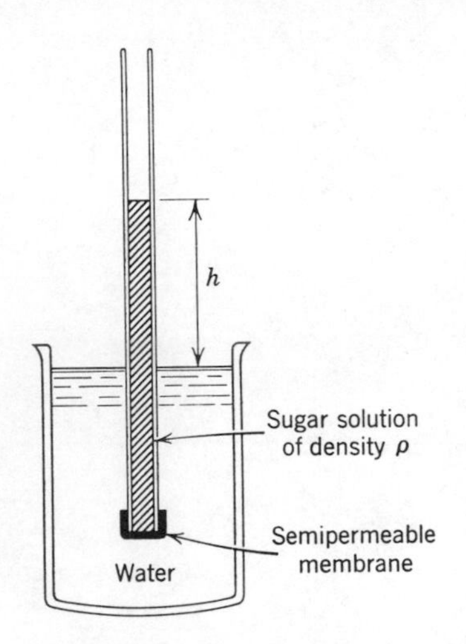

Fig. 2.12. Osmotic pressure.

2.4 OSMOTIC PRESSURE

If we cover one end of an open glass tube with a "semipermeable membrane" that is permeable to water but not to sugar in solution, fill the tube with a sugar solution, and then dip this end of the tube into a beaker of water, we find that the sugar solution rises to a height h above the level of the water, as illustrated in Fig. 2.12. This indicates that the sugar solution has a pressure ρgh higher than that of pure water at the same temperature. This pressure must be due to the presence of the sugar, and is called the *osmotic pressure* exerted by the sugar in solution. It is by virtue of this pressure that a living cell, which is mostly water, can absorb sugar when it is immersed in a sugar solution. The osmotic pressure P' exerted by n_1 moles of solute in a very dilute solution of temperature T and volume V is experimentally given by

$$P' = \frac{n_1 RT}{V} \tag{2.35}$$

We derive this result with the help of the second law of thermodynamics.

Consider a solution containing n_0 moles of solvent and n_1 moles of solute, with

$$n_1/n_0 \ll 1$$

The free energy of the solution can be obtained from the definition $A = U - TS$. To this end we first discuss the internal energy of the solution. It is a function of T, P, n_0, n_1. Further, it is assumed to be a homogeneous function of n_0 and n_1. That is, if n_0 and n_1 are simultaneously increased by a certain factor, U increases by the same factor. By making a Taylor series expansion we can write

$$U(T, P, n_0, n_1) = U(T, P, n_0, 0) + n_1(\partial U/\partial n_1)_{n_1=0} + \cdots$$

Up to first-order terms, with the assumption of homogeneity in mind, we rewrite this in the form

$$U(T, P, n_0, n_1) = n_0 u_0(T, P) + n_1 u_1(T, P) \tag{2.36}$$

Similarly the volume occupied by the solution is

$$V(T, P, n_0, n_1) = n_0 v_0(T, P) + n_1 v_1(T, P) \tag{2.37}$$

where u_0 and v_0 are the internal energy per mole and volume per mole of the pure solvent respectively. On the other hand, u_1 and v_1 have no comparably simple interpretation.

Let us now consider the entropy of the solution. Imagine that the solution undergoes an infinitesimal reversible transformation, with n_0 and n_1 held fixed. The change in entropy is the exact differential

$$dS = \frac{dQ}{T} = \frac{1}{T}(dU + P\,dV) = n_0\left[\frac{1}{T}(du_0 + P\,dv_0) + \frac{n_1}{n_0}\frac{1}{T}(du_1 + P\,dv_1)\right]$$

Since n_1/n_0 is arbitrary, the two differentials

$$ds_0 = \frac{1}{T}(du_0 + P\,dv_0), \qquad ds_1 = \frac{1}{T}(du_1 + P\,dv_1)$$

must be separately exact. Therefore the entropy has the form

$$S(T, P, n_0, n_1) = n_0 s_0(T, P) + n_1 s_1(T, P) + \lambda(n_0, n_1) \tag{2.38}$$

where the constant of integration $\lambda(n_0, n_1)$ does not depend on P, T. Accordingly we can find $\lambda(n_0, n_1)$ by making T so high and P so low that the solution completely evaporates and becomes a mixture of two ideal gases, the entropy of which we can explicitly calculate. It will be seen that the osmotic pressure arises solely from the term $\lambda(n_0, n_1)$.

The entropy of 1 mole of ideal gas at given T and P is

$$s(T, P) = c_P \log T - R \log P + K \tag{2.39}$$

where K is a numerical constant. The entropy of a mixture of two ideal gases of n_0 and n_1 moles respectively is

$$S_{\text{ideal}}(T, P, n_0, n_1) = (n_0 c_{P_0} + n_1 c_{P_1}) \log T - n_0 R \log P_0 - n_1 R \log P_1 + n_0 K_0 + n_1 K_1 \tag{2.40}$$

where P_0 and P_1 are the partial pressures of the two gases. To express the entropy in terms of the total pressure P, we make use of the facts

$$P = P_0 + P_1$$
$$P_0/n_0 = P_1/n_1 \tag{2.41}$$

which imply

$$P_0 = \frac{n_0 P}{n_0 + n_1} \approx P$$
$$P_1 = \frac{n_1 P}{n_0 + n_1} \approx \frac{n_1}{n_0} P \tag{2.42}$$

Thus

$$S_{\text{ideal}}(T, P, n_0, n_1) = (n_0 c_{P_0} + n_1 c_{P_1}) \log T - (n_0 + n_1) R \log P - n_1 R \log (n_1/n_0) + n_0 K_0 + n_1 K_1 \tag{2.43}$$

Comparison of this with (2.38) yields

$$\lambda(n_0, n_1) = -n_1 R \log (n_1/n_0) + n_0 K_0 + n_1 K_1 \tag{2.44}$$

The first term on the right-hand side is known as the *entropy of mixing*, since it arises solely from the mixing of the two gases without interaction. It is the entropy of mixing that gives rise to osmotic pressure. The free energy of the solution can now be written in the form

$$A(T, P, n_0, n_1) = n_0 a_0(T, P) + n_1 a_1(T, P) + n_1 RT \log (n_1/n_0) \tag{2.45}$$

where a_0 is the free energy per mole of the pure solvent. The explicit forms of a_1 and a_0 are irrelevant to our purpose.

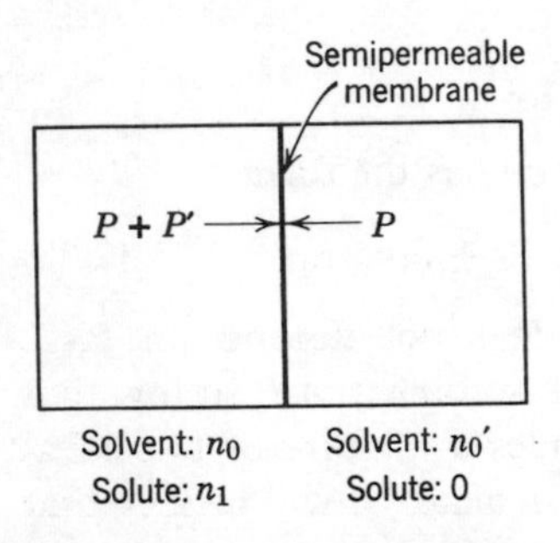

Fig. 2.13. Thought experiment in the derivation of the osmotic pressure.

To find the osmotic pressure, consider a solution separated from the pure solvent by a semipermeable membrane, as shown in Fig. 2.13. The pressure of the solution is by definition higher than that of the pure solvent by P', the osmotic pressure. The total free energy of this composite system is

$$A = (n_0 + n_0')a_0 + n_1 a_1 + n_1 RT \log (n_1/n_0) \tag{2.46}$$

Suppose the semipermeable membrane is displaced reversibly, with the temperature and the total volume of the composite system held fixed. Then n_0 suffers a change dn_0, and n_0' suffers a change $-dn_0$, for, as far as the pure solvent is concerned, the membrane is nonexistent. The volume of the *solution* changes by the amount $v_0\, dn_0$. The work done by the entire composite system is

$$dW = P' v_0\, dn_0 \tag{2.47}$$

According to the second law this is equal to the negative of the change in free energy dA, given by

$$-dA = \frac{n_1}{n_0} RT\, dn_0 \tag{2.48}$$

Therefore

$$P' = \frac{n_1 RT}{n_0 v_0}$$

Since $n_1/n_0 \ll 1$, however, $V = n_0 v_0$ is just the volume occupied by the solution. Hence

$$P' = \frac{n_1 RT}{V} \tag{2.49}$$

It is easy to see that the boiling point of a solution is higher than that of the pure solvent, on account of osmotic pressure. To deduce the change in boiling point, let us first find the difference between the vapor pressure of a solution and that of a pure solvent. This can be done by considering the arrangement shown in Fig. 2.14, which is self-explanatory. Under equilibrium conditions the difference between the pressures is the difference between the pressures of the vapor at B and at C. The pressure at C, however, is the same as that at A, because the vapor is at rest. Hence

$$\Delta P_{\text{vapor}} = P_B - P_C = P_B - P_A = \rho' g h \tag{2.50}$$

where ρ' is the mass density of the vapor. On the other hand, the osmotic

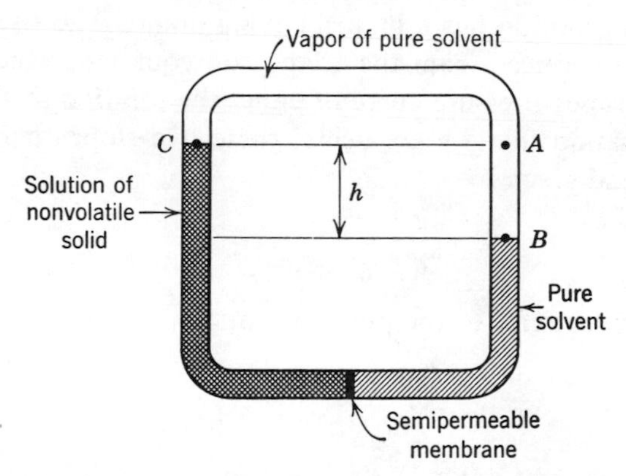

Fig. 2.14. Aid in the derivation of the difference in vapor pressure of a solution and the pure solvent.

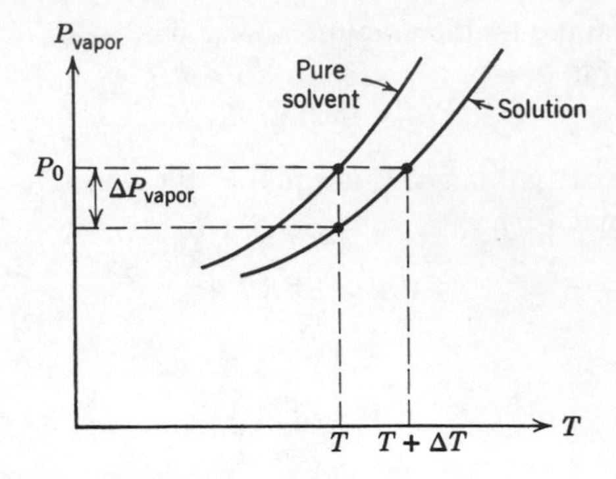

Fig. 2.15. Difference in boiling point between a solution and the pure solvent.

pressure is by definition equal to the pressure exerted by the column of solution of height h:

$$P' = \rho g h \tag{2.51}$$

where ρ is the density of the solution. Dividing (2.50) by (2.51) we have

$$\frac{\Delta P_{\text{vapor}}}{P'} = \frac{\rho'}{\rho} \tag{2.52}$$

which reduces, on using (2.49), to

$$\Delta P_{\text{vapor}} = \frac{\rho'}{\rho} \frac{n_1}{n_0} \frac{RT}{v_0} \tag{2.53}$$

where v_0 is the volume per mole of the solvent. Thus, at a given temperature, the solution has a lower vapor pressure than the pure solvent by the amount (2.53). The meaning of this formula may be made vivid by the qualitative plot of the vapor pressures in Fig. 2.15, from which we immediately see that the solution has a higher boiling point. The rise in boiling point ΔT can be deduced from the Clapeyron equation, which gives us the slope of the vapor pressure curve of either the solution or the solvent. In the approximation that we are using, these two slopes may be taken to be the same and given by

$$\frac{dP}{dT} = \frac{l}{T \Delta v}$$

where l and Δv both refer to the pure solvent. Therefore

$$\Delta T = \frac{\Delta P_{\text{vapor}}}{(dP/dT)}$$

or

$$\frac{\Delta T}{T} = \frac{\Delta v}{l} \frac{\rho'}{\rho} \frac{n_1}{n_0} \frac{RT}{v_0}$$

We may further make the approximation that the volume per mole of the solvent is negligible compared to that of its vapor, and that the vapor is an ideal gas.

$$\Delta v \approx \frac{RT}{P}$$

$$\frac{\rho'}{\rho} = \frac{Pv_0}{RT}$$

We obtain, with these approximations,

$$\frac{\Delta T}{T} = \frac{n_1}{n_0} \frac{RT}{l} \tag{2.54}$$

where l is the heat of vaporization per mole of the pure solvent.

We have seen that a substance in solution exerts osmotic pressure. A solution, in the derivation we have given, is any mixture of substances for which the entropy is greater than the sum of the entropies of the individual substances before they were mixed. Thermodynamics itself does not tell us what entropy really is, however, and therefore it does not tell us what constitutes a solution and what does not. For example, on purely thermodynamic grounds there is no way to answer the question, "Does a suspension of small particles in water exert osmotic pressure?" To answer this question we would have to form a definite opinion concerning the constitution of matter. Depending on our faiths, we can arrive at opposite answers. The following dialogue might be imagined between two believers of different faiths, who, for the present purpose, shall be identified as 𝔈 and 𝔇.

𝔈. The atomicity of matter points to the inescapable conclusion that a suspension of small particles in water exerts osmotic pressure. There is, in fact, no qualitative distinction between such a suspension and a sugar solution, for the latter is nothing but a suspension of individual molecules. If thermodynamics predicts osmotic pressure for a sugar solution, it predicts the same for a suspension of particles.

𝔇. It is evident, as I shall demonstrate, that a suspension of small particles cannot exert osmotic pressure. Your argument therefore precisely proves that the hypothesis of the atomicity of matter is incorrect. If a rock is placed into a glass of water, no one can argue that the pressure of the water is thereby increased. Break the rock in two. The pressure of the water will still be the same. Now keep breaking the rock into smaller and smaller pieces until you have a suspension of small rock particles. The pressure of the water against the wall of the glass should be completely unaffected, because the situation is not qualitatively different from the original one.

𝔈. In the atomic theory, water is made up of molecules too. A rock placed in a glass of water will suffer constant bombardments by water molecules. In any finite time interval, there is a small but finite probability that the rock will be pushed by the water molecules against the side of the glass. Therefore, in principle, a rock placed in a glass of water increases the pressure against the wall of the glass, although this increase is imperceptibly small. We merely have to replace the rock by a sugar molecule, and the description I have given furnishes the mechanism that gives rise to the observable osmotic pressure in a sugar solution.

𝔇. But this is contrary to common sense.

𝔈. To settle the argument, I suggest that an experiment be performed to detect the osmotic pressure of a suspension of small particles.

It is, of course, difficult to measure directly the osmotic pressure exerted by a suspension of particles, even if this pressure exists. For example, a suspension of 5×10^{10} particles/cc at room temperature would exert an osmotic pressure of 10^{-9} atm. Accordingly, indirect methods of detection have to be used. In 1905 Einstein proposed to measure the density $n(x)$ of a vertical column of suspended particles as a function of height x. If there were no osmotic pressure, all the suspended particles would eventually sink to the bottom. Assuming that there is osmotic pressure, we can deduce $n(x)$ as follows. The osmotic pressure at height x is

$$P'(x) = n(x)kT \tag{2.55}$$

If $n(x)$ is not a constant, there will be a net force per unit volume acting on the particles at height x, given by

$$F_{\text{osmotic}}(x) = -kT\frac{dn(x)}{dx} \tag{2.56}$$

The force per unit volume due to gravity, on the other hand, is

$$F_{\text{gravity}}(x) = -mgn(x) \tag{2.57}$$

where m is the mass of a suspended particle. Under equilibrium conditions these two forces must cancel.* Therefore

$$\frac{dn(x)}{dx} + \frac{mg}{kT}\,n(x) = 0 \tag{2.58}$$

from which follows immediately

$$n(x) = n(0)\,e^{-mgx/kT} \tag{2.59}$$

This formula was verified by experiments.

* We assume that the force of buoyancy can be neglected.

Equation (2.59) can also be derived from purely kinetic considerations. If the viscosity of the medium is η and the radius of a suspended particle is r, a suspended particle falling under gravity will eventually reach the terminal velocity $mg/6\pi r\eta$, according to Stokes' law. The flux of particles falling because of gravity is therefore

$$-\frac{n(x)mg}{6\pi r\eta}$$

On the other hand, when $n(x)$ is not a constant, these particles are expected from kinetic theory to diffuse, giving rise to a net upward flux of

$$D\frac{dn(x)}{dx}$$

where D is the coefficient of diffusion. In equilibrium these two fluxes must be equal. Therefore

$$\frac{dn(x)}{dx} + \frac{mg}{6\pi r\eta D}n(x) = 0 \tag{2.60}$$

Comparison of this with (2.58) yields

$$D = \frac{kT}{6\pi r\eta} \tag{2.61}$$

Experimental verification of this relation also constitutes a demonstration of the existence of osmotic pressure in a suspension.

Finally, from experiments involving suspensions, we can deduce atomic constants such as Avogadro's number. These experiments have to do with the motion of a single suspended particle (Brownian motion) and are beyond the scope of the present discussion.

PROBLEMS

2.1. What is the boiling point of water on Mt. Evans, Colorado, where the atmospheric pressure is two-thirds that at sea level?

2.2. A substance whose state is specified by P, V, T can exist in two distinct phases. At a given temperature T the two phases can coexist if the pressure is $P(T)$. The following information is known about the two phases. At the temperatures and pressures where they can coexist in equilibrium,
(i) there is no difference in the specific volume of the two phases;
(ii) there is no difference in the specific entropy of the two phases;
(iii) the specific heat c_P and the volume expansion coefficient α are different for the two phases.
(*a*) Find $dP(T)/dT$ as a function of T.
(*b*) What is the qualitative shape of the transition region in the P-V diagram?

In what way is it different from that of an ordinary gas-liquid transition?

The phase transition we have described is a *second-order transition.*

2.3. A cloud chamber contains water vapor at its equilibrium vapor pressure $P_\infty(T_0)$ corresponding to an absolute temperature T_0. Assume that
(i) the water vapor may be treated as an ideal gas;
(ii) the specific volume of water may be neglected compared to that of the vapor;
(iii) the latent heat l of condensation and $\gamma = c_P/c_V$ may be taken to be constants: $l = 540$ cal/g, $\gamma = \frac{3}{2}$.
(*a*) Calculate the equilibrium vapor pressure $P_\infty(T)$ as a function of the absolute temperature T.
(*b*) The water vapor is expanded adiabatically until the temperature is T, $T < T_0$. Assume the vapor is now supersaturated. If a small number of droplets of water is formed (catalyzed, e.g., by the presence of ions produced by the passage of an α-particle), what is the equilibrium radius of these droplets?
(*c*) In the approximations considered, does adiabatic expansion always lead to supersaturation?

2.4. Show that the heat capacity at constant volume C_V of a Van der Waals gas is a function of the temperature alone.

2.5. Derive the Maxwell construction by considerations involving the minimization of the Gibbs potential instead of the Helmholtz free energy.

2.6. Consider an open tank partitioned in two by a vertical semipermeable membrane that is permeable to water but not to sugar in solution. Fill the tank with water and dissolve sugar on the left side of the partition. The level of the sugar solution will be higher than that of the pure water because of osmotic pressure. Since the partition is permeable to water, will the water in the sugar solution leak out through the partition?

chapter 3

THE PROBLEM OF KINETIC THEORY

3.1 FORMULATION OF THE PROBLEM

The system under consideration in the classical kinetic theory of gases is a dilute gas of N molecules enclosed in a box of volume V. The temperature is sufficiently high and the density is sufficiently low for the molecules to be localized wave packets whose extensions are small compared to the average intermolecular distance. For this to be realized the average de Broglie wavelength of a molecule must be much smaller than the average interparticle separation:

$$\frac{\hbar}{\sqrt{2mkT}}\left(\frac{N}{V}\right)^{1/3} \ll 1 \tag{3.1}$$

Under such conditions each molecule may be considered a classical particle with a rather well-defined position and momentum. Furthermore, two molecules may be considered to be distinguishable from each other. The molecules interact with each other through collisions whose nature is specified through a given differential scattering cross section σ. Throughout our discussion of kinetic theory only the special case of a system of one kind of molecule will be considered.

An important simplification of the problem is made by ignoring the atomic structure of the walls containing the gas under consideration. That is, the physical walls of the container are replaced by idealized surfaces which act on an impinging gas molecule in a simple way, e.g., reflecting it elastically.

We are not interested in the motion of each molecule in detail. Rather, we are interested in the distribution function $f(\mathbf{r}, \mathbf{v}, t)$, so defined that

$$f(\mathbf{r}, \mathbf{v}, t)\, d^3r\, d^3v \tag{3.2}$$

is the number of molecules which, at time t, have positions lying within a volume element d^3r about $\mathbf{r}$ and velocities lying within a velocity-space element d^3v about $\mathbf{v}$. The volume elements d^3r and d^3v are not to be taken literally as mathematically infinitesimal quantities. They are finite volume elements which are large enough to contain a very large number of molecules and yet small enough so that compared to macroscopic dimensions they are essentially points. That such a choice is possible can be seen by an example. Under standard conditions there are about 3×10^{19} molecules/cc in a gas. If we choose $d^3r \sim 10^{-10}$ cc, which to us is small enough to be called a point, there are still on the order of 3×10^9 molecules in d^3r.

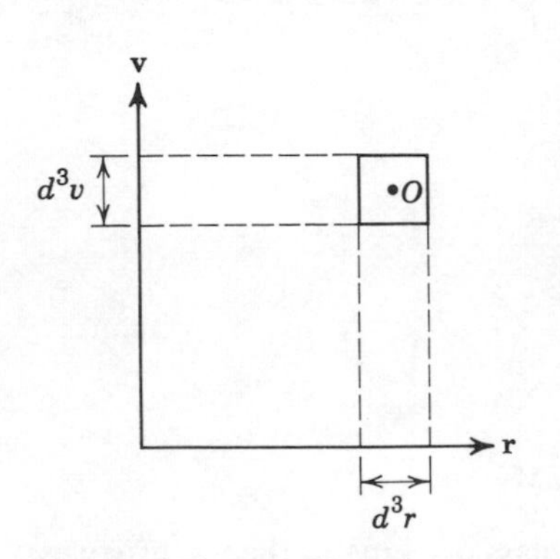

Fig. 3.1. The six-dimensional μ-space of a molecule.

To make the definition of $f(\mathbf{r}, \mathbf{v}, t)$ more precise, let us consider the six-dimensional space, called the μ-space, spanned by the coordinates* $(\mathbf{r}, \mathbf{v})$ of a molecule. The μ-space is schematically represented in Fig. 3.1. A point in this space represents a state of a molecule. At any instant of time, the state of the entire system of N molecules is represented by N points in μ-space. Let a volume element $d^3r\, d^3v$ be constructed about each point in μ-space, such as that shown about the point O in Fig. 3.1. If we count the number of points in this volume element, the result is by definition $f(\mathbf{r}, \mathbf{v}, t)\, d^3r\, d^3v$. If the sizes of these volume elements are chosen so that each of them contains a very large number of points, such as 10^9, and if the density of these points does not vary rapidly from one element to a neighboring element, then $f(\mathbf{r}, \mathbf{v}, t)$ may be regarded as a continuous function of its arguments. If we cover the entire μ-space with such volume elements, we can make the approximation

$$\sum f(\mathbf{r}, \mathbf{v}, t)\, d^3v\, d^3r \approx \int f(\mathbf{r}, \mathbf{v}, t)\, d^3v\, d^3r \tag{3.3}$$

* For brevity, the collection of spatial and velocity coordinates $(\mathbf{r}, \mathbf{v})$ is referred to as the coordinates of a molecule.

where the sum on the left extends over all the centers of the volume elements, and the integral on the right-hand side is taken in the sense of calculus. Such an approximation shall always be understood.

Having defined the distribution function, we can express the information that there are N molecules in the volume V through the normalization condition

$$\int f(\mathbf{r}, \mathbf{v}, t)\, d^3r\, d^3v = N \tag{3.4}$$

If the molecules are uniformly distributed in space, so that f is independent of $\mathbf{r}$, then

$$\int f(\mathbf{r}, \mathbf{v}, t)\, d^3v = \frac{N}{V} \tag{3.5}$$

The aim of kinetic theory is to find the distribution function $f(\mathbf{r}, \mathbf{v}, t)$ for a given form of molecular interaction. The limiting form of $f(\mathbf{r}, \mathbf{v}, t)$ as $t \to \infty$ would then contain all the equilibrium properties of the system. The aim of kinetic theory therefore includes the derivation of the thermodynamics of a dilute gas.

To fulfill this aim, our first task is to find the equation of motion for the distribution function. The distribution function changes with time, because molecules constantly enter and leave a given volume element in μ-space. Suppose there were no molecular collisions (i.e., $\sigma = 0$). Then a molecule with the coordinates $(\mathbf{r}, \mathbf{v})$ at the instant t will have the coordinates $(\mathbf{r} + \mathbf{v}\,\delta t, \mathbf{v} + (\mathbf{F}/m)\,\delta t)$ at the instant $t + \delta t$, where $\mathbf{F}$ is the external force acting on a molecule, and m is the mass of a molecule. We may take δt to be a truly infinitesimal quantity. Thus all the molecules contained in a μ-space element $d^3r\, d^3v$, at $(\mathbf{r}, \mathbf{v})$, at the instant t, will all be found in an element $d^3r'\, d^3v'$, at $(\mathbf{r} + \mathbf{v}\,\delta t, \mathbf{v} + (\mathbf{F}/m)\,\delta t)$, at the instant $t + \delta t$. Hence in the absence of collisions we have the equality

$$f\left(\mathbf{r} + \mathbf{v}\,\delta t, \mathbf{v} + \frac{\mathbf{F}}{m}\,\delta t, t + \delta t\right) d^3r'\, d^3v' = f(\mathbf{r}, \mathbf{v}, t)\, d^3r\, d^3v$$

which reduces to

$$f\left(\mathbf{r} + \mathbf{v}\,\delta t, \mathbf{v} + \frac{\mathbf{F}}{m}\,\delta t, t + \delta t\right) = f(\mathbf{r}, \mathbf{v}, t) \tag{3.6}$$

because $d^3r\, d^3v = d^3r'\, d^3v'$. The last fact is easily established if we assume that the external force $\mathbf{F}$ depends on position only. At any instant t, we may choose $d^3r\, d^3v$ to be a six-dimensional cube. It is sufficient to show that the area of any projection of this cube, say, $dx\, dv_x$, does not change. A simple calculation will show that this projection, originally a square, becomes a parallelogram of the same area in the time δt, as illustrated in

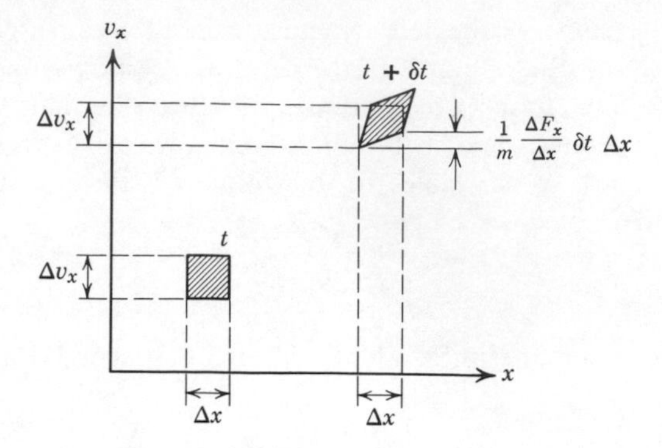

Fig. 3.2. The invariance of the volume element in μ-space under dynamical evolution in time.

Fig. 3.2. This invariance is valid under more general conditions if we describe the system by generalized coordinates and momenta instead of by **r**, **v**.

When there are collisions (i.e., $\sigma > 0$), equality (3.6) must be modified. We write

$$f\left(\mathbf{r} + \mathbf{v}\,\delta t, \mathbf{v} + \frac{\mathbf{F}}{m}\,\delta t, t + \delta t\right) = f(\mathbf{r}, \mathbf{v}, t) + \left(\frac{\partial f}{\partial t}\right)_{\text{coll}} \delta t \tag{3.7}$$

which *defines* $(\partial f/\partial t)_{\text{coll}}$. Expanding the left-hand side to the first order in δt, we obtain the equation of motion for the distribution function as we let $\delta t \to 0$:

$$\left(\frac{\partial}{\partial t} + \mathbf{v}\cdot\nabla_{\mathbf{r}} + \frac{\mathbf{F}}{m}\cdot\nabla_{\mathbf{v}}\right) f(\mathbf{r}, \mathbf{v}, t) = \left(\frac{\partial f}{\partial t}\right)_{\text{coll}} \tag{3.8}$$

where $\nabla_{\mathbf{r}}$, $\nabla_{\mathbf{v}}$ are, respectively, the gradient operators with respect to **r** and **v**. This equation is not meaningful until we explicitly specify $(\partial f/\partial t)_{\text{coll}}$. It is in specifying this term that the assumption that the system is a dilute gas becomes relevant.

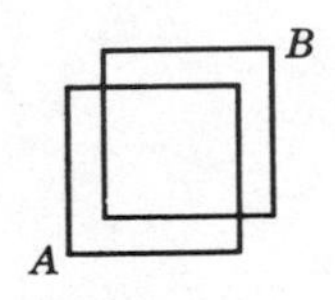

Fig. 3.3. A volume element in μ-space at the times t and $t + \delta t$.

An explicit form for $(\partial f/\partial t)_{\text{coll}}$ can be obtained by going back to its definition (3.7). Consider Fig. 3.3, where the square labeled A represents the μ-space volume element at $\{\mathbf{r}, \mathbf{v}, t\}$ and the one labeled B represents that at $\{\mathbf{r} + \mathbf{v}\,\delta t, \mathbf{v} + (\mathbf{F}/m)\,\delta t, t + \delta t\}$, where δt eventually tends to zero. During the time interval δt, some molecules in A will be removed from A by collision. We regard A as so small that *any* collision that a molecule in A suffers will knock it out of A. Such a molecule will not reach B. On the other hand, there are molecules outside A which, through collisions, will get into A during the time interval δt. These will be in B. Therefore the number of molecules in B at $t + \delta t$, as $\delta t \to 0$, equals the original number of molecules in A at time t plus the *net* gain of molecules in A due to collisions during the time interval δt. This statement is the content of (3.7), and may be expressed in the form

$$\left(\frac{\partial f}{\partial t}\right)_{\text{coll}} \delta t = (\bar{R} - R)\,\delta t \tag{3.9}$$

where $R\,\delta t\,d^3r\,d^3v$ = no. of collisions occurring during the time between t and $t + \delta t$ in which one of the *initial* molecules is in $d^3r\,d^3v$ about $(\mathbf{r}, \mathbf{v})$ (3.10)

$\bar{R}\,\delta t\,d^3r\,d^3v$ = no. of collisions occurring during the time between t and $t + \delta t$, in which one of the *final* molecules is in $d^3r\,d^3v$ about $(\mathbf{r}, \mathbf{v})$ (3.11)

Strictly speaking, we make a small error here. For example, in (3.10), we are implicitly assuming that if a molecule qualifies under the description, none of its partners in collision qualifies. This error is negligible because of the smallness of d^3v.

To proceed further, we assume that the gas is extremely dilute, so that we may consider only binary collisions and ignore the possibility that three or more molecules may collide simultaneously. This considerably simplifies the evaluation of R and $\bar{R}$. It is thus natural to study the nature of binary collisions next.

3.2 BINARY COLLISIONS

We consider an elastic collision in free space between two spinless molecules of equal mass. The molecules, as we have assumed, are wave packets with sufficiently well-defined position and velocity for us to be able to describe the initial and final states of the collision classically.

Let the velocities of the incoming molecules be $\mathbf{v}_1$, $\mathbf{v}_2$ and the velocities of the outgoing molecules be $\mathbf{v}_1'$, $\mathbf{v}_2'$. From the conservation of momentum and energy we have

$$\begin{aligned} \mathbf{v}_1 + \mathbf{v}_2 &= \mathbf{v}_1' + \mathbf{v}_2' \\ |\mathbf{v}_1|^2 + |\mathbf{v}_2|^2 &= |\mathbf{v}_1'|^2 + |\mathbf{v}_2'|^2 \end{aligned} \tag{3.12}$$

Introducing the new variables

$$\begin{aligned} \mathbf{V} &= \tfrac{1}{2}(\mathbf{v}_1 + \mathbf{v}_2) \\ \mathbf{u} &= \mathbf{v}_2 - \mathbf{v}_1 \end{aligned} \tag{3.13}$$

and variables $\mathbf{V}'$, $\mathbf{u}'$ similarly defined, we can rewrite (3.12) in the form

$$\begin{aligned} \mathbf{V} &= \mathbf{V}' \\ |\mathbf{u}| &= |\mathbf{u}'| \end{aligned} \tag{3.14}$$

It is convenient to represent these conditions geometrically, as shown in Fig. 3.4. We see that the collision merely rotates $\mathbf{u}$ into $\mathbf{u}'$, without changing its magnitude. Specifying $\mathbf{V}$, $\mathbf{u}$, and the angles θ, ϕ of $\mathbf{u}'$ with respect to $\mathbf{u}$ completely determines the collision. The angles θ, ϕ are called the scattering angles.

Imagine $\mathbf{V}$ and $\mathbf{u}$ slightly changed, to $\mathbf{V} + d\mathbf{V}$ and $\mathbf{u} + d\mathbf{u}$ respectively, with θ, ϕ kept fixed.* That is, consider a collision where the initial state is slightly different from the original one. Then $\mathbf{V}'$ and $\mathbf{u}'$ change respectively to $\mathbf{V}' + d\mathbf{V}'$ and $\mathbf{u}' + d\mathbf{u}'$. Since $\mathbf{V} = \mathbf{V}'$, we have $d\mathbf{V} = d\mathbf{V}'$. Since θ is

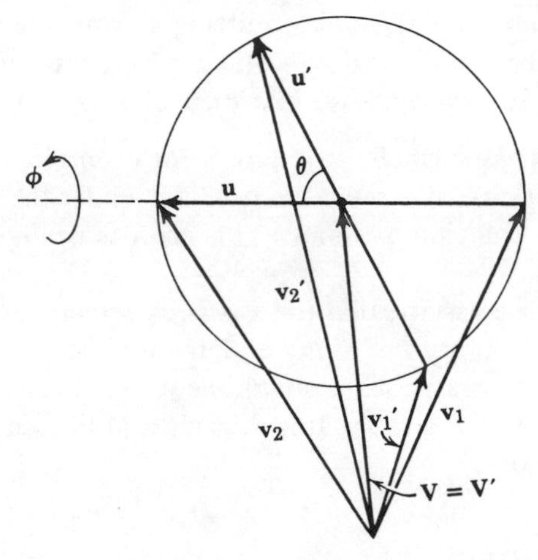

Fig. 3.4. Geometrical interpretation of the conservation of energy and momentum in a binary collision. The vectors $\mathbf{u}$ and $\mathbf{u}'$ always lie along a diameter of the sphere.

fixed, we have $|d\mathbf{u}| = |d\mathbf{u}'|$. Therefore we have

$$d^3V\, d^3u = d^3V'\, d^3u' \tag{3.15}$$

where $d^3V = dV_x\, dV_y\, dV_z$, etc. From (3.13) we readily verify that $d^3V\, d^3u = d^3v_1\, d^3v_2$, with a similar equality for the primed quantities. Therefore, if we change $\mathbf{v}_1$ and $\mathbf{v}_2$ slightly, keeping the scattering angles fixed, then $\mathbf{v}_1'$ and $\mathbf{v}_2'$ change in such a way that

$$d^3v_1\, d^3v_2 = d^3v_1'\, d^3v_2' \tag{3.16}$$

The total velocity $\mathbf{V}$ is not a very interesting quantity. In fact, if we translate our coordinate system with a uniform velocity $\mathbf{V}$, then in the new coordinate system only the relative velocities $\mathbf{u}$ and $\mathbf{u}'$ need be considered. Such a coordinate system is called the center-of-mass system. The collision process as viewed in the laboratory coordinate system and that as viewed in the center-of-mass coordinate system are shown in Fig. 3.5.

* This means that in the center-of-mass system the impact parameter is fixed. In terms of Fig. 3.4, it means that θ and the angle between $\mathbf{u} \times \mathbf{v}_2$ and $\mathbf{u}' \times \mathbf{v}_2'$ are both fixed.

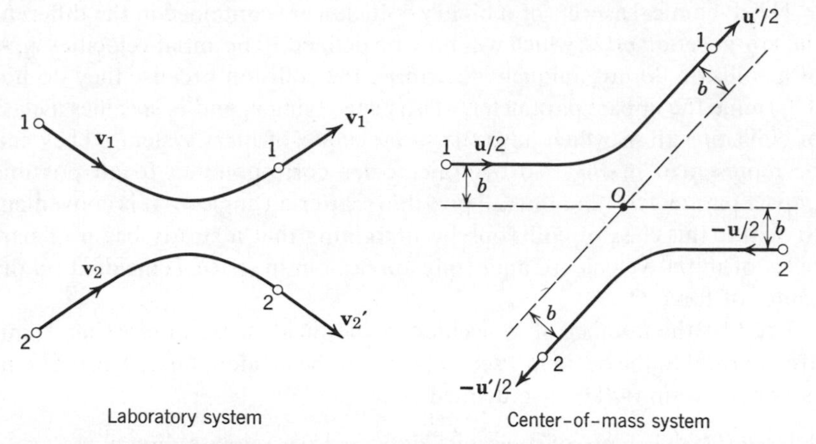

Fig. 3.5. A binary collision as viewed in the laboratory coordinate system and in the center-of-mass coordinate system.

In the center-of-mass system it suffices to focus our attention on one of the molecules, because its partner always moves oppositely. Thus the problem reduces to an equivalent problem of the scattering of a molecule by a fictitious fixed center of force, represented by the point O in Fig. 3.5. This molecule approaches O with velocity $\mathbf{u}$, whose perpendicular distance to O is called the impact parameter b. Let a frame of reference be chosen with O located at the origin of the coordinate system with the z axis parallel to $\mathbf{u}$. Since $|\mathbf{u}'| = |\mathbf{u}|$, the final state is specified by the two scattering angles θ and ϕ, collectively denoted by Ω, with θ the angle between $\mathbf{u}'$ and the z axis and ϕ the azimuth angle of $\mathbf{u}'$ about the z axis, as shown in Fig. 3.6. This completes the kinematic description of a binary collision.

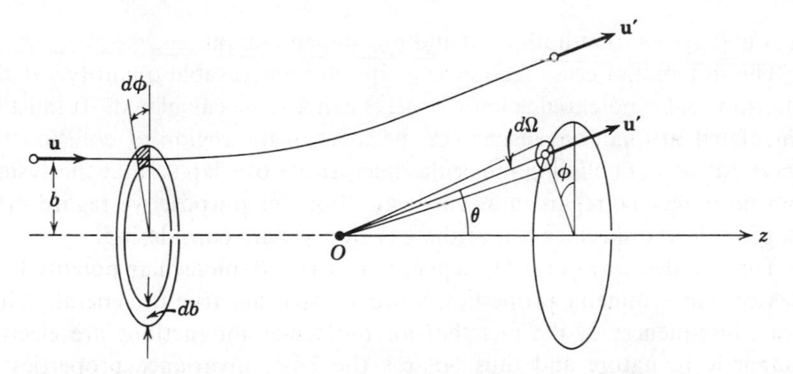

Fig. 3.6. Scattering of a molecule by a fixed center of force O.

The dynamical aspects of a binary collision are contained in the differential cross section $\sigma(\Omega)$, which will now be defined. The initial velocities $\mathbf{v}_1, \mathbf{v}_2$ of a collision do not uniquely determine the collision because they do not determine the impact parameter. Thus specifying $\mathbf{v}_1$ and $\mathbf{v}_2$ specifies a class of collisions all of which have the same center-of-mass system. They can be represented in Fig. 3.6 by trajectories corresponding to all possible impact parameters (and thus all possible scattering angles). It is convenient to picture this class of collisions by imagining that a steady beam of particles of initial velocity $\mathbf{u}$, uniformly spread out in space, is incident on the center of force O.

Let I be the number of molecules in the incident beam crossing a unit area normal to the beam in 1 sec. I is called the incident flux. The differential cross section $\sigma(\Omega)$ is so defined that

$$I\sigma(\Omega)\, d\Omega = \text{no. of molecules deflected per sec in a direction lying within a solid angle element } d\Omega \tag{3.17}$$

It is obvious from Fig. 3.6 that

$$I\sigma(\Omega)\, d\Omega = Ib\, db\, d\phi \tag{3.18}$$

The differential cross section $\sigma(\Omega)$ has the dimension of area and has the following geometrical meaning:

$$\begin{array}{c}\text{no. of molecules deflected}\\ \text{into } d\Omega \text{ per sec}\end{array} = \begin{array}{l}\text{no. of molecules crossing}\\ \text{an area equal to } \sigma(\Omega)\, d\Omega\\ \text{per sec, in incident beam}\end{array} \tag{3.19}$$

The total cross section σ_{tot} is the integral of $\sigma(\Omega)$ over all solid-angle elements:

$$\sigma_{tot} = \int \sigma(\Omega)\, d\Omega \tag{3.20}$$

It is understood that both $\sigma(\Omega)$ and σ_{tot} depend on $|\mathbf{u}|$.

The differential cross section is a directly measurable quantity. If the intermolecular potential is known, $\sigma(\Omega)$ can also be calculated. It must be calculated in quantum mechanics, because in the region of collision the wave packets of colliding molecules necessarily overlap so that the system can no longer be regarded as classical. For our purpose we regard $\sigma(\Omega)$ as given in the specification of the gas that we are considering.

The detailed form of $\sigma(\Omega)$ depends on the intermolecular potential. It has certain symmetry properties, however, that are true in general. They are consequences of the fact that all molecular interactions are electromagnetic in nature and thus possess the basic invariance properties of electromagnetic interactions. We now enumerate the relevant symmetries.

For this purpose it is convenient to introduce the notation

$$\sigma(\mathbf{v}_1, \mathbf{v}_2 \mid \mathbf{v}_1', \mathbf{v}_2') \equiv \sigma(\Omega) \tag{3.21}$$

where $\mathbf{v}_1$, $\mathbf{v}_2$ and $\mathbf{v}_1'$, $\mathbf{v}_2'$ have the same meaning as in (3.12), and

$$\Omega \equiv \text{the angles between } \mathbf{v}_2 - \mathbf{v}_1 \text{ and } \mathbf{v}_2' - \mathbf{v}_1' \tag{3.22}$$

(*a*) *Invariance under Time Reversal*

$$\sigma(\mathbf{v}_1, \mathbf{v}_2 \mid \mathbf{v}_1', \mathbf{v}_2') = \sigma(-\mathbf{v}_1', -\mathbf{v}_2' \mid -\mathbf{v}_1, -\mathbf{v}_2) \tag{3.23}$$

This expresses the property that if we reverse the sense of time each molecule will retrace its original course.

(*b*) *Invariance under Rotation and Reflection*

$$\sigma(\mathbf{v}_1, \mathbf{v}_2 \mid \mathbf{v}_1', \mathbf{v}_2') = \sigma(\mathbf{v}_1^*, \mathbf{v}_2^* \mid \mathbf{v}_1'^*, \mathbf{v}_2'^*) \tag{3.24}$$

Here $\mathbf{v}^*$ denotes the vector obtained from $\mathbf{v}$ after performing a given rotation in space or a reflection with respect to a given plane, or a combination of both.

Let the *inverse collision* be defined as the collision that differs from a given collision by interchanging the initial and the final state. As a consequence of the symmetries we have stated, the inverse collision has the same differential cross section as the collision:

$$\sigma(\mathbf{v}_1, \mathbf{v}_2 \mid \mathbf{v}_1', \mathbf{v}_2') = \sigma(\mathbf{v}_1', \mathbf{v}_2' \mid \mathbf{v}_1, \mathbf{v}_2) \tag{3.25}$$

To prove this let us represent a collision by the schematic drawing A of Fig. 3.7, which is self-explanatory. The diagram A' beneath it has the same meaning as Fig. 3.4. The cross section for this collision is the same as the time-reversed collision represented by B. Now rotate the coordinate system through 180° about a suitable axis $\mathbf{n}$ perpendicular to the total momentum, and then reflect with respect to a plane pp' perpendicular to $\mathbf{n}$. As a result we obtain the collision D, which is the inverse of the original collision and which has the same cross section because of (3.23) and (3.24).

The symmetry laws (3.23) and (3.24) require amendment if the molecules have spin. By making use of the transformation properties of the spin quantum number under time reversal, rotation, and reflection, we can arrive at formally the same conditions as (3.23) and (3.24) if $\mathbf{v}$ is understood to denote collectively the velocity and the spin quantum number. It is then clear that (3.25) remains valid, if each cross section in (3.25) is replaced by the sum of the cross section over all spin quantum numbers in the initial and the final state.

If collisions were treated classically, we might encounter features that are irrelevant to molecular collisions. In particular, in the classical collision between two macroscopic objects, (3.25) can be easily violated.

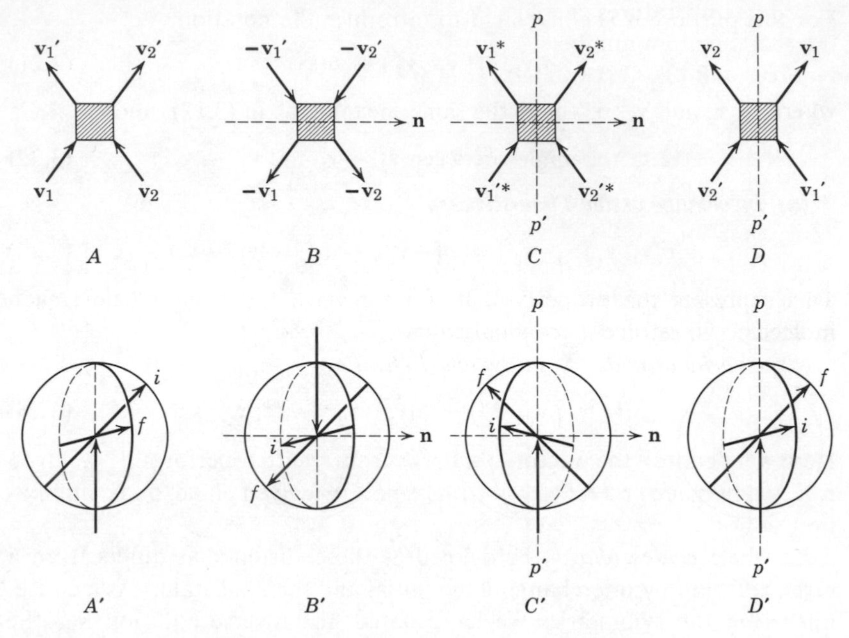

Fig. 3.7. Symmetry operations that take a collision to the inverse collision. In A', B', C', and D', i and f respectively denote initial and final relative velocities.

As a concrete example consider the collision between a sphere and a wedge.* A glance at Fig. 3.8 shows that the inverse collision can be very different from the original collision. This point, however, is not relevant to molecular collisions, because no molecule is a concrete wedge. The state of a molecule is completely specified by the total momentum, the spin quantum number, and the internal quantum numbers. In collisions

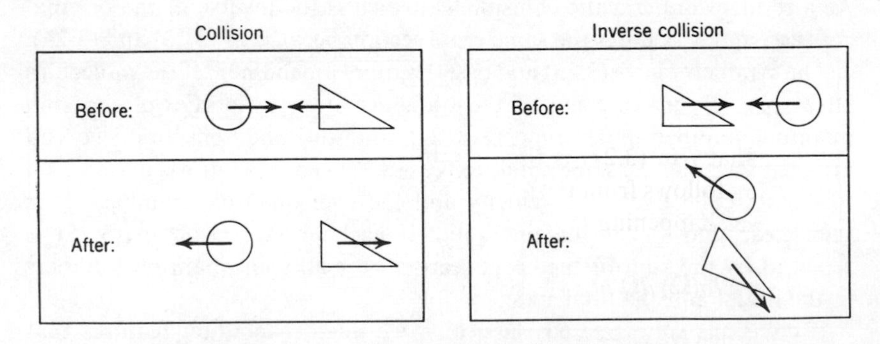

Fig. 3.8. Classical collision between macroscopic objects.

* Both made of concrete.

at room temperatures energies are generally not high enough to change the internal quantum numbers. The internal quantum numbers may hence be ignored, and (3.25) is valid.

3.3 BOLTZMANN TRANSPORT EQUATION

To derive an explicit formula for $(\partial f/\partial t)_{\text{coll}}$, we make the following approximations:

(*a*) Only binary collisions are taken into account. This is valid if the gas is sufficiently dilute.

(*b*) The walls of the container are ignored. This is shown later to be justified.

(*c*) The effect of the external force on the collision cross section is ignored.

(*d*) The velocity of a molecule is uncorrelated with its position.

Assumption (*d*) is known as the *assumption of molecular chaos.* In more precise terms, it states that in a spatial volume element d^3r the number of pairs of molecules with respective velocities lying in the velocity volume elements d^3v_1 about $\mathbf{v}_1$, and d^3v_2 about $\mathbf{v}_2$, is

$$[f(\mathbf{r}, \mathbf{v}_1, t)\, d^3r\, d^3v_1][f(\mathbf{r}, \mathbf{v}_2, t)\, d^3r\, d^3v_2] \tag{3.26}$$

This assumption is introduced as a mathematical convenience for the description of the state of the gas under consideration. It is clearly a *possible* condition for the gas, but it is not clear whether it is a general condition. We analyze its meaning in Sec. 4.4.

We now proceed to find an explicit expression for $(\partial f/\partial t)_{\text{coll}}$. The rate of decrease of $f(\mathbf{r}, \mathbf{v}_1, t)$ owing to collisions, denoted by R in (3.10), can be obtained by directing our attention to a molecule whose velocity lies in d^3v_1 about $\mathbf{v}_1$, in the spatial volume d^3r about $\mathbf{r}$. In the same spatial volume there are molecules of any velocity $\mathbf{v}_2$ that pose as an incident beam of molecules incident on the molecule $\mathbf{v}_1$. The flux of this incident beam is

$$I = [f(\mathbf{r}, \mathbf{v}_2, t)\, d^3v_2]\, |\mathbf{v}_1 - \mathbf{v}_2| \tag{3.27}$$

The fact that f in (3.27) is the same distribution function we have been considering follows from the assumption of molecular chaos. The number of collisions happening in d^3r during δt of the type* $\{\mathbf{v}_1, \mathbf{v}_2\} \to \{\mathbf{v}_1', \mathbf{v}_2'\}$ is given by

$$I\sigma(\Omega)\, d\Omega\, \delta t = f(\mathbf{r}, \mathbf{v}_2, t)\, d^3v_2\, |\mathbf{v}_2 - \mathbf{v}_1|\, \sigma(\Omega)\, d\Omega\, \delta t \tag{3.28}$$

* The notation $\{\mathbf{v}_1, \mathbf{v}_2\} \to \{\mathbf{v}_1', \mathbf{v}_2'\}$ denotes the binary collision in which the velocities of the colliding molecules are $\mathbf{v}_1$ and $\mathbf{v}_2$ and the velocities of the product molecules are $\mathbf{v}_1'$ and $\mathbf{v}_2'$.

where $\sigma(\Omega)$ is the differential cross section in the center-of-mass system and Ω is the angles between $\mathbf{v}_2 - \mathbf{v}_1$ and $\mathbf{v}_2' - \mathbf{v}_1'$. The rate R is obtained by summing (3.28) over all $\mathbf{v}_2$ and then multiplying the result by the spatial density of molecules in d^3v_1:

$$R = f(\mathbf{r}, \mathbf{v}_1, t) \int d^3v_2 \int d\Omega\sigma(\Omega)\, |\mathbf{v}_1 - \mathbf{v}_2|\, f(\mathbf{r}, \mathbf{v}_2, t) \tag{3.29}$$

The integration over Ω can be immediately effected to yield the total cross section. We prefer, however, to leave (3.29) in its present form.

In a similar fashion we can calculate $\bar{R}$, defined in (3.11). We are now interested in collisions of the type $\{\mathbf{v}_1', \mathbf{v}_2'\} \rightarrow \{\mathbf{v}_1, \mathbf{v}_2\}$ where $\mathbf{v}_1$ is fixed. Consider a molecule $\mathbf{v}_1'$ with a beam of molecules $\mathbf{v}_2'$ incident upon it. The incident flux is

$$f(\mathbf{r}, \mathbf{v}_2', t)\, d^3v_2'\, |\mathbf{v}_2' - \mathbf{v}_1'| \tag{3.30}$$

The number of collisions of this type during δt is

$$f(\mathbf{r}, \mathbf{v}_2', t)\, d^3v_2'\, |\mathbf{v}_2' - \mathbf{v}_1'|\, \sigma'(\Omega)\, d\Omega\, \delta t \tag{3.31}$$

The rate $\bar{R}$ is given by

$$\bar{R}\, d^3v_1 = \int d^3v_2' \int d\Omega\sigma'(\Omega)\, |\mathbf{v}_2' - \mathbf{v}_1'|\, [f(\mathbf{r}, \mathbf{v}_1', t)\, d^3v_1'] f(\mathbf{r}, \mathbf{v}_2', t) \tag{3.32}$$

Now the vectors $\mathbf{v}_1, \mathbf{v}_2, \mathbf{v}_1', \mathbf{v}_2'$ refer to collisions that are inverses of each other, hence $\sigma'(\Omega) = \sigma(\Omega)$. From (3.14) and (3.16) we have

$$|\mathbf{v}_1 - \mathbf{v}_2| = |\mathbf{v}_1' - \mathbf{v}_2'|$$
$$d^3v_1\, d^3v_2 = d^3v_1'\, d^3v_2'$$

Therefore

$$\bar{R} = \int d^3v_2 \int d\Omega\sigma(\Omega)\, |\mathbf{v}_1 - \mathbf{v}_2|\, f(\mathbf{r}, \mathbf{v}_1', t) f(\mathbf{r}, \mathbf{v}_2', t) \tag{3.33}$$

It is to be noted that here $\mathbf{v}_1$ is fixed, whereas $\mathbf{v}_1'$ and $\mathbf{v}_2'$ are functions of $\mathbf{v}_1$, $\mathbf{v}_2$ and Ω.

Combining the results for R and $\bar{R}$ we obtain

$$\left[\frac{\partial f(\mathbf{r}, \mathbf{v}_1, t)}{\partial t}\right]_{\text{coll}} = \bar{R} - R = \int d^3v_2 \int d\Omega\sigma(\Omega)\, |\mathbf{v}_1 - \mathbf{v}_2|\, (f_1' f_2' - f_1 f_2) \tag{3.34}$$

where $\sigma(\Omega)$ is the differential cross section for the collision

$$\{\mathbf{v}_1, \mathbf{v}_2\} \rightarrow \{\mathbf{v}_1', \mathbf{v}_2'\}$$

and the following abbreviations have been used:

$$\begin{aligned} f_1 &\equiv f(\mathbf{r}, \mathbf{v}_1, t) \\ f_2 &\equiv f(\mathbf{r}, \mathbf{v}_2, t) \\ f_1' &\equiv f(\mathbf{r}, \mathbf{v}_1', t) \\ f_2' &\equiv f(\mathbf{r}, \mathbf{v}_2', t) \end{aligned} \tag{3.35}$$

Substituting (3.34) into (3.8) we obtain the Boltzmann transport equation

$$\left(\frac{\partial}{\partial t} + \mathbf{v}_1 \cdot \nabla_r + \frac{\mathbf{F}}{m} \cdot \nabla_{\mathbf{v}_1}\right) f_1 = \int d\Omega \int d^3v_2 \sigma(\Omega) \, |\mathbf{v}_1 - \mathbf{v}_2| \, (f_2' f_1' - f_2 f_1) \tag{3.36}$$

which is a nonlinear integro-differential equation for f. If the molecules have spin, the same equation holds, provided all spin states are equally populated, so that f is independent of the spin quantum number. The problem of the kinetic theory of gases is now reduced to the mathematical problem of solving this equation.

If we did not make use of the assumption of molecular chaos, the quantity $(\partial f/\partial t)_{\text{coll}}$ would not bc expressible in terms of f itself. Instead, it would involve a two-particle correlation function which is independent of f. In place of (3.36), therefore, we would not have an equation for f, but an equation relating f to a two-particle correlation function. In general, we can obtain equations relating an n-particle correlation function to an $(n + 1)$-particle correlation function. In the general case, therefore, (3.36) is replaced by a set of N coupled equations.*

PROBLEMS

3.1. Give a few numerical examples to show that the condition (3.1) is fulfilled for physical gases at room temperatures.

3.2. Explain qualitatively why all molecular interactions are electromagnetic in origin.

3.3. For the collision between perfectly elastic spheres of diameter a,
(*a*) calculate the differential cross section with classical mechanics in the coordinate system in which one of the spheres is initially at rest;
(*b*) compare your answer with the quantum mechanical result. Consider both the low-energy and the high-energy limit. [See, e.g., L. I. Schiff, *Quantum Mechanics*, 2nd ed. (McGraw-Hill Book Co., New York, 1955), p. 110].

3.4. Consider a mixture of two gases whose molecules have masses m and M respectively and which are subjected to external forces $\mathbf{F}$ and $\mathbf{Q}$, respectively. Denote the respective distribution functions by f and g. Assuming that only binary collisions between molecules are important, derive the Boltzmann transport equation for the system.

* See N. N. Bogolubov in J. de Boer and G. E. Uhlenbeck, *Studies in Statistical Mechanics*, Vol. I (North-Holland Publishing Co., Amsterdam, 1962).

chapter 4

THE EQUILIBRIUM STATE OF A DILUTE GAS

4.1 BOLTZMANN'S *H* THEOREM

We define the equilibrium distribution function as the solution of the Boltzmann transport equation that is independent of time. We shall see that it is also the limiting form of the distribution function as the time tends to infinity. Assume that there is no external force. It is then consistent to assume further that the distribution function is independent of $\mathbf{r}$ and hence can be denoted by $f(\mathbf{v}, t)$. The equilibrium distribution function, denoted by $f_0(\mathbf{v})$, is the solution to the equation $\partial f(\mathbf{v}, t)/\partial t = 0$. According to the Boltzmann transport equation (3.36), $f_0(\mathbf{v})$ satisfies the integral equation

$$0 = \int d^3v_2 \int d\Omega\sigma(\Omega)\,|\mathbf{v}_2 - \mathbf{v}_1|\,[f_0(\mathbf{v}_2')f_0(\mathbf{v}_1') - f_0(\mathbf{v}_2)f_0(\mathbf{v}_1)] \tag{4.1}$$

where $\mathbf{v}_1$ is a given velocity.

A sufficient condition for $f_0(\mathbf{v})$ to solve (4.1) is

$$f_0(\mathbf{v}_2')f_0(\mathbf{v}_1') - f_0(\mathbf{v}_2)f_0(\mathbf{v}_1) = 0 \tag{4.2}$$

where $\{\mathbf{v}_1, \mathbf{v}_2\} \to \{\mathbf{v}_1', \mathbf{v}_2'\}$ is any possible collision (i.e., one with non-vanishing cross section). We show that this condition is also necessary, and we thus arrive at the interesting conclusion that $f_0(\mathbf{v})$ is independent of $\sigma(\Omega)$, as long as the latter is nonzero.

To show the necessity of (4.2) we define with Boltzmann the functional

$$H(t) \equiv \int d^3v\, f(\mathbf{v}, t) \log f(\mathbf{v}, t) \tag{4.3}$$

where $f(\mathbf{v}, t)$ is the distribution function at time t, satisfying

$$\frac{\partial f(\mathbf{v}_1, t)}{\partial t} = \int d^3v_2 \int d\Omega\sigma(\Omega)\, |\mathbf{v}_2 - \mathbf{v}_1|\, (f_2' f_1' - f_2 f_1) \tag{4.4}$$

Differentiation of (4.3) yields

$$\frac{dH(t)}{dt} = \int d^3v\, \frac{\partial f(\mathbf{v}, t)}{\partial t}\, [1 + \log f(\mathbf{v}, t)] \tag{4.5}$$

Therefore $\partial f/\partial t = 0$ implies $dH/dt = 0$. This means that a necessary condition for $\partial f/\partial t = 0$ is $dH/dt = 0$. We now show that the statement

$$\frac{dH}{dt} = 0 \tag{4.6}$$

is the same as (4.2). It would then follow that (4.2) is also a necessary condition for the solution of (4.1). To this end we prove the following theorem.

BOLTZMANN'S *H* THEOREM. If f satisfies the Boltzmann transport equation, then

$$\frac{dH(t)}{dt} \leq 0 \tag{4.7}$$

Proof. Substituting (4.4) into the integrand of (4.5) we have*

$$\frac{dH}{dt} = \int d^3v_1 \int d^3v_2 \int d\Omega\sigma(\Omega)\, |\mathbf{v}_1 - \mathbf{v}_2|\, (f_2' f_1' - f_2 f_1)(1 + \log f_1) \tag{4.8}$$

Interchanging $\mathbf{v}_1$ and $\mathbf{v}_2$ in this integrand leaves the integral invariant because $\sigma(\Omega)$ is invariant under such an interchange. Making this change of variables of integration and taking one-half of the sum of the new expression and (4.8), we obtain

$$\frac{dH}{dt} = \frac{1}{2} \int d^3v_1 \int d^3v_2 \int d\Omega\sigma(\Omega)\, |\mathbf{v}_2 - \mathbf{v}_1|\, (f_2' f_1' - f_2 f_1)[2 + \log (f_1 f_2)] \tag{4.9}$$

This integral is invariant under the interchange of $\{\mathbf{v}_1, \mathbf{v}_2\}$ and $\{\mathbf{v}_1', \mathbf{v}_2'\}$ because for every collision there is an inverse collision with the same cross

* Note that the use of (4.4) presupposes that the state of the system under consideration satisfies the assumption of molecular chaos.

section. Hence

$$\frac{dH}{dt} = \frac{1}{2} \int d^3v_1' \int d^3v_2' \int d\Omega \sigma'(\Omega) \, |\mathbf{v}_2' - \mathbf{v}_1'| \, (f_2 f_1 - f_2' f_1')[2 + \log (f_1' f_2')] \tag{4.10}$$

Noting that $d^3v_1' \, d^3v_2' = d^3v_1 \, d^3v_2$, $|\mathbf{v}_2' - \mathbf{v}_1'| = |\mathbf{v}_2 - \mathbf{v}_1|$, and $\sigma'(\Omega) = \sigma(\Omega)$, we take half the sum of (4.9) and (4.10) and obtain

$$\frac{dH}{dt} = \frac{1}{4} \int d^3v_1 \int d^3v_2 \int d\Omega \sigma(\Omega) \, |\mathbf{v}_2 - \mathbf{v}_1| \, (f_2' f_1' - f_2 f_1)[\log (f_1 f_2) - \log(f_1' f_2')] \tag{4.11}$$

The integrand of the integral in (4.11) is never positive. (QED)

As a byproduct of the proof, we deduce from (4.11) that $dH/dt = 0$ if and only if the integrand of (4.11) identically vanishes. This proves that the statement (4.6) is identical with (4.2). It also shows that under an arbitrary initial condition $f(\mathbf{v}, t) \xrightarrow[t\to\infty]{} f_0(\mathbf{v})$.

4.2 THE MAXWELL-BOLTZMANN DISTRIBUTION

It has been shown that the equilibrium distribution function $f_0(\mathbf{v})$ is a solution of (4.2). It will be called the Maxwell-Boltzmann distribution. To find it, let us take the logarithm of both sides of (4.2):

$$\log f_0(\mathbf{v}_1) + \log f_0(\mathbf{v}_2) = \log f_0(\mathbf{v}_1') + \log f_0(\mathbf{v}_2') \tag{4.12}$$

Since $\{\mathbf{v}_1, \mathbf{v}_2\}$ and $\{\mathbf{v}_1', \mathbf{v}_2'\}$ are respectively the initial and final velocities of *any* possible collision, (4.12) has the form of a conservation law. If $\chi(\mathbf{v})$ is any quantity associated with a molecule of velocity $\mathbf{v}$, such that $\chi(\mathbf{v}_1) + \chi(\mathbf{v}_2)$ is conserved in a collision between molecules $\mathbf{v}_1$ and $\mathbf{v}_2$, a solution of (4.12) is

$$\log f_0(\mathbf{v}) = \chi(\mathbf{v})$$

The most general solution of (4.12) is

$$\log f_0(\mathbf{v}) = \chi_1(\mathbf{v}) + \chi_2(\mathbf{v}) + \cdots$$

where the list $\chi_1, \chi_2, \ldots$ exhausts all independently conserved quantities. For spinless molecules these are the energy and the momentum of a molecule, and, of course, a constant. Hence $\log f$ is a linear combination of v^2 and the three components of $\mathbf{v}$ plus an arbitrary constant:

$$\log f_0(\mathbf{v}) = -A(\mathbf{v} - \mathbf{v}_0)^2 + \log C$$

or

$$f_0(\mathbf{v}) = Ce^{-A(\mathbf{v}-\mathbf{v}_0)^2} \tag{4.13}$$

where C, A, and the three components of $\mathbf{v}_0$ are five arbitrary constants.

We can determine these constants in terms of observed properties of the system.

Applying the condition (3.5), and denoting the particle density N/V by n we have

$$n = C\int d^3v\, e^{-A(\mathbf{v}-\mathbf{v}_0)^2} = C\int d^3v\, e^{-Av^2} = C\left(\frac{\pi}{A}\right)^{3/2}$$

from which we conclude that $A > 0$ and

$$C = \left(\frac{A}{\pi}\right)^{3/2} n \tag{4.14}$$

Let the average velocity $\langle \mathbf{v} \rangle$ of a gas molecule be defined by

$$\langle \mathbf{v} \rangle \equiv \frac{\int d^3v\, \mathbf{v}\, f_0(\mathbf{v})}{\int d^3v\, f_0(\mathbf{v})} \tag{4.15}$$

Then

$$\langle \mathbf{v} \rangle = \frac{C}{n}\int d^3v\, \mathbf{v}e^{-A(\mathbf{v}-\mathbf{v}_0)^2} = \frac{C}{n}\int d^3v\, (\mathbf{v}+\mathbf{v}_0)e^{-Av^2} = \mathbf{v}_0 \tag{4.16}$$

Thus we must take $\mathbf{v}_0 = 0$, if the gas has no translational motion as a whole.

Next we calculate the average energy ϵ of a molecule, defined by

$$\epsilon \equiv \frac{\int d^3v\, \frac{1}{2}mv^2 f_0(\mathbf{v})}{\int d^3v\, f_0(\mathbf{v})} \tag{4.17}$$

We have, setting $\mathbf{v}_0 = 0$:

$$\epsilon = \frac{mC}{2n}\int d^3v\, v^2 e^{-Av^2} = \frac{2\pi mC}{n}\int_0^\infty dv\, v^4 e^{-Av^2} = \frac{3}{4}\frac{m}{A}$$

The constant A is therefore related to the average energy by

$$A = \frac{3}{4}\frac{m}{\epsilon} \tag{4.18}$$

Substituting this into (4.14) we obtain for the constant C the expression

$$C = n\left(\frac{3m}{4\pi\epsilon}\right)^{3/2} \tag{4.19}$$

To relate the average energy ϵ to a directly measurable quantity, let us find the equation of state corresponding to the equilibrium distribution function. We do this by calculating the pressure, which is defined as the average force per unit area exerted by the gas on one face of a perfectly reflecting plane exposed to the gas. Let the disk shown in Fig. 4.1 represent such a unit area, and let us call the axis normal to it the x axis. A molecule

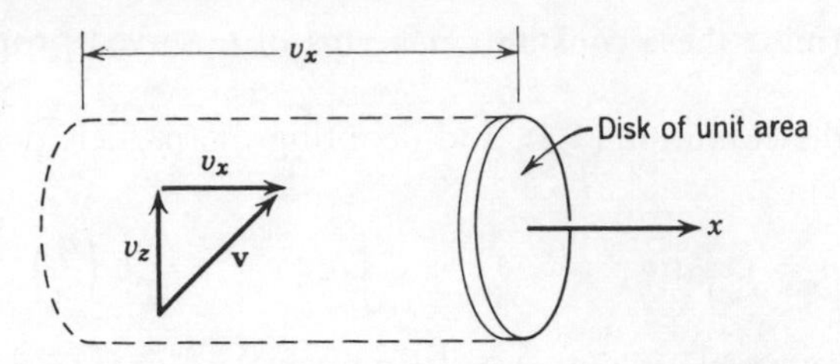

Fig. 4.1. Illustration for the calculation of the pressure.

can hit this disk only if the x component of its velocity v_x is positive. Then it loses an amount of momentum $2mv_x$ upon reflection from this disk. The number of molecules reflected by the disk per second is the number of molecules contained in the cylinder shown in Fig. 4.1 with $v_x > 0$. This number is $v_x f_0(\mathbf{v})\, d^3v$, with $v_x > 0$. Therefore the pressure is, for a gas with zero average velocity

$$\begin{aligned} P &= \int_{v_x>0} (2mv_x)v_x f_0(\mathbf{v})\, d^3v \\ &= 2mC \int_0^\infty dv_x v_x^2 e^{-Av_x^2} \int_{-\infty}^\infty dv_y e^{-Av_y^2} \int_{-\infty}^\infty dv_z e^{-Av_z^2} \\ &= mC \int d^3v\, v_x^2 e^{-Av^2} = \tfrac{1}{3} mC \int d^3v\, v^2 e^{-Av^2} \end{aligned} \tag{4.20}$$

where the last step comes about because $f_0(\mathbf{v})$ depends only on $|\mathbf{v}|$ so that the average values of v_x^2, v_y^2, and v_z^2 are all equal to one-third of the average of $v^2 = v_x^2 + v_y^2 + v_z^2$. Finally we notice that

$$P = \tfrac{2}{3} C \int d^3v\, \tfrac{1}{2} mv^2 e^{-Av^2} = \tfrac{2}{3} n\epsilon \tag{4.21}$$

This is the equation of state. Experimentally we define the temperature T by $P = nkT$, where k is Boltzmann's constant. Hence

$$\epsilon = \tfrac{3}{2} kT \tag{4.22}$$

In terms of the temperature T, the average velocity $\mathbf{v}_0$, and the particle density n the equilibrium distribution function for a dilute gas in the absence of external force is

$$f_0(\mathbf{v}) = n \left(\frac{m}{2\pi kT} \right)^{3/2} e^{-m(\mathbf{v}-\mathbf{v}_0)^2/2kT} \tag{4.23}$$

This is the Maxwell-Boltzmann distribution, the probability of finding a molecule with velocity $\mathbf{v}$ in the gas, under equilibrium conditions.*

* We have assumed, in accordance with experimental facts, that the temperature T is independent of the translational velocity $\mathbf{v}_0$.

If a perfectly reflecting wall is introduced into the gas, $f_0(\mathbf{v})$ will remain unchanged because $f_0(\mathbf{v})$ depends only on the magnitude of $\mathbf{v}$, which is unchanged by reflection from the wall.

For a gas with $\mathbf{v}_0 = 0$ it is customary to define the most probable speed $\bar{v}$ of a molecule by the value of v at which $4\pi v^2 f(\mathbf{v})$ attains a maximum. We easily find

$$\bar{v} = \sqrt{\frac{2kT}{m}} \tag{4.24}$$

The root mean square speed v_{rms} is defined by

$$v_{\text{rms}} \equiv \left[\frac{\int d^3v\, v^2 f_0(\mathbf{v})}{\int d^3v\, f_0(\mathbf{v})}\right]^{1/2} = \sqrt{\frac{3kT}{m}} \tag{4.25}$$

At room temperatures these speeds for an O_2 gas are of the order of magnitude of 10^5 cm/sec.

A plot of $4\pi v^2 f_0(\mathbf{v})$ against v is shown in Fig. 4.2. We notice that $f_0(\mathbf{v})$ does not vanish, as it should, when v exceeds the velocity of light c. This is because we have used Newtonian dynamics for the molecules instead of the more correct relativistic dynamics. The error is negligible at room temperatures, because $\bar{v} \ll c$. The temperature above which relativistic dynamics must be used can be roughly estimated by putting $\bar{v} = c$, from which we obtain $kT \approx mc^2$. Hence $T \approx 10^{13}$ °K for H_2.

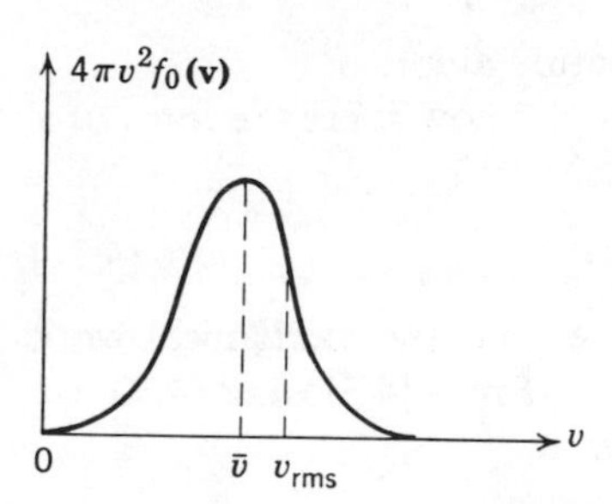

Fig. 4.2. The Maxwell-Boltzmann distribution.

Let us now consider the equilibrium distribution for a dilute gas in the presence of an external conservative force field given by

$$\mathbf{F} = -\nabla\phi(\mathbf{r}) \tag{4.26}$$

We assert that the equilibrium distribution function is now

$$f(\mathbf{r}, \mathbf{v}) = f_0(\mathbf{v})e^{-\phi(\mathbf{r})/kT} \tag{4.27}$$

where $f_0(\mathbf{v})$ is given by (4.23). To prove this we show that (4.27) satisfies Boltzmann's equation. We see immediately that $\partial f/\partial t = 0$ because (4.27) is independent of the time. Furthermore $(\partial f/\partial t)_{\text{coll}} = 0$ because $\phi(\mathbf{r})$ is independent of $\mathbf{v}$:

$$\left(\frac{\partial f}{\partial t}\right)_{\text{coll}} = e^{-2\phi(\mathbf{r})/kT}\int d^3v_2 \int d\Omega\,\sigma(\Omega)\,|\mathbf{v}_2 - \mathbf{v}_1|\,[f_0(\mathbf{v}_2')f_0(\mathbf{v}_1') - f_0(\mathbf{v}_2)f_0(\mathbf{v}_1)] = 0$$

Hence it is only necessary to verify that

$$\left(\mathbf{v}\cdot\nabla_{\mathbf{r}} + \frac{\mathbf{F}}{m}\cdot\nabla_{\mathbf{v}}\right) f(\mathbf{r}, \mathbf{v}) = 0$$

and this is trivial. We may absorb the factor $\exp(-\phi/kT)$ in (4.27) into the density n and write

$$f(\mathbf{r}, \mathbf{v}) = n(\mathbf{r})\left(\frac{m}{2\pi kT}\right)^{3/2} e^{-m(\mathbf{v}-\mathbf{v}_0)^2/2kT} \tag{4.28}$$

where

$$n(\mathbf{r}) = \int d^3v\, f(\mathbf{r}, \mathbf{v}) = ne^{-\phi(\mathbf{r})/kT} \tag{4.29}$$

Finally we derive the thermodynamics of a dilute gas. We have defined the temperature by (4.22) and we have obtained the equation of state. By the very definition of the pressure, the work done by the gas when its volume increases by dV is $P\,dV$. The internal energy is defined by

$$U(T) = N\epsilon = \tfrac{3}{2}NkT \tag{4.30}$$

which is obviously a function of the temperature alone.

The analog of the first law of thermodynamics now takes the form of a definition for the heat absorbed by the system:

$$dQ = dU + P\,dV \tag{4.31}$$

It tells us that heat added to the system goes into the mechanical work $P\,dV$ and the energy of molecular motion dU. From (4.31) and (4.30) we obtain for the heat capacity at constant volume

$$C_V = \tfrac{3}{2}Nk \tag{4.32}$$

The analog of the second law of thermodynamics is Boltzmann's H theorem, where we identify H with the negative of the entropy per unit volume divided by Boltzmann's constant:

$$H = -\frac{S}{Vk} \tag{4.33}$$

Thus the H theorem states that for a fixed volume (i.e., for an isolated gas) the entropy never decreases, which is a statement of the second law.

To justify (4.33) we calculate H in equilibrium:

$$H_0 = \int d^3v\, f_0 \log f_0 = n\left\{\log\left[n\left(\frac{m}{2\pi kT}\right)^{3/2}\right] - \frac{3}{2}\right\}$$

Using the equation of state we can rewrite this as

$$-kVH_0 = \tfrac{3}{2}Nk \log(PV^{5/3}) + \text{constant} \tag{4.34}$$

We recognize that the right-hand side is the entropy of an ideal gas in thermodynamics. It follows from (4.34), (4.33), and (4.31) that $dS = dQ/T$.

Thus we have derived all of classical thermodynamics for a dilute gas; and moreover, we were able to calculate the equation of state and the specific heat. The third law of thermodynamics cannot be derived here because we have used classical mechanics and thus are obliged to confine our considerations to high temperatures.

4.3 THE METHOD OF THE MOST PROBABLE DISTRIBUTION

We have noted the interesting fact that the Maxwell-Boltzmann distribution is independent of the detailed form of molecular interactions, as long as they exist. This fact endows the Maxwell-Boltzmann distribution with universality. We might suspect that as long as we are interested only in the *equilibrium* behavior of a gas there is a way to derive the Maxwell-Boltzmann distribution without explicitly mentioning molecular interactions. Such a derivation is now supplied. Through it we shall understand better the meaning of the Maxwell-Boltzmann distribution. The conclusion we reach will be the following. If we choose a state of the gas at random from among all its possible states consistent with certain macroscopic conditions, the probability that we shall choose a Maxwell-Boltzmann distribution is overwhelmingly greater than that for any other distribution. But first, some preliminary notions.

A state of the gas under consideration can be specified by the $3N$ canonical coordinates $\mathbf{q}_1, \ldots, \mathbf{q}_N$ and the $3N$ canonical momenta $\mathbf{p}_1, \ldots, \mathbf{p}_N$ of the N molecules. The $6N$-dimensional space spanned by the vectors $\{\mathbf{p}_1, \ldots, \mathbf{p}_N;\ \mathbf{q}_1, \ldots, \mathbf{q}_N\}$ shall be called the Γ-space of the system. A point in Γ-space represents a state of the entire gas. We call it a *representative point.*

It is obvious that a very large (in fact, infinite) number of states of the gas corresponds to a given macroscopic condition of the gas. For example, the condition that the gas is contained in a box of volume 1 cc is consistent with an infinite number of ways to distribute the molecules in space. Through macroscopic measurements we would not be able to distinguish between two gases existing in different states (thus corresponding to two distinct representative points) but satisfying the same macroscopic conditions. Thus when we speak of a gas under certain macroscopic conditions, we are in fact referring not to a single state, but to an infinite number of states. In other words, we refer not to a single system, but to a collection of systems, identical in composition and macroscopic condition but existing in different states. With Gibbs, we call such a collection of systems an *ensemble*, which is geometrically represented by a distribution of representative points in Γ-space, usually a continuous distribution. It may be conveniently described by a density function $\rho(p, q, t)$, where (p, q) is an abbreviation for $(\mathbf{p}_1, \ldots, \mathbf{p}_N; \mathbf{q}_1, \ldots, \mathbf{q}_N)$, so defined that

$$\rho(p, q, t)\, d^{3N}p\, d^{3N}q \tag{4.35}$$

is the number of representative points which at time t are contained in the infinitesimal volume element $d^{3N}p\, d^{3N}q$ of Γ-space centered about the point (p, q). An ensemble is completely specified by $\rho(p, q, t)$. It is to be emphasized that members of an ensemble are mental copies of a system and do not interact with one another.

Given $\rho(p, q, t)$ at any time t, its subsequent values are determined by the dynamics of molecular motion. Let the Hamiltonian of a system in the ensemble be $H(p_1, \ldots, p_{3N}; q_1, \ldots, q_{3N})$. The equations of motion for a system are given by

$$\begin{aligned} \dot{p}_i &= -\frac{\partial H}{\partial q_i} \qquad (i = 1, \ldots, 3N) \\ \dot{q}_i &= \frac{\partial H}{\partial p_i} \qquad (i = 1, \ldots, 3N) \end{aligned} \tag{4.36}$$

These will tell us how a representative point moves in Γ-space as time evolves. We assume that the Hamiltonian does not depend on any time derivative of p and q. It is then clear that (4.36) is invariant under time reversal and that (4.36) uniquely determines the motion of a representative point for all times, when the position of the representative point is given at any time. It follows immediately from these observations that the locus of a representative point is either a simple closed curve or a curve that never intersects itself. Furthermore, the loci of two distinct representative points never intersect.

We now prove the following theorem.

LIOUVILLE'S THEOREM

$$\frac{\partial \rho}{\partial t} + \sum_{i=1}^{3N} \left(\frac{\partial \rho}{\partial p_i} \dot{p}_i + \frac{\partial \rho}{\partial q_i} \dot{q}_i \right) = 0 \tag{4.37}$$

Proof. Since the total number of systems in an ensemble is conserved, the number of representative points leaving any volume in Γ-space per second must be equal to the rate of decrease of the number of representative points in the same volume. Let ω be an arbitrary volume in Γ-space and let S be its surface. If we denote by $\mathbf{v}$ the $6N$-dimensional vector whose components are

$$\mathbf{v} \equiv (\dot{p}_1, \dot{p}_2, \ldots, \dot{p}_{3N};\ \dot{q}_1, \dot{q}_2, \ldots, \dot{q}_{3N})$$

and **n** the vector locally normal to the surface S, then

$$-\frac{d}{dt}\int_\omega d\omega\, \rho = \int_S dS\, \mathbf{n}\cdot\mathbf{v}\rho$$

With the help of the divergence theorem in $6N$-dimensional space, we convert this to the equation

$$\int_\omega d\omega \left[\frac{\partial \rho}{\partial t} + \nabla\cdot(\mathbf{v}\rho)\right] = 0 \tag{4.38}$$

where ∇ is the $6N$-dimensional gradient operator:

$$\nabla \equiv \left(\frac{\partial}{\partial p_1}, \frac{\partial}{\partial p_2}, \ldots, \frac{\partial}{\partial p_{3N}}; \frac{\partial}{\partial q_1}, \frac{\partial}{\partial q_2}, \ldots, \frac{\partial}{\partial q_{3N}}\right)$$

Since ω is an arbitrary volume the integrand of (4.38) must identically vanish. Hence

$$-\frac{\partial \rho}{\partial t} = \nabla\cdot(\mathbf{v}\rho) = \sum_{i=1}^{3N}\left[\frac{\partial}{\partial p_i}(\dot{p}_i\rho) + \frac{\partial}{\partial q_i}(\dot{q}_i\rho)\right]$$

$$= \sum_{i=1}^{3N}\left(\frac{\partial \rho}{\partial p_i}\dot{p}_i + \frac{\partial \rho}{\partial q_i}\dot{q}_i\right) + \sum_{i=1}^{3N}\rho\left(\frac{\partial \dot{p}_i}{\partial p_i} + \frac{\partial \dot{q}_i}{\partial q_i}\right)$$

By the equations of motion (4.36) we have

$$\frac{\partial \dot{p}_i}{\partial p_i} + \frac{\partial \dot{q}_i}{\partial q_i} = 0 \qquad (i = 1, \ldots, 3N)$$

Therefore

$$-\frac{\partial \rho}{\partial t} = \sum_{i=1}^{3N}\left(\frac{\partial \rho}{\partial p_i}\dot{p}_i + \frac{\partial \rho}{\partial q_i}\dot{q}_i\right) \tag{QED}$$

Liouville's theorem is equivalent to the statement

$$\frac{d\rho}{dt} = 0 \tag{4.39}$$

since by virtue of the equations of motion p_i and q_i are functions of the time. Its geometrical interpretation is as follows. If we follow the motion of a representative point in Γ-space, we find that the density of representative points in its neighborhood is constant. Hence the distribution of representative points moves in Γ-space like an incompressible fluid.

What is the relationship between an ensemble representing a gas and the distribution function f of the gas? To examine this question we remind ourselves that $f(\mathbf{p}, \mathbf{q}, t)$ is the density of molecules in μ-space. That is, the number of molecules in the gas which at time t are located in the volume element at $(\mathbf{p}, \mathbf{q})$ in μ-space is equal to $f(\mathbf{p}, \mathbf{q}, t)\, d^3p\, d^3q$. If the state of the gas is given, then f is uniquely determined; but if f is given, the state of the gas is not uniquely determined. For example, consider a state of the gas in which molecule 1 is located at $\mathbf{x}$ and molecule 2 is located at $\mathbf{y}$. This state is distinct from that in which $\mathbf{x}$ and $\mathbf{y}$ are interchanged. These two states correspond to two distinct representative points in Γ-space, but these two states obviously possess the same distribution function. A given distribution function therefore corresponds not to a point but to a volume in Γ-space. The volume in Γ-space corresponding to a given distribution function f is called the *volume occupied by* f.

The relevance of these considerations to physics is based on the following assumptions.

(*a*) If we know nothing about the system under consideration, apart from a few macroscopic conditions, the system is equally likely to exist in any state satisfying these macroscopic conditions. That is, to a given macroscopic situation there corresponds a definite ensemble of systems.

(*b*) The equilibrium distribution function is the *most probable distribution function*, which is defined as the distribution function that occupies the maximum volume in Γ-space.

The definition of the most probable distribution is invariant under a canonical transformation of the variables (p, q), because both the macroscopic conditions and the volume element $d^{3N}p\, d^{3N}q$ are invariant under canonical transformations.

Assumptions (*a*) and (*b*) have not been proved from the basic laws of molecular dynamics, from which they should follow logically. Their justification lies in the experimental correctness of the results deduced from them.

The practical procedure for finding the equilibrium distribution function is as follows.

(*a*) Determine the ensemble that corresponds to the macroscopic situation under consideration.

(*b*) Pick an arbitrary distribution function and calculate the volume it occupies in Γ-space by counting the systems in the ensemble that possess this distribution function.

(*c*) Vary the distribution function in an arbitrary way until we find the one that occupies the maximum volume. This one is the equilibrium distribution function.

This procedure is now carried out for a dilute gas. Let a gas of N molecules be enclosed in a box of volume V with perfectly reflecting walls.

Let the total energy of the gas lie between E and $E + \Delta$, where $\Delta \ll E$. The ensemble that corresponds to the macroscopic conditions consists of a uniform distribution of representative points in a region in Γ-space bounded by the surfaces of constant energy E and $E + \Delta$ and the surfaces corresponding to the physical boundary of the container, as shown schematically in Fig. 4.3. Since the walls of the box are perfectly reflecting, energy is conserved. A representative point never leaves this region. By Liouville's theorem the distribution of representative points moves about in this region like an incompressible fluid. The density of representative points of the ensemble remains uniform for all times. This ensemble is therefore the same for all times.

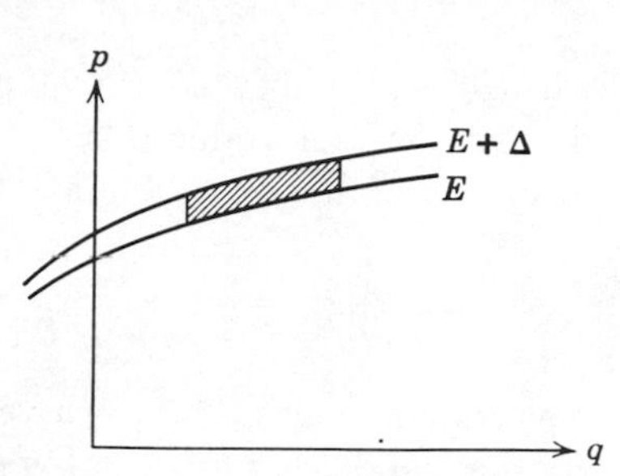

Fig. 4.3. The ensemble corresponding to a gas contained in a finite volume with energy between E and $E + \Delta$.

Next consider an arbitrary distribution function of a gas. A molecule in the gas is confined to a finite region of μ-space because the values of p and q are restricted by the macroscopic conditions. Cover this finite region of μ-space with volume elements of volume $\omega = d^3p\, d^3q$, and number them from 1 to K, where K is a very large number which eventually will be made to approach infinity. We refer to these volume elements as cells. An arbitrary distribution function is defined if we specify the number of molecules n_i found in the ith cell. These are called occupation numbers, and they satisfy the conditions

$$\sum_{i=1}^{K} n_i = N \tag{4.40}$$

$$\sum_{i=1}^{K} \epsilon_i n_i = E \tag{4.41}$$

where ϵ_i is the energy of a molecule in the ith cell:

$$\epsilon_i = \frac{p_i^2}{2m}$$

where $\mathbf{p}_i$ is the momentum of the ith cell. It is in (4.41), and only in (4.41), that the assumption of a dilute gas enters. An arbitrary set of integers $\{n_i\}$ satisfying (4.40) and (4.41) defines an arbitrary distribution function. The value of the distribution function in the ith cell, denoted by f_i, is

$$f_i = \frac{n_i}{\omega} \tag{4.42}$$

We now calculate the volume in Γ-space occupied by the distribution function corresponding to $\{n_i\}$. Let this volume be denoted by $\Omega\{n_i\}$, which is a function of the set of integers $\{n_i\}$. It is proportional to the number of ways of distributing N distinguishable molecules among K cells so that there are n_i of them in the ith cell ($i = 1, 2, \ldots, K$). Therefore

$$\Omega\{n_i\} \propto \frac{N!}{n_1!\, n_2!\, n_3! \cdots n_K!} g_1^{\,n_1} g_2^{\,n_2} \cdots g_K^{\,n_K} \tag{4.43}$$

where g_i is a number that we will put equal to unity at the end of the calculation but that is introduced here for mathematical convenience. Taking the logarithm of (4.43) we obtain

$$\log \Omega\{n_i\} = \log N! - \sum_{i=1}^{K} \log n_i! + \sum_{i=1}^{K} n_i \log g_i + \text{constant}$$

Now assume that each n_i is a very large integer, so we can use Stirling's approximation, $\log n_i! \approx n_i \log n_i$. We then have

$$\log \Omega\{n_i\} = N \log N - \sum_{i=1}^{K} n_i \log n_i! + \sum_{i=1}^{K} n_i \log g_i + \text{constant} \tag{4.44}$$

To find the equilibrium distribution we vary the set of integers $\{n_i\}$ subject to the conditions (4.40) and (4.41) until $\log \Omega$ attains a maximum. Let $\{\bar{n}_i\}$ denote the set of occupation numbers that maximizes $\log \Omega$. By the well-known method of Lagrange multipliers we have

$$\delta[\log \Omega\{n_i\}] - \delta\left(\alpha \sum_{i=1}^{K} n_i + \beta \sum_{i=1}^{K} \epsilon_i n_i\right) = 0 \qquad (n_i = \bar{n}_i) \tag{4.45}$$

where α, β are Lagrangian multipliers. Now the n_i can be considered independent of one another. Substituting (4.44) into (4.45) we obtain

$$\sum_{i=1}^{K} [-(\log n_i + 1) + \log g_i - \alpha - \beta\epsilon_i]\, \delta n_i = 0 \qquad (n_i = \bar{n}_i)$$

Since δn_i are independent variations, we obtain the equilibrium condition by setting the summand equal to zero:

$$\log \bar{n}_i = -1 + \log g_i - \alpha - \beta\epsilon_i$$

$$\bar{n}_i = g_i e^{-\alpha-\beta\epsilon_1-1} \tag{4.46}$$

The most probable distribution function is, by (4.42) and (4.46),

$$f_i = Ce^{-\beta\epsilon_i} \tag{4.47}$$

where C is a constant. The determination of the constants C and β

proceeds in the same way as for (4.13). Writing $f_i \equiv f(\mathbf{v}_i)$, we see that $f(\mathbf{v})$ is the Maxwell-Boltzmann distribution (4.23) for $\mathbf{v}_0 = 0$. To show that (4.46) actually corresponds to a maximum of $\log \Omega\{n_i\}$ we calculate the second variation. It is easily shown that the second variation of the quantity on the left-hand side of (4.45), for $n_i = \bar{n}_i$, is

$$-\sum_{i=1}^{K} \frac{1}{n_i} (\delta n_i)^2 < 0$$

We have obtained the Maxwell-Boltzmann distribution as the most probable distribution, in the sense that among all the systems satisfying the macroscopic conditions the Maxwell-Boltzmann distribution is the distribution common to the largest number of them. The question remains: What fraction of these systems have the Maxwell-Boltzmann distributions? In other words, how probable is the most probable distribution? The probability for the occurrence of any set of occupation numbers $\{n_i\}$ is given by

$$P\{n_i\} = \frac{\Omega\{n_i\}}{\sum_{\{n_j'\}} \Omega\{n_j'\}} \tag{4.48}$$

where the sum in the denominator extends over all possible sets of integers $\{n_j'\}$ satisfying (4.40) and (4.41). The probability for finding the system in the Maxwell-Boltzmann distribution is therefore $P\{\bar{n}_i\}$. A direct calculation of $P\{\bar{n}_i\}$ is not easy. We shall be satisfied with an estimate, which, however, becomes an exact evaluation if this probability approaches unity.

We define the ensemble average $\langle n_i \rangle$ of the occupation number n_i by

$$\langle n_i \rangle = \frac{\sum_{\{n_j\}} n_i \Omega\{n_j\}}{\sum_{\{n_j\}} \Omega\{n_j\}} \tag{4.49}$$

It is obvious from (4.43) that

$$\langle n_i \rangle = g_i \frac{\partial}{\partial g_i} \log \left[\sum_{\{n_j\}} \Omega\{n_j\} \right] \tag{4.50}$$

if we let $g_i \to 1$. The deviations from the average value can be estimated by calculating the mean square fluctuation $\langle n_i^2 \rangle - \langle n_i \rangle^2$. We can express $\langle n_i^2 \rangle$ in terms of $\langle n_i \rangle$ as follows:

$$\langle n_i^2 \rangle \equiv \frac{\sum n_i^2 \Omega}{\sum \Omega} = \frac{g_i \dfrac{\partial}{\partial g_i} \left(g_i \dfrac{\partial}{\partial g_i} \sum \Omega \right)}{\sum \Omega} \tag{4.51}$$

where the sum Σ extends over all allowed $\{n_j\}$. Through the series of steps given next we obtain the desired results:

$$\langle n_i^2\rangle = g_i \frac{\partial}{\partial g_i}\left(\frac{1}{\sum\Omega} g_i \frac{\partial}{\partial g_i}\sum\Omega\right) - g_i\left(\frac{\partial}{\partial g_i}\frac{1}{\sum\Omega}\right) g_i \frac{\partial}{\partial g_i}\sum\Omega$$

$$= g_i \frac{\partial}{\partial g_i}\left(g_i \frac{\partial}{\partial g_i}\log\sum\Omega\right) + \left(\frac{1}{\sum\Omega} g_i \frac{\partial}{\partial g_i}\sum\Omega\right)^2$$

$$= g_i \frac{\partial}{\partial g_i}\langle n_i\rangle + \langle n_i\rangle^2 \tag{4.52}$$

Therefore the mean square fluctuation is

$$\langle n_i^2\rangle - \langle n_i\rangle^2 = g_i \frac{\partial}{\partial g_i}\langle n_i\rangle \tag{4.53}$$

where we must let $g_i \to 1$ at the end of the calculation.

If the mean square fluctuation is large compared to $\langle n_i\rangle^2$, then $\langle n_i\rangle$ may differ considerably from $\bar{n}_i$; but then neither of them will be physically meaningful. If the mean square fluctuation is small compared to $\langle n_i\rangle^2$, we may expect $\langle n_i\rangle$ and $\bar{n}_i$ to be almost equal. We assume the latter is so, and we shall see that this is a consistent assumption.* Putting

$$\langle n_i\rangle \approx \bar{n}_i$$

we find from (4.46) and (4.53) that

$$\langle n_i^2\rangle - \langle n_i\rangle^2 \approx \bar{n}_i$$

or

$$\sqrt{\left\langle\left(\frac{n_i}{N}\right)^2\right\rangle - \left\langle\frac{n_i}{N}\right\rangle^2} \approx \frac{\sqrt{\bar{n}_i/N}}{\sqrt{N}} \tag{4.54}$$

Since $\bar{n}_i/N$ is less than one, the right-hand side of (4.54) becomes vanishingly small if N is the number of molecules in 1 mole of gas, namely $N \approx 10^{23}$. This result implies that the probability $P\{n_i\}$ defined by (4.48) has an extremely sharp peak at $\{n_i\} = \{\bar{n}_i\}$. The width of the peak is such that $P\{n_i\}$ is essentially reduced to zero when any n_i/N differ from $\bar{n}_i/N$ by a number of the order of $1/\sqrt{N}$. A schematic plot of $P\{n_i\}$ is shown in Fig. 4.4. We shall call the distributions lying within the peak "essentially Maxwell-Boltzmann" distributions. They are physically indistinguishable from the strict Maxwell-Boltzmann distribution. From these considerations we conclude that in a physical gas any state picked out at random from among all those satisfying the given macroscopic

* This assumption can be proved by the method described in Chap. 10. The desired result is essentially stated in (10.29).

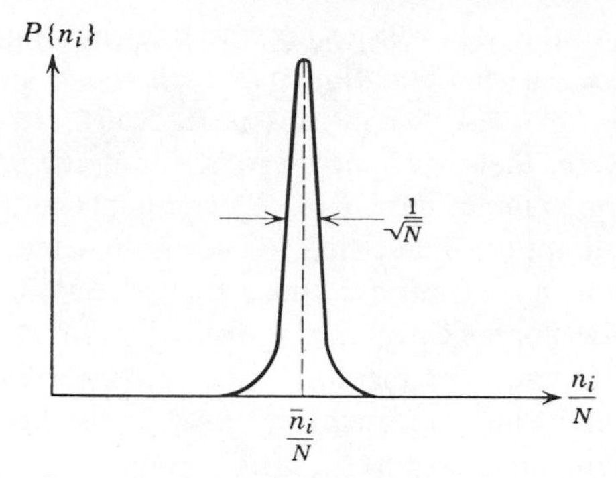

Fig. 4.4. Probability of a gas having the occupation numbers $\{n_i\}$. The most probable occupation numbers $\{\bar{n}_i\}$ correspond to the Maxwell-Boltzmann distribution. Occupation numbers $\{n_i\}$ for which $P\{n_i\}$ lies within the peak are called "essentially Maxwell-Boltzmann" distributions.

conditions will almost certainly have a distribution function that is Maxwell-Boltzmann.

The meaning of the Maxwell-Boltzmann distribution is therefore as follows. If a dilute gas is prepared in an arbitrary initial state, and if there exist interactions to enable the gas to go into states other than the initial state, the gas will in time almost certainly become Maxwell-Boltzmann, because among all possible states of the gas satisfying the macroscopic conditions (which are conserved by the interactions), almost all of them have the Maxwell-Boltzmann distribution. This, however, does not tell us how long it will take for the gas to reach the equilibrium situation. Nor does it rule out the possibility that the gas may never reach the equilibrium situation, nor that of leaving the equilibrium situation after attaining it. From this point of view, we see that the laws of thermodynamics are not rigorously true but only overwhelmingly probable.

To illustrate these ideas, consider a gas enclosed in a cubical box with perfectly reflecting walls. Suppose initially the gas molecules are distributed in an arbitrary way within the box, and all have exactly the same velocity parallel to one edge of the box. If there are no interactions, this distribution will be maintained indefinitely, and the system never becomes Maxwell-Boltzmann. For such a gas thermodynamics is invalid. If there is molecular interaction, *no matter how small*, the initial distribution will, through collisions, change with time. Since almost any state of the gas will have a Maxwell-Boltzmann distribution, it is reasonable that the distribution after a *sufficiently long time*, depending on the collision cross section, will become Maxwell-Boltzmann. The considerations we have made cannot tell us how long this time must be. They only tell us what the equilibrium situation is, if it is reached.

The derivation of the Maxwell-Boltzmann distribution presented here is independent of the earlier derivation based on the Boltzmann transport equation. Neither of these derivations is rigorous. In the present one there are assumptions that we did not justify, and in the previous one there was the assumption of molecular chaos, which remains unproved and is not related to the assumptions made here. The present method seems to be more satisfactory as a derivation of the Maxwell-Boltzmann distribution because it reveals more clearly the statistical nature of the Maxwell-Boltzmann distribution. The method of the most probable distribution, however, does not furnish information about a gas not in equilibrium, whereas the Boltzmann transport equation does. Hence the main value of the Boltzmann equation lies in its application to nonequilibrium phenomena.

4.4 ANALYSIS OF THE *H* THEOREM

We now discuss the physical implication of Boltzmann's H Theorem. For a given distribution function $f(\mathbf{v}, t)$, H is defined by

$$H = \int d^3v\, f(\mathbf{v}, t) \log f(\mathbf{v}, t) \tag{4.55}$$

The time evolution of H is determined by the time evolution of $f(\mathbf{v}, t)$, which does not in general satisfy the Boltzmann transport equation. It satisfies the Boltzmann transport equation only at the instant when the assumption of molecular chaos happens to be valid.

The H theorem states that *if at a given instant t the state of the gas satisfies the assumption of molecular chaos*, then at the instant $t + \epsilon$ $(\epsilon \to 0)$,

(*a*) $\dfrac{dH}{dt} \leq 0$

(*b*) $\dfrac{dH}{dt} = 0$ if and only if $f(\mathbf{v}, t)$ is the Maxwell-Boltzmann distribution.

The proof of the theorem given earlier is rigorous in the limiting case of an infinitely dilute gas. Therefore an inquiry into the validity of the H theorem can only be an inquiry into the validity of the assumption of molecular chaos.

We recall that the assumption of molecular chaos states the following: If $f(\mathbf{v}, t)$ is the probability of finding a molecule with velocity $\mathbf{v}$ at time t, the probability of simultaneously finding a molecule with velocity $\mathbf{v}$ and a molecule with velocity $\mathbf{v}'$ at time t is $f(\mathbf{v}, t) f(\mathbf{v}', t)$. This assumption concerns the correlation between two molecules and has nothing to say about the form of the distribution function. Thus a state of the gas possessing a given distribution function may or may not satisfy the assumption of molecular chaos. For brevity we call a state of the gas a state of "molecular chaos" if it satisfies the assumption of molecular chaos.

We now show that when the gas is in a state of "molecular chaos" H is at a local peak. Consider a dilute gas, in the absence of external force, prepared with an initial condition that is invariant under time reversal.* Under these conditions, the distribution function depends on the magnitude but not the direction of $\mathbf{v}$. Let the gas be in a state of "molecular chaos" and be non-Maxwell-Boltzmann at time $t = 0$. According to the H theorem $dH/dt \leq 0$ at $t = 0^+$. Now consider another gas, which at $t = 0$ is precisely the same as the original one except that all molecular velocities are reversed in direction. This gas must have the same H and must also be in a state of "molecular chaos." Therefore for this new gas $dH/dt \leq 0$ at $t = 0^+$. On the other hand, according to the invariance of the equations of motion under time reversal, the future of the new gas is the past of the original gas. Therefore for the original gas we must have

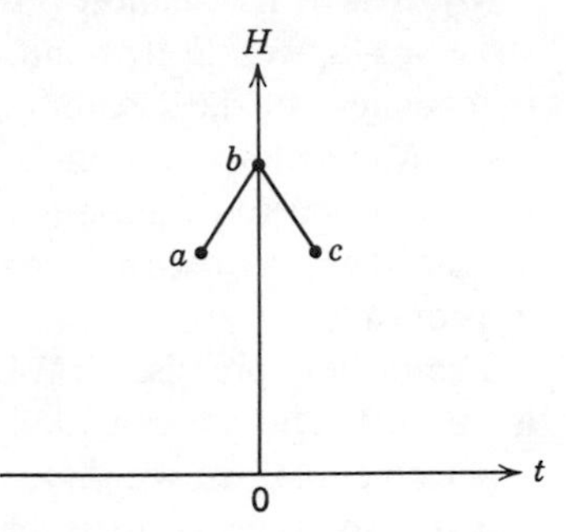

Fig. 4.5. H is at a local peak when the gas is in a state of "molecular chaos."

$$\frac{dH}{dt} \leq 0 \qquad \text{at} \qquad t = 0^+$$

$$\frac{dH}{dt} \geq 0 \qquad \text{at} \qquad t = 0^-$$

Thus H is at a local peak,† as illustrated in Fig. 4.5.

When H is not at a local peak, such as at the points a and c in Fig. 4.5, the state of the gas is not a state of "molecular chaos." Hence molecular collisions, which are responsible for the change of H with time, can create "molecular chaos" when there is none and destroy "molecular chaos" once established.

It is important to note that dH/dt is not necessarily a continuous function of time; it can be changed abruptly by molecular collisions. Overlooking this fact might lead us to conclude, erroneously, that the H theorem is

* These simplifying features are introduced to avoid the irrelevant complications arising from the time reversal properties of the external force and the agent preparing the system.

† The foregoing argument is due to F. E. Low (unpublished).

inconsistent with the invariance under time reversal. A statement of the H theorem that is manifestly invariant under time reversal is the following. If there is "molecular chaos" now, then $dH/dt \leq 0$ in the next instant. If there will be "molecular chaos" in the next instant, then $dH/dt \geq 0$ now.

We now discuss the general behavior of H as a function of time. Our discussion rests on the following premises.

(*a*) H is at its smallest possible value when the distribution function is strictly Maxwell-Boltzmann. This easily follows from (4.55), and it is independent of the assumption of molecular chaos.*

(*b*) Molecular collisions happen at random, i.e., the time sequence of the states of a gas is a sequence of states chosen at random from those that satisfy the macroscopic conditions. This assumption is plausible but unproved.

From these premises it follows that the distribution function of the gas is almost always essentially Maxwell-Boltzmann, i.e., a distribution function contained within the peak shown in Fig. 4.4. The curve of H as a function of time consists mostly of microscopic fluctuations above the minimum value. Between two points at which H is at the minimum value there is likely to be a small peak.

If at any instant the gas has a distribution function appreciably different from the Maxwell-Boltzmann distribution, then H is appreciably larger than the minimum value. Since collisions are assumed to happen at random, it is overwhelmingly probable that after the next collision the distribution will become essentially Maxwell-Boltzmann and H will decrease to essentially the minimum value. By time reversal invariance it is overwhelmingly probable that before the last collision H was at essentially the minimum value. Thus H is overwhelmingly likely to be at a sharp peak when the gas is in an improbable state. The more improbable the state, the sharper the peak.

A very crude model of the curve of H as a function of time is shown in Fig. 4.6. The duration of a fluctuation, large or small, should be of the order of the time between two successive collisions of a molecule, i.e., 10^{-11} sec for a gas under ordinary conditions. The large fluctuations, such as that labeled *a* in Fig. 4.6, almost never occur spontaneously.† We can, of course, prepare a gas in an improbable state, e.g., by suddenly removing a wall of the container of the gas, so that H is initially at a peak. But it is overwhelmingly probable that within a few collision times the distribution would be reduced to an essentially Maxwell-Boltzmann distribution.

Most of the time the value of H fluctuates within a small range above

* See Problem 4.9.

† See Problems 4.5 and 4.6.

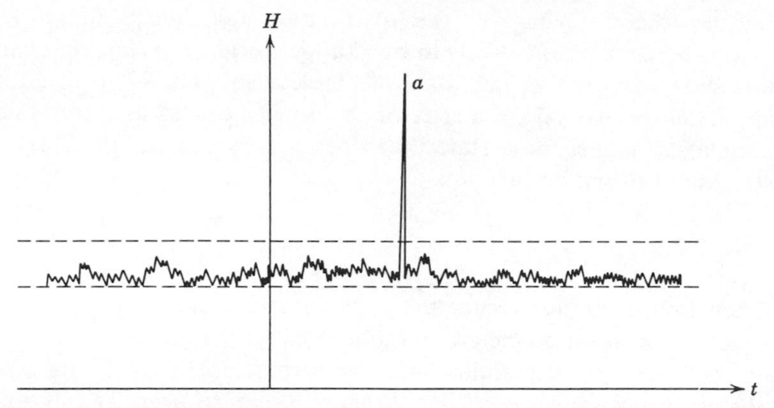

Fig. 4.6. H as a function of time. The range of values of H lying between the two horizontal dashed lines is called the "noise range."

the minimum value. This range, shown enclosed by the dashed lines in Fig. 4.6, corresponds to states of the gas with distribution functions that are essentially Maxwell-Boltzmann, i.e., distribution functions contained within the peak of Fig. 4.4. We call this range the "noise range." These features of the curve of H have been deduced only through plausibility arguments, but they are in accord with experience. We can summarize them as follows.

(*a*) For all practical purposes H never fluctuates spontaneously above the noise range. This corresponds to the observed fact that a system in thermodynamic equilibrium never spontaneously goes out of equilibrium.

(*b*) If at an instant H has a value above the noise range, then, for all practical purposes, H always decreases after that instant. In a few collision times its value will be within the noise range. This corresponds to the observed fact that if a system is initially not in equilibrium (the initial state being brought about by external agents), it always tends to equilibrium. In a few collision times it will be in equilibrium. This feature, together with (*a*), constitutes the second law of thermodynamics.

(*c*) Most of the time the value of H fluctuates in the noise range, in which dH/dt is as frequently positive as negative. (This is not a contradiction to the H theorem, because the H theorem merely requires that when the system is in a state of "molecular chaos," then $dH/dt \leq 0$ in the next instant.) These small fluctuations produce no observable change in the equation of state and other thermodynamic quantities. When H is in the noise range, the system is, for all practical purposes, in thermodynamic equilibrium. These fluctuations, however, do lead to observable effects, e.g., the fluctuation scattering of light. We witness it in the blue of the sky.

Finally let us say something about the meaning of the assumption of molecular chaos. Whenever the distribution function is not strictly Maxwell-Boltzmann, H is likely to be at a local peak. On the other hand, it was shown earlier that in a state of "molecular chaos" H *is* at a local peak. Thus we may regard a state of "molecular chaos" as a convenient mathematical model for a state that does not have a strictly Maxwell-Boltzmann distribution function.

4.5 TWO "PARADOXES"

When Boltzmann announced the H theorem a century ago, objections were raised against it on the ground that it led to "paradoxes." These are the so-called "reversal paradox" and "recurrence paradox," both based on the erroneous statement of the H theorem that $dH/dt \leq 0$ at all times. The correct statement of the H theorem, as given in the last section, is free from such objections. We mention these "paradoxes" purely for historical interest.

The "reversal paradox" is as follows. "The H theorem singles out a preferred direction of time. It is therefore inconsistent with time reversal invariance." This is not a paradox, because the statement of the alleged paradox is false. We have seen in the last section that time reversal invariance is consistent with the H theorem, because dH/dt need not be a continuous function of time. In fact, we have made use of time reversal invariance to deduce interesting properties of the curve of H.

The "recurrence paradox" is based on the following true theorem.

POINCARÉ'S THEOREM. A system having a finite energy and confined to a finite volume will, after a sufficiently long time, return to an arbitrarily small neighborhood of almost any given initial state.

By "almost any state" is meant any state of the system, except for a set of measure zero (i.e., a set that has no volume, e.g., a discrete point set). A neighborhood of a state has an obvious definition in terms of the Γ-space of the system.

A proof of Poincaré's theorem is given at the end of this section. This theorem implies that H is an almost periodic function of time. The "recurrence paradox" arises in an obvious way, if we take the statement of the H theorem to be $dH/dt \leq 0$ at all times. Since this is not the statement of the H theorem, there is no paradox. In fact, Poincaré's theorem furnishes further information concerning the curve of H.

Most of the time H lies in the noise range. Poincaré's theorem implies that the small fluctuations in the noise range repeat themselves. This is only to be expected.

For the rare spontaneous fluctuations above the noise range, Poincaré's theorem requires that if one such fluctuation occurs another one must occur after a sufficiently long time. The time interval between two large fluctuations is called a *Poincaré cycle.* A crude estimate (see Problem 4.7) shows that a Poincaré cycle is of the order of e^N, where N is the total number of molecules in the system. Since $N \approx 10^{23}$, a Poincaré cycle is enormously long. In fact, it is essentially the same number, be it $10^{10^{23}}$ sec or $10^{10^{23}}$ ages of the universe, (the age of the universe being a mere 10^{10} years.) Thus it has nothing to do with physics.

Proof of Poincaré's Theorem. Let a state of the system be represented by a point in Γ-space. As time goes on, any point in Γ-space traces out a locus that is uniquely determined by any given point on the locus. Let g_0 be an arbitrary volume element in Γ-space of volume ω_0. After time t all the points in g_0 will be in another volume element g_t, of volume ω_t, which is uniquely determined by g_0. By Liouville's theorem, $\omega_t = \omega_0$.

Let Γ_0 denote the subspace that is the union of all g_t for $0 \leq t < \infty$. Let its volume be Ω_0. Similarly, let Γ_τ denote the subspace that is the union of all g_t for $\tau \leq t < \infty$. Let its volume be Ω_τ. The numbers Ω_0 and Ω_τ are finite because, since the energy of the system and the spatial extension of the system are finite, a representative point is confined to a finite region of Γ-space. The definitions immediately imply that Γ_0 contains Γ_τ.

We may think of Γ_0 and Γ_τ in a different way. Imagine the region Γ_0 to be filled uniformly with representative points. As time goes on, Γ_0 will evolve into some other regions that are uniquely determined. It is clear, from the definitions, that after a time τ, Γ_0 will become Γ_τ. Hence, by Liouville's theorem,

$$\Omega_0 = \Omega_\tau$$

We recall that Γ_0 contains all the future destinations of the points in g_0, and Γ_τ contains all the future destinations of the points in g_τ, which in turn is evolved from g_0 after the time τ. It has been shown that Γ_0 has the same volume as Γ_τ. Therefore Γ_0 and Γ_τ must contain the same set of points except for a set of measure zero.

In particular, Γ_τ contains all of g_0 except for a set of measure zero. But, by definition, all points in Γ_τ are future destinations of the points in g_0. Therefore all points in g_0, except for a set of measure zero, must return to g_0 after a sufficiently long time. Since g_0 can be made as small as we wish, Poincaré's theorem follows.

4.6 VALIDITY OF THE BOLTZMANN TRANSPORT EQUATION

In the light of the analysis in the last section we examine the validity of the Boltzmann transport equation. Without any detailed discussion we can see that the Boltzmann transport equation is not a rigorous consequence of molecular dynamics; the latter is invariant under time reversal, but the former is not. We shall discuss where the Boltzmann transport equation fails and in what sense it may be considered approximately valid.

The Boltzmann transport equation is strictly valid for a dilute gas at an instant when the gas is in a state of "molecular chaos." But we have seen that collisions can destroy "molecular chaos," once established. The Boltzmann transport equation therefore cannot be rigorously valid for all times. In fact, if the Boltzmann transport equation were rigorously valid for all times, it would imply that a distribution which is initially Maxwell-Boltzmann would remain forever Maxwell-Boltzmann. It would imply that a chair may never spontaneously fly up in the air as a result of statistical fluctuations—a conclusion we are willing to accept; but it would also imply that there is no Brownian motion—a conclusion that is wrong. Therefore we must ask, "In what sense can we regard the Boltzmann transport equation as valid?"

An answer is provided by the argument given earlier, that a state of "molecular chaos" may be regarded as a convenient mathematical model for a state not strictly in equilibrium. If "molecular chaos" is the condition most of the time, then H will be at a local peak most of the time. This result is a valid description of the curve of H in a statistical sense. Hence the Boltzmann transport equation may be regarded as valid in a statistical sense.

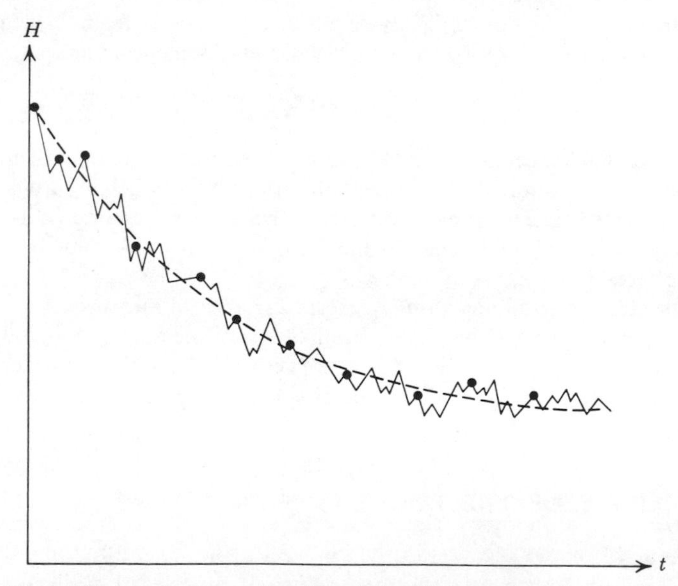

Fig. 4.7. The solid curve is H as a function of time for a gas initially in an improbable state. The dots are the points at which there is "molecular chaos." The dashed curve is that predicted by the Boltzmann transport equation.

To illustrate the statistical validity of the Boltzmann transport equation let us imagine that a gas is prepared in an improbable initial state. The curve of H as a function of time might look like the solid curve in Fig. 4.7. Let us mark with a dot a point on this curve at which the gas is in a state of "molecular chaos." All these dots must be at a local peak of H (but not all local peaks are marked with a dot). By assumption of the randomness of the time sequence of states, they are likely to be evenly distributed in time. The distribution of dots might look like that illustrated in Fig. 4.7.

A solution to the Boltzmann transport equation would yield a smooth curve of negative slope that tries to fit these dots, as shown by the dashed curve in Fig. 4.7. It is in this sense that the Boltzmann transport equation provides a description of the approach to equilibrium.

These arguments make it only plausible that the Boltzmann transport equation is useful for the description of the approach to equilibrium. The final test lies in the comparison of results with experiments.

PROBLEMS

4.1. Describe an experimental method for the verification of the Maxwell-Boltzmann distribution.

4.2. A cylindrical column of gas of given temperature rotates about a fixed axis with constant angular velocity. Find the equilibrium distribution function.

4.3. (*a*) What fraction of the H_2 gas at sea level and at a temperature of 300°K can escape from the earth's gravitational field?
(*b*) Why do we still have H_2 gas in the atmosphere at sea level?

4.4. Using relativistic dynamics for gas molecules find, for a dilute gas of zero total momentum,
(*a*) the equilibrium distribution function;
(*b*) the equation of state. *Answer:* PV is independent of the volume. Hence it is NkT by definition of T.

4.5. (*a*) *Estimate* the probability that a 7¢ airmail stamp (mass = 0.1 g) resting on a desk top at room temperature (300°K) will spontaneously fly up to a height of 10^{-8} cm above the desk top.
Hint: Think not of one stamp but of an infinite number of noninteracting stamps placed side-by-side. Formulate an argument showing that these stamps obey the Maxwell-Boltzmann distribution.
Answer: Let m = mass of stamps, h = height, g = acceleration of gravity. Probability $\approx e^{-mgh/kT}$

4.6. A room of volume 3 × 3 × 3 cubic meters is under standard conditions (atmospheric pressure and 300°K).
(*a*) *Estimate* the probability that at any instant of time a 1-cc volume

anywhere within this room becomes totally devoid of air because of spontaneous statistical fluctuations.

(*b*) Estimate the same for a 1-Å^3 volume.

Answer. Let N = total number of air molecules, V = volume of room, v = the volume devoid of air. Probability $\approx e^{-N(v/V)}$

4.7. Suppose the situation referred to in Problem 4.6*a* has occurred. Describe qualitatively the behavior of the distribution function thereafter. Estimate the time it takes for such a situation to occur again, under the assumption that molecular collisions are such that the time sequence of the state of the system is a random sequence of states.

4.8. (*a*) Explain why in (4.47) we arrived at the formula for the Maxwell-Boltzmann distribution for a gas with no average velocity ($\mathbf{v}_0 = 0$), although average velocity was not mentioned as a macroscopic condition in (4.40) and (4.41).

(*b*) Derive the Maxwell-Boltzmann distribution for a gas with average velocity $\mathbf{v}_0$, using the method of the most probable distribution.

4.9. Let

$$H = \int d^3v\, f(\mathbf{v}, t) \log f(\mathbf{v}, t)$$

when $f(\mathbf{v}, t)$ is arbitrary except for the conditions

$$\int d^3v\, f(\mathbf{v}, t) = n$$

$$\int d^3v \tfrac{1}{2}mv^2 f(\mathbf{v}, t) = \epsilon$$

Show that H is minimum when f is the Maxwell-Boltzmann distribution.

chapter 5

TRANSPORT PHENOMENA

5.1 THE MEAN FREE PATH

To begin our discussion on the approach to equilibrium of a gas initially not in equilibrium, we introduce the qualitative concept of the mean free path and related quantities.

A gas is not in equilibrium when the distribution function is different from the Maxwell-Boltzmann distribution. The most common case of a nonequilibrium situation is that in which the temperature, density, and average velocity are not constant throughout the gas. To approach equilibrium, these nonuniformities have to be ironed out through the transport of energy, mass, and momentum from one part of the gas to another. The mechanism of transport is molecular collision, and the average distance over which molecular properties can be transported in one collision is the *mean free path*. It is the average distance traveled by a molecule between successive collisions. We give an estimate of its order of magnitude.

The number of collisions happening per second per unit volume at the point $\mathbf{r}$ in a gas is given by

$$Z = \int d^3v_1 \int d^3v_2 \int d\Omega \, \sigma(\Omega) \, |\mathbf{v}_1 - \mathbf{v}_2| \, f(\mathbf{r}, \mathbf{v}_1, t) f(\mathbf{r}, \mathbf{v}_2, t)$$

where $f(\mathbf{r}, \mathbf{v}, t)$ is the distribution function. The integration over the scattering angles Ω can be immediately effected to yield

$$Z = \int d^3v_1 \int d^3v_2 \, \sigma_{\text{tot}} \, |\mathbf{v}_1 - \mathbf{v}_2| \; f(\mathbf{r}, \mathbf{v}_1, t) f(\mathbf{r}, \mathbf{v}_2, t) \tag{5.1}$$

A free path is defined as the distance traveled by a molecule between two successive collisions. Since it takes two molecules to make a collision, every collision terminates two free paths. The total number of free paths occurring per second per unit volume is therefore $2Z$. Since there are n molecues per unit volume, the average number of free paths traveled by a molecule per second is $2Z/n$. The *mean free path*, which is the average length of a free path, is given by

$$\lambda = \frac{n}{2Z} \bar{v} \tag{5.2}$$

where $\bar{v} = \sqrt{2kT/m}$ is the most probable speed of a molecule. The average duration of a free path is called the *collision time* and is given by

$$\tau = \frac{\lambda}{\bar{v}} \tag{5.3}$$

For a gas in equilibrium, $f(\mathbf{r}, \mathbf{v}, t)$ is the Maxwell-Boltzmann distribution. Assume for an order-of-magnitude estimate that σ_{tot} is insensitive to the energy of the colliding molecules and may be replaced by a constant of the order of πa^2 where a is the molecular diameter. Then we have

$$\begin{aligned} Z &= \sigma_{\text{tot}} \int d^3v_1 \int d^3v_2 |\mathbf{v}_1 - \mathbf{v}_2| \; f(\mathbf{v}_1) f(\mathbf{v}_2) \\ &= \sigma_{\text{tot}} n^2 \left(\frac{m}{2\pi kT}\right)^3 \int d^3v_1 \int d^3v_2 \, |\mathbf{v}_1 - \mathbf{v}_2| \exp\left[-\frac{m}{2kT}(v_1^2 + v_2^2)\right] \\ &= \sigma_{\text{tot}} n^2 \left(\frac{m}{2\pi kT}\right)^3 \int d^3V \int d^3v \, |\mathbf{v}| \exp\left[-\frac{m}{2kT}(2V^2 + \tfrac{1}{2}v^2)\right] \end{aligned}$$

where $\mathbf{V} = \frac{1}{2}(\mathbf{v}_1 + \mathbf{v}_2)$, $\mathbf{v} = \mathbf{v}_2 - \mathbf{v}_1$. The integrations are elementary and give

$$Z = 4n^2 \sigma_{\text{tot}} \sqrt{\frac{kT}{\pi m}} = 4n^2 \, \sigma_{\text{tot}} \frac{\bar{v}}{\sqrt{2\pi}} \tag{5.4}$$

Therefore

$$\lambda = \frac{1}{4} \sqrt{\frac{\pi}{2}} \frac{1}{n\sigma_{\text{tot}}} \tag{5.5}$$

$$\tau = \frac{1}{4} \sqrt{\frac{\pi}{2}} \frac{1}{n\sigma_{\text{tot}} \bar{v}} \tag{5.6}$$

where $\bar{v} = \sqrt{2kT/m}$. We see that the mean free path is independent of the temperature and is inversely proportional to the density times the total cross section. The numbers (5.5) and (5.6) are also good estimates for a gas not far from equilibrium, which is the only case we discuss further.

The following are some numerical estimates. For H_2 gas at its critical point,

$$\lambda \approx 10^{-7} \text{ cm}$$
$$\tau \approx 10^{-11} \text{ sec}$$

For H_2 gas in inter-stellar space, where the density is about 1 molecule/cc,

$$\lambda \approx 10^{15} \text{ cm}$$

The diameter of H_2 has been taken to be about 1 Å.

From these qualitative estimates, it is expected that in H_2 gas under normal conditions, for example, any nonuniformity in density or temperature over distances of order 10^{-7} cm will be ironed out in the order of 10^{-11} sec. Variations in density or temperature over macroscopic distances may persist for a long time.

5.2 THE CONSERVATION LAWS

To investigate nonequilibrium phenomena, we must solve the Boltzmann transport equation, with given initial conditions, to obtain the distribution function as a function of time. Some rigorous properties of any solution to the Boltzmann equation may be obtained from the fact that in any molecular collision there are dynamical quantities that are rigorously conserved.

Let $\chi(\mathbf{r}, \mathbf{v})$ be any quantity associated with a molecule of velocity $\mathbf{v}$ located at $\mathbf{r}$, such that in any collision $\{\mathbf{v}_1, \mathbf{v}_2\} \to \{\mathbf{v}_1', \mathbf{v}_2'\}$ taking place at $\mathbf{r}$, we have

$$\chi_1 + \chi_2 = \chi_1' + \chi_2' \tag{5.7}$$

where $\chi_1 = \chi(\mathbf{r}_1, \mathbf{v}_1)$, etc. We call χ a conserved property. The following theorem holds.

THEOREM

$$\int d^3v\, \chi(\mathbf{r}, \mathbf{v}) \left[\frac{\partial f(\mathbf{r}, \mathbf{v}, t)}{\partial t}\right]_{\text{coll}} = 0 \tag{5.8}$$

where $(\partial f/\partial t)_{\text{coll}}$ is the right-hand side of (3.36).*

Proof. By definition of $(\partial f/\partial t)_{\text{coll}}$ we have

$$\int d^3v\, \chi \left(\frac{\partial f}{\partial t}\right)_{\text{coll}} = \int d^3v_1 \int d^3v_2 \int d\Omega\, \sigma(\Omega)\, |\mathbf{v}_2 - \mathbf{v}_1|\, \chi_1 (f_1' f_2' - f_1 f_2) \tag{5.9}$$

Making use of the properties of $\sigma(\Omega)$ discussed in Section 3.2, and proceeding in a manner similar to the proof of the H theorem, we make each

* Note that it is not required that f be a solution of the Boltzmann transport equation.

of the following interchanges of integration variables.

First: $\mathbf{v}_1 \rightleftarrows \mathbf{v}_2$

Next: $\mathbf{v}_1 \rightleftarrows \mathbf{v}_1'$ and $\mathbf{v}_2 \rightleftarrows \mathbf{v}_2'$

Next: $\mathbf{v}_1 \rightleftarrows \mathbf{v}_2'$ and $\mathbf{v}_2 \rightleftarrows \mathbf{v}_1'$

For each case we obtain a different form for the same integral. Adding the three new formulas so obtained to (5.9) and dividing the result by 4 we get

$$\int d^3v\, \chi \left(\frac{\partial f}{\partial t}\right)_{\text{coll}} = \frac{1}{4}\int d^3v_1 \int d^3v_2 \int d\Omega\, \sigma(\Omega)\, |\mathbf{v}_1 - \mathbf{v}_2| \times (f_2'f_1' - f_2f_1)(\chi_1 + \chi_2 - \chi_1' - \chi_2') \equiv 0 \quad \text{(QED)}$$

The conservation theorem relevant to the Boltzmann transport equation is obtained by multiplying the Boltzmann transport equation on both sides by χ and then integrating over $\mathbf{v}$. The collision term vanishes by virtue of (5.8), and we have*

$$\int d^3v\, \chi(\mathbf{r}, \mathbf{v}) \left(\frac{\partial}{\partial t} + v_i \frac{\partial}{\partial x_i} + \frac{1}{m} F_i \frac{\partial}{\partial v_i}\right) f(\mathbf{r}, \mathbf{v}, t) = 0 \tag{5.10}$$

We may rewrite (5.10) in the form

$$\frac{\partial}{\partial t}\int d^3v\, \chi f + \frac{\partial}{\partial x_i}\int d^3v\, \chi v_i f - \int d^3v \frac{\partial \chi}{\partial x_i} v_i f + \frac{1}{m}\int d^3v \frac{\partial}{\partial v_i}(\chi F_i f) - \frac{1}{m}\int d^3v \frac{\partial \chi}{\partial v_i} F_i f - \frac{1}{m}\int d^3v\, \chi \frac{\partial F_i}{\partial v_i} f = 0 \tag{5.11}$$

The fourth term vanishes if $f(\mathbf{r}, \mathbf{v}, t)$ is assumed to vanish when $|\mathbf{v}| \to \infty$. Defining the average value $\langle A \rangle$ by

$$\langle A \rangle \equiv \frac{\int d^3v\, Af}{\int d^3v\, f} = \frac{1}{n}\int d^3v\, Af \tag{5.12}$$

where

$$n(\mathbf{r}, t) \equiv \int d^3v\, f(\mathbf{r}, \mathbf{v}, t) \tag{5.13}$$

we obtain finally the desired theorem.

CONSERVATION THEOREM

$$\frac{\partial}{\partial t}\langle n\chi \rangle + \frac{\partial}{\partial x_i}\langle n v_i \chi \rangle - n\left\langle v_i \frac{\partial \chi}{\partial x_i}\right\rangle - \frac{n}{m}\left\langle F_i \frac{\partial \chi}{\partial v_i}\right\rangle - \frac{n}{m}\left\langle \frac{\partial F_i}{\partial v_i}\chi \right\rangle = 0 \tag{5.14}$$

* The summation convention, whereby a repeated vector index is understood to be summed from 1 to 3, is used.

where χ is any conserved property. Note that $\langle nA \rangle = n\langle A \rangle$ because n is independent of $\mathbf{v}$. From now on we restrict our attention to velocity-independent external forces so that the last term of (5.14) may be dropped.

For simple molecules the independent conserved properties are mass, momentum, and energy. For charged molecules we also include the charge, but this extension is trivial. Accordingly we set successively

$$\chi = m \qquad \text{(mass)}$$

$$\chi = mv_i \quad (i = 1, 2, 3) \qquad \text{(momentum)}$$

$$\chi = \tfrac{1}{2}m\,|\mathbf{v} - \mathbf{u}(\mathbf{r}, t)|^2 \qquad \text{(thermal energy)}$$

where $\mathbf{u}(\mathbf{r}, t) \equiv \langle \mathbf{v} \rangle$

We should then have three independent conservation theorems.

For $\chi = m$ we have immediately

$$\frac{\partial}{\partial t}(mn) + \frac{\partial}{\partial x_i}\langle mnv_i \rangle = 0$$

or, introducing the mass density

$$\rho(\mathbf{r}, t) \equiv mn(\mathbf{r}, t)$$

we obtain

$$\frac{\partial \rho}{\partial t} + \nabla \cdot (\rho \mathbf{u}) = 0 \tag{5.15}$$

Next we put $\chi = mv_i$, obtaining

$$\frac{\partial}{\partial t}\langle \rho v_i \rangle + \frac{\partial}{\partial x_j}\langle \rho v_i v_j \rangle - \frac{1}{m}\rho F_i = 0 \tag{5.16}$$

To reduce this further let us write

$$\begin{aligned} \langle v_i v_j \rangle &= \langle (v_i - u_i)(v_j - u_j) \rangle + \langle v_i \rangle u_j + u_i \langle v_j \rangle - u_i u_j \\ &= \langle (v_i - u_i)(v_j - u_j) \rangle + u_i u_j \end{aligned}$$

Substituting this into (5.16) we obtain

$$\rho\left(\frac{\partial u_i}{\partial t} + u_j \frac{\partial u_i}{\partial x_j}\right) = \frac{1}{m}\rho F_i - \frac{\partial}{\partial x_j}\langle \rho(v_i - u_i)(v_j - u_j) \rangle \tag{5.17}$$

Introducing the abbreviation

$$P_{ij} \equiv \rho \langle (v_i - u_i)(v_j - u_j) \rangle$$

which is called the *pressure tensor*, we finally have

$$\left(\frac{\partial}{\partial t} + u_j \frac{\partial}{\partial x_j}\right) u_i = \frac{1}{m} F_i - \frac{1}{\rho} \frac{\partial}{\partial x_j} P_{ij} \tag{5.18}$$

Finally we set $\chi = \frac{1}{2} m |\mathbf{v} - \mathbf{u}|^2$. Then

$$\frac{1}{2} \frac{\partial}{\partial t} \langle \rho |\mathbf{v} - \mathbf{u}|^2 \rangle + \frac{1}{2} \frac{\partial}{\partial x_i} \langle \rho v_i |\mathbf{v} - \mathbf{u}|^2 \rangle - \frac{1}{2} \rho \left\langle v_i \frac{\partial}{\partial x_i} |\mathbf{v} - \mathbf{u}|^2 \right\rangle = 0 \tag{5.19}$$

We define the *temperature* by

$$kT \equiv \theta \equiv \tfrac{1}{3} m \langle |\mathbf{v} - \mathbf{u}|^2 \rangle$$

and the *heat flux* by

$$\mathbf{q} \equiv \tfrac{1}{2} m \rho \langle (\mathbf{v} - \mathbf{u}) \, |\mathbf{v} - \mathbf{u}|^2 \rangle$$

We then have

$$\begin{aligned} \tfrac{1}{2} m \rho \langle v_i |\mathbf{v} - \mathbf{u}|^2 \rangle &= \tfrac{1}{2} m \rho \langle (v_i - u_i) \, |\mathbf{v} - \mathbf{u}|^2 \rangle + \tfrac{1}{2} m \rho u_i \langle |\mathbf{v} - \mathbf{u}|^2 \rangle \\ &= q_i + \tfrac{3}{2} \rho \theta u_i \end{aligned}$$

and

$$\rho \langle v_i (v_j - u_j) \rangle = \rho \langle (v_i - u_i)(v_j - u_j) \rangle + \rho u_i \langle v_j - u_j \rangle = P_{ij}$$

Thus (5.19) can be written

$$\frac{3}{2} \frac{\partial}{\partial t} (\rho \theta) + \frac{\partial q_i}{\partial x_i} + \frac{3}{2} \frac{\partial}{\partial x_i} (\rho \theta u_i) + m P_{ij} \frac{\partial u_j}{\partial x_i} = 0$$

Since $P_{ij} = P_{ji}$

$$m P_{ij} \frac{\partial u_j}{\partial x_i} = P_{ij} \frac{m}{2} \left(\frac{\partial u_j}{\partial x_i} + \frac{\partial u_i}{\partial x_j} \right) \equiv P_{ij} \Lambda_{ij}$$

The final form is then obtained after a few straightforward steps:

$$\rho \left(\frac{\partial}{\partial t} + u_i \frac{\partial}{\partial x_i} \right) \theta + \frac{2}{3} \frac{\partial}{\partial x_i} q_i = -\frac{2}{3} \Lambda_{ij} P_{ij} \tag{5.20}$$

The three conservation theorems are summarized in (5.21), (5.22), and (5.23).

$$\frac{\partial \rho}{\partial t} + \nabla \cdot (\rho \mathbf{u}) = 0 \qquad \text{(conservation of mass)} \tag{5.21}$$

$$\rho \left(\frac{\partial}{\partial t} + \mathbf{u} \cdot \nabla \right) \mathbf{u} = \frac{\rho}{m} \mathbf{F} - \nabla \cdot \overleftrightarrow{P} \qquad \text{(conservation of momentum)} \tag{5.22}$$

$$\rho \left(\frac{\partial}{\partial t} + \mathbf{u} \cdot \nabla \right) \theta = -\tfrac{2}{3} \nabla \cdot \mathbf{q} - \tfrac{2}{3} \overleftrightarrow{P} \cdot \overleftrightarrow{\Lambda} \qquad \text{(conservation of energy)} \tag{5.23}$$

where $\overleftrightarrow{P}$ is a dyadic whose components are P_{ij}, $\nabla \cdot \overleftrightarrow{P}$ is a vector whose

ith component is $\partial P_{ij}/\partial x_j$, and $\overleftrightarrow{P} \cdot \overleftrightarrow{\Lambda}$ is the scalar $P_{ij}\Lambda_{ij}$. The auxiliary quantities are defined as follows.

$$\rho(\mathbf{r}, t) \equiv m \int d^3v\, f(\mathbf{r}, \mathbf{v}, t) \qquad \text{(mass density)} \qquad (5.24)$$

$$\mathbf{u}(\mathbf{r}, t) \equiv \langle \mathbf{v} \rangle \qquad \text{(average velocity)} \qquad (5.25)$$

$$\theta(\mathbf{r}, t) \equiv \tfrac{1}{3} m \langle |\mathbf{v} - \mathbf{u}|^2 \rangle \qquad \text{(temperature)} \qquad (5.26)$$

$$\mathbf{q}(\mathbf{r}, t) \equiv \tfrac{1}{2} m\rho \langle (\mathbf{v} - \mathbf{u})\, |\mathbf{v} - \mathbf{u}|^2 \rangle \qquad \text{(heat flux vector)} \qquad (5.72)$$

$$P_{ij} \equiv \rho \langle (v_i - u_i)(v_j - u_j) \rangle \qquad \text{(pressure tensor)} \qquad (5.28)$$

$$\Lambda_{ij} \equiv \tfrac{1}{2} m \left(\frac{\partial u_i}{\partial x_j} + \frac{\partial u_j}{\partial x_i} \right) \qquad (5.29)$$

Although the conservation theorems are exact, they have no practical value unless we can actually solve the Boltzmann transport equation and use the distribution function so obtained to evaluate the quantities (5.24)–(5.29). Despite the fact that these quantities have been given rather suggestive names, their physical meaning, if any, can only be ascertained after the distribution function is known. We shall see that when it is known these conservation theorems become the physically meaningful equations of hydrodynamics.

5.3 THE ZERO-ORDER APPROXIMATION

We assume that we are dealing with a gas that, although not in equilibrium, is not far from it. In particular, we assume that in the neighborhood of any point in the gas, the distribution function is locally Maxwell-Boltzmann, and that the density, temperature, and average velocity vary only slowly in space and time. For such a gas it is natural that we try the approximation

$$f(\mathbf{r}, \mathbf{v}, t) \approx f^{(0)}(\mathbf{r}, \mathbf{v}, t) \qquad (5.30)$$

where

$$f^{(0)}(\mathbf{r}, \mathbf{v}, t) = n \left(\frac{m}{2\pi\theta} \right)^{3/2} \exp\left[-\frac{m}{2\theta} (\mathbf{v} - \mathbf{u})^2 \right] \qquad (5.31)$$

where n, θ, $\mathbf{u}$ are all slowly varying functions of $\mathbf{r}$ and t. It is obvious that (5.30) cannot be an exact solution of the Boltzmann transport equation. It is obvious that

$$\left(\frac{\partial f^{(0)}}{\partial t} \right)_{\text{coll}} = 0 \qquad (5.32)$$

because n, θ, $\mathbf{u}$ do not depend on $\mathbf{v}$, but it is also clear that in general

$$\left(\frac{\partial}{\partial t} + \mathbf{v} \cdot \nabla_{\mathbf{r}} + \frac{\mathbf{F}}{m} \cdot \nabla_{\mathbf{v}}\right) f^{(0)}(\mathbf{r}, \mathbf{v}, t) \neq 0 \tag{5.33}$$

We postpone the discussion of the accuracy of the approximation (5.30). For the moment let us assume that it is a good approximation and discuss the physical consequences.

If (5.30) is a good approximation, the left-hand side of (5.33) must be approximately equal to zero. This in turn would mean that n, θ, $\mathbf{u}$ are such that the conservation theorems (5.21)–(5.23) are approximately satisfied. The conservation theorems then become the equations restricting the behavior of n, θ, $\mathbf{u}$. To see what they are, we must calculate $\mathbf{q}$ and P_{ij} to the lowest order. The results are denoted respectively by $\mathbf{q}^{(0)}$ and $P_{ij}^{(0)}$. Let $C(\mathbf{r}, t) = n(m/2\pi\theta)^{3/2}$ and $A(\mathbf{r}, t) = m/2\theta$. We easily obtain

$$\begin{aligned}
\mathbf{q}^{(0)} &= \frac{1}{2}\frac{m\rho}{n} \int d^3v\, (\mathbf{v} - \mathbf{u})\, |\mathbf{v} - \mathbf{u}|^2\, C(\mathbf{r}, t) e^{-A(\mathbf{r},t)|\mathbf{v}-\mathbf{u}|^2} \\
&= \tfrac{1}{2} m^2 C(\mathbf{r}, t) \int d^3U\, \mathbf{U} U^2 e^{-A(\mathbf{r},t)U^2} = 0
\end{aligned} \tag{5.34}$$

$$\begin{aligned}
P_{ij}^{(0)} &= \frac{\rho}{n}\, C(\mathbf{r}, t) \int d^3v\, (v_i - u_i)(v_j - u_j) e^{-A(\mathbf{r},t)|\mathbf{v}-\mathbf{u}|^2} \\
&= mC(\mathbf{r}, t) \int d^3U U_i U_j e^{-A(\mathbf{r},t)U^2} = \delta_{ij} P
\end{aligned} \tag{5.35}$$

where

$$P = \tfrac{1}{3}\rho\left(\frac{m}{2\pi\theta}\right)^{3/2} \int d^3\mathbf{U}\, U^2 e^{-A(\mathbf{r},t)U^2} = n\theta \tag{5.36}$$

which is the local hydrostatic pressure.

Substituting these into (5.21) and (5.23), and noting that

$$\nabla \cdot \overleftrightarrow{P}^{(0)} = \nabla P$$

$$\overleftrightarrow{P}^{(0)} \cdot \overleftrightarrow{\Lambda} = P \sum_{i=1}^{3} \Lambda_{ii} = mP\nabla \cdot \mathbf{u}$$

We obtain the equations

$$\frac{\partial \rho}{\partial t} + \nabla \cdot (\rho \mathbf{u}) = 0 \qquad \text{(continuity equation)} \tag{5.37}$$

$$\left(\frac{\partial}{\partial t} + \mathbf{u} \cdot \nabla\right)\mathbf{u} + \frac{1}{\rho}\nabla P = \frac{\mathbf{F}}{m} \qquad \text{(Euler's equation)} \tag{5.38}$$

$$\left(\frac{\partial}{\partial t} + \mathbf{u} \cdot \nabla\right)\theta + \frac{1}{c_V}(\nabla \cdot \mathbf{u})\theta = 0 \tag{5.39}$$

where $c_V = \frac{3}{2}$. These are the hydrodynamic equations for the nonviscous flow of a gas. They possess solutions describing flow patterns that persist indefinitely. Thus, in this approximation, the local Maxwell-Boltzmann distribution never decays to the true Maxwell-Boltzmann distribution. This is in rough accord with experience, for we know that a hydrodynamic flow, left to itself, takes a long time to die out.

Although derived for dilute gases, (5.37)–(5.39) are also used for liquids, because these equations can also be derived through heuristic arguments which indicate that they are of a more general validity.

We shall now briefly point out some of the consequences of (5.37)–(5.39) that are of practical interest.

The quantity $\left(\frac{\partial}{\partial t} + \mathbf{u} \cdot \nabla\right) X$ is known as the "material derivative of X," because it is the time rate of change of X to an observer moving with the local average velocity $\mathbf{u}$. Such an observer is said to be moving along a streamline. We now show that in the zero-order approximation a dilute gas undergoes only adiabatic transformations to an observer moving along a streamline. Equations (5.37) and (5.39) may be rewritten as

$$\left(\frac{\partial}{\partial t} + \mathbf{u} \cdot \nabla\right) \rho = -\rho \nabla \cdot \mathbf{u}$$

$$-\frac{3}{2}\frac{\rho}{\theta}\left(\frac{\partial}{\partial t} + \mathbf{u} \cdot \nabla\right) \theta = \rho \nabla \cdot \mathbf{u}$$

Adding these two equations we obtain

$$\left(\frac{\partial}{\partial t} + \mathbf{u} \cdot \nabla\right) \rho - \frac{3}{2}\frac{\rho}{\theta}\left(\frac{\partial}{\partial t} + \mathbf{u} \cdot \nabla\right) \theta = 0$$

or

$$\left(\frac{\partial}{\partial t} + \mathbf{u} \cdot \nabla\right)(\rho \theta^{-3/2}) = 0 \tag{5.40}$$

Using the equation of state $P = \rho\theta/m$ we can convert (5.40) to the condition

$$P\rho^{-5/3} = \text{constant} \quad \text{(along a streamline)} \tag{5.41}$$

This is the condition for adiabatic transformation for an ideal gas, since $c_P/c_V = \frac{5}{3}$.

Next we derive the linear equation for a sound wave. Let us restrict ourselves to the case in which $\mathbf{u}$ and all the space and time derivatives of

$\mathbf{u}$, ρ, and θ are small quantities of the first order. For $\mathbf{F} = 0$, (5.37) and (5.38) may be replaced by

$$\frac{\partial \rho}{\partial t} + \rho \nabla \cdot \mathbf{u} = 0 \tag{5.42}$$

$$\rho \frac{\partial \mathbf{u}}{\partial t} + \nabla P = 0 \tag{5.43}$$

$$\frac{3}{2} \rho \frac{\partial \theta}{\partial t} - \theta \frac{\partial \rho}{\partial t} = 0 \tag{5.44}$$

where quantities smaller than first-order ones are neglected. Note that (5.44) is none other than (5.40) or (5.41). Taking the divergence of (5.43) and the time derivative of (5.42), and subtracting one resulting equation from the other, we obtain

$$\nabla^2 P - \frac{\partial^2 \rho}{\partial t^2} = 0 \tag{5.45}$$

in which higher-order quantities are again neglected. Now P is a function of ρ and θ, but the latter are not independent quantities, being related to each other through the condition of adiabatic transformation (5.44). Hence we may regard P as a function of ρ alone, and write

$$\nabla^2 P = \nabla \cdot \left[\left(\frac{\partial P}{\partial \rho} \right)_S \nabla \rho \right] \approx \left(\frac{\partial P}{\partial \rho} \right)_S \nabla^2 \rho$$

where $(\partial P/\partial \rho)_S$ is the adiabatic derivative, related to the adiabatic compressibility κ_S by

$$\kappa_S = \frac{1}{\rho} \left(\frac{\partial \rho}{\partial P} \right)_S = \frac{3}{5} \frac{m}{\rho \theta} \tag{5.46}$$

Thus (5.45) can be written in the form

$$\nabla^2 \rho - \rho \kappa_S \frac{\partial^2 \rho}{\partial t^2} = 0 \tag{5.47}$$

which is a wave equation for ρ, describing a sound wave with a velocity of propagation c given by

$$c = \frac{1}{\sqrt{\rho \kappa_S}} = \sqrt{\frac{5}{3} \frac{\theta}{m}} = \sqrt{\frac{5}{6}}\, \bar{v} \tag{5.48}$$

It is hardly surprising that the adiabatic compressibility enters here, because in the present approximation there can be no heat conduction in the gas, as (5.34) indicates.

Finally consider the case of steady flow under the influence of a conservative external force field, i.e., under the conditions

$$\mathbf{F} = -\nabla\phi \tag{5.49}$$

$$\frac{\partial \mathbf{u}}{\partial t} = 0$$

Using the vector identity

$$(\mathbf{u}\cdot\nabla)\mathbf{u} = \tfrac{1}{2}\nabla(u^2) - \mathbf{u}\times(\nabla\times\mathbf{u}) \tag{5.50}$$

we can rewrite (5.38) as follows

$$\nabla\left(\tfrac{1}{2}u^2 + \frac{1}{\rho}P + \frac{1}{m}\phi\right) = \mathbf{u}\times(\nabla\times\mathbf{u}) - \frac{\theta}{m}\frac{\nabla\rho}{\rho} \tag{5.51}$$

Two further specializations are of interest. First, in the case of uniform density and irrotational flow, namely, $\nabla\rho = 0$ and $\nabla\times\mathbf{u} = 0$, we have

$$\nabla\left(\tfrac{1}{2}u^2 + \frac{1}{\rho}P + \frac{1}{m}\phi\right) = 0 \tag{5.52}$$

which is *Bernoulli's equation*. Second, in the case of uniform temperature and irrotational flow, namely, $\nabla\theta = 0$ and $\nabla\times\mathbf{u} = 0$, we have

$$\nabla\left(\tfrac{1}{2}u^2 + \frac{1}{m}\phi\right) = -\frac{\theta}{m}\nabla(\log\rho)$$

which may be immediately integrated to yield

$$\rho \equiv \rho_0 \exp\left[-\frac{1}{\theta}\left(\frac{1}{2}mu^2 + \phi\right)\right] \tag{5.53}$$

where ρ_0 is an arbitrary constant.

5.4 THE FIRST-ORDER APPROXIMATION

We now give an estimate of the error incurred in the zero-order approximation (5.30). Let $f(\mathbf{r}, \mathbf{v}, t)$ be the exact distribution function, and let

$$g(\mathbf{r}, \mathbf{v}, t) \equiv f(\mathbf{r}, \mathbf{v}, t) - f^{(0)}(\mathbf{r}, \mathbf{v}, t) \tag{5.54}$$

We are interested in the magnitude of g as compared to $f^{(0)}$. First let us estimate the order of magnitude of $(\partial f/\partial t)_{\text{coll}}$. We have, by definition,

$$\begin{aligned}\left(\frac{\partial f}{\partial t}\right)_{\text{coll}} &= \int d^3v_2 \int d\Omega\,\sigma(\Omega)\,|\mathbf{v}_2 - \mathbf{v}_1|\,(f_2'f_1' - f_2f_1)\\ &\approx \int d^3v_2 \int d\Omega\,\sigma(\Omega)\,|\mathbf{v}_2 - \mathbf{v}_1|\,(f_2^{(0)\prime}g_1' - f_2^{(0)}g_1 + g_2'f_1^{(0)\prime} - g_2f_1^{(0)})\end{aligned} \tag{5.55}$$

where we have used (5.54), the fact that $(\partial f^{(0)}/\partial t)_{\text{coll}} = 0$, and the assumption that g is a small quantity whose square can be neglected. An order-of-magnitude estimate of (5.55) may be obtained by calculating the second term of the right-hand side of (5.55), which is

$$-g(\mathbf{r}, \mathbf{v}_1, t) \int d^3v_2 \sigma_{\text{tot}} \, |\mathbf{v}_2 - \mathbf{v}_1| \, f_2^{(0)} = -\frac{g(\mathbf{r}, \mathbf{v}_1, t)}{\tau} \tag{5.56}$$

where τ is a number of the order of magnitude of the collision time. Thus if we put

$$\left(\frac{\partial f}{\partial t}\right)_{\text{coll}} \approx -\frac{f - f^{(0)}}{\tau} \tag{5.57}$$

we obtain results that are qualitatively correct. With (5.57) the Boltzmann transport equation becomes

$$\left(\frac{\partial}{\partial t} + \mathbf{v} \cdot \nabla_{\mathbf{r}} + \frac{\mathbf{F}}{m} \cdot \nabla_{\mathbf{v}}\right)(f^{(0)} + g) \approx -\frac{g}{\tau} \tag{5.58}$$

Assuming $g \ll f^{(0)}$, we can neglect g on the left-hand side of (5.58). Assume further that $f^{(0)}$ varies by a significant amount (i.e., of the order of itself) only when $|\mathbf{r}|$ varies by a distance L. Then (5.58) furnishes the estimate

$$\bar{v}\frac{f^{(0)}}{L} \approx -\frac{g}{\tau}$$

or

$$\frac{g}{f^{(0)}} \approx -\frac{\lambda}{L} \tag{5.59}$$

where λ is a length of the order of the mean free path. From these considerations we conclude that $f^{(0)}$ is a good approximation if the local density, temperature, and velocity have characteristic wavelengths L much larger than the mean free path λ. The corrections to $f^{(0)}$ would be of the order of λ/L.

A systematic expansion of f in powers of λ/L is furnished by the Chapman-Enskog expansion, which is somewhat complicated. In order not to lose sight of the physical aspects of the problem, we first give a qualitative discussion of the first-order approximation based on the approximate equation (5.58). The precise value of τ cannot be ascertained. For the present we have to be content with the knowledge that τ is of the order of the collision time. Thus we put

$$f = f^{(0)} + g \tag{5.60}$$

where, with (5.58), we take

$$g = -\tau\left(\frac{\partial}{\partial t} + \mathbf{v} \cdot \nabla_{\mathbf{r}} + \frac{\mathbf{F}}{m} \cdot \nabla_{\mathbf{v}}\right) f^{(0)} \tag{5.61}$$

To calculate g, note that $f^{(0)}$ depends on $\mathbf{r}$ and t only through the functions ρ, θ, and $\mathbf{u}$. Thus we need the derivatives

$$
\begin{aligned}
\frac{\partial f^{(0)}}{\partial \rho} &= \frac{f^{(0)}}{\rho} \\
\frac{\partial f^{(0)}}{\partial \theta} &= \frac{1}{\theta}\left(\frac{m}{2\theta}U^2 - \frac{3}{2}\right) f^{(0)} \\
\frac{\partial f^{(0)}}{\partial u_i} &= \frac{m}{\theta} U_i f^{(0)} \\
\frac{\partial f^{(0)}}{\partial v_i} &= -\frac{m}{\theta} U_i f^{(0)}
\end{aligned}
\tag{5.62}
$$

where

$$
\mathbf{U} \equiv \mathbf{v} - \mathbf{u}(\mathbf{r}, t) \tag{5.63}
$$

Hence

$$
\begin{aligned}
g &= -\tau\left(\frac{\partial}{\partial t} + v_i \frac{\partial}{\partial x_i} + \frac{F_i}{m}\frac{\partial}{\partial v_i}\right) f^{(0)} \\
&= -\tau f^{(0)}\left[\frac{1}{\rho} D(\rho) + \frac{1}{\theta}\left(\frac{m}{2\theta}U^2 - \frac{3}{2}\right) D(\theta) + \frac{m}{\theta} U_j D(u_j) - \frac{1}{\theta}\mathbf{F}\cdot\mathbf{U}\right]
\end{aligned}
\tag{5.64}
$$

where

$$
D(X) \equiv \left(\frac{\partial}{\partial t} + v_i \frac{\partial}{\partial x_i}\right) X \tag{5.65}
$$

Using the zero-order hydrodynamic equations (5.37)–(5.39), we can show that

$$
\begin{aligned}
D(\rho) &= -\rho(\nabla\cdot\mathbf{u}) + \mathbf{U}\cdot\nabla\rho \\
D(\theta) &= -\tfrac{2}{3}\theta\nabla\cdot\mathbf{u} + \mathbf{U}\cdot\nabla\theta \\
D(u_j) &= -\frac{1}{\rho}\frac{\partial P}{\partial x_j} + \frac{F_j}{m} + U_i \frac{\partial u_j}{\partial x_i}
\end{aligned}
\tag{5.66}
$$

where $P = \rho\theta/m$. Substituting these into (5.64) we obtain

$$
\begin{aligned}
g = -\tau f^{(0)}\Bigg[&-(\nabla\cdot\mathbf{u}) + \mathbf{U}\cdot\frac{\nabla\rho}{\rho} + \frac{1}{\theta}\left(\frac{m}{2\theta}U^2 - \frac{3}{2}\right)\left(-\frac{2}{3}\theta\nabla\cdot\mathbf{u} + \mathbf{U}\cdot\nabla\theta\right) \\
&+ \frac{m}{\theta}\left(-\mathbf{U}\cdot\frac{\nabla P}{\rho} + \mathbf{U}\cdot\frac{\mathbf{F}}{m} + U_i U_j \frac{\partial u_j}{\partial x_i}\right) - \frac{1}{\theta}\mathbf{F}\cdot\mathbf{U}\Bigg]
\end{aligned}
$$

which, after some rearrangement and cancellation of terms, becomes

$$
g = -\tau\left[\frac{1}{\theta}\frac{\partial\theta}{\partial x_i} U_i \left(\frac{m}{2\theta}U^2 - \frac{5}{2}\right) + \frac{1}{\theta}\Lambda_{ij}(U_i U_j - \tfrac{1}{3}\delta_{ij}U^2)\right] f^{(0)} \tag{5.67}
$$

where Λ_{ij} is defined by (5.29).

It is now necessary to calculate $\mathbf{q}$ and P_{ij} with the help of (5.60), in order to obtain the equations of hydrodynamics to the first order. We have

$$\mathbf{q} = \frac{m\rho}{2n}\int d^3v\,(\mathbf{v}-\mathbf{u})\,|\mathbf{v}-\mathbf{u}|^2\,g$$

Noting that the second term of (5.67) does not contribute to this integral, we obtain

$$\mathbf{q} = -\frac{\tau m^2}{2}\int d^3U\,\mathbf{U}U^2\left(\frac{m}{2\theta}U^2-\frac{5}{2}\right)\frac{1}{\theta}U_i\frac{\partial\theta}{\partial x_i}f^{(0)}$$

or
$$\mathbf{q} = -K\nabla\theta \tag{5.68}$$

where
$$K = \frac{m^2\tau}{6\theta}\int d^3U\,U^4\left(\frac{m}{2\theta}U^2-\frac{5}{2}\right)f^{(0)} = \tfrac{5}{2}\tau\theta n \tag{5.69}$$

It is clear from (5.68) that K is to be identified as the coefficient of thermal conductivity. It is also clear that $|\mathbf{q}|$ is a small quantity of the first order, being of the order of λ/L.

For the pressure tensor P_{ij}, only the second term of (5.67) contributes:

$$P_{ij} = \frac{\rho}{n}\int d^3v\,(v_i-u_i)(v_j-u_j)(f^{(0)}+g) = \delta_{ij}P + P_{ij}' \tag{5.70}$$

where $P = \rho\theta/m$ and

$$P_{ij}' = -\frac{\tau\rho}{\theta n}\Lambda_{kl}\int d^3U\,U_iU_j(U_kU_l - \tfrac{1}{3}\delta_{kl}U^2)f^{(0)} \tag{5.71}$$

To evaluate this, note that P_{ij}' is a symmetric tensor of zero trace (i.e., $\sum_{i=1}^{3}P_{ii}'=0$), and it depends linearly on the symmetric tensor Λ_{ij}. Therefore P_{ij}' must have the form

$$P_{ij}' = -\frac{2\mu}{m}\left(\Lambda_{ij}-\frac{m}{3}\delta_{ij}\nabla\cdot\mathbf{u}\right) \tag{5.72}$$

where $m\nabla\cdot\mathbf{u}$ is none other than the trace of Λ_{ij}:

$$\sum_{i=1}^{3}\Lambda_{ii} = m\sum_{i=1}^{3}\frac{\partial u_i}{\partial x_i} = m\nabla\cdot\mathbf{u} \tag{5.73}$$

and μ is a constant. It remains to calculate μ. For this purpose it suffices to calculate any component of P_{ij}', e.g., P_{12}'. From (5.71) we have

$$P_{12}' = -\frac{\tau m}{\theta}\Lambda_{kl}\int d^3U\,U_1U_2(U_kU_l-\tfrac{1}{3}\delta_{kl}U^2)f^{(0)}$$

$$= -2\frac{\tau m}{\theta}\Lambda_{12}\int d^3U\,U_1^2U_2^2f^{(0)}$$

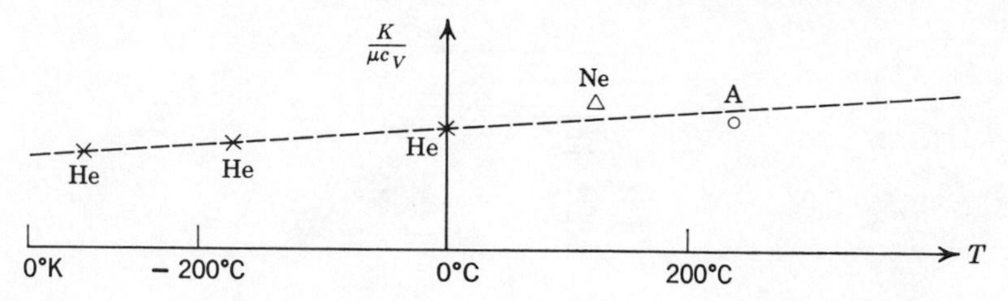

Fig. 5.1. Ratio of thermal conductivity to the product of viscosity and specific heat for different dilute gases.

Therefore
$$\mu = \frac{\tau m^2}{\theta}\int d^3U\, U_1{}^2 U_2{}^2 f^{(0)} = \tau n\theta \tag{5.74}$$

With this we have

$$P_{ij} = \delta_{ij}P - \frac{2\mu}{m}\left(\Lambda_{ij} - \frac{m}{3}\,\delta_{ij}\nabla\cdot\mathbf{u}\right) \tag{5.75}$$

The second term is of the order of λ/L. The coefficient μ turns out to be the coefficient of viscosity, as we show shortly.

A comparison of (5.74) with (5.69) shows that

$$\frac{K}{\mu} = \tfrac{5}{2} = \tfrac{5}{3}c_V \tag{5.76}$$

Since the unknown collision time τ drops out in this relation, we might expect (5.76) to be of quantitative significance. A plot of some experimental data for different dilute gases in Fig. 5.1 shows that it is indeed so.

Let us put, with (5.6),

$$\tau \approx \sqrt{\frac{m}{kT}}\,\frac{1}{na^2} \tag{5.77}$$

where a is the molecular diameter. Then we find that

$$\mu \approx K \approx \frac{\sqrt{mkT}}{a^2} \tag{5.78}$$

5.5 VISCOSITY

To show that (5.74) is the coefficient of viscosity, we independently calculate the coefficient of viscosity using its experimental definition. Consider a gas of uniform and constant density and temperature, with an average velocity given by

$$\begin{aligned} u_x &= A + By \\ u_y &= u_z = 0 \end{aligned} \tag{5.79}$$

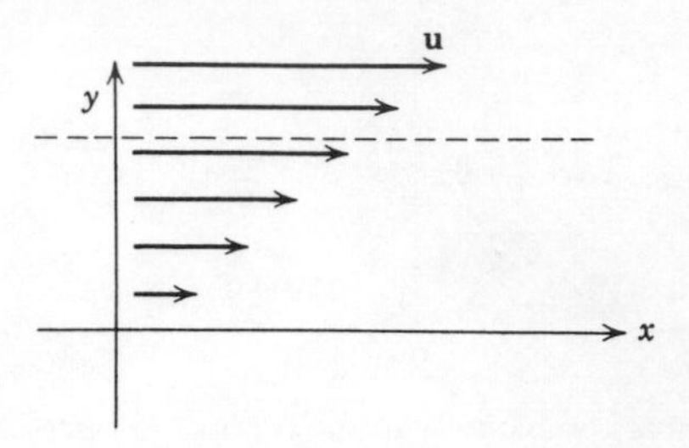

Fig. 5.2. Horizontal flow of a gas with average velocity increasing linearly with height.

where A, B are constants. The gas may be thought of as being composed of different layers sliding over each other, as shown in Fig. 5.2. Draw any plane perpendicular to the y axis, as shown by the dotted line in Fig. 5.2. Let F' be the frictional force experienced by the gas above this plane, per unit area of the plane. Then the coefficient of viscosity μ is experimentally defined by the relation

$$F' = -\mu \frac{\partial u_x}{\partial y} \tag{5.80}$$

The gas above the plane experiences a frictional force by virtue of the fact that it suffers a net loss of "x component of momentum" to the gas below. Thus

$$F' \equiv \textit{net} \text{ amount of "}x\text{ component of momentum" transported per sec across unit area in the } y \text{ direction} \tag{5.81}$$

The quantity being transported is $m(v_x - u_x)$, whereas the flux effective in the transport is $n(v_y - u_y)$. Hence we have

$$F' = mn\langle(v_x - u_x)(v_y - u_y)\rangle = m \int d^3v\,(v_x - u_x)(v_y - u_y)(f^{(0)} + g) \tag{5.82}$$

We easily see that the term $f^{(0)}$ does not contribute to the integral in (5.82). The first correction g may be obtained directly from the approximate Boltzmann transport equation

$$\mathbf{v} \cdot \nabla f^{(0)} = -\frac{g}{\tau}$$

$$g = -\frac{\tau m}{\theta} v_y(v_x - u_x)Bf^{(0)} = -\frac{\tau m}{\theta} U_y U_x \frac{\partial u_x}{\partial y} f^{(0)} \tag{5.83}$$

where $\mathbf{U} \equiv \mathbf{v} - \mathbf{u}$. Thus

$$F' = -\frac{\partial u_x}{\partial y} \frac{\tau m^2}{\theta} \int d^3U\, U_x^{\,2} U_y^{\,2} f^{(0)} \tag{5.84}$$

A comparison between this and (5.80) yields

$$\mu = \frac{\tau m^2}{\theta} \int d^3U \; U_x{}^2 U_y{}^2 f^{(0)} \tag{5.85}$$

which is identical with (5.74).

From the nature of this derivation it is possible to understand physically why μ has the order of magnitude given by (5.78). Across the imaginary plane mentioned previously, a net transport of momentum exists, because molecules constantly cross this plane in both directions. The flux is the same in both directions, being of the order of $n\sqrt{kT/m}$. On the average, however, those that cross from above to below carry more "x component of momentum" than the opposite ones, because the average velocity u_x is greater above than below. Since most molecules that cross the plane from above originated within a mean free path λ above the plane, their u_x is in excess of the local u_x below the plane by the amount $\lambda(\partial u_x/\partial y)$. Hence the net amount of "x component of momentum" transported per second from above to below, per unit area of the plane, is

$$\lambda nm \sqrt{\frac{kT}{m}} \frac{\partial u_x}{\partial y} = \frac{\sqrt{mkT}}{a^2} \frac{\partial u_x}{\partial y} \tag{5.86}$$

Therefore

$$\mu \approx \frac{\sqrt{mkT}}{a^2} \tag{5.87}$$

It is interesting to note that according to (5.87) μ is independent of the density for a given temperature. When Maxwell first derived this fact, he was so surprised that he put it to experimental test by observing the rate of damping of a pendulum suspended in gases of different densities. To his satisfaction, it was verified.

According to (5.87) the coefficient of viscosity increases as the molecular diameter decreases, everything else being constant. This is physically easy to understand, because the mean free path λ increases with decreasing molecular diameter. For a given gradient $\partial u_x/\partial y$, the momentum transported across any plane normal to the y axis obviously increases as λ increases. When λ becomes so large that it is comparable to the size of the container of the gas, the whole method adopted here breaks down, and the coefficient of viscosity ceases to be a meaningful concept.

As a topic related to the concept of viscosity we consider the boundary condition for a gas flowing past a wall. A gas, unlike a liquid, does not stick to the wall of its container. Rather, it slips by with an average velocity u_0. To determine u_0, it is necessary to know how the gas molecules interact with the wall. We make the simplifying assumption that a fraction $1 - \alpha$ of the molecules striking the wall is reflected elastically while

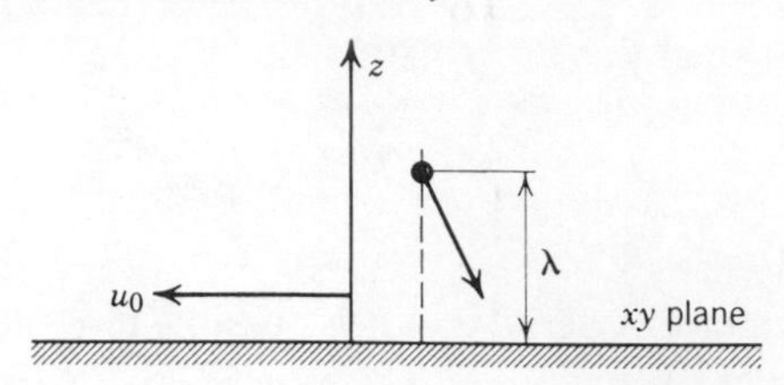

Fig. 5.3. A gas slipping past a wall.

the remaining fraction α is absorbed by the wall, only to return to the gas later with thermal velocity. The number α is called the *coefficient of accommodation*. Suppose the wall is the xy plane, as shown in Fig. 5.3. Then the downward flux of particles is given by

$$\int_{-\infty}^{\infty} dv_x \int_{-\infty}^{\infty} dv_y \int_{0}^{\infty} dv_z \, nv_z f^{(0)} = n\sqrt{\frac{\theta}{2\pi m}} \tag{5.88}$$

The particles that reach the wall came from a mean free path λ above the wall. Thus the gas loses to a unit area of the wall an amount of momentum per second equal to

$$F' = -\alpha nm \sqrt{\frac{\theta}{2\pi m}} \left[u_0 + \lambda \left(\frac{\partial u}{\partial z} \right)_0 \right] \tag{5.89}$$

where $(\partial u / \partial z)_0$ is the normal gradient of u at the wall. This is the force of friction per unit area that the wall exerts on the gas, and must equal $-\mu(\partial u / \partial z)_0$. Hence the boundary condition at the wall is

$$\alpha nm \sqrt{\frac{\theta}{2\pi m}} \left[u_0 + \lambda \left(\frac{\partial u}{\partial z} \right)_0 \right] = \mu \left(\frac{\partial u}{\partial z} \right)_0$$

or

$$u_0 = \left(\sqrt{\frac{2\pi}{m\theta}} \frac{\mu}{n\alpha} - \lambda \right) \left(\frac{\partial u}{\partial z} \right)_0 \tag{5.90}$$

Using $\mu = \tau n \theta$ and $\lambda = \beta\tau\sqrt{2\pi\theta / m}$, where β is a constant of the order of unity, we obtain the boundary condition

$$u_0 = s\lambda \left(\frac{\partial u}{\partial z} \right)_0 \tag{5.91}$$

where

$$s = \frac{1 - \alpha\beta}{\alpha\beta}$$

is an empirical constant which may be called the "slipping coefficient." When $s = 0$ there is no slipping at the wall. In general the velocity of slip is equal to the velocity in the gas at a distance of s mean free paths from the wall. Usually $s\lambda$ is a few mean free paths.

5.6 VISCOUS HYDRODYNAMICS

The equations of hydrodynamics in the first-order approximation can be obtained by substituting **q** and P_{ij}, given respectively in (5.68) and (5.75), into the conservation theorems (5.21)–(5.23). We first evaluate a few relevant quantities.

$$\nabla \cdot \mathbf{q} = -\nabla(K\nabla\theta) = -K\nabla^2\theta - \nabla K \cdot \nabla\theta \tag{5.92}$$

$$\frac{\partial P_{ij}}{\partial x_j} = \frac{\partial P}{\partial x_i} - \mu\left[\nabla^2 u_i + \frac{1}{3}\frac{\partial}{\partial x_i}(\nabla \cdot \mathbf{u})\right] - \frac{2}{m}\frac{\partial \mu}{\partial x_j}\left(\Lambda_{ij} - \frac{m}{3}\delta_{ij}\nabla \cdot \mathbf{u}\right) \tag{5.93}$$

$$P_{ij}\Lambda_{ij} = mP(\nabla \cdot \mathbf{u}) - \frac{2\mu}{m}\Lambda_{ij}\Lambda_{ij} + \tfrac{2}{3}\mu m(\nabla \cdot \mathbf{u})^2 \tag{5.94}$$

The quantity $\Lambda_{ij}\Lambda_{ij}$ can be further reduced:

$$\Lambda_{ij}\Lambda_{ij} = \frac{m^2}{4}\left(\frac{\partial u_i}{\partial x_j} + \frac{\partial u_j}{\partial x_i}\right)\left(\frac{\partial u_i}{\partial x_j} + \frac{\partial u_j}{\partial x_i}\right) = \frac{m^2}{2}\frac{\partial u_i}{\partial x_j}\left(\frac{\partial u_i}{\partial x_j} + \frac{\partial u_j}{\partial x_i}\right)$$

Now we reduce the two terms above separately:

$$\frac{\partial u_i}{\partial x_j}\frac{\partial u_i}{\partial x_j} = \frac{\partial}{\partial x_j}\left(u_i\frac{\partial u_i}{\partial x_j}\right) - u_i\frac{\partial^2 u_i}{\partial x_j \partial x_j} = \frac{1}{2}\nabla^2(u^2) - \mathbf{u} \cdot \nabla^2\mathbf{u}$$

$$\begin{aligned}\frac{\partial u_i}{\partial x_j}\frac{\partial u_j}{\partial x_i} &= \left(\frac{\partial u_i}{\partial x_j} - \frac{\partial u_j}{\partial x_i}\right)\left(\frac{\partial u_j}{\partial x_i} - \frac{\partial u_i}{\partial x_j}\right) + \frac{\partial u_i}{\partial x_j}\frac{\partial u_i}{\partial x_j} + \frac{\partial u_j}{\partial x_i}\frac{\partial u_j}{\partial x_i} - \frac{\partial u_j}{\partial x_i}\frac{\partial u_i}{\partial x_j} \\ &= -2(\nabla \times \mathbf{u})^2 + 2\frac{\partial u_i}{\partial x_j}\frac{\partial u_i}{\partial x_j} - \frac{\partial u_j}{\partial x_i}\frac{\partial u_i}{\partial x_j}\end{aligned}$$

Hence

$$\frac{\partial u_i}{\partial x_j}\frac{\partial u_j}{\partial x_i} = -(\nabla \times \mathbf{u})^2 + \frac{\partial u_i}{\partial x_j}\frac{\partial u_i}{\partial x_j}$$

and finally

$$\Lambda_{ij}\Lambda_{ij} = \frac{m^2}{2}\left[\nabla^2(u^2) - 2\mathbf{u} \cdot \nabla^2\mathbf{u} - |\nabla \times \mathbf{u}|^2\right] \tag{5.95}$$

Substituting (5.92)–(5.94) into (5.21)–(5.23) we obtain

$$\frac{\partial \rho}{\partial t} + \nabla \cdot (\rho\mathbf{u}) = 0 \tag{5.96}$$

$$\rho\left(\frac{\partial}{\partial t} + \mathbf{u} \cdot \nabla\right)\mathbf{u} = \frac{\mathbf{F}}{m} - \nabla\left(P - \frac{\mu}{3}\nabla \cdot \mathbf{u}\right) + \mu\nabla^2\mathbf{u} + \mathbf{R} \tag{5.97}$$

$$\begin{aligned}\rho\left(\frac{\partial}{\partial t} + \mathbf{u} \cdot \nabla\right)\theta = \frac{K}{c_V}\nabla^2\theta + \frac{1}{c_V}\nabla K \cdot \nabla\theta - \frac{1}{c_V}\big[m\rho(\nabla \cdot \mathbf{u}) + \tfrac{2}{3}\mu m(\nabla \cdot \mathbf{u})^2 \\ - \mu m\{\nabla^2(u^2) - 2\mathbf{u} \cdot \nabla^2\mathbf{u} - |\nabla \times \mathbf{u}|^2\}\big]\end{aligned} \tag{5.98}$$

where $c_V = \frac{3}{2}$ and $\mathbf{R}$ is a vector whose components are given by

$$R_i = \frac{2}{m}\frac{\partial \mu}{\partial x_j}\left(\Lambda_{ij} - \frac{m}{3}\delta_{ij}\nabla \cdot \mathbf{u}\right) \tag{5.99}$$

In these equations the quantities of first-order smallness are μ, K, $\mathbf{u}$ and the derivatives of ρ, θ, and $\mathbf{u}$. Keeping only quantities of first-order smallness, we can neglect all terms involving derivatives of μ and K and the last four terms on the right side of (5.98). We then have the equations of hydrodynamics to the first order:

$$\frac{\partial \rho}{\partial t} + \nabla \cdot (\rho \mathbf{u}) = 0 \qquad \text{(continuity equation)} \tag{5.100}$$

$$\left(\frac{\partial}{\partial t} + \mathbf{u} \cdot \nabla\right)\mathbf{u} = \frac{\mathbf{F}}{m} - \frac{1}{\rho}\nabla\left(P - \frac{\mu}{3}\nabla \cdot \mathbf{u}\right) + \frac{\mu}{\rho}\nabla^2 \mathbf{u} \tag{5.101}$$

(Navier-Stokes equation)

$$\left(\frac{\partial}{\partial t} + \mathbf{u} \cdot \nabla\right)\theta = -\frac{1}{c_V}(\nabla \cdot \mathbf{u})\theta + \frac{K}{\rho c_V}\nabla^2 \theta \tag{5.102}$$

(heat conduction equation)

where $c_V = \frac{3}{2}$. The boundary condition to be used when a wall is present is the slip boundary condition (5.91).

If $\mathbf{u} = 0$, (5.102) reduces to

$$\rho c_V \frac{\partial \theta}{\partial t} - K\nabla^2 \theta = 0 \tag{5.103}$$

which is the familiar diffusion equation governing heat conduction. This equation can be derived intuitively from the fact that $\mathbf{q} = -K\nabla\theta$. Although we have proved this fact only for a dilute gas, it is experimentally correct for liquids and solids as well. For this reason (5.103) is often applied to systems other than a dilute gas.

The Navier-Stokes equation can also be derived on an intuitive basis provided we take the meaning of viscosity from experiments. We discuss this derivation in the next section.

5.7 THE NAVIER-STOKES EQUATION

We give a phenomenological derivation of the Navier-Stokes equation in order to show why it is expected to be valid even for liquids. Some examples of its use are then discussed.

Consider a small element of fluid whose volume is $dx_1\, dx_2\, dx_3$ and whose velocity is $\mathbf{u}(\mathbf{r}, t)$. According to Newton's second law the equation of motion of this element of fluid is

$$m \frac{d\mathbf{u}}{dt} = \mathscr{L}$$

where m is the mass of the fluid element and $\mathscr{L}$ is the total force acting on the fluid element. Let the mass density of the fluid be ρ and let there be two forces acting on any element of fluid: A force due to agents external to the fluid, and a force due to neighboring fluid elements. These forces *per unit volume* will be respectively denoted by $\mathbf{F}_1$ and $\mathbf{G}$. Thus we can write

$$m = \rho \, dx_1 \, dx_2 \, dx_3$$
$$\mathscr{L} = (\mathbf{F}_1 + G) \, dx_1 \, dx_2 \, dx_3$$

Therefore Newton's second law for a fluid element takes the form

$$\rho\left(\frac{\partial}{\partial t} + \mathbf{u} \cdot \nabla\right)\mathbf{u} = \mathbf{F}_1 + \mathbf{G} \tag{5.104}$$

Thus the derivation of the Navier-Stokes equation reduces to the derivation of a definite expression for $\mathbf{G}$.

Let us choose a coordinate system such that the fluid element under consideration is a cube with edges along the three coordinate axes, as shown in Fig. 5.4. The six faces of this cube are subjected to forces exerted by neighboring fluid elements. The force on each face is such that its direction is determined by the direction of the normal vector to the face. That is, its direction depends on which side of the face is considered the "outside." This is physically obvious if we remind ourselves that this force arises from hydrostatic pressure and viscous drag. Let $\mathbf{T}_i$ be the force per unit area acting on the face whose normal lies along the x_i

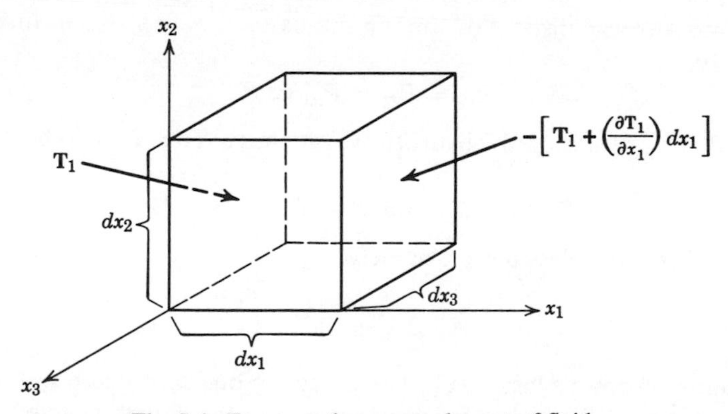

Fig. 5.4. Forces acting on an element of fluid.

axis. Then the forces *per unit area* acting on the two faces normal to the x_i axis are respectively (see Fig. 5.4)

$$\mathbf{T}_i, \quad -\left(\mathbf{T}_i + \frac{\partial \mathbf{T}_i}{\partial x_i}\, dx_i\right) \qquad (i = 1, 2, 3) \tag{5.105}$$

The total force acting on the cube by neighboring fluid elements is then given by

$$\mathbf{G}\, dx_1\, dx_2\, dx_3 = -\left(\frac{\partial \mathbf{T}_1}{\partial x_1} + \frac{\partial \mathbf{T}_2}{\partial x_2} + \frac{\partial \mathbf{T}_3}{\partial x_3}\right) dx_1\, dx_2\, dx_3 \tag{5.106}$$

We denote the components of the vectors $\mathbf{T}_1$, $\mathbf{T}_2$, $\mathbf{T}_3$ as follows:

$$\begin{aligned} \mathbf{T}_1 &= (P_{11}, P_{12}, P_{13}) \\ \mathbf{T}_2 &= (P_{21}, P_{22}, P_{23}) \\ \mathbf{T}_3 &= (P_{31}, P_{32}, P_{33}) \end{aligned} \tag{5.107}$$

Then

$$G_i = -\frac{\partial P_{ji}}{\partial x_j} \tag{5.108}$$

or

$$\mathbf{G} = -\nabla \cdot \overleftrightarrow{P} \tag{5.109}$$

With this, (5.104) becomes

$$\rho\left(\frac{\partial}{\partial t} + \mathbf{u} \cdot \nabla\right)\mathbf{u} = \mathbf{F}_1 - \nabla \cdot \overleftrightarrow{P} \tag{5.110}$$

which is of the same form as (5.22) if we set $\mathbf{F}_1 = \rho\mathbf{F}/m$, where $\mathbf{F}$ is the external force per molecule and m is the mass of a molecule. To derive the Navier-Stokes equation, we only have to deduce a more explicit form for P_{ij}. We postulate that (5.110) is valid, whatever the coordinate system we choose. It follows that P_{ij} is a tensor.

We assume the fluid under consideration to be isotropic, so that there can be no intrinsic distinction among the axes x_1, x_2, x_3. Accordingly we must have

$$P_{11} = P_{22} = P_{33} \equiv P \tag{5.111}$$

where P is by definition the hydrostatic pressure. Thus P_{ij} can be written in the form

$$P_{ij} = \delta_{ij}P + P_{ij}' \tag{5.112}$$

where P_{ij}' is a traceless tensor, namely,

$$\sum_{i=1}^{3} P_{ii}' = 0 \tag{5.113}$$

This follows from the fact that (5.113) is true in one coordinate system and that the trace of a tensor is independent of the coordinate system.

Next we make the physically reasonable assumption that the fluid element under consideration, which is really a point in the fluid, has no intrinsic angular momentum. This assumption implies that P_{ij}, and hence P_{ij}', is a symmetric tensor:

$$P_{ij}' = P_{ji}' \tag{5.114}$$

To see this we need only remind ourselves of the meaning of, for example, P_{12}'. A glance at Fig. 5.5*a* makes (5.114) obvious.

Finally we incorporate into P_{ij} the empirical connection between the shear force applied to a fluid element and the rate of deformation of the same fluid element. A shear force F' per unit area acting parallel to a face of a cube of fluid tends to stretch the cube into a parallelopiped at a rate given by $R' = \mu(d\phi/dt)$, where μ is the coefficient of viscosity and ϕ is the angle shown in Fig. 5.5*b*.

Consider now the effect of P_{12}' on one fluid element. It can be seen from Fig. 5.5*c*, where P_{12}' is indicated in its positive sense in accordance with (5.105), that

$$P_{21}' = -\mu\left(\frac{d\phi_1}{dt} + \frac{d\phi_2}{dt}\right) = -\mu\left(\frac{\partial u_2}{\partial x_1} + \frac{\partial u_1}{\partial x_2}\right) \tag{5.115}$$

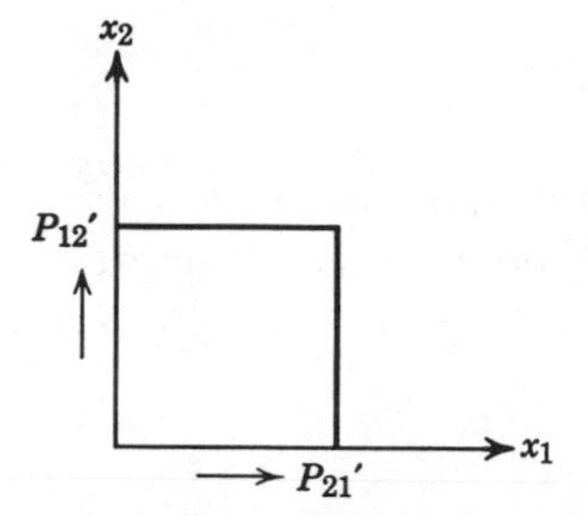

Fig. 5.5*a*. Nonrotation of fluid element implies $P_{12}' = P_{21}'$.

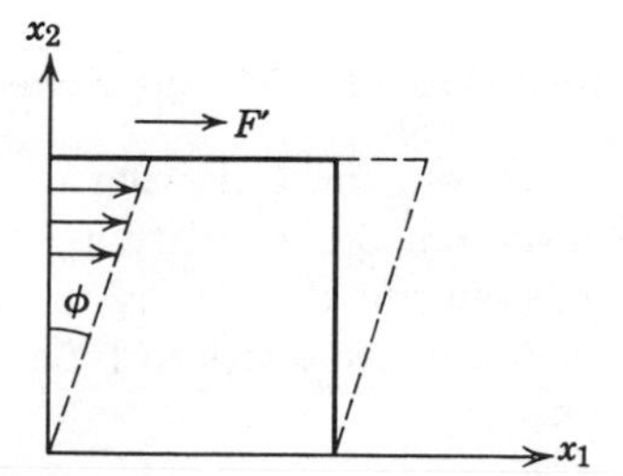

Fig. 5.5*b*. Deformation of fluid element due to shear force.

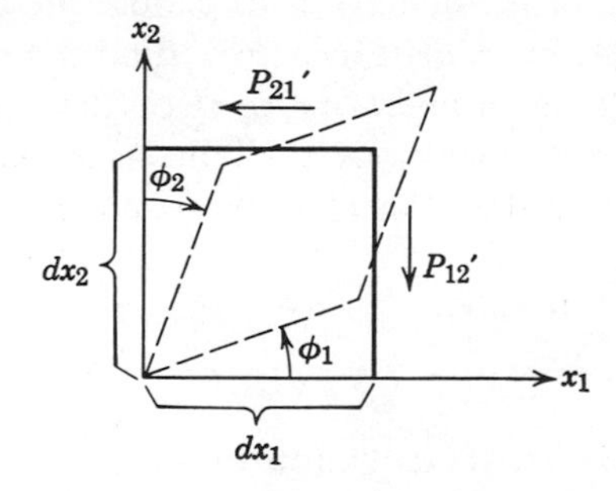

Fig. 5.5*c*. P_{12}' as shear force.

In general we have

$$P_{ij}' = -\mu\left(\frac{\partial u_i}{\partial x_j} + \frac{\partial u_j}{\partial x_i}\right) \qquad (i \neq j) \tag{5.116}$$

To make P_{ij}' traceless we must take

$$P_{ij}' = -\mu\left[\left(\frac{\partial u_i}{\partial x_j} + \frac{\partial u_j}{\partial x_i}\right) - \frac{2}{3}\delta_{ij}\nabla \cdot \mathbf{u}\right] \tag{5.117}$$

Therefore

$$P_{ij} = \delta_{ij}P - \mu\left[\left(\frac{\partial u_i}{\partial x_j} + \frac{\partial u_j}{\partial x_i}\right) - \frac{2}{3}\delta_{ij}\nabla \cdot \mathbf{u}\right] \tag{5.118}$$

which is identical in form to (5.75). This completes the phenomenological derivation, which makes it plausible that the Navier-Stokes equation is valid for dilute gas and dense liquid alike.

5.8 EXAMPLES IN HYDRODYNAMICS

To illustrate the mathematical techniques of dealing with the equations of hydrodynamics (5.100)–(5.102), we consider two examples of the application of the Navier-Stokes equation to a liquid.

Effective Mass of a Moving Sphere

We consider the following problem: A sphere of radius r is moving with instantaneous velocity $\mathbf{u}_0$ in an infinite, nonviscous, incompressible fluid of constant density, in the absence of external force. The Navier-Stokes equation reduces to Euler's equation:

$$\rho\left(\frac{\partial}{\partial t} + \mathbf{u} \cdot \nabla\right)\mathbf{u} = -\nabla P \tag{5.119}$$

where $\mathbf{u}$ is the velocity field of the liquid and P the pressure as given by the equation of state of the fluid. Let us choose the center of the sphere to be the origin of the coordinate system and label any point in space by either the rectangular coordinates (x, y, z) or the spherical coordinates (r, θ, ϕ). The boundary conditions shall be such that the normal component of $\mathbf{u}$ vanishes on the surface of the sphere and that the liquid is at rest at infinity:

$$\begin{aligned} &[\mathbf{r} \cdot \mathbf{u}(\mathbf{r})]_{r=a} - (\mathbf{r} \cdot \mathbf{u}_0)_{r=a} = 0 \\ &\mathbf{u}(\mathbf{r}) \xrightarrow[r=\infty]{} 0 \end{aligned} \tag{5.120}$$

It is tacitly assumed that the equation of state of the fluid, though unknown, is consistent with these boundary conditions. Thus the sphere can only push or pull the fluid along but cannot drag it. Since we assume no source for the fluid, we must have everywhere

$$\nabla \cdot \mathbf{u} = 0 \tag{5.121}$$

Taking the curl of both sides of (5.119), remembering that ρ is a constant, and neglecting terms of the form $(\partial \mathbf{u}/\partial x_i)(\partial \mathbf{u}/dx_j)$, we find that

$$\left(\frac{\partial}{\partial t} + \mathbf{u} \cdot \nabla\right)(\nabla \times \mathbf{u}) = 0 \tag{5.122}$$

i.e., that $\nabla \times \mathbf{u}$ is constant along a streamline. Since very far from the sphere we have $\nabla \times \mathbf{u} = 0$, it follows that everywhere

$$\nabla \times \mathbf{u} = 0 \tag{5.123}$$

This means that $\mathbf{u}$ is the gradient of some function:

$$\mathbf{u} = \nabla \Phi \tag{5.124}$$

where Φ is called the velocity potential. By (5.120) and (5.121) the equation and boundary conditions for Φ are

$$\begin{aligned} &\nabla^2 \Phi(\mathbf{r}) = 0 \\ &\left(\frac{\partial \Phi}{\partial r}\right)_{r=a} = u_0 \cos\theta \\ &\Phi(\mathbf{r}) \xrightarrow[r\to\infty]{} 0 \end{aligned} \tag{5.125}$$

where θ is the angle between u_0 and $\mathbf{r}$, as shown in Fig. 5.6. It is fortunate that to solve the problem we do not need the equation of state of the fluid apart from the assumption that it is consistent with the boundary conditions.

The most general solution to $\nabla^2 \Phi = 0$ is a superposition of solid harmonics.* Since the boundary condition involves $\cos\theta$, we try the solution

$$\Phi(\mathbf{r}) = A \frac{\cos\theta}{r^2} \qquad (r \geq a) \tag{5.126}$$

which is a solid harmonic of order 1 and is the potential that would be set up if a dipole source were placed at the center of the sphere. Choosing $A = -\frac{1}{2} u_0 a^3$ satisfies the boundary conditions. Therefore

$$\Phi(r) = -\tfrac{1}{2} u_0 a^3 \frac{\cos\theta}{r^2} \qquad (r \geq a) \tag{5.127}$$

* A solid harmonic is $r^l Y_{lm}$ or $r^{-l-1} Y_{lm}$, where Y_{lm} is a spherical harmonic.

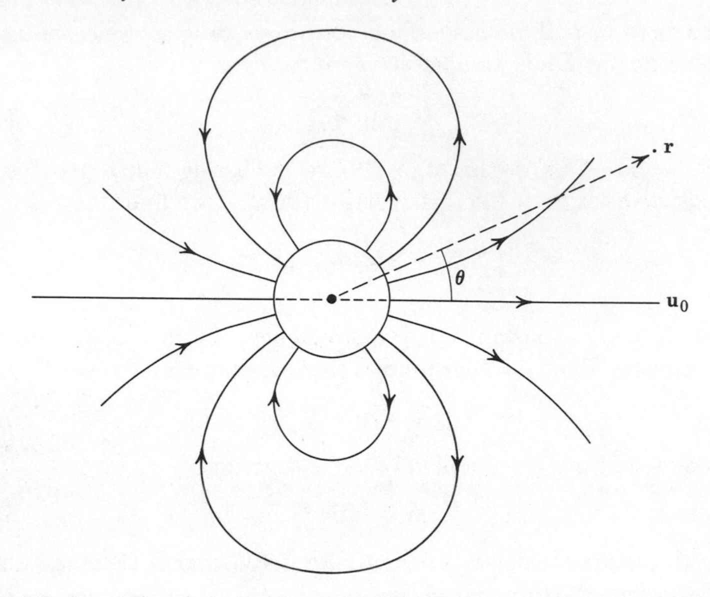

Fig. 5.6. Streamlines in a nonviscous liquid in the presence of a moving sphere.

This is the only solution of (5.125), by the well-known uniqueness theorem of the Laplace equation. The velocity field of the fluid is then given by

$$\mathbf{u}(\mathbf{r}) = -\tfrac{1}{2}u_0 a^3 \nabla \frac{\cos\theta}{r^2} \qquad (r \geq a) \tag{5.128}$$

The streamlines can be sketched immediately, and they look like the electric field due to a dipole, as shown in Fig. 5.6.

Let us calculate the kinetic energy of the fluid. It is given by the integral

$$\begin{aligned}
\text{K.E.} &= \int d^3r\, \tfrac{1}{2}\rho\, |\mathbf{u}|^2 = \frac{\rho}{2}\left(\frac{u_0 a^3}{2}\right)^2 \int_{r\geq a} d^3r\, \nabla\frac{\cos\theta}{r^2}\cdot\nabla\frac{\cos\theta}{r^2} \\
&= \frac{\rho}{2}\left(\frac{u_0 a^3}{2}\right)^2 \int_{r\geq a} d^3r\, \nabla\cdot\left[\frac{\cos\theta}{r^2}\nabla\frac{\cos\theta}{r^2}\right] \\
&= -\frac{\rho}{2}\left(\frac{u_0 a^3}{2}\right)^2 \int_{r=a} d\mathbf{S}\cdot\left[\frac{\cos\theta}{r^2}\nabla\frac{\cos\theta}{r^2}\right] \\
&= -\frac{\rho}{2}\left(\frac{u_0 a^3}{2}\right)^2 a^2 \int_0^{2\pi} d\phi \int_{-1}^{+1} d(\cos\theta)\left(\frac{\cos\theta}{r^2}\frac{\partial}{\partial r}\frac{\cos\theta}{r^2}\right)_{r=a} = \tfrac{1}{2}m' u_0^{\,2}
\end{aligned} \tag{5.129}$$

where m' is half the mass of the displaced fluid:

$$m' = \tfrac{1}{2}(\tfrac{4}{3}\pi a^3 \rho) \tag{5.130}$$

If the sphere has a mechanical mass m, the total kinetic energy of the system of liquid plus sphere is

$$E = \tfrac{1}{2}(m_0 + m')u_0^2 \tag{5.131}$$

The mass $m_0 + m'$ may be interpreted to be the effective mass of the sphere, since (5.131) is the total energy that has to be supplied for the sphere to move with velocity u_0.

Stokes' Law

We proceed to solve the same problem when the fluid has a nonvanishing coefficient of viscosity μ. The Navier-Stokes equation will be taken to be

$$0 = -\nabla\left(P - \frac{\mu}{3}\nabla \cdot \mathbf{u}\right) + \mu\nabla^2\mathbf{u} \tag{5.132}$$

on the assumption that the material derivative of $\mathbf{u}$, which gives rise to the effective mass, is small compared to the viscous terms. We return to examine the validity of this approximation later. Since there is no source for the fluid, we still require $\nabla \cdot \mathbf{u} = 0$, and (5.132) becomes the simultaneous equations

$$\begin{aligned} \nabla^2\mathbf{u} &= \frac{1}{\mu}\nabla P \\ \nabla \cdot \mathbf{u} &= 0 \end{aligned} \tag{5.133}$$

with the boundary condition that the fluid sticks to the sphere. Let us translate the coordinate system so that the sphere is at rest at the origin while the fluid at infinity flows with uniform constant velocity $\mathbf{u}_0$. The equations (5.133) remain invariant under the translation, whereas the boundary conditions become

$$\begin{aligned} [\mathbf{u}(\mathbf{r})]_{r=a} &= 0 \\ \mathbf{u}(\mathbf{r}) &\xrightarrow[r\to\infty]{} \mathbf{u}_0 \end{aligned} \tag{5.134}$$

Taking the divergence of both sides of the first equation of (5.133), we obtain

$$\nabla^2 P = 0 \tag{5.135}$$

Thus the pressure, whatever it is, must be a linear superposition of solid harmonics. A systematic way to proceed would be to write P as the most general superposition of solid harmonics and to determine the coefficient by requiring that (5.133) be satisfied. We take a short cut, however, and *guess* that P is, apart from an additive constant, a pure solid harmonic of order 1:

$$P = P_0 + \mu P_1 \frac{\cos\theta}{r^2} \tag{5.136}$$

where P_0 and P_1 are constants to be determined later. With this, the problem reduces to solving the inhomogeneous Laplace equation

$$\nabla^2 \mathbf{u} = P_1 \nabla \frac{\cos\theta}{r^2} \tag{5.137}$$

subject to the conditions

$$\begin{gathered} \nabla \cdot \mathbf{u} = 0 \\ [\mathbf{u}(\mathbf{r})]_{r=a} = 0 \\ \mathbf{u}(\mathbf{r}) \xrightarrow[r\to\infty]{} \mathbf{u}_0 \end{gathered} \tag{5.138}$$

A particular solution of (5.137) is

$$\mathbf{u}_1 = -\frac{P_1}{6} r^2 \nabla \frac{\cos\theta}{r^2} = -\frac{P_1}{6}\left(\frac{\hat{z}}{r} - 3\mathbf{r}\frac{z}{r^3}\right) \tag{5.139}$$

where $\hat{z}$ denotes the unit vector along the z-axis, which lies along $\mathbf{u}$. It is easily verified that (5.139) solves (5.137), if we note that $1/r$ and z/r^3 are both solid harmonics. Thus,

$$\nabla^2 \mathbf{u}_1 = -\frac{P_1}{6}\left[-3\nabla^2\left(\frac{\mathbf{r}z}{r^3}\right)\right] = P_1 \nabla\left(\frac{z}{r^3}\right) = P_1 \nabla \frac{\cos\theta}{r^2} \tag{5.140}$$

The complete solution is obtained by adding an appropriate homogeneous solution to (5.139) in order to satisfy (5.138). By inspection we see that the complete solution is

$$\mathbf{u} = \mathbf{u}_0\left(1 - \frac{a}{r}\right) + \frac{1}{4} u_0 a (r^2 - a^2) \nabla \frac{\cos\theta}{r^2} \tag{5.141}$$

where we have set

$$P_1 = -\tfrac{3}{2} u_0 a \tag{5.142}$$

in order to have $\nabla \cdot \mathbf{u} = 0$.

We now calculate the force acting on the sphere by the fluid. By definition the force per unit area acting on a surface whose normal point along the x_j axis is $-\mathbf{T}_j$ of (5.107). It follows that the force per unit area acting on a surface element of the sphere is

$$\mathbf{f} = -\left(\frac{x}{r}\mathbf{T}_1 + \frac{y}{r}\mathbf{T}_2 + \frac{z}{r}\mathbf{T}_3\right) = -\hat{r} \cdot \overleftrightarrow{P} \tag{5.143}$$

where $\hat{r}$ is the unit vector in the radial direction and $\overleftrightarrow{P}$ is given by (5.118). The total force experienced by the sphere is

$$\mathbf{F}' = \int dS\, \mathbf{f} \tag{5.144}$$

where dS is a surface element of the sphere and the integral extends over the entire surface of the sphere. Thus it is sufficient to calculate $\mathbf{f}$ for $r = a$.

The vector $\hat{r} \cdot \overleftrightarrow{P}$ has the components

$$(\hat{r} \cdot \overleftrightarrow{P})_i = \frac{1}{r} x_j P_{ji} = \frac{1}{r} x_j \left[\delta_{ji} P - \mu \left(\frac{\partial u_j}{\partial x_i} + \frac{\partial u_i}{\partial x_j} \right) \right]$$

$$= \frac{x_i}{r} P - \frac{\mu}{r} \left[\frac{\partial}{\partial x_i} (x_j u_j) - u_i + x_j \frac{\partial}{\partial x_j} u_i \right]$$

Hence

$$\mathbf{f} = -\hat{r}P + \frac{\mu}{r} [\nabla(\mathbf{r} \cdot \mathbf{u}) - \mathbf{u} + (\mathbf{r} \cdot \nabla)\mathbf{u}] \tag{5.145}$$

where P is given by (5.136) and (5.142), and $\mathbf{u}$ is given by (5.141). Since $\mathbf{u} = 0$ when $r = a$, we only need to consider the first and the last terms in the bracket. The first term is zero at $r = a$ by a straightforward calculation. At $r = a$ the second term is found to be

$$\frac{1}{r} [(\mathbf{r} \cdot \nabla)\mathbf{u}]_{r=a} = \left(\frac{\partial \mathbf{u}}{\partial r} \right)_{r=a} = \frac{3}{2} \frac{\mathbf{u}_0}{a} - \frac{3}{2} \hat{r} u_0 \frac{\cos \theta}{a} \tag{5.146}$$

When this is substituted into (5.145), the second term exactly cancels the dipole part of $\hat{r}P$, and we obtain

$$(\mathbf{f})_{r=a} = -\hat{r}P_0 + \frac{3}{2} \frac{\mu}{a} \mathbf{u}_0$$

The constant P_0 is unknown, but it does not contribute to the force on the sphere. From (5.144) we obtain

$$\mathbf{F} = 6\pi\mu a \mathbf{u}_0 \tag{5.147}$$

which is Stokes' law.

The validity of (5.141) depends on the smallness of the material derivative of $\mathbf{u}$ as compared to $\mu \nabla^2 \mathbf{u}$. Both these quantities can be computed from (5.141). It is then clear that we must require

$$\frac{\rho u_0 a}{\mu} \ll 1 \tag{5.148}$$

Thus Stokes' law holds only for small velocities and small radii of the sphere. A more elaborate treatment shows that a more accurate formula for $\mathbf{F}'$ is

$$\mathbf{F}' = 6\pi\mu a \mathbf{u}_0 \left(1 + \frac{3}{8} \frac{\rho u_0 a}{\mu} + \cdots \right) \tag{5.149}$$

The pure number $\rho u_0 a/\mu$ is called the Reynolds number. When the Reynolds number becomes large, turbulence sets in and streamline motion completely breaks down.

PROBLEMS

5.1. Make order-of-magnitude estimates for the mean free path and the collision time for

(*a*) H_2 molecules in a hydrogen gas in standard condition (diameter of $H_2 = 2.9$ Å);

(*b*) protons in a plasma (gas of totally ionized H_2) at $T = 3 \times 10^5$°K, $n = 10^{15}$ protons/cc, $\sigma = \pi r^2$, where $r = e^2/kT$;

(*c*) protons in a plasma at the same density as (*b*) but at $T = 10^7$°K, where thermonuclear reactions occur;

(*d*) protons in the sun's corona, which is a plasma at $T = 10^6$°K, $n = 10^6$ protons/cc;

(*e*) slow neutrons of energy 0.5 Mev in U^{238} ($\sigma \approx \pi r^2$, $r \approx 10^{-13}$ cm).

5.2. (*a*) Explain why it is meaningless to speak of a sound wave in a gas of strictly noninteracting molecules.

(*b*) In view of (*a*), explain the meaning of a sound wave in an ideal gas.

5.3. Show that the velocity of sound in a real substance is to a good approximation given by $c = 1/\sqrt{\rho\kappa_S}$, where ρ is the mass density and κ_S the adiabatic compressibility, by the following steps.

(*a*) Show that in a sound wave the density oscillates adiabatically if

$$K \ll \frac{l\rho c_V}{c}$$

where K = coefficient of thermal conductivity
l = wavelength of sound wave
ρ = mass density
c_V = specific heat
c = velocity of sound

(*b*) Show by numerial examples, that the criterion stated in (*a*) is well satisfied in most practical situations.

5.4. A flat disk of unit area is placed in a dilute gas at rest with initial temperature T. Face A of the disk is at temperature T, and face B is at temperature $T_1 > T$ (see sketch). Molecules striking face A reflect elastically. Molecules striking face B are absorbed by the disk, only to re-emerge from the same face with a Maxwellian distribution of temperature T_1.

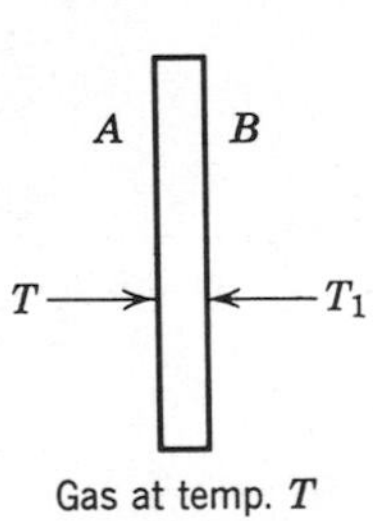

Problem 5.4

(*a*) Assume that the mean free path in the gas is much smaller than the dimension of the disk. Present an argument to show that after a few collision times the gas can be described by the hydrodynamic equations, with face B replaced by a boundary condition for the temperature.

(*b*) Write down the first-order hydrodynamic equations for (*a*), neglecting the flow of the gas. Show that there is no net force acting on the disk.

(*c*) Assume that the mean free path is much larger than the dimensions of the disk, and find the net force acting on the disk.

5.5. A square vane, of area 1 cm^2, painted white on one side, black on the other, is attached to a vertical axis and can rotate freely about it (see the sketch). Suppose the arrangement is placed in He gas at room temperature and sunlight is allowed to shine on the vane. Explain qualitatively why

(*a*) at high density of the gas the vane does not move;

(*b*) at extremely small densities the vane rotates;

(*c*) at some intermediate density the vane rotates in a sense opposite to that in (*b*). *Estimate* this intermediate density and the corresponding pressure.

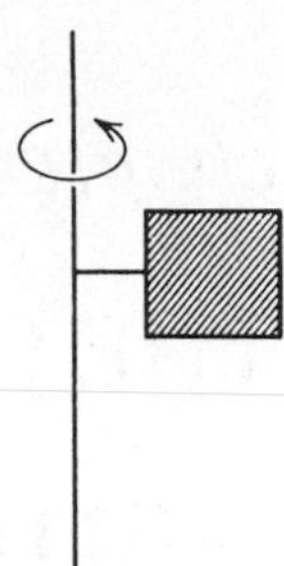

Problem 5.5

5.6. A dilute gas, infinite in extension and composed of charged molecules, each of charge e and mass m, comes to equilibrium in an infinite lattice of fixed ions. In the absence of an external electric field the equilibrium distribution function is

$$f^{(0)}(\mathbf{r}) = n\left(\frac{2\pi kT}{m}\right)^{-3/2} e^{-mv^2/kT}$$

where n, T are constants. A weak uniform electric field $\mathbf{E}$ is then turned on, leading to a new equilibrium distribution function. Assume that a collision term of the form

$$\left(\frac{\partial f}{\partial t}\right)_{\text{coll}} = -\frac{f - f^{(0)}}{\tau}$$

where τ is a collision time, adequately takes into account the effect of collisions among molecules and between molecules and lattice. Calculate

(*a*) the new equilibrium distribution function f, to the first order;

(*b*) the electrical conductivity σ, defined by the relation

$$ne\langle\mathbf{v}\rangle = \sigma\mathbf{E}$$

chapter 6

THE CHAPMAN-ENSKOG METHOD

6.1 PURPOSE OF THE METHOD

The purpose of the Chapman-Enskog method is to solve the Boltzmann transport equation (3.36) by successive approximation. This method will not yield the most general solution of the Boltzmann equation. It only yields a particular type of solution, namely, those that depend on the time implicitly through the local density, velocity, and temperature. In Chapter 5 we discussed this type of solution under the simplifying assumption that $(\partial f/\partial t)_{\text{coll}} \approx -(f - f^{(0)})/\tau$. This assumption is not made here.

Before we proceed, it is appropriate to point out the special importance attached to this type of solution. Its importance lies in the fact that a solution initially not of this type will become one of the type in a time of the order of a collision time. Since the collision time for ordinary gases is of the order of 10^{-11} sec, solutions of the type mentioned are the relevant ones in most physical applications. This claim may be justified as follows. Let f be a solution of the Boltzmann transport equation. It may depend on the time explicitly as well as implicitly through the time dependence of the local density, velocity, and temperature. Thus we may decompose the time derivative of f into two terms:

$$\frac{\partial f}{\partial t} = \left(\frac{\partial f}{\partial t}\right)_{\text{explicit}} + \left(\frac{\partial f}{\partial t}\right)_{\text{implicit}} \tag{6.1}$$

Restoring temporarily the simplifying assumption* for $(\partial f/\partial t)_{\text{coll}}$, we may write the Boltzmann transport equation in the form

$$\left(\frac{\partial f}{\partial t}\right)_{\text{explicit}} + L(f) = -\frac{f - f^{(0)}}{\tau} \tag{6.2}$$

where

$$L(f) \equiv \left(\frac{\partial f}{\partial t}\right)_{\text{implicit}} + \left(\mathbf{v}\cdot\nabla_r + \frac{\mathbf{F}}{m}\cdot\nabla_v\right) f \tag{6.3}$$

Define f_1 to be the solution of the equation

$$L(f_1) = -\frac{f_1 - f^{(0)}}{\tau} \tag{6.4}$$

Then f_1 has no explicit dependence on the time. Defining f_2 by

$$f = f_1 + f_2 \tag{6.5}$$

we easily find that f_2 satisfies the equation

$$\left(\frac{\partial f_2}{\partial t}\right)_{\text{explicit}} + L(f_2) = \left(\frac{\partial f_2}{\partial t}\right)_{\text{coll}} \tag{6.6}$$

If we can neglect $L(f_2)$, this becomes

$$\frac{\partial f_2}{\partial t} = -\frac{f_2}{\tau} \tag{6.7}$$

which immediately leads to

$$f_2 \propto e^{-t/\tau} \tag{6.8}$$

Therefore, after a time τ, the solution essentially reduces to f_1, which has no explicit dependence on the time.

6.2 THE CHAPMAN-ENSKOG EXPANSION

Let the Boltzmann transport equation be written in the form

$$\frac{\partial f}{\partial t} + Df = J(f\,|\,f) \tag{6.9}$$

where

$$Df \equiv \left(\mathbf{v}\cdot\nabla_r + \frac{\mathbf{F}}{m}\cdot\nabla_v\right) f \tag{6.10}$$

and where, for any functions f and g of $\mathbf{v}$,

$$J(f\,|\,g) \equiv \int d^3v_1 \int d\Omega\,\sigma(\Omega)\,|\mathbf{v} - \mathbf{v}_1|\,[f(\mathbf{v}')g(\mathbf{v}_1') - f(\mathbf{v})g(\mathbf{v}_1)] \tag{6.11}$$

* This is permissible because we intend to draw only qualitative conclusions.

in which $\sigma(\Omega)$ is the differential cross section for the binary collision $\{\mathbf{v}, \mathbf{v}_1\} \rightarrow \{\mathbf{v}', \mathbf{v}_1'\}$. We wish to find the solutions of (6.9) that depend on the time only implicitly through the time dependence of the quantities

$$n(\mathbf{r}, t) \equiv \int d^3v\, f(\mathbf{r}, \mathbf{v}, t) \tag{6.12}$$

$$\mathbf{u}(\mathbf{r}, t) \equiv \frac{1}{n} \int d^3v\, f(\mathbf{r}, \mathbf{v}, t)\mathbf{v} \tag{6.13}$$

$$\theta(\mathbf{r}, t) \equiv \frac{1}{n} \int d^3v\, f(\mathbf{r}, \mathbf{v}, t) \frac{m}{3} |\mathbf{v} - \mathbf{u}|^2 \tag{6.14}$$

It is further assumed that these functions vary negligibly over a spatial distance comparable to a mean free path and also vary negligibly over a time interval comparable to the collision time. The method of solution is one of successive approximation.

Let us formally introduce a parameter ζ, which eventually will be set equal to unity, by writing

$$f = \frac{1}{\zeta}(f^{(0)} + \zeta f^{(1)} + \zeta^2 f^{(2)} + \zeta^3 f^{(3)} + \cdots) \tag{6.15}$$

where the functions $f^{(n)}$ are defined in such a way that $f^{(n)}$ gets smaller and smaller as n increases. The parameter ζ has no physical significance and is introduced only to keep track of the order of the terms in the series. The main achievement of the Chapman-Enskog expansion is to provide a way of defining $f^{(n)}$ that is both consistent and practicable.

We require that

$$\int d^3v\, f^{(0)} = n \tag{6.16}$$

$$\frac{1}{n} \int d^3v\, f^{(0)}\, \mathbf{v} = \mathbf{u} \tag{6.17}$$

$$\frac{1}{n} \int d^3v\, f^{(0)} \frac{m}{3} |\mathbf{v} - \mathbf{u}|^2 = \theta \tag{6.18}$$

and that

$$\int d^3v\, f^{(n)} \begin{Bmatrix} 1 \\ \mathbf{v} \\ v^2 \end{Bmatrix} = 0 \qquad (n \neq 0) \tag{6.19}$$

Thus (6.12)–(6.14) are satisfied. This is not the only way to satisfy (6.12)–(6.14), but it is clearly a possible way.

Corresponding to the expansion (6.15), the conservation theorems (5.21)–(5.23) can be written in the following forms:

$$\frac{\partial \rho}{\partial t} + \frac{\partial \rho}{\partial x_i}(\rho u_i) = 0 \tag{6.20}$$

$$\rho\left(\frac{\partial u_i}{\partial t} + u_j \frac{\partial u_i}{\partial x_j}\right) = \frac{\rho}{m} F_i - \sum_{n=0}^{\infty} \zeta^n \frac{\partial P_{ij}^{(n)}}{\partial x_j} \tag{6.21}$$

$$\rho\left(\frac{\partial \theta}{\partial t} + u_j \frac{\partial \theta}{\partial x_j}\right) = -\frac{2}{3} \sum_{n=0}^{\infty} \zeta^n \left(\frac{\partial q_i^{(n)}}{\partial x_i} + \Lambda_{ij} P_{ij}^{(n)}\right) \tag{6.22}$$

where

$$P_{ij}^{(n)} = m \int d^3v\, f^{(n)}(v_i - u_i)(v_j - u_j) \tag{6.23}$$

$$q_i^{(n)} = \frac{m^2}{2} \int d^3v\, f^{(n)}(v_i - u_i)|\mathbf{v} - \mathbf{u}|^2 \tag{6.24}$$

To obtain a consistent scheme of successive approximation we define $f^{(n)}$ in such a way that if all $f^{(k)}$, $P_{ij}^{(k)}$, $q_i^{(k)}$ are neglected for $k > n$, then we have the nth approximation to the distribution function and to the hydrodynamic equations. To find such a definition we decompose the Boltzmann transport equation into successive equations for $f^{(n)}$ in the manner described next.

We first write

$$Df = \frac{1}{\zeta}(Df^{(0)} + \zeta Df^{(1)} + \zeta^2 Df^{(2)} + \cdots) \tag{6.25}$$

That this is consistent with (6.15) follows from the linearity of the operator D. Next we consider $\partial f/\partial t$. By assumption, f depends on t only through ρ, $\mathbf{u}$, and θ. Therefore

$$\frac{\partial f}{\partial t} = \frac{\partial f}{\partial \rho}\frac{\partial \rho}{\partial t} + \frac{\partial f}{\partial u_i}\frac{\partial u_i}{\partial t} + \frac{\partial f}{\partial \theta}\frac{\partial \theta}{\partial t} \tag{6.26}$$

To expand (6.26) into an infinite series in powers of ζ we must expand $\partial f/\partial \rho$, $\partial \rho/\partial t$, etc. The following expansions are straightforward:

$$\frac{\partial f}{\partial \rho} = \frac{1}{\zeta}\left(\frac{\partial f^{(0)}}{\partial \rho} + \zeta \frac{\partial f^{(1)}}{\partial \rho} + \cdots\right) \tag{6.27}$$

$$\frac{\partial f}{\partial u_i} = \frac{1}{\zeta}\left(\frac{\partial f^{(0)}}{\partial u_i} + \zeta \frac{\partial f^{(1)}}{\partial u_i} + \cdots\right) \tag{6.28}$$

$$\frac{\partial f}{\partial \theta} = \frac{1}{\zeta}\left(\frac{\partial f^{(0)}}{\partial \theta} + \zeta \frac{\partial f^{(1)}}{\partial \theta} + \cdots\right) \tag{6.29}$$

The expansions for $\partial\rho/\partial t$, $\partial u_i/\partial t$ and $\partial\theta/\partial t$ must be so defined as to be consistent with the conservation theorems. We shall formally prescribe that, whenever the operator $\partial/\partial t$ is applied to ρ, u_i, or θ, it shall be considered to be the following infinite series of operators:

$$\frac{\partial}{\partial t} \equiv \frac{\partial_0}{\partial t} + \zeta\frac{\partial_1}{\partial t} + \zeta\frac{\partial_2}{\partial t} + \cdots \tag{6.30}$$

The definition of $\partial_n/\partial t$ is taken from the nth approximation to the conservation theorems:

$$\begin{aligned} \frac{\partial_0}{\partial t}\rho &\equiv -\frac{\partial}{\partial x_i}(\rho u_i) \\ \frac{\partial_n}{\partial t}\rho &\equiv 0 \qquad (n > 0) \end{aligned} \tag{6.31}$$

$$\begin{aligned} \frac{\partial_0}{\partial t}u_i &\equiv -u_j\frac{\partial u_i}{\partial x_j} + \frac{F_i}{m} - \frac{1}{\rho}\frac{\partial P_{ij}^{(0)}}{\partial x_j} \\ \frac{\partial_n}{\partial t}u_i &\equiv -\frac{1}{\rho}\frac{\partial P_{ij}^{(n)}}{\partial x_j} \qquad (n > 0) \end{aligned} \tag{6.32}$$

$$\begin{aligned} \frac{\partial_0}{\partial t}\theta &\equiv -u_j\frac{\partial\theta}{\partial x_j} - \frac{2}{3}\frac{1}{\rho}\frac{\partial q_i^{(0)}}{\partial x_i} - \frac{2}{3}\frac{1}{\rho}\Lambda_{ij}P_{ij}^{(0)} \\ \frac{\partial_n}{\partial t}\theta &\equiv -\frac{2}{3}\frac{1}{\rho}\left(\frac{\partial q_i^{(n)}}{\partial x_i} + \Lambda_{ij}P_{ij}^{(n)}\right) \qquad (n > 0) \end{aligned} \tag{6.33}$$

With these definitions we obtain the expansion of $\partial f/\partial t$ into a power series in ζ by formally writing

$$\frac{\partial f}{\partial t} = \frac{1}{\zeta}\left(\frac{\partial_0}{\partial t} + \zeta\frac{\partial_1}{\partial t} + \zeta^2\frac{\partial_2}{\partial t} + \cdots\right)(f^{(0)} + \zeta f^{(1)} + \zeta^2 f^{(2)} + \cdots) \tag{6.34}$$

Next we examine the collision term in the Boltzmann transport equation. We have, from (6.11) and (6.15),

$$\begin{aligned} J(f\,|\,f) &= \frac{1}{\zeta^2}\sum_{n=0}^{\infty}\sum_{m=0}^{\infty}\zeta^{m+n}\int d^3v_1\int d\Omega(\Omega)\,|\mathbf{v} - \mathbf{v}_1| \\ &\quad\times [f^{(n)}(\mathbf{v}')f^{(m)}(\mathbf{v}_1') - f^{(n)}(\mathbf{v})f^{(m)}(\mathbf{v}_1')] \\ &= \frac{1}{\zeta^2}\sum_{n=0}^{\infty}\sum_{m=0}^{\infty}\zeta^{m+n}J(f^{(n)}\,|\,f^{(m)}) \end{aligned} \tag{6.35}$$

Let us introduce the definition

$$J^{(n)}(f^{(0)}, f^{(1)}, \ldots, f^{(n)}) \equiv \sum_{\substack{r,s \\ r+s=n}} J(f^{(r)}\,|\,f^{(s)}) \tag{6.36}$$

where the sum extends over all sets of integers (r, s) for which $r + s = n$. Then we can write

$$J(f\,|\,f) = \frac{1}{\zeta^2}\sum_{n=0}^{\infty}\zeta^n J^{(n)} = \frac{1}{\zeta^2}[J^{(0)}(f^{(0)}) + \zeta J^{(1)}(f^{(0)}, f^{(1)}) + \zeta^2 J^{(2)}(f^{(0)}, f^{(1)}, f^{(2)}) + \cdots] \quad (6.37)$$

The Boltzmann transport equation can be written in the form

$$\frac{1}{\zeta}\left[\left(\frac{\partial_0}{\partial t} + \zeta\frac{\partial_1}{\partial t} + \zeta^2\frac{\partial_2}{\partial t} + \cdots\right) + D\right](f^{(0)} + \zeta f^{(1)} + \zeta^2 f^{(2)} + \cdots) - \frac{1}{\zeta^2} \times [J^{(0)}(f^{(0)}) + \zeta J^{(1)}(f^{(0)}, f^{(1)}) + \zeta^2 J^{(2)}(f^{(0)}, f^{(1)}, f^{(2)}) + \cdots] = 0 \quad (6.38)$$

We now define $f^{(n)}$ uniquely by requiring that in (6.38) the coefficient of each power of ζ vanish separately. Accordingly the equations that are to be successively solved to yield all the $f^{(n)}$ are

$$\begin{aligned}
&(0)\quad J^{(0)}(f^{(0)}) = 0\\
&(1)\quad \frac{\partial_0}{\partial t}f^{(0)} + Df^{(0)} = J^{(1)}(f^{(0)}, f^{(1)})\\
&(2)\quad \frac{\partial_0}{\partial t}f^{(1)} + \frac{\partial_1}{\partial t}f^{(0)} + Df^{(1)} = J^{(2)}(f^{(0)}, f^{(1)}, f^{(2)}) \qquad (6.39)\\
&\qquad\vdots \qquad\qquad\qquad \vdots
\end{aligned}$$

The zero equation is familiar. We have arranged it to be so by starting the expansion (6.15) with ζ^{-1}.

The zero equation in (6.39) was solved earlier and gives for $f^{(0)}$ a local Maxwell-Boltzmann distribution. This solution determines ρ, u_i, and θ, by (6.18)–(6.19). The nth equation involves only $f^{(n)}$ and $f^{(k)}$ for $k < n$. Therefore we can hope to solve these equations successively, starting from the local Maxwell-Boltzmann distribution. To complete the formal scheme of Chapman and Enskog we only need to show that a solution of the nth equation exists.

6.3 EXISTENCE OF SOLUTIONS

That $f^{(0)}$ exists has been previously demonstrated by explicitly displaying it. We indicate the proof that there is a solution to the nth equation of (6.39). The equation is

$$\frac{\partial_0}{\partial t}f^{(n-1)} + \frac{\partial_1}{\partial t}f^{(n-2)} + \cdots + \frac{\partial_{n-1}}{\partial t}f^{(0)} + Df^{(n-1)} = J^{(n)}(f^{(0)}, f^{(1)}, \ldots, f^{(n)}) \quad (6.40)$$

where $f^{(0)}, f^{(1)}, \ldots, f^{(n-1)}$ are assumed to be known functions. The unknown function $f^{(n)}$ appears only in the argument of $J^{(n)}$. By definition,

$$J^{(n)}(f^{(0)}, \ldots, f^{(n)}) = J(f^{(n)}\,|\,f^{(0)}) + J(f^{(n-1)}\,|\,f^{(1)}) + \cdots + J(f^{(0)}\,|\,f^{(n)}) \tag{6.41}$$

Thus we can rewrite (6.40) in the form

$$J(f^{(n)}\,|\,f^{(0)}) + J(f^{(0)}\,|\,f^{(n)}) = F(\mathbf{v}) - G(\mathbf{v}) \tag{6.42}$$

where

$$F(\mathbf{v}) \equiv \frac{\partial_0}{\partial t} f^{(n-1)} + \frac{\partial_1}{\partial t} f^{(n-2)} + \cdots + \frac{\partial_{n-1}}{\partial t} f^{(0)} + D f^{(n-1)} \tag{6.43}$$

$$G(\mathbf{v}) \equiv J(f^{(n-1)}\,|\,f^{(1)}) + J(f^{(n-2)}\,|\,f^{(2)}) + \cdots + J(f^{(1)}\,|\,f^{(n-1)}) \tag{6.44}$$

which are known functions of $\mathbf{v}$. (The dependences on $\mathbf{r}$ and t are understood.) Using the definition (6.11) and defining $\Phi(\mathbf{v})$ by

$$f^{(n)}(\mathbf{v}) \equiv f^{(0)}(\mathbf{v})\Phi(\mathbf{v}) \tag{6.45}$$

we have

$$\int d^3v_1 \int d\Omega\,\sigma(\Omega)\,|\mathbf{v} - \mathbf{v}_1|\,f^{(0)}(\mathbf{v})\,f^{(0)}(\mathbf{v}_1)[\Phi(\mathbf{v}') + \Phi(\mathbf{v}_1') - \Phi(\mathbf{v}) - \Phi(\mathbf{v}_1)] = F(\mathbf{v}) - G(\mathbf{v}) \tag{6.46}$$

This can be put into the form

$$K_0(\mathbf{v})\Phi(\mathbf{v}) = [G(\mathbf{v}) - F(\mathbf{v})] + \int d^3v_1\, K(\mathbf{v}, \mathbf{v}_1)\Phi(\mathbf{v}_1) \tag{6.47}$$

where

$$K_0(\mathbf{v}) = f^{(0)}(\mathbf{v}) \int d^3v_1 \int d\Omega\,\sigma(\Omega)\,|\mathbf{v} - \mathbf{v}_1|\,f^{(0)}(\mathbf{v}_1) \tag{6.48}$$

and

$$\int d^3v_1\, K(\mathbf{v}, \mathbf{v}_1)\Phi(\mathbf{v}_1) = \int d^3v_1 \int d\Omega\,\sigma(\Omega)\,|\mathbf{v} - \mathbf{v}_1|\,f^{(0)}(\mathbf{v})\,f^{(0)}(\mathbf{v}) \times [\Phi(\mathbf{v}') + \Phi(\mathbf{v}_1') - \Phi(\mathbf{v}_1)] \tag{6.49}$$

It will be shown that

$$K(\mathbf{v}, \mathbf{v}_1) = K(\mathbf{v}_1, \mathbf{v}) \tag{6.50}$$

Hence (6.47) is a linear inhomogeneous integral equation with symmetric kernel. The existence theorem* for such integral equations states that a solution $\Phi(\mathbf{v})$ exists if and only if the inhomogeneous term of the equation, namely $G(\mathbf{v}) - F(\mathbf{v})$, is orthogonal to all the homogeneous solutions of (6.47).

* R. Courant and D. Hilbert, *Methods of Mathematical Physics*, Vol. I (Interscience Publishers, New York, 1953), pp. 136–137.

The homogeneous solutions of (6.47) are simply the solutions to the equation

$$J(f^{(n)}\,|\,f^{(0)}) + J(f^{(0)}\,|\,f^{(n)}) = 0 \tag{6.51}$$

or

$$2\int d^3v_1 \int d\Omega\sigma(\Omega)\,|\mathbf{v}-\mathbf{v}_1|\,f^{(0)}(\mathbf{v})f^{(0)}(\mathbf{v}_1)[\Phi(\mathbf{v}') + \Phi(\mathbf{v}) - \Phi(\mathbf{v}) - \Phi(\mathbf{v}_1)] = 0 \tag{6.52}$$

The only independent solutions are 1, $\mathbf{v}$, v^2. Therefore a solution to (6.47) exists if and only if

$$\int d^3v\,[F(\mathbf{v}) - G(\mathbf{v})]\begin{Bmatrix}1\\ \mathbf{v}\\ v^2\end{Bmatrix} = 0 \tag{6.53}$$

That this is true is easily shown. From (6.43) it is seen that $F(\mathbf{v})$ is the coefficient of ζ^{n-1} in the expansion of $(\partial/\partial t + D)f$ in powers of ζ. Now the conservation theorems are just the statements

$$\int d^3v \begin{Bmatrix}1\\ \mathbf{v}\\ v^2\end{Bmatrix}\left(\frac{\partial}{\partial t} + D\right) f = 0 \tag{6.54}$$

The definitions (6.31)–(6.33) for $\partial_n/\partial t$ are just such as to make (6.54) true to each order, i.e.,

$$\int d^3v \begin{Bmatrix}1\\ \mathbf{v}\\ v^2\end{Bmatrix} F(\mathbf{v}) = 0 \tag{6.55}$$

It is straightforward to show that

$$\int d^3v \begin{Bmatrix}1\\ \mathbf{v}\\ v^2\end{Bmatrix} G(\mathbf{v}) = 0 \tag{6.56}$$

Therefore (6.53) is true.

To complete the proof that a solution of (6.47) exists, we only have to prove (6.50). First note that in the integral on the right-hand side of (6.49) the first two terms are identical, because we can interchange $\mathbf{v}'$ and $\mathbf{v}_1'$ without affecting the integral. Thus

$$\begin{aligned}\int d^3v_1\,K(\mathbf{v},\mathbf{v}_1)\Phi(\mathbf{v}_1) &= 2\int d^3v_1 \int d\Omega\sigma(\Omega)\,|\mathbf{v}-\mathbf{v}_1|f^{(0)}(\mathbf{v})f^{(0)}(\mathbf{v}_1)\Phi(\mathbf{v}_1')\\ &\quad - \int d^3v_1 \int d\Omega\sigma(\Omega)\,|\mathbf{v}-\mathbf{v}_1|f^{(0)}(\mathbf{v})f^{(0)}(\mathbf{v}_1)\Phi(\mathbf{v}_1)\end{aligned} \tag{6.57}$$

The second term on the right-hand side already contains a symmetric kernel. We only need to examine further the first term. For this purpose, note that

$$\int d^3v_1 \int d\Omega\sigma(\Omega)\, |\mathbf{v}-\mathbf{v}_1|\, f(\mathbf{v},\mathbf{v}_1,\mathbf{v}',\mathbf{v}_1') =$$

$$\int d^3v_1 \int d^3v' \int d^3v_1'\, \delta(\mathbf{v}+\mathbf{v}_1-\mathbf{v}'-\mathbf{v}_1')\, \delta(v^2+v_1^2-v'^2-v_1'^2) \\ \times R(\mathbf{v},\mathbf{v}_1 \,|\, \mathbf{v}',\mathbf{v}_1') f(\mathbf{v},\mathbf{v}_1,\mathbf{v}',\mathbf{v}_1') \quad (6.58)$$

where $R(\mathbf{v}, \mathbf{v}_1 \,|\, \mathbf{v}', \mathbf{v}_1')$ is the rate of the transition $\{\mathbf{v}, \mathbf{v}_1\} \to \{\mathbf{v}', \mathbf{v}_1'\}$ and differs from $\sigma(\Omega)\,|\mathbf{v} - \mathbf{v}_1|$ only by a factor representing the density of final states. The integration on the right-hand side, which is an integration over nine variables, reduces to an integration over five variables in agreement with the left-hand side, by virtue of the δ functions representing the four conditions of momentum and energy conservation. With the help of (6.58) it is easy to show that the first term on the right-hand side of (6.57) also contains a symmetric kernel, after judicious changes of variables of integration. This completes the proof.

6.4 THE FIRST-ORDER APPROXIMATION

The zero-order approximation to the distribution function is

$$f^{(0)} = \frac{\rho}{m}\left(\frac{m}{2\pi\theta}\right)^{3/2} \exp\left(-\frac{m}{2\theta}|\mathbf{v}-\mathbf{u}|^2\right) \quad (6.59)$$

The first-order correction $f^{(1)}$ is a solution of the equation

$$\left(\frac{\partial_0}{\partial t} + D\right) f^{(0)} = J^{(1)}(f^{(0)}, f^{(1)}) \quad (6.60)$$

subject to the conditions

$$\int d^3v\, f^{(1)}(\mathbf{v}) \begin{Bmatrix} 1 \\ \mathbf{v} \\ v^2 \end{Bmatrix} = 0 \quad (6.61)$$

From (6.31)–(6.33) we obtain, after a calculation identical to that made in arriving at (5.67), the formula

$$\left(\frac{\partial_0}{\partial t} + D\right) f^{(0)} = f^{(0)}\left[\frac{1}{\theta}\frac{\partial\theta}{\partial x_i} U_i\left(\frac{m}{2\theta}U^2 - \frac{5}{2}\right) + \frac{1}{\theta}\Lambda_{ij}(U_iU_j - \tfrac{1}{3}\delta_{ij}U^2)\right] \quad (6.62)$$

where $\mathbf{U} \equiv \mathbf{v} - \mathbf{u}$. Let

$$f^{(1)}(\mathbf{v}) \equiv f^{(0)}(\mathbf{v})\Phi(\mathbf{v}) \quad (6.63)$$

and let $\mathscr{I}(\Phi)$ be defined by

$$\mathscr{I}(\Phi) \equiv \frac{J^{(1)}(f^{(0)},f^{(1)})}{f^{(0)}} = \int d^3v \int d\Omega\sigma(\Omega)\, |\mathbf{v}-\mathbf{v}_1|\, f^{(0)}(\mathbf{v}_1) \times [\Phi(\mathbf{v}_1') + \Phi(\mathbf{v}') - \Phi(\mathbf{v}_1) - \Phi(\mathbf{v})] \tag{6.64}$$

Then $\Phi(\mathbf{v})$ is the solution of the equation

$$\mathscr{I}(\Phi) = \frac{1}{\theta}\frac{\partial \theta}{\partial x_i}\, U_i\left(\frac{m}{2\theta}\, U^2 - \frac{5}{2}\right) + \frac{1}{\theta}\Lambda_{ij}(U_iU_j - \tfrac{1}{3}\,\delta_{ij}U^2) \tag{6.65}$$

subject to the requirements

$$\int d^3v\, f^{(0)}\Phi \begin{Bmatrix} 1 \\ \mathbf{v} \\ v^2 \end{Bmatrix} = 0 \tag{6.66}$$

Since $\mathscr{I}(\Phi)$ is linear in Φ, there must be a particular solution which is linear in $\partial\theta/\partial x_i$ and Λ_{ij}. Furthermore, Φ is a scalar. Hence there must be a particular solution of the form

$$\Phi = \frac{1}{\theta}\, R_i\, \frac{\partial \theta}{\partial x_i} + \frac{1}{\theta}\, S_{ij}\Lambda_{ij} \tag{6.67}$$

where R_i and S_{ij} are respectively vector and tensor functions of θ, ρ, and U. Substituting this into (6.65) we find that R_i and S_{ij} must satisfy the equations

$$\begin{aligned} \mathscr{I}(R_i) &= \left(\frac{m}{2\theta}\, U^2 - \frac{5}{2}\right) U_i \\ \mathscr{I}(S_{ij}) &= (U_iU_j - \tfrac{1}{3}\,\delta_{ij}U^2) \end{aligned} \tag{6.68}$$

Particular solutions to (6.68) are of the form

$$\begin{aligned} R_i &= -U_iF(U^2, \rho, \theta) \\ S_{ij} &= -(U_iU_j - \tfrac{1}{3}\,\delta_{ij}U^2)G(U^2, \rho, \theta) \end{aligned} \tag{6.69}$$

where F and G are scalar functions. The independent homogeneous solutions to (6.65) are $\Phi = 1$, $\mathbf{U}$, U^2. Therefore the most general solution to (6.65) is

$$\Phi = -\left[\frac{1}{\theta}\frac{\partial\theta}{\partial x_i}\, U_iF + \frac{1}{\theta}\Lambda_{ij}(U_iU_j - \tfrac{1}{3}\,\delta_{ij})G + \alpha + \beta_iU_i + \gamma U^2\right] \tag{6.70}$$

where α, β_i, γ are five arbitrary constants, to be determined by the five requirements (6.66).

Noting that

$$\int d^3v\, f^{(0)}(U_iU_j - \tfrac{1}{3}\,\delta_{ij}U^2) \begin{Bmatrix} 1 \\ \mathbf{U} \\ U^2 \end{Bmatrix} = 0 \tag{6.71}$$

we may reduce the requirements (6.65) to the following conditions:

$$\int d^3U\, f^{(0)}(\alpha + \gamma U^2) = 0 \tag{6.72}$$

$$\int d^3U\, f^{(0)}\left(\frac{1}{\theta}\frac{\partial\theta}{\partial x_i} F + \beta_i\right) U^2 = 0 \tag{6.73}$$

$$\int d^3U\, f^{(0)}(\alpha + \gamma U^2) U^2 = 0 \tag{6.74}$$

Multiplying (6.72) by α, (6.74) by γ, and adding the result, we obtain

$$(\alpha + \gamma) \int d^3U\, (\alpha + \gamma U^2) f^{(0)} = 0 \tag{6.75}$$

Since the integrand is non-negative, we must have $\alpha + \gamma U^2 = 0$ for all U. Hence

$$\alpha = 0, \qquad \gamma = 0 \tag{6.76}$$

From (6.73) we obtain

$$\beta_i = -\frac{1}{\theta}\frac{\partial\theta}{\partial x_i} \frac{\int d^3U\, U^2 f^{(0)} F}{\int d^3U\, U^2 f^{(0)}} \tag{6.77}$$

The term in (6.70) proportional to β_i shall be absorbed into the first term by redefining F. We can now state the complete first-order correction to the distribution function:

$$f^{(1)} = -f^{(0)}\left[\frac{1}{\theta}\frac{\partial\theta}{\partial x_i} U_i F(U^2, \theta, \rho) + \frac{1}{\theta}\Lambda_{ij}(U_i U_j - \tfrac{1}{3}\delta_{ij}U^2) G(U^2, \theta, \rho)\right] \tag{6.78}$$

where F and G are solutions of

$$\mathscr{I}(U_i F) = -U_i\left[\frac{m}{2\theta} U^2 - \frac{5}{2}\right] \tag{6.79}$$

$$\mathscr{I}((U_i U_j - \tfrac{1}{3}\delta_{ij}U^2)G) = -(U_i U_j - \tfrac{1}{3}\delta_{ij}U^2) \tag{6.80}$$

Now the solutions to $\mathscr{I}(X) = 0$ are $X = 1$, U, U^2. Thus the most general solution of (6.79) is any particular solution plus a constant, which must be so chosen as to make $\int d^3v\, f^{(1)} U = 0$. On the other hand, the only homogeneous solution to (6.80) is $G = 0$. Hence G is uniquely determined by (6.80).

Having obtained $f^{(1)}$, we straightforwardly find, in a manner similar to that described in Section 5.6, the coefficient of thermal conductivity K and the coefficient of viscosity μ:

$$K = \frac{m^2}{6\theta}\int d^3U\, U^4 f^{(0)} F(\rho, \theta, U^2) \tag{6.81}$$

$$\mu = \frac{m^2}{15\theta}\int d^3U\, U^4 f^{(0)} G(\rho, \theta, U^2) \tag{6.82}$$

From (6.79) and (6.80) we can verify that both F and G are of the dimension of $(\text{time})^{-1}$ and are of the order of magnitude of $(\text{cross section})^{-1}$. Our previous simplified treatment replaced both F and G by the inverse of a collision time:

$$F \approx G \approx \frac{1}{\tau} \tag{6.83}$$

This is now seen to be qualitatively correct.

The smallness of the first correction $f^{(1)}$ depends on the smallness of the mean free path compared to the "wavelengths" of ρ, θ, U, as discussed in the last chapter. In practice, the first approximation yields results that are in excellent agreement with experiments. To obtain numerical values for K and μ in specific cases, it would be necessary to solve the equations (6.79) and (6.80) with a given collision cross section $\sigma(\Omega)$. This is a difficult task even for simple intermolecular potentials, e.g., the hard-sphere potential. We forego any discussion of the actual solution of (6.79) and (6.80).*

* The techniques useful for solving (6.79) and (6.80), together with results for a few simple intermolecular potentials, may be found in S. Chapman and T. G. Cowling, *The Mathematical Theory of Non-Uniform Gases*, 2nd ed. (Cambridge University Press, Cambridge, 1952).

B

STATISTICAL MECHANICS

chapter 7

CLASSICAL STATISTICAL MECHANICS

7.1 THE POSTULATE OF CLASSICAL STATISTICAL MECHANICS

Statistical mechanics is concerned with the properties of matter in equilibrium, with equilibrium meaning the empirical notion of equilibrium introduced in thermodynamics.

The aim of statistical mechanics is to derive all the equilibrium properties of a macroscopic molecular system from the laws of molecular dynamics. Thus it aims to derive not only the general laws of thermodynamics but also the specific thermodynamic functions of a given system. Statistical mechanics, however, does not describe how a system approaches equilibrium, nor does it determine whether a system can ever be found to be in equilibrium. It merely states what the equilibrium situation is, for a given system.

We recall that in the kinetic theory of gases the process of the approach to equilibrium is rather complicated, but the equilibrium situation, the Maxwell-Boltzmann distribution, is simple. Furthermore, the Maxwell-Boltzmann distribution can be derived in a simple way, independent of the details of molecular interactions. We might suspect that a slight generalization of the method used—the method of the most probable distribution—would enable us to discuss the equilibrium situation of not only a dilute gas but also any macroscopic system. This indeed is true. The generalization is classical statistical mechanics.

We consider a classical system composed of a large number N of molecules occupying a large volume V. Typical magnitudes of N and V are

$$N \approx 10^{23} \quad \text{molecules}$$
$$V \approx 10^{23} \quad \text{molecular volumes}$$

These being enormously large numbers, it would be convenient to consider the system in the limit

$$\begin{aligned} N &\to \infty \\ V &\to \infty \\ \frac{V}{N} &= v \end{aligned} \tag{7.1}$$

where the specific volume v is a given finite number.

The system shall be regarded as isolated in the sense that the energy is a constant of the motion. This is clearly an idealization, for we never deal with truly isolated systems in the laboratory. The very fact that measurements can be performed on the system necessitates some interaction between the system and the external world. If the interactions with the external world, however, are sufficiently weak, so that the energy of the system remains approximately constant, we shall consider the system isolated. The walls of the container containing the system (if present) shall be idealized as perfectly reflecting walls.

A state of the system is completely and uniquely defined by $3N$ canonical coordinates $q_1, q_2, \ldots, q_{3N}$ and $3N$ canonical momenta $p_1, p_2, \ldots, p_{3N}$. These $6N$ variables are denoted collectively by the abbreviation (p, q). The dynamics of the system is completely contained in the Hamiltonian $H(p, q)$, from which we may obtain the canonical equations of motion

$$\begin{aligned} \frac{\partial H(p, q)}{\partial p_i} &= \dot{q}_i \\ \frac{\partial H(p, q)}{\partial q_i} &= -\dot{p}_i \end{aligned} \tag{7.2}$$

It is convenient to introduce, as we did in Chapter 4, the $6N$-dimensional Γ-space of the system, in which each point represents a state of the system, and *vice versa*. The locus of all points in Γ-space satisfying the condition $H(p, q) = E$ defines a surface called the energy surface of energy E. As the state of the system evolves in time according to (7.2) the representative point traces out a path in Γ-space. This path always stays on the same energy surface because by definition energy is conserved.

For a macroscopic system, we have no means, nor desire, to ascertain the state at every instant. We are interested only in a few macroscopic properties of the system. Specifically, we only require that the system has N particles, a volume V, and an energy lying between the values E and $E + \Delta$. An infinite number of states satisfy these conditions. Therefore we think not of a single system, but of an infinite number of mental copies of the same system, existing in all possible states satisfying the given conditions. Any one of these systems can be the system we are dealing with. The mental picture of such a collection of systems is called an *ensemble*. It is represented by a distribution of points in Γ-space characterized by a density function $\rho(p, q, t)$, defined in such a way that

$$\rho(p, q, t)\, d^{3N}p\, d^{3N}q = \begin{array}{l}\text{no. of representative points contained}\\ \text{in the volume element } d^{3N}p\, d^{3N}q \text{ lo-}\\ \text{cated at } (p, q) \text{ in } \Gamma\text{-space at the instant } t\end{array} \tag{7.3}$$

The concept of the ensemble and the density function $\rho(p, q, t)$ were first introduced in Chapter 4. It is recalled that from (7.2) and the fact that N is a constant follows Liouville's theorem:

$$\frac{\partial \rho}{\partial t} + \sum_{i=1}^{3N} \left(\frac{\partial \rho}{\partial q_i} \frac{\partial H}{\partial p_i} - \frac{\partial H}{\partial q_i} \frac{\partial \rho}{\partial p_i} \right) = 0 \tag{7.4}$$

In geometrical language it states that the distribution of points in Γ-space moves like an incompressible fluid. We restrict our considerations to ensembles whose density function does not depend explicitly on the time and depends on (p, q) only through the Hamiltonian. That is,

$$\rho(p, q) = \rho'(H(p, q)) \tag{7.5}$$

where $\rho'(H)$ is a given function of H. It follows immediately that the second term on the left-hand side of (7.4) is identically zero. Therefore

$$\frac{\partial}{\partial t} \rho(p, q) = 0 \tag{7.6}$$

Hence the ensemble described by $\rho(p, q)$ is the same for all times.

Classical statistical mechanics is founded on the following postulate.

Postulate of Equal a Priori Probability. When a macroscopic system is in thermodynamic equilibrium, its state is equally likely to be any state satisfying the macroscopic conditions of the system.

This postulate implies that in thermodynamic equilibrium the system under consideration is a member of an ensemble, called the *microcanonical ensemble*, with the density function

$$\rho(p, q) = \begin{cases} 1 & \text{if } E < H(p, q) < E + \Delta \\ 0 & \text{otherwise} \end{cases} \tag{7.7}$$

It is understood that all members of the ensemble have the same number of particles N and the same volume V. By virtue of (7.6) this ensemble is the same for all times. The postulate is therefore consistent with the idea of equilibrium.

Suppose $f(p, q)$ is a measurable property of the system, such as energy or momentum. When the system is in equilibrium, the observed value of $f(p, q)$ must be the result obtained by averaging $f(p, q)$ over the microcanonical ensemble in some manner. If the postulate of equal *a priori* probability is to be useful, all manners of averaging must yield essentially the same answer.

Two kinds of average values are commonly introduced: The most probable value and the ensemble average. The *most probable value* of $f(p, q)$ is the value of $f(p, q)$ that is possessed by the largest number of systems in the ensemble. The *ensemble average* of $f(p, q)$ is defined by

$$\langle f \rangle \equiv \frac{\int d^{3N}p\, d^{3N}q\, f(p, q)\rho(p, q)}{\int d^{3N}p\, d^{3N}q\, \rho(p, q)} \tag{7.8}$$

The ensemble average and the most probable value are nearly equal if the *mean square fluctuation* is small, i.e., if

$$\frac{\langle f^2 \rangle - \langle f \rangle^2}{\langle f \rangle^2} \ll 1 \tag{7.9}$$

If this condition is not satisfied, there is no unique way to determine how the observed value of f may be calculated. When it is not, we should question the validity of statistical mechanics. In all physical cases we shall find that mean square fluctuations are of the order of $1/N$. Thus in the limit as $N \to \infty$ the ensemble average and the most probable value became identical.

Strictly speaking, systems in nature do not obey classical mechanics. They obey quantum mechanics, which contains classical mechanics as a special limiting case. Logically we should start with quantum statistical mechanics and then arrive at classical statistical mechanics as a special case. We do this later. It is only for pedagogical reasons that we begin with classical statistical mechanics.

From a purely logical point of view there is no room for an independent postulate of classical statistical mechanics. It would not be logically satisfactory even if we could show that the postulate introduced here follows from the equations of motion (7.2), for, the world being quantum mechanical, the foundation of statistical mechanics lies not in classical mechanics but in quantum mechanics. At present we take this postulate to be a working hypothesis whose justification lies in the agreement between results derived from it and experimental facts.

7.2 MICROCANONICAL ENSEMBLE

The microcanonical ensemble has been defined, and its connection with physics has been postulated. We now work out the equilibrium properties of a system in the microcanonical ensemble.

In the microcanonical ensemble it is trivially true that in equilibrium the system has N molecules, a volume V, and an energy between E and $E + \Delta$. It is also evident, from a moment's reflection, that the average total momentum of the system is zero. We show that it is possible to define quantities that correspond to thermodynamic quantities.

The fundamental quantity that furnishes the connection between the microcanonical ensemble and thermodynamics is the entropy. It is the main task of the present section to define the entropy and to show that it possesses all the properties attributed to it in thermodynamics.

Let $\Gamma(E)$ denote the volume in Γ-space occupied by the microcanonical ensemble:

$$\Gamma(E) \equiv \int_{E<H(p,q)<E+\Delta} d^{3N}p\, d^{3N}q\, \rho(p, q) \tag{7.10}$$

The dependence of $\Gamma(E)$ on N, V, and Δ is understood. Let $\Sigma(E)$ denote the volume in Γ-space enclosed by the energy surface of energy E:

$$\Sigma(E) = \int_{H(p,q)<E} d^{3N}p\, d^{3N}q \tag{7.11}$$

Then

$$\Gamma(E) = \Sigma(E + \Delta) - \Sigma(E) \tag{7.12}$$

If Δ is so chosen that $\Delta \ll E$, then

$$\Gamma(E) = \omega(E)\,\Delta \tag{7.13}$$

where $\omega(E)$ is called the density of states of the system at the energy E and is defined by

$$\omega(E) = \frac{\partial \Sigma(E)}{\partial E} \tag{7.14}$$

The entropy is defined by

$$S(E, V) \equiv k \log \Gamma(E) \tag{7.15}$$

where k is a universal constant eventually shown to be Boltzmann's constant. To justify this definition we show that (7.15) possesses all the properties of the entropy function in thermodynamics, namely,

(*a*) S is an extensive quantity: If a system is composed of two subsystems whose entropies are respectively S_1 and S_2, the entropy of the total system is $S_1 + S_2$, when the subsystems are sufficiently large.

(*b*) S satisfies the properties of the entropy as required by the second law of the thermodynamics.

To show the extensive property, let the system be divided into two subsystems which have N_1 and N_2 particles and the volumes V_1 and V_2 respectively.* The energy of molecular interaction between the two subsystems is negligible compared to the total energy of each subsystem, if the intermolecular potential has a finite range, and if the surface-to-volume ratio of each subsystem is negligibly small. The total Hamiltonian of the composite system accordingly may be taken to be the sum of the Hamiltonians of the two subsystems:

$$H(p, q) = H_1(p_1, q_1) + H_2(p_2, q_2) \tag{7.16}$$

where (p_1, q_1) and (p_2, q_2) denote respectively the coordinates and momenta of the particles contained in the two subsystems.

Let us first imagine that the two subsystems are isolated from each other and consider the microcanonical ensemble for each taken alone. Let the energy of the first subsystem lie between E_1 and $E_1 + \Delta$ and the energy of the second subsystem lie between E_2 and $E_2 + \Delta$. The entropies of the subsystems are respectively

$$S_1(E_1, V_1) = k \log \Gamma_1(E_1)$$
$$S_2(E_2, V_2) = k \log \Gamma_2(E_2)$$

where $\Gamma_1(E_1)$ and $\Gamma_2(E_2)$ are the volumes occupied by the two ensembles in their respective Γ-spaces. They are schematically represented in Fig. 7.1 by the volumes of the shaded regions, which lie between successive energy surfaces that differ in energy by Δ.

Now consider the microcanonical ensemble of the composite system made up of the two subsystems, and let the total energy lie between E and $E + 2\Delta$. We choose Δ such that $\Delta \ll E$. This ensemble contains all copies of the composite system for which

(*a*) the N_1 particles whose momenta and coordinates are (p_1, q_1) are contained in the volume V_1,

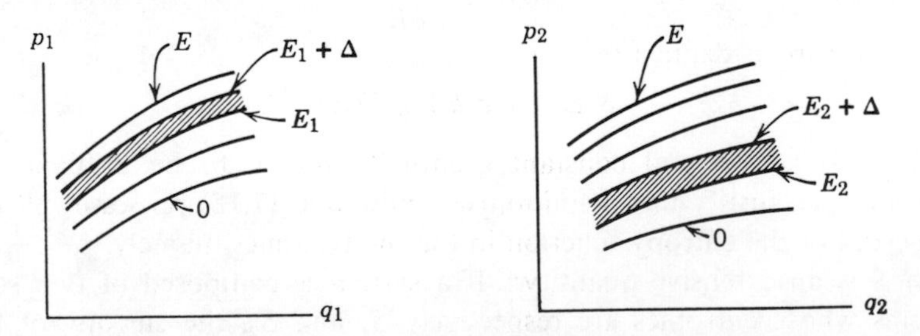

Fig. 7.1. The microcanonical ensemble of the two subsystems.

* For simplicity we assume that the same N_1, N_2 particles are always confined respectively to the volumes V_1, V_2. The proof is therefore invalid for a gas, for which S has to be modified (See Section 7.6).

(*b*) the N_2 particles whose momenta and coordinates are (p_2, q_2) are contained in the volume V_2,

(*c*) the energies E_1, E_2 of the subsystems have values satisfying the condition

$$E < (E_1 + E_2) < E + 2\Delta \tag{7.17}$$

Obviously, the volume of the region of Γ-space that corresponds to conditions (*a*) and (*b*) with a total energy lying between $E_1 + E_2$ and $E_1 + E_2 + 2\Delta$ is

$$\Gamma_1(E_1)\,\Gamma_2(E_2)$$

To obtain the total volume of the ensemble specified by (*a*), (*b*), and (*c*), we only have to take the sum of $\Gamma_1(E_1)\Gamma_2(E_2)$ over values of E_1 and E_2 consistent with (*c*). Since E_1 and E_2 are possible values of the Hamiltonians $H_1(p_1, q_1)$ and $H_2(p_2, q_2)$, their spectra of values must be bounded from below, for otherwise the subsystems would not be stable. For simplicity we take the lower bounds for both spectra to be 0. If we divide each of the energy spectra E_1 and E_2 into intervals of size Δ, then between 0 and E there are E/Δ intervals in each spectrum. Thus, since $\Delta \ll E$, we can write

$$\Gamma(E) = \sum_{i=1}^{E/\Delta} \Gamma_1(E_i)\Gamma_2(E - E_i) \tag{7.18}$$

where E_i is the energy lying in the center of each energy interval.

The entropy of the composite system of N particles and of volume V, with

$$N = N_1 + N_2$$

$$V = V_1 + V_2$$

is given by

$$S(E, V) = k \log \sum_{i=1}^{E/\Delta} \Gamma_1(E_i)\Gamma_2(E - E_i) \tag{7.19}$$

It will now be shown that when $N_1 \to \infty$ and $N_2 \to \infty$ a single term in the sum of (7.18) dominates the sum. The sum in (7.18) is a sum of E/Δ positive terms. Let the largest term in the sum be $\Gamma_1(\bar{E}_1)\Gamma_2(\bar{E}_2)$, where

$$\bar{E}_1 + \bar{E}_2 = E \tag{7.20}$$

Then it is obvious that

$$\Gamma_1(\bar{E}_1)\Gamma_2(\bar{E}_2) \le \Gamma(E) \le \frac{E}{\Delta}\,\Gamma_1(\bar{E}_1)\Gamma_2(\bar{E}_2)$$

or

$$k \log [\Gamma_1(\bar{E}_1)\Gamma_2(\bar{E}_2)] \le S(E, V) \le k \log [\Gamma_1(\bar{E}_1)\Gamma_2(\bar{E}_2)] + k \log \frac{E}{\Delta} \tag{7.21}$$

If the subsystems are molecular systems with N_1 and N_2 particles respectively, we expect that as $N_1 \to \infty$ and $N_2 \to \infty$,

$$\begin{aligned} \log \Gamma_1 &\propto N_1 \\ \log \Gamma_2 &\propto N_2 \\ E &\propto N_1 + N_2 \end{aligned} \tag{7.22}$$

Thus the term $\log(E/\Delta)$ in (7.21) may be neglected, because Δ is a constant independent of N. Therefore

$$S(E, V) = S_1(\bar{E}_1, V_1) + S_2(\bar{E}_2, V_2) + O(\log N) \tag{7.23}$$

which proves the extensive property of the entropy.

We have actually proved more than the extensive property of the entropy, because (7.23) also implies that the energies of subsystems have the definite values $\bar{E}_1$ and $\bar{E}_2$, respectively. They are the values of E_1 and E_2 that maximize the function $\Gamma_1(E_1)\Gamma_2(E_2)$ under the restriction $E_1 + E_2 = E$. That is,

$$\delta[\Gamma_1(E_1)\Gamma_2(E_2)] = 0, \qquad \delta E_1 + \delta E_2 = 0$$

This leads to the condition

$$\left[\frac{\partial}{\partial E_1} \log \Gamma_1(E_1)\right]_{E_1=\bar{E}_1} = \left[\frac{\partial}{\partial E_2} \log \Gamma_2(E_2)\right]_{E_2=\bar{E}_2}$$

or

$$\left[\frac{\partial S_1(E_1)}{\partial E_1}\right]_{E_1=\bar{E}_1} = \left[\frac{\partial S_2(E_2)}{\partial E_2}\right]_{E_2=\bar{E}_2} \tag{7.24}$$

We define the temperature of any system by

$$\frac{\partial S(E, V)}{\partial E} \equiv \frac{1}{T} \tag{7.25}$$

Then $\bar{E}_1$ and $\bar{E}_2$ are such that the two subsystems have the same temperature:

$$T_1 = T_2 \tag{7.26}$$

The temperature defined by (7.25) is precisely the absolute temperature in thermodynamics; not only is it a parameter associated with the condition for equilibrium, it is also related to the entropy by (7.25), which is one of the Maxwell relations in thermodynamics. Choosing the standard temperature interval to be the conventional Centrigrade degree defines the constant k in (7.15) to be Boltzmann's constant. Thus the proof of the extensive property of the entropy also reveals the meaning of the temperature for an isolated system: *The temperature of an isolated system is the parameter governing the equilibrium between one part of the system and another.*

Although the condition (7.17) allows a range of values of (E_1, E_2) to occur among members of the microcanonical ensemble, the result (7.21) shows that as the number of particles becomes very large almost all members of the ensemble have the values $(\bar{E}_1, \bar{E}_2)$. This fact is fundamental to the success of statistical mechanics as a theory of matter.

A calculation similar to that leading to (7.23) shows that the following definitions of S are equivalent, up to additive constant terms of order $\log N$ or smaller:

$$S = k \log \Gamma(E) \tag{7.27}$$

$$S = k \log \omega(E) \tag{7.28}$$

$$S = k \log \Sigma(E) \tag{7.29}$$

In fact, if these definitions were not equivalent, the validity of statistical mechanics would be in doubt.

To show that S possesses the properties of the entropy as required by the second law of thermodynamics, let us first state the form of the second law that is most convenient for the present purpose. The entropy in thermodynamics, just as S here, is defined only for equilibrium situations. The second law states that if an isolated system undergoes a change of thermodynamic state such that the initial and final states are equilibrium states, the entropy of the final state is not smaller than that of the initial state. For the system we are considering, the only independent macroscopic parameters are N, V, and E. By definition N and E cannot change, for the system is isolated. Thus only V can change. Now V cannot decrease without compressing the system, thereby disturbing its isolation. Hence V can only increase. (An example is the free expansion of a gas when one of the containing walls is suddenly removed.) For our purpose the second law states that the entropy is a nondecreasing function of V.

Let us use the definition (7.29), namely

$$S(E, V) = k \log \Sigma(E)$$

It is obvious that $\Sigma(E)$ is a nondecreasing function of V, for if $V_1 > V_2$, then the integral (7.11) for $V = V_1$ extends over a domain of integration which includes that for $V = V_2$. This shows that $S(E, V)$ is a nondecreasing function of V.

We conclude that the function $S(E, V)$, as defined by any one of the formulas (7.27)–(7.29), is the entropy of a system of volume V and internal energy E. This conclusion furnishes the connection between the microcanonical ensemble and thermodynamics.

7.3 DERIVATION OF THERMODYNAMICS

We have defined the entropy of a system and have shown that the second law of thermodynamics holds. The complete thermodynamics of a system can now be obtained.

First we discuss the analog of quasi-static thermodynamic transformations. A quasi-static thermodynamic transformation corresponds to a slow variation of E and V, induced by coupling the system to external agents. During such a transformation the ensemble is represented by a collection of representative points uniformly distributed over a slowly changing region in Γ-space. The change is so slow that at every instant we have a microcanonical ensemble. Accordingly, the change in the entropy in an infinitesimal transformation is given by

$$dS(E, V) = \left(\frac{\partial S}{\partial E}\right)_V dE + \left(\frac{\partial S}{\partial V}\right)_E dV \tag{7.30}$$

The coefficient of dE has been defined earlier as the inverse absolute temperature T^{-1}. We now define the pressure of the system to be

$$P \equiv T\left(\frac{\partial S}{\partial V}\right)_E \tag{7.31}$$

Hence

$$dS = \frac{1}{T}(dE + P\,dV) \tag{7.32}$$

or

$$dE = T\,dS - P\,dV \tag{7.33}$$

This is the first law of thermodynamics.

Thus we have succeeded not only in deriving the first and second laws of thermodynamics, but also in finding means to calculate all thermodynamic functions in terms of molecular interactions. The third law of thermodynamics cannot be obtained in classical statistical mechanics, because it is quantum mechanical.

We summarize by giving a practical recipe for finding all the thermodynamic functions of a system.

RECIPE. Consider an isolated system that occupies volume V and has an energy E within a small uncertainty $\Delta \ll E$. The Hamiltonian is presumed known. To find all thermodynamic functions of the system, proceed as follows:

(*a*) Calculate the density of states $\omega(E)$ of the system from the Hamiltonian.

(*b*) Find the entropy up to an arbitrary additive constant by the formula

$$S(E, V) = k \log \omega(E)$$

where k is Boltzmann's constant. Alternatively we can use the formula (7.27) or (7.29).

(*c*) Solve for E in terms of S and V. The resulting function is the thermodynamic internal energy of the system

$$U(S, V) \equiv E(S, V)$$

(*d*) Find other thermodynamic functions from the following formulas:

$$T = \left(\frac{\partial U}{\partial S}\right)_V \qquad \text{(absolute temperature)}$$

$$P = -\left(\frac{\partial U}{\partial V}\right)_S \qquad \text{(pressure)*}$$

$$A = U - TS \qquad \text{(Helmholtz free energy)}$$

$$G = U + PV - TS \qquad \text{(Gibbs potential)}$$

$$C_V = \left(\frac{\partial U}{\partial T}\right)_V \qquad \text{(heat capacity at constant volume)}$$

(*e*) To study any equilibrium behavior of the system, use thermodynamics.

7.4 EQUIPARTITION THEOREM

Let x_i be either p_i or q_i $(i = 1, \ldots, 3N)$. We calculate the ensemble average of $x_i(\partial H/\partial x_j)$, where H is the Hamiltonian. Using the abbreviation $dp\,dq \equiv d^{3N}p\,d^{3N}q$, we can write

$$\left\langle x_i \frac{\partial H}{\partial x_j}\right\rangle = \frac{1}{\Gamma(E)} \int_{E<H<E+\Delta} dp\,dq\,x_i \frac{\partial H}{\partial x_j} = \frac{\Delta}{\Gamma(E)} \frac{\partial}{\partial E} \int_{H<E} dp\,dq\,x_i \frac{\partial H}{\partial x_j}$$

Noting that $\partial E/\partial x_j = 0$, we may calculate the last integral in the following manner:

$$\int_{H<E} dp\,dq\,x_i \frac{\partial H}{\partial x_j} = \int_{H<E} dp\,dq\,x_i \frac{\partial}{\partial x_j}(H - E)$$

$$= \int_{H<E} dp\,dq \frac{\partial}{\partial x_j}\, x_i(H - E) - \delta_{ij} \int_{H<E} dp\,dq(H - E)$$

The first integral on the right-hand side vanishes, because it reduces to a surface integral over the boundary of the region defined by $H < E$, and on this boundary $H - E = 0$. Substituting the latest result into the

* This is equivalent to (7.31) by the chain relation.

previous equation, and noting that $\Gamma(E) = \omega(E)\Delta$, we obtain

$$\left\langle x_i \frac{\partial H}{\partial x_j} \right\rangle = \frac{\delta_{ij}}{\omega(E)} \frac{\partial}{\partial E} \int_{H<E} dp\, dq(E - H) = \frac{\delta_{ij}}{\omega(E)} \int_{H<E} dp\, dq = \frac{\delta_{ij}}{\omega(E)} \Sigma(E)$$

$$= \delta_{ij} \frac{\Sigma(E)}{\partial \Sigma(E)/\partial E} = \delta_{ij} \left[\frac{\partial}{\partial E} \log \Sigma(E) \right]^{-1} = \delta_{ij} \frac{k}{\partial S/\partial E}$$

that is,

$$\left\langle x_i \frac{\partial H}{\partial x_j} \right\rangle = \delta_{ij} kT \tag{7.34}$$

This is the *generalized equipartition theorem.*

For the special case $i = j$, $x_i = p_i$, we have

$$\left\langle p_i \frac{\partial H}{\partial p_i} \right\rangle = kT \tag{7.35}$$

For $i = j$ and $x_i = q_i$, we have

$$\left\langle q_i \frac{\partial H}{\partial q_i} \right\rangle = kT \tag{7.36}$$

According to the canonical equations of motion, $\partial H / \partial q_i = -\dot{p}_i$. Hence (7.36) leads to the statement

$$\left\langle \sum_{i=1}^{3N} q_i \dot{p}_i \right\rangle = -3NkT \tag{7.37}$$

which is known as the *virial theorem*, because $\Sigma q_i \dot{p}_i$—the sum of the ith coordinate times the ith component of the generalized force—is known in classical mechanics as the virial.

Many physical systems have Hamiltonians that, through a canonical transformation, can be cast in the form

$$H = \sum_i A_i P_i^2 + \sum_i B_i Q_i^2 \tag{7.38}$$

where P_i, Q_i are canonically conjugate variables and A_i, B_i are constants. For such systems we have

$$\sum_i \left(P_i \frac{\partial H}{\partial P_i} + Q_i \frac{\partial H}{\partial Q_i} \right) = 2H \tag{7.39}$$

Suppose f of the constants A_i and B_i are nonvanishing. Then (7.35) and (7.36) imply that

$$\langle H \rangle = \tfrac{1}{2} f kT \tag{7.40}$$

That is, each harmonic term in the Hamiltonian contributes $\frac{1}{2}kT$ to the average energy of the system. This is known as the *theorem of equipartition of energy*. But (7.40) is the internal energy of the system. Therefore

$$\frac{C_V}{k} = \frac{f}{2} \tag{7.41}$$

Thus the heat capacity is directly related to the number of degrees of freedom of the system.

A paradox arises from the theorem of equipartition of energy. In classical physics every system must in the last analysis have an infinite number of degrees of freedom, for after we have resolved matter into atoms we must continue to resolve an atom into its constituents and the constituents of the constituents, *ad infinitum*. Therefore the heat capacity of any system is infinite. This is a real paradox in classical physics and is resolved by quantum mechanics. Quantum mechanics possesses the feature that the degrees of freedom of a system are manifest only when there is sufficient energy to excite them, and that those degrees of freedom that are not excited can be forgotten. Thus the formula (7.41) is valid only when the temperature is sufficiently high.

7.5 CLASSICAL IDEAL GAS

To illustrate the method of calculation in the microcanonical ensemble we consider the classical ideal gas. This has been considered earlier in our discussion of the kinetic theory of gases. In that discussion we also introduced the microcanonical ensemble, but we obtained all the thermodynamic properties of the ideal gas via the distribution function. For the sake of illustration, we now derive the same results using the recipe given in Sec. 7.3.

The Hamiltonian is

$$H = \frac{1}{2m}\sum_{i=1}^{N} p_i^2 \tag{7.42}$$

We first calculate

$$\Sigma(E) = \frac{1}{h^{3N}}\int_{H<E} d^3p_1 \cdots d^3p_N\, d^3q_1 \cdots d^3q_N \tag{7.43}$$

where h is a constant of the dimension of momentum $\times$ distance, introduced to make $\Sigma(E)$ dimensionless. The integration over q_i can be immediately carried out, giving a factor of V^N. Let

$$R = \sqrt{2mE} \tag{7.44}$$

Then

$$\Sigma(E) = \left(\frac{V}{h^3}\right)^N \Omega_{3N}(R) \tag{7.45}$$

where Ω_n is the volume of an n-sphere of radius R:

$$\Omega_n(R) = \int_{x_1^2+x_2^2+\cdots+x_n^2<R^2} dx_1\, dx_2 \cdots dx_n \tag{7.46}$$

Clearly,

$$\Omega_n(R) = C_n R^n \tag{7.47}$$

where C_n is a constant. To find C_n, consider the identity

$$\int_{-\infty}^{+\infty} dx_1 \cdots \int_{-\infty}^{+\infty} dx_n e^{-(x_1^2+\cdots+x_n^2)} = \left(\int_{-\infty}^{+\infty} dx\, e^{-x^2}\right)^n = \pi^{n/2} \tag{7.48}$$

The left-hand side of (7.48) can be re-expressed as follows. Let $S_n(R) \equiv d\Omega_n(R)/dR$ be the surface area of an n-sphere of radius R. Then

$$\begin{aligned}\int_{-\infty}^{+\infty} dx_1 \cdots \int_{-\infty}^{+\infty} dx_n e^{-(x_1^2+\cdots+x_n^2)} &= \int_0^\infty dR\, S_n(R) e^{-R^2} \\ &= nC_n \int_0^\infty dR\, R^{n-1} e^{-R^2} \\ &= \tfrac{1}{2} n C_n \int_0^\infty dt\, t^{(n/2)-1} e^{-t} = \tfrac{1}{2} n C_n \left(\frac{n}{2} - 1\right)!\end{aligned} \tag{7.49}$$

Comparison of (7.49) and (7.48) yields

$$C_n = \frac{\pi^{n/2}}{(n/2-1)!} \tag{7.50}$$

$$\log C_n \xrightarrow[n\to\infty]{} \frac{n}{2}\log \pi - \frac{n}{2}\log\frac{n}{2} + \frac{n}{2} \tag{7.51}$$

Hence

$$\Sigma(E) = C_{3N}\left[\frac{V}{h^3}(2mE)^{3/2}\right]^N \tag{7.52}$$

The entropy of the ideal gas is

$$S(E, V) = k\left[\log C_{3N} + N\log\frac{V}{h^3} + \frac{3}{2}N\log(2mE)\right] \tag{7.53}$$

By (7.51), this reduces to

$$S(E, V) = Nk\log\left[V\left(\frac{4\pi m}{3h^2}\frac{E}{N}\right)^{3/2}\right] + \frac{3}{2}Nk \tag{7.54}$$

Solving for E in terms of S and V, and calling the resulting function $U(S, V)$ the internal energy, we obtain

$$U(S, V) = \left(\frac{3}{4\pi}\frac{h^2}{m}\right)\frac{N}{V^{2/3}}\exp\left(\frac{2}{3}\frac{S}{Nk} - 1\right) \tag{7.55}$$

The temperature is

$$T = \left(\frac{\partial U}{\partial S}\right)_V = \frac{2}{3}\frac{U}{Nk} \tag{7.56}$$

from which follows

$$C_V = \tfrac{3}{2}Nk \tag{7.57}$$

Finally the equation of state is

$$P = -\left(\frac{\partial U}{\partial V}\right)_S = \frac{2}{3}\frac{U}{V} = \frac{NkT}{V} \tag{7.58}$$

This calculation shows that the microcanonical ensemble is clumsy to use. There seems little hope that we can straightforwardly carry out the recipe of the microcanonical ensemble for any system but the ideal gas. We later introduce the canonical ensemble, which gives results equivalent to those of the microcanonical ensemble but which is more convenient for practical calculations.

7.6 GIBBS PARADOX

According to (7.54), the entropy of an ideal gas is

$$S = Nk \log (Vu^{3/2}) + Ns_0 \tag{7.59}$$

where

$$u = \tfrac{3}{2}kT$$
$$s_0 = \frac{3k}{2}\left(1 + \log\frac{4\pi m}{3h^2}\right) \tag{7.60}$$

Consider two ideal gases, with N_1 and N_2 particles respectively, kept in two separate volumes V_1 and V_2 at the same temperature. Let us find the change in entropy of the combined system after the gases are allowed to mix in a volume $V = V_1 + V_2$. The temperature will be the same after the mixing process. Hence u remains unchanged. From (7.59) we find that the change in entropy is

$$\frac{\Delta S}{k} = N_1 \log\frac{V}{V_1} + N_2 \log\frac{V}{V_2} > 0 \tag{7.61}$$

which is the entropy of mixing. If the two gases are different (e.g., argon and neon), this result is experimentally correct.

The Gibbs paradox presents itself if we consider the case in which the two mixing gases are of the same kind. Since the derivation of (7.61) does not depend on the identity of the gases, we would obtain the same increase of entropy (7.61). This is a disastrous result, because it implies that the entropy of a gas depends on the history of the gas, and thus cannot be a function of the thermodynamic state alone. Worse, the entropy does not exist, because we can always imagine that the existing state of a gas is arrived at by pulling off any number of partitions that initially divided the gas into any number of compartments. Hence S is larger than any number.

Gibbs resolved the paradox in an empirical fashion by postulating that we have made an error in calculating $\Sigma(E)$, the number of states of the gas with energy less than E. Gibbs assumed that the correct answer is $N!$ times smaller than we thought it was. By this assumption we should subtract from (7.59) the term $\log N! \approx N \log N - N$ and obtain

$$S = Nk \log\left(\frac{V}{N} u^{3/2}\right) + \tfrac{3}{2}Nk\left(\tfrac{5}{3} + \log \frac{4\pi m}{3h^2}\right) \tag{7.62}$$

This formula does not affect the equation of state and other thermodynamic functions of a system, because the subtracted term is independent of T and V. For the mixing of two different gases (7.62) still predicts (7.61), because N_1 and N_2 are the same constants before and after the mixing. For the mixing of gases that are of the same kind, however, it gives no entropy of mixing because the specific volume V/N is the same before and after mixing.

The formula (7.62) has been experimentally verified as the correct entropy of an ideal gas at high temperatures, if h is numerically set equal to Planck's constant. It is known as the *Sackur-Tetrode equation*.

It is not possible to understand classically why we must divide $\Sigma(E)$ by $N!$ to obtain the correct counting of states. The reason is inherently quantum mechanical. Quantum mechanically, atoms are inherently indistinguishable in the following sense: A state of the gas is described by an N-particle wave function, which is either symmetric or antisymmetric with respect to the interchange of any two particles. A permutation of the particles can at most change the wave function by a sign, and it does not produce a new state of the system. From this fact it seems reasonable that the Γ-space volume element $dp\,dq$ corresponds to not one but only $dp\,dq/N!$ states of the system. Hence we should divide $\Sigma(E)$ by $N!$. This rule of counting is known as the "correct Boltzmann counting." It is something that we must append to classical mechanics in order to get right answers.

The foregoing discussion contains the correct reason for, but is not a derivation of, the "correct Boltzmann counting," because in classical mechanics there is no consistent way in which we can regard the particles as indistinguishable. In all classical considerations other than the counting of states we must continue to regard the particles in a gas as distinguishable.

We may derive the "correct Boltzmann counting" by showing that in the limit of high temperatures quantum statistical mechanics reduces to classical statistical mechanics with "correct Boltzmann counting." This is done in Sec. 10.2.

PROBLEMS

7.1. Show that the formulas (7.27), (7.28), and (7.29) are equivalent to one another.

7.2. Let the "uniform" ensemble of energy E be defined as the ensemble of all systems of the given type with energy less than E. The equivalence between (7.29) and (7.27) means that we should obtain the same thermodynamic functions from the "uniform" ensemble of energy E as from the microcanonical ensemble of energy E. In particular, the internal energy is E in both ensembles. Explain why this seemingly paradoxical result is true.

7.3. Consider a system of N free particles in which the energy of each particle can assume two and only two distinct values, 0 and E ($E > 0$). Denote by n_0 and n_1 the occupation numbers of the energy level 0 and E, respectively. The total energy of the system is U.

(*a*) Find the entropy of such a system.

(*b*) Find the most probable values of n_0 and n_1, and find the mean square fluctuations of these quantities.

(*c*) Find the temperature as a function of U, and show that it can be negative.

(*d*) What happens when a system of negative temperature is allowed to exchange heat with a system of positive temperature?

Reference: N. F. Ramsey, *Phys. Rev.*, **103**, 20 (1956).

chapter 8

CANONICAL ENSEMBLE AND GRAND CANONICAL ENSEMBLE

8.1 CANONICAL ENSEMBLE

We wish to consider the question, "What ensemble is appropriate for the description of a system not in isolation, but in thermal equilibrium with a larger system?" To answer it we must find the probability that the system has energy E, because this probability is proportional to the density in Γ-space for the ensemble we want.

We investigated a similar problem in Sec. 7.2, when we considered the energies of the component parts of a composite system. What we do in the following is to discuss the case in which one component part is much smaller than the other.

Consider an isolated composite system made up of two subsystems whose Hamiltonians are respectively $H_1(p_1, q_1)$ and $H_2(p_2, q_2)$, with number of particles N_1 and N_2 respectively. We assume that $N_2 \gg N_1$ but that both N_1 and N_2 are macroscopically large. We are interested in system 1 only. Consider a microcanonical ensemble of the composite system with total energy between E and $E + 2\Delta$. The energies E_1 and E_2 of the subsystems accordingly can have any values satisfying

$$E < (E_1 + E_2) < E + 2\Delta \tag{8.1}$$

Although this includes a range of values of E_1, E_2, the analysis of Sec. 7.2 shows that only one set of values, namely $\bar{E}_1$, $\bar{E}_2$, is important. We assume that $\bar{E}_2 \gg \bar{E}_1$. Let $\Gamma_2(E_2)$ be the volume occupied by system 2 in its own Γ-space. The probability of finding system 1 in a state within $dp_1\,dq_1$ of (p_1, q_1), regardless of the state of system 2, is proportional to $dp_1\,dq_1\Gamma_2(E_2)$, where $E_2 = E - E_1$. Therefore up to a proportionality constant the density in Γ-space for system 1 is

$$\rho(p_1, q_1) \propto \Gamma_2(E - E_1) \tag{8.2}$$

Since only the values near $E_1 = \bar{E}_1$ are expected to be important, and $\bar{E}_1 \ll E$, we may perform the expansion

$$\begin{aligned} k \log \Gamma_2(E - E_1) = S_2(E - E_1) &= S_2(E) - E_1\left[\frac{\partial S_2(E_2)}{\partial E_2}\right]_{E_2=E} + \cdots \\ &\approx S_2(E) - \frac{E_1}{T} \end{aligned} \tag{8.3}$$

where T is the temperature of the larger subsystem. Hence

$$\Gamma_2(E - E_1) \approx \exp\left[\frac{1}{k}\, S_2(E)\right] \exp\left(-\frac{E_1}{kT}\right) \tag{8.4}$$

The first factor is independent of E_1 and is thus a constant as far as the small subsystem is concerned. Owing to (8.2) and the fact that $E_1 = H_1(p_1, q_1)$, we may take the ensemble density for the small subsystem to be

$$\rho(p, q) = e^{-H(p,q)/kT} \tag{8.5}$$

Where the subscript 1 labeling the subsystem has been omitted, since we may now forget about the larger subsystem, apart from the information that its temperature is T. The larger subsystem in fact behaves like a heat reservoir in thermodynamics. The ensemble defined by (8.5), appropriate for a system whose temperature is determined through contact with a heat reservoir, is called the *canonical ensemble.*

The volume in Γ-space occupied by the canonical ensemble is called the *partition function*:

$$Q_N(V, T) \equiv \int \frac{d^{3N}p\, d^{3N}q}{N!\, h^{3N}}\, e^{-\beta H(p,q)} \tag{8.6}$$

where $\beta = 1/kT$, and where we have introduced a constant h, of the dimension of *momentum* × *distance*, in order to make Q_N dimensionless. The factor $1/N!$ appears, in accordance with the rule of "correct Boltzmann counting." These constants are of no importance for the equation of state.

Strictly speaking we should not integrate over the entire Γ-space in (8.6), because (8.2) requires that $\rho(p_1, q_1)$ vanish if $E_1 > E$. The justification for ignoring such a restriction is that in the integral (8.6) only one value of the energy $H(p, q)$ contributes to the integral and that this value will lie in the range where the approximation (8.4) is valid. We prove this contention in Sec. 8.2.

The thermodynamics of the system is to be obtained from the formula

$$Q_N(V, T) = e^{-\beta A(V, T)} \tag{8.7}$$

where $A(V, T)$ is the Helmholtz free energy. To justify this identification we show that

(*a*) A is an extensive quantity,

(*b*) A is related to the internal energy $U \equiv \langle H \rangle$ and the entropy $S \equiv -(\partial A/\partial T)_V$ by the thermodynamic relation

$$A = U - TS$$

That A is an extensive quantity follows from (8.6), because if the system is made up of two subsystems whose mutual interaction can be neglected, then Q_N is a product of two factors. To prove the relation (*b*), we first convert (*b*) into the following differential equation for A:

$$\langle H \rangle = A - T\left(\frac{\partial A}{\partial T}\right)_V \tag{8.8}$$

To prove (8.8), note the identity

$$\frac{1}{N!\, h^{3N}} \int dp\, dq e^{\beta[A(V, T) - H(p, q)]} = 1 \tag{8.9}$$

Differentiating with respect to β on both sides, we obtain

$$\frac{1}{N!\, h^{3N}} \int dp\, dq e^{\beta[A(V, T) - H(p, q)]} \left[A(V, T) - H(p, q) + \beta\left(\frac{\partial A}{\partial \beta}\right)_V\right] = 0$$

This is the same as

$$A(V, T) - U(V, T) - T\left(\frac{\partial A}{\partial T}\right)_V = 0$$

All other thermodynamic functions may be found from $A(V, T)$ by the Maxwell relations in thermodynamics:

$$P = -\left(\frac{\partial A}{\partial V}\right)_T$$

$$S = -\left(\frac{\partial A}{\partial T}\right)_V$$

$$G = A + PV$$

$$U = \langle H \rangle = A + TS$$

Therefore all calculations in the canonical ensembles begin (and nearly end) with the calculation of the partition function (8.6).

8.2 ENERGY FLUCTUATIONS IN THE CANONICAL ENSEMBLE

We now show that the canonical ensemble is mathematically equivalent to the microcanonical ensemble in the sense that although the canonical ensemble contains systems of all energies the overwhelming majority of them have the same energy. To do this we calculate the mean square fluctuation of energy in the canonical ensemble. The average energy is

$$U = \langle H \rangle = \frac{\int dp\, dq H e^{-\beta H}}{\int dp\, dq e^{-\beta H}} \tag{8.10}$$

Hence

$$\int dp\, dq\, [U - H(p, q)] e^{\beta[A(V,T) - H(p,q)]} = 0 \tag{8.11}$$

Differentiating both sides with respect to β, we obtain

$$\frac{\partial U}{\partial \beta} + \int dp\, dq e^{\beta(A-H)} (U - H)\left(A - H - T\frac{\partial A}{\partial T}\right) = 0 \tag{8.12}$$

By (8.8) this can be rewritten in the form

$$\frac{\partial U}{\partial \beta} + \langle (U - H)^2 \rangle = 0 \tag{8.13}$$

Therefore the mean square fluctuation of energy is

$$\langle H^2 \rangle - \langle H \rangle^2 = \langle (U - H)^2 \rangle = -\frac{\partial U}{\partial \beta} = kT^2 \frac{\partial U}{\partial T}$$

or

$$\langle H^2 \rangle - \langle H \rangle^2 = kT^2 C_V \tag{8.14}$$

For a macroscopic system $\langle H \rangle \propto N$ and $C_V \propto N$. Hence (8.14) is a normal fluctuation. As $N \to \infty$, almost all systems in the ensemble have the energy $\langle H \rangle$, which is the internal energy. Therefore the canonical ensemble is equivalent to the microcanonical ensemble.

It is instructive to calculate the fluctuations in another way. We begin by calculating the partition function in the following manner:

$$\frac{1}{N!\, h^{3N}} \int dp\, dq e^{-\beta H(p,q)} = \int_0^\infty dE \omega(E) e^{-\beta E} = \int_0^\infty dE e^{-\beta E + \log \omega(E)}$$

$$= \int_0^\infty dE e^{\beta[TS(E) - E]} \tag{8.15}$$

where S is the entropy defined in the microcanonical ensemble. Since both S and U are proportional to N, the exponent in the last integrand is enormous. We expect that as $N \to \infty$ the integral receives contribution only from the neighborhood of the maximum of the integrand. The maximum of the integrand occurs at $E = \bar{E}$, where $\bar{E}$ satisfies the conditions

$$T\left(\frac{\partial S}{\partial E}\right)_{E=\bar{E}} = 1 \tag{8.16}$$

$$\left(\frac{\partial^2 S}{\partial E^2}\right)_{E=\bar{E}} < 0 \tag{8.17}$$

The first condition implies $\bar{E} = U$, the internal energy. Next we note that

$$\left(\frac{\partial^2 S}{\partial E^2}\right)_{E=\bar{E}} = \left(\frac{\partial}{\partial E}\frac{1}{T}\right)_{E=\bar{E}} = -\frac{1}{T^2}\left(\frac{\partial T}{\partial E}\right)_{E=\bar{E}} = -\frac{1}{T^2 C_V} \tag{8.18}$$

Thus the condition (8.17) is satisfied if $C_V > 0$, which is true for physical systems. Now let us expand the exponent in (8.15) about $E = \bar{E}$:

$$\begin{aligned} TS(E) - E &= [TS(\bar{E}) - \bar{E}] + \tfrac{1}{2}(E - \bar{E})^2 T\left(\frac{\partial^2 S}{\partial E^2}\right)_{E=\bar{E}} + \cdots \\ &= [TS(U) - U] - \frac{1}{2TC_V}(E - U)^2 + \cdots \end{aligned} \tag{8.19}$$

Hence

$$\frac{1}{N!\,h^{3N}}\int dp\,dq e^{-\beta H(p,q)} \approx e^{\beta(TS-U)}\int_0^\infty dE e^{-(E-U)^2/2kT^2C_V} \tag{8.20}$$

showing that in the canonical ensemble the distribution in *energy* is a Gaussian distribution centered about the value $E = U$ with a width equal to

$$\Delta E = \sqrt{2kT^2C_V} \tag{8.21}$$

Since $U \propto N$ and $C_V \propto N$, $\Delta E/U$ is negligibly small. As $N \to \infty$ the Gaussian approaches a δ-function. Finally, let us perform the integral in (8.20). It is elementary.

$$\int_0^\infty dE e^{-(E-U)^2/2kT^2C_V} = \int_{-U}^\infty dx e^{-x^2/2kT^2C_V} \approx \int_{-\infty}^{+\infty} dx e^{-x^2/2kT^2C_V} = \sqrt{2\pi kT^2C_V}$$

Therefore

$$\frac{1}{N!\,h^{3N}}\int dp\,dq e^{-\beta H(p,q)} \approx e^{+\beta(TS-U)}\sqrt{2\pi kT^2C_V} \tag{8.22}$$

$$A \approx (U - TS) - \tfrac{1}{2}kT\log(C_V) \tag{8.23}$$

This last term is negligible when $N \to \infty$. In that limit we have exactly $A = U - TS$. Statement (8.23) shows that the entropy as defined in the canonical and microcanonical ensemble differs only by terms of the order of $\log N$.

We have shown that almost all systems in the canonical ensemble have the same energy—namely, the energy that is equal to the internal energy of a system at the given temperature T. The reason for this is easy to see, both mathematically and physically.

In the canonical ensemble we distribute systems in Γ-space according to the density function $\rho(p, q) = \exp[-\beta H(p, q)]$, which is represented in Fig. 8.1. The density of points fall off exponentially as we go away from the origin of Γ-space. The distribution in energy is obtained by "counting" the number of points on energy surfaces. As we go away from the origin, the energy increases and the area of the energy surface increases. This is why we get a peak in the distribution in energy. The sharpness of the peak is due to the rapidity with which the area of the energy surface increases as E increases. For an N-body system this area increases like e^E, where $E \propto N$.

From a physical point of view, a microcanonical ensemble must be equivalent to a canonical ensemble, otherwise we would seriously doubt the utility of either. A macroscopic substance has the extensive property, i.e., any part of the substance has the same thermodynamic property as the whole substance. If we consider a piece of substance isolated from everything, it is still true that any part of the substance must be in equilibrium with the rest; and the rest serves as a heat reservoir that defines a temperature for the part on which we focus our attention. Therefore the whole substance must have a well-defined temperature.

We have seen earlier that in the microcanonical ensemble it matters little whether we take the entropy to be k times the logarithm of the density of states at the energy E, the number of states with energies between E, $E + \Delta$, or all the states with energy below E. In all these cases we arrive at the same thermodynamic behavior. Now we see that it matters little

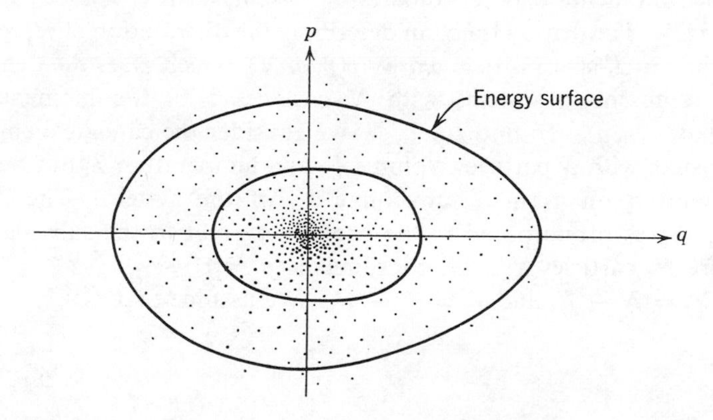

Fig. 8.1. Distribution of representative points in Γ-space for the canonical ensemble.

whether we specify the energy of the system or the temperature of the system, for specifying one fixes the other, and we find the same thermodynamic behavior in both cases. All these examples illustrate the insensitivity of thermodynamic results to methods of derivation. The reasons behind this insensitivity are, in all cases, the facts that

(*a*) density of states $\propto e^E$
(*b*) $E \propto N$
(*c*) $N \to \infty$

On these facts depends the validity of statistical mechanics.

8.3 GRAND CANONICAL ENSEMBLE

Although the canonical and the microcanonical ensemble give equivalent results, it may be argued that conceptually the canonical ensemble corresponds more closely to physical situations. In experiments we never deal with a completely isolated system, nor do we ever directly measure the total energy of a macroscopic system. We usually deal with systems with a given temperature—a parameter that we can control in experiments.

By the same attitude we would consider it unphysical to specify the number of particles of a macroscopic system, for that is never precisely known. All we can find out from experiments is the average number of particles. This is the motivation for introducing the *grand canonical ensemble*, in which the systems can have any number of particles, with the average number determined by conditions external to the system. This is analogous to the situation in the canonical ensemble, where the average energy of a system is determined by the temperature of the heat reservoir with which it is in contact.

The Γ-space for the grand canonical ensemble is spanned by all the canonical momenta and coordinates of systems with 0, 1, 2, . . . number of particles. The density function describing the distribution of representative points in Γ-space is denoted by $\rho(p, q, N)$, which gives the density of points representing systems with N particles with the momenta and coordinates (p, q). To find $\rho(p, q, N)$ we consider the canonical ensemble for a system with N particles, volume V, and temperature T, but we focus our attention on a small subvolume V_1 of the system. The density $\rho(p_1, q_1, N_1)$ is proportional to the probability that in the subvolume V_1 there are N_1 particles with the coordinates (p_1, q_1).

Let $N_2 = N - N_1$ and $V_2 = V - V_1$. We assume that

$$N_2 \gg N_1$$
$$V_2 \gg V_1$$

If there are N_1 particles in V_1, there must be N_2 particles in the remaining volume V_2. Hence, neglecting molecular interactions across the surface separating V_2 and V_1, we must have

$$\rho(p_1, q_1, N_1) \propto e^{-\beta H(p_1, q_1, N_1)} \int_{V_2} dp_2\, dq_2 e^{-\beta H(p_2, q_2, N_2)} \tag{8.24}$$

where the integral in (8.24) extends over all p_2, but only over such values of q_2 as to keep the N_2 particles always in the volume V_2. The Hamiltonians $H(p_1, q_1, N_1)$ and $H(p_2, q_2, N_2)$ have the same functional form but refer respectively to N_1 particles and N_2 particles.* We arbitrarily choose the proportionality constant in (8.24) so as to give

$$\rho(p_1, q_1, N_1) = \frac{N!}{N_1!\, N_2!} \frac{e^{-\beta H(p_1, q_1, N_1)} \displaystyle\int_{V_2} dp_2\, dq_2 e^{-\beta H(p_2, q_2, N_2)}}{\displaystyle\int_V dp\, dq e^{-\beta H(p, q, N)}} \tag{8.25}$$

We can also write

$$\rho(p_1, q_1, N_1) = \frac{Q_{N_2}(V_2, T)}{Q_N(V, T)} \frac{e^{-\beta H(p_1, q_1, N_1)}}{N_1!\, h^{3N_1}} \tag{8.26}$$

where $Q_N(V, T)$ is the partition function defined in (8.6). It follows from (8.25) that

$$\sum_{N_1=0}^{N} \int dp_1\, dq_1 \rho(p_1, q_1, N_1) = 1 \tag{8.27}$$

The proof of this statement is left as an exercise (see Problem 8.4.)

Using (8.7) we can write

$$\frac{Q_{N_2}(V_2, T)}{Q_N(V, T)} = e^{-\beta[A(N_2, V_2, T) - A(N, V, T)]} = e^{-\beta[A(N - N_1, V - V_1, T) - A(N, V, T)]} \tag{8.28}$$

where $A(N, V, T)$ is the Helmholtz free energy. Since $N \gg N_1$ and $V \gg V_1$, we may use the approximation

$$A(N - N_1, V - V_1, T) - A(N, V, T) \approx -N_1\mu + V_1 P \tag{8.29}$$

where μ and P are respectively the chemical potential and the pressure of the part of the system external to the small volume V_1:

$$\mu = \left[\frac{\partial A(N_2, V, T)}{\partial N_2}\right]_{N_2 = N} \tag{8.30}$$

$$P = -\left[\frac{\partial A(N, V_2, T)}{\partial V_2}\right]_{V_2 = V} \tag{8.31}$$

* Which particle belongs to which group depends on the instant of time, but the numerical values of the Hamiltonians are independent of time.

We introduce the *fugacity*, defined by

$$z = e^{\beta\mu} \tag{8.32}$$

Substituting (8.32) and (8.29) into (8.28), and then substituting (8.28) into (8.26), we obtain

$$\rho(p, q, N) = \frac{z^N}{N!\, h^{3N}} e^{-\beta PV - \beta H(p;\, q)} \tag{8.33}$$

where the subscript 1 identifying the volume under consideration has been omitted because the system external to the volume can now be forgotten, apart from the information that it has the temperature T, pressure P, and chemical potential μ. We now allow the system external to the volume under consideration to become infinite in size. Then the range of N in (8.33) becomes

$$0 \le N < \infty$$

The thermodynamic functions for the volume under consideration may be found as follows. First of all, the internal energy shall be the ensemble average of $H(p, q)$. Second, the temperature, pressure, and chemical potential shall be respectively equal to T, P, μ. To show that this is a correct recipe, it suffices to remind ourselves that thermodynamics has been derived from the canonical ensemble. It is an elementary thermodynamic exercise to show that if a system is in equilibrium any part of the system must have the same T, P, μ as any other part; but this is the desired result.

To obtain a convenient formal recipe for finding all the thermodynamic functions we define the *grand partition function* as follows:

$$\mathcal{Q}(z, V, T) \equiv \sum_{N=0}^{\infty} z^N Q_N(V, T) \tag{8.34}$$

which in principle can be calculated from a knowledge of the Hamiltonian. Integrating both sides of (8.33) over all (p, q) for a given N, and then summing N from 0 to ∞, we find that

$$\frac{PV}{kT} = \log \mathcal{Q}(z, V, T) \tag{8.35}$$

Thus the grand partition function directly gives the pressure as a function of z, V, and T. The average number N of particles in the volume V is by definition the ensemble average

$$N = \frac{\sum_{N'=0}^{\infty} N' z^{N'} Q_{N'}(V, T)}{\sum_{N'=0}^{\infty} z^{N'} Q_{N'}(V, T)} = z \frac{\partial}{\partial z} \log \mathcal{Q}(z, V, T) \tag{8.36}$$

The equation of state, which is the equation expressing P as a function of N, V, and T, is obtained by eliminating z between (8.35) and (8.36).

All other thermodynamic functions may be obtained from the internal energy:

$$U = -\frac{\partial}{\partial \beta} \log \mathcal{Q}(z, V, T) \tag{8.37}$$

After eliminating z with the help of (8.36), U becomes a function of N, V, and T. We can then use the formulas

$$C_V = \left(\frac{\partial U}{\partial T}\right)_V$$

$$S = \int_0^T dT \frac{C_V}{T}$$

$$A = U - TS$$

Alternatively, all thermodynamic functions may be found from the Helmholtz free energy, which can be shown to be directly obtainable from $\log \mathcal{Q}$ through the formula

$$A = NkT \log z - kT \log \mathcal{Q}(z, V, T) \tag{8.38}$$

Again it is necessary to eliminate z with the help of (8.36) in order to obtain A as a function of N, V, and T. The justification for (8.38) is given in the following two sections.

8.4 DENSITY FLUCTUATIONS IN THE GRAND CANONICAL ENSEMBLE

We begin a study of the equivalence between the grand canonical ensemble and the canonical ensemble. These two are trivially equivalent to each other, if almost all systems in the grand canonical ensemble have the same number of particles. Since all systems have exactly the same volume, this means that the fluctuation of density is small. We first find the condition under which the fluctuation of density is small.

The probability that a system in the grand canonical ensemble has N' particles is proportional to

$$W(N') = z^{N'} Q_{N'}(V, T) = e^{\beta\mu N' - \beta A(N', V, T)} \tag{8.39}$$

where $A(N, V, T)$ is the Helmholtz free energy, calculated from the canonical ensemble for N particles in volume V at temperature T. For the fluctuation of density in the grand canonical ensemble to be small, it is necessary and sufficient that $W(N')$ be essentially zero except in the neighborhood of some point $N' = N$, where $W(N')$ should have a sharp maximum. That is, we require that there be a value N for which

$$\left[\frac{\partial A(N', V, T)}{\partial N'}\right]_{N'=N} = \mu \tag{8.40}$$

$$\gamma \equiv \left[\frac{\partial^2 A(N', V, T)}{\partial N'^2}\right]_{N'=N} > 0 \tag{8.41}$$

The first of these conditions requires that the system have the same chemical potential μ as the external system. It is identical with the condition (8.36). To find out the meaning of the second condition we first express γ in terms of measurable quantities in the following manner.

Since the Helmholtz free energy is an extensive quantity it may be written in the form

$$A(N', V, T) = N'a(v') \tag{8.42}$$

where $v' \equiv V/N'$, and where the dependence on the temperature is understood. Then

$$\begin{aligned}\frac{\partial A}{\partial N'} &= a(v') - v'\,\frac{\partial a(v')}{\partial v'}\\ \frac{\partial^2 A}{\partial N'^2} &= \frac{1}{N'}\,v'^2\,\frac{\partial^2 a(v')}{\partial v'^2}\end{aligned} \tag{8.43}$$

On the other hand the pressure of the system is

$$P(v') = -\,\frac{\partial a(v')}{\partial v'}$$

Hence
$$\frac{\partial P(v')}{\partial v'} = -\,\frac{\partial^2 a(v')}{\partial v'^2} \tag{8.44}$$

Comparing (8.44) with (8.43) and (8.41) we obtain

$$\gamma \equiv \left(\frac{\partial^2 A}{\partial N'^2}\right)_{N'=N} = -\,\frac{v^2}{N}\frac{\partial P(v)}{\partial v} \tag{8.45}$$

where $v \equiv V/N$. Thus (8.41) is the same as the requirement

$$\frac{\partial P(v)}{\partial v} < 0 \tag{8.46}$$

Experimentally the equation of state of a substance is always such that $\partial P(v)/\partial v \leq 0$. Hence (8.46) can be fulfilled.

We now expand $W(N')$ about N:

$$W(N') \approx W(N)e^{-\frac{1}{2}\beta\gamma(N'-N)^2} \tag{8.47}$$

This leads to a Gaussian distribution of N', centered about N, with a width equal to

$$\Delta N = \sqrt{\frac{2}{\beta\gamma}} = \sqrt{\frac{2kTN}{v^2(-\partial P/\partial v)}} \tag{8.48}$$

Thus $\Delta N/N \to 0$ as $N \to \infty$, if (8.46) is fulfilled.

We may also directly calculate the mean square fluctuation of N' and find that

$$\frac{\sum_{N'=0}^{\infty} (N' - N)^2 W(N')}{\sum_{N'=0}^{\infty} W(N')} = \frac{kTN}{v^2(-\partial P/\partial v)} \tag{8.49}$$

We have seen that if $\partial P/\partial v < 0$ then almost all systems in the grand canonical ensemble have the same number of particles N. Then the grand canonical ensemble is trivially equivalent to the canonical ensemble for N particles. We must then have

$$\mathcal{Q}(z, V, T) \approx z^N Q_N(V, T) \tag{8.50}$$

or

$$\begin{cases} -\beta A(N, V, T) = \log \mathcal{Q}(z, V, T) - N \log z \\ N = z\dfrac{\partial}{\partial z} \log \mathcal{Q}(z, V, T) \end{cases} \tag{8.51}$$

Eliminating z between the two equations of (8.51) yields the Helmholtz free energy, from which all thermodynamic functions can be obtained. In particular we recover the equation of state in the form (8.35) by using the formula $P = -\partial A/\partial V$ from the canonical ensemble. There is still the question whether there always exists a z such that the second equation of (8.51) gives any desired value of N. We postpone the question until the next section, where it is answered in the affirmative.

Since we derive the grand canonical ensemble from the canonical ensemble by focusing our attention on a volume within the system, the grand canonical ensemble cannot contain more information than the canonical ensemble. The grand canonical ensemble does, however, make it more convenient for us to consider density fluctuations. These fluctuations give rise to physically observable effects, e.g., the fluctuation scattering of light. The formula (8.49) indicates that near the critical point of a gas, where $\partial P/\partial v = 0$, the density fluctuations become abnormally large. This is experimentally borne out by the phenomenon of critical opalescence.

8.5 EQUIVALENCE OF THE CANONICAL ENSEMBLE AND THE GRAND CANONICAL ENSEMBLE

To complete our investigation of the equivalence between the canonical and the grand canonical ensemble it is necessary to consider values of v for which $\partial P/\partial v = 0$. It will be shown that in such cases the function $W(N)$ given in (8.39) will no longer have a sharp maximum; the equation of state as given by the recipe in the grand canonical ensemble nevertheless still agrees with that given by the recipe in the canonical ensemble. In this sense the two ensembles are always equivalent.

Physically the values of v for which $\partial P/\partial v = 0$ correspond to the transition region of a first-order phase transition. In this region, (8.49) leads us to expect that the fluctuations of density in a given volume of the system will be large. This is also expected physically, for in such a region the system is composed of two or more phases of *different* densities. Therefore the number of particles in any given volume can have a whole range of values, depending on the amounts of each phase present. At the critical point of a gas-liquid system fluctuations in density are also expected to be large, because throughout the system molecules are spontaneously forming large clusters and breaking up. It is clear that under these conditions the grand canonical ensemble must continue to yield thermodynamic predictions that are in agreement with those obtained by the canonical ensemble. Otherwise the validity of either as a description of matter would be in doubt, for it is a basic experimental fact that we can obtain the same thermodynamic information whether we look at the whole system or at only a subvolume of the system.

The mathematical questions that we try to answer are as follows. Suppose $Q_N(V, T)$ is given, and we wish to calculate

$$\mathcal{Q}(z, V, T) \equiv \sum_{N'=0}^{\infty} z^{N'} Q_{N'}(V, T) \tag{8.52}$$

for given values of z, V, and T.

(*a*) For a given value of z is the following true for some N?

$$\mathcal{Q}(z, V, T) \approx z^N Q_N(V, T) \tag{8.53}$$

(*b*) Does there always exist a value of z for which N has any given positive value?

The answers are obviously no, if $Q_N(V, T)$ is *any* function of N, V, T. We are only interested, however, in the answers when $Q_N(V, T)$ is the partition function of a physical system. Thus we must first make some assumptions about $Q_N(V, T)$.

In order to incorporate the salient features of a physical system into our considerations, and yet keep the mathematics simple, we assume that we are dealing with a system

(*a*) whose molecules interact through an intermolecular potential that contains a hard-sphere repulsion of finite diameter plus a finite potential of finite range, and

(*b*) whose Helmholtz free energy has the form

$$A(N, V) \equiv -\frac{1}{\beta}\log Q_N(V) = -\frac{V}{\beta}f(v) \tag{8.54}$$

where $v \equiv V/N$, $\beta = 1/kT$, and $f(v)$ is finite. The temperature shall be fixed throughout our discussions and will not be displayed unless necessary. The function $f(v)$ is related to the pressure $P(v)$ of the canonical ensemble by

$$f(v) = \frac{1}{v}\int_{v_0}^{v} dv'\beta P(v') \tag{8.55}$$

where the integration is carried out along an isotherm and v_0 is an arbitrary constant corresponding to an arbitrary additive constant in the Helmholtz free energy.

(*c*) We further assume that $f(v)$ is such that

$$\frac{\partial P}{\partial v} \le 0 \tag{8.56}$$

This immediately implies that

$$\frac{\partial^2 f(v)}{\partial(1/v)^2} \le 0 \tag{8.57}$$

With these assumptions the grand partition function may be written in the form

$$\mathcal{Q}(z, V) = \sum_{N=0}^{\infty} \exp\left[V\phi\left(\frac{V}{N}, z\right)\right] \tag{8.58}$$

where z is an arbitrary fixed number and

$$\phi(v, z) \equiv f(v) + \frac{1}{v}\log z \tag{8.59}$$

Using (8.55), we obtain

$$\phi(v, z) = \frac{1}{v}\log z + \frac{1}{v}\int_{v_0}^{v} dv'\beta P(v') \tag{8.60}$$

By (8.57), we have $\partial^2\phi/\partial(1/v)^2 \le 0$, or

$$\frac{\partial^2\phi}{\partial v^2} + \frac{2}{v}\frac{\partial\phi}{\partial v} \le 0 \tag{8.61}$$

We now calculate the grand partition function. For a fixed volume V the partition function $Q_N(V)$ vanishes whenever

$$N > N_0(V)$$

where $N_0(V)$ is the maximum number of particles that can be accommodated in the volume V, such that no two particles are separated by a distance less than the diameter of the hard sphere in the interparticle potential. Therefore $\mathcal{Q}(z, V)$ is a polynomial of degree $N_0(V)$. For large V it is clear that

$$N_0(V) = aV \tag{8.62}$$

where a is a constant. Let the largest value among the terms in this polynomial be $\exp [V\phi_0(z)]$, where

$$\phi_0(z) = \text{Max}\left[\phi\left(\frac{V}{N}, z\right)\right] \qquad (N = 0, 1, 2, \ldots) \tag{8.63}$$

Then the following inequality holds:

$$e^{V\phi_0(z)} \leq \mathcal{Q}(z, v) \leq N_0(V)e^{V\phi_0(z)}$$

Using (8.62) we obtain

$$e^{V\phi_0(z)} \leq \mathcal{Q}(z, V) \leq aVe^{V\phi_0(z)}$$

or

$$\phi_0(z) \leq \frac{1}{V}\log \mathcal{Q}(z, V) \leq \phi_0(z) + \frac{\log (aV)}{V} \tag{8.64}$$

Therefore

$$\lim_{V\to\infty} \frac{1}{V}\log \mathcal{Q}(z, V) = \phi_0(z) \tag{8.65}$$

This shows that (8.50) is true.

Let $\bar{v}$ be a value of v at which $\phi(v, z)$ assumes its largest possible value. Since $\phi(v, z)$ is differentiable, $\bar{v}$ is determined by the conditions

$$\left(\frac{\partial \phi}{\partial v}\right)_{v=\bar{v}} = 0 \tag{8.66}$$

$$\left(\frac{\partial^2 \phi}{\partial v^2}\right)_{v=\bar{v}} \leq 0 \tag{8.67}$$

By virtue of (8.61) the first condition implies the second. Therefore $\bar{v}$ is determined by (8.66) alone. By (8.59) and (8.55) we may rewrite it in the form

$$\int_{v_0}^{\bar{v}} dv'P(v') - \bar{v}P(\bar{v}) = -kT \log z$$

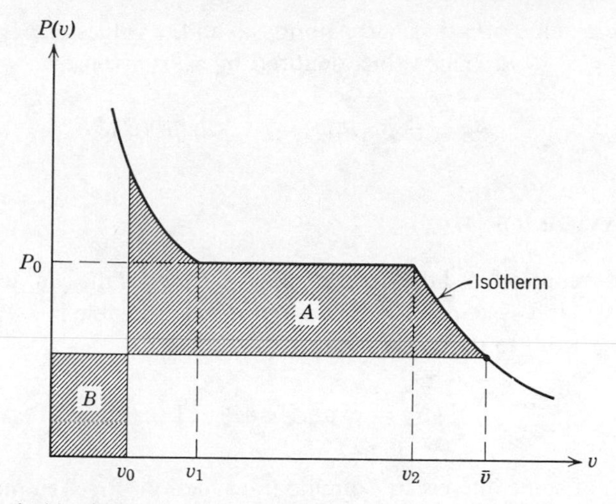

Fig. 8.2. Typical isotherm of a substance in the transition region of a first-order phase transition.

or

$$\left[\int_{v_0}^{\bar{v}} dv' P(v') - (\bar{v} - v_0)P(\bar{v})\right] - v_0 P(\bar{v}) = -kT \log z \qquad (8.68)$$

A geometrical representation of this condition is shown in Fig. 8.2. The value of $\bar{v}$ is such that the difference between the area of the region A and that of the region B is numerically equal to $-kT \log z$. The result is shown in Fig. 8.3. It is seen that to every value of $\bar{v}$ greater than the close-packing volume there corresponds a value of z. This answers question (*b*) in the affirmative.

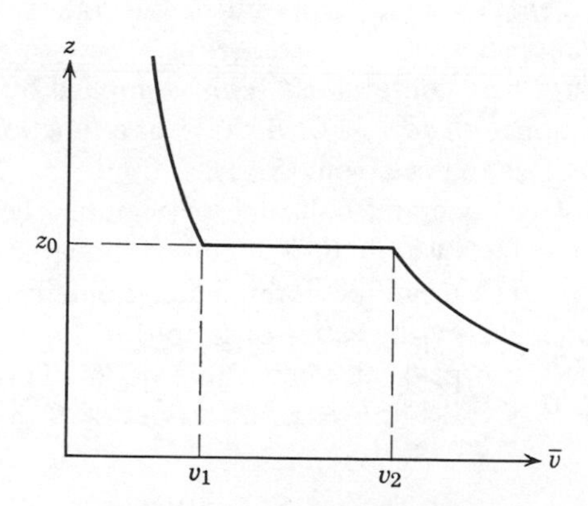

Fig. 8.3. z as a function of $\bar{v}$.

There is a value of z that corresponds to all the values of $\bar{v}$ lying in the interval $v_1 \leq \bar{v} \leq v_2$. This value, denoted by z_0, is given by

$$\log z_0 = \beta v_1 P(v_1) - \int_{v_0}^{v_1} dv' \beta P(v') \tag{8.69}$$

8.6 BEHAVIOR OF $W(N)$

In (8.39) we introduced the quantity $W(N)$, which is the (un-normalized) probability that a system in the grand canonical ensemble has N particles. Comparing (8.39) to (8.58) we see that

$$W(N) = \exp\left[V\phi\left(\frac{V}{N}, z\right)\right] \tag{8.70}$$

Hence it is of some interest to examine the function $\phi(v, z)$ in more detail. Suppose $P(v)$ has the form shown in the $P - v$ diagram of Fig. 8.2. For values of v lying in the range $v_1 \leq v \leq v_2$, P has the constant value P_0. For this range of v we have

$$\phi(v, z) = \frac{1}{v}\left[\log z + \int_{v_0}^{v_1} dv' \beta P(v') - \beta P_0 v_1\right] + \beta P_0$$

which is the same as

$$\phi(v, z) = \frac{1}{v} \log\left(\frac{z}{z_0}\right) + \beta P_0 \qquad (v_1 \leq v \leq v_2) \tag{8.71}$$

where z_0 is defined by (8.69). Hence we can immediately make a qualitative sketch of a family of curves, one for each z, for the function $\phi(v, z)$ in the interval $v_1 \leq v \leq v_2$. The result is shown in Fig. 8.4.

To deduce the behavior of $\phi(v, z)$ outside the interval just discussed we use the following facts:

(*a*) $\partial\phi/\partial v$ is everywhere continuous. This is implied by (8.60).

(*b*) $\partial\phi/\partial v = 0$ implies $\partial^2\phi/\partial v^2 \leq 0$. That is, as a function of v, ϕ cannot have a minimum. This follows from (8.61).

(*c*) For $z \neq z_0$, ϕ has one and only one maximum. This follows from (*b*). Guided by these facts we obtain the curves shown in Fig. 8.4.

The behavior of $W(N)$ can be immediately obtained from that of $\phi(v, z)$. It is summarized by the series of graphs in Fig. 8.5. For $z \neq z_0$, $W(N)$ has a single sharp peak at some value of N. This peak becomes infinitely sharp as $V \to \infty$. For $z = z_0$, all values of N in the interval

$$v_1 \leq \frac{N}{V} \leq v_2 \tag{8.72}$$

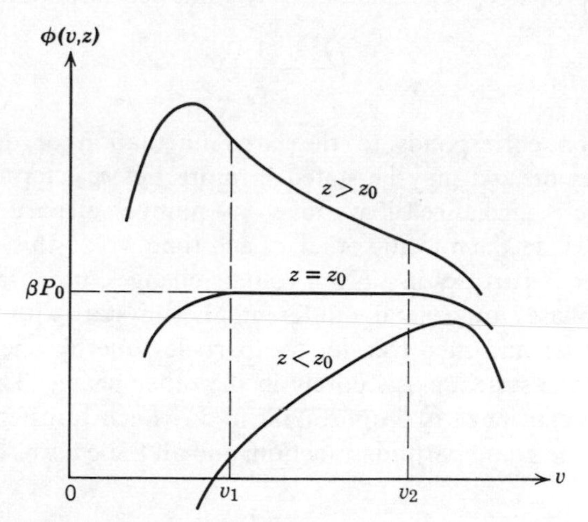

Fig. 8.4. Qualitative form of $\phi(v, z)$ for a physical substance.

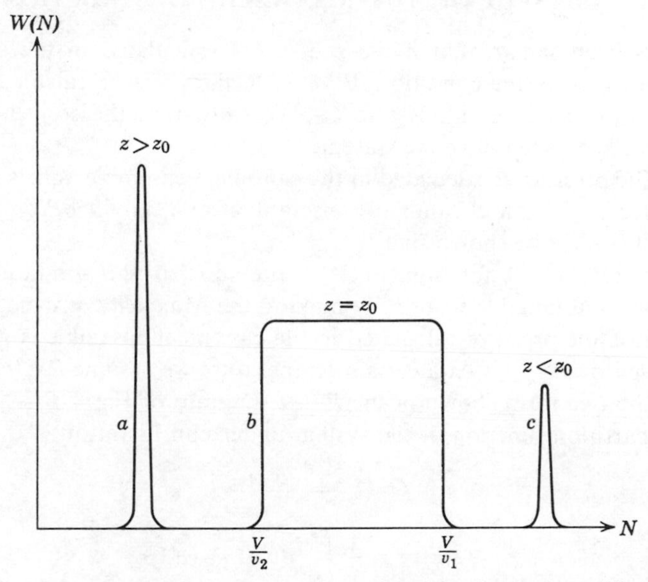

Fig. 8.5. The function $W(N)$ for three different fugacities (hence three different densities). For curves a and c the system is in a single pure phase. For curve b the system is undergoing a first-order phase transition.

are equally probable. The number of N values corresponding to (8.72) is

$$\left(\frac{1}{v_1} - \frac{1}{v_2}\right) V \tag{8.73}$$

This situation corresponds to the large fluctuation of density in the transition region and may be stated in more physical terms as follows: The pressure is unchanged if we take any number of particles from one phase and deliver them to the other. Each time we do this, however, the total number of particles in a given volume changes, because the densities of the two phases are generally different. Let us start with the system in one pure phase and then transfer the particles one by one to the other phase, until the system exists purely in the other phase. The number of transfers we can make is proportional to V. Each transfer corresponds to a term in the grand partition function, and all these terms have the same value.

8.7 THE MEANING OF THE MAXWELL CONSTRUCTION

It has been shown that if the pressure P calculated in the canonical ensemble satisfies the condition $\partial P/\partial v \leq 0$, the pressure calculated in the grand canonical ensemble is also P. We show that the converse is also true. We shall then have the statement

(*a*) The pressure P calculated in the canonical ensemble agrees with that calculated in the grand canonical ensemble if and only if $\partial P/\partial v \leq 0$.

It will further be shown that

(*b*) If $\partial P/\partial v > 0$ for some v, the pressure in the grand canonical ensemble is obtainable from P by making the Maxwell construction.

Suppose the pressure calculated in the canonical ensemble is given and is denoted by $P_{\text{can}}(v)$. At a certain temperature we assume $P_{\text{can}}(v)$ to have the qualitative form shown in the $P - v$ diagram of Fig. 8.6.

The partition function of the system under consideration is

$$Q_N(V) = e^{VF(v)} \tag{8.74}$$

where

$$F(v) = \frac{1}{v}\int_{v_0}^{v} dv'\beta P_{\text{can}}(v') \tag{8.75}$$

It is easily seen that

$$\beta P_{\text{can}}(v) = F(v) + v\,\frac{\partial F(v)}{\partial v} \tag{8.76}$$

Let

$$\Phi(v, z) \equiv F(v) + \frac{1}{v}\log z \tag{8.77}$$

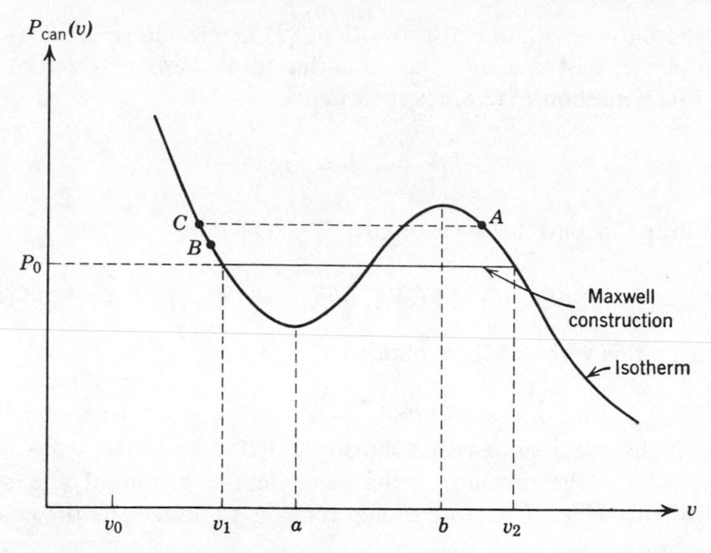

Fig. 8.6. Isotherm with $\partial P_{\mathrm{can}}/\partial v > 0$ for v lying in the range $a < v < b$.

It is easily verified that

$$\frac{\partial^2 \Phi}{\partial v^2} + \frac{2}{v}\frac{\partial \Phi}{\partial v} = \frac{\beta}{v}\frac{\partial P_{\mathrm{can}}}{\partial v} \begin{cases} > 0 & (a < v < b) \\ \leq 0 & (\text{otherwise}) \end{cases} \tag{8.78}$$

To calculate the grand partition function we recall that the derivation of (8.65) is independent of the sign of $\partial P/\partial v$. Hence, in analogy with (8.65), we have in the present case

$$\lim_{V \to \infty} \frac{1}{V} \log \mathcal{Q}(z, V) = \Phi(\bar{v}, z) \tag{8.79}$$

where

$$\Phi(\bar{v}, z) = \mathrm{Max}\,[\Phi(v, z)] \tag{8.80}$$

This determines $\bar{v}$ in terms of z, or vice versa. The pressure in the grand canonical ensemble, denoted by $P_{\mathrm{gr}}(\bar{v})$, is given by

$$\beta P_{\mathrm{gr}}(\bar{v}) = \Phi(\bar{v}, z) \tag{8.81}$$

From (8.77) and (8.75) we see that both Φ and $\partial\Phi/\partial v$ are continuous functions of v. Hence (8.80) is equivalent to the conditions

$$\left(\frac{\partial \Phi}{\partial v}\right)_{v=\bar{v}} = 0$$
$$\left(\frac{\partial^2 \Phi}{\partial v^2}\right)_{v=\bar{v}} \leq 0 \tag{8.82}$$

with the following additional rule: If (8.82) determines more than one value of $\bar{v}$, we must take only the value that gives the largest $\Phi(\bar{v}, z)$.

The first condition of (8.82) is the same as

$$\left(v^2 \frac{\partial F}{\partial v}\right)_{v=\bar{v}} = \log z \tag{8.83}$$

Substituting this into (8.76) we obtain

$$\beta P_{\text{can}}(\bar{v}) = F(\bar{v}) + \frac{1}{\bar{v}} \log z = \Phi(\bar{v}, z) \tag{8.84}$$

Comparing this with (8.81) we obtain

$$P_{\text{can}}(\bar{v}) = P_{\text{gr}}(\bar{v}) \tag{8.85}$$

That is, if there is a value $\bar{v}$ that satisfies (8.82), then at this value of the specific volume the pressure is the same in the canonical and grand canonical ensemble. Therefore it only remains to investigate the possible values of $\bar{v}$.

It is obvious that $\bar{v}$ can never lie between the values a and b shown in Fig. 8.6, because, as we can see from (8.78), in that region $\partial\Phi/\partial v = 0$ implies $\partial^2\Phi/\partial v^2 > 0$, in contradiction to (8.82). On the other hand, outside this region, $\partial\Phi/\partial v = 0$ implies $\partial^2\Phi/\partial v^2 \leq 0$. Hence the first condition of (8.82) alone determines $\bar{v}$. Using (8.75) we can write this condition in the form

$$\int_{v_0}^{\bar{v}} dv' P_{\text{can}}(v') - \bar{v} P_{\text{can}}(\bar{v}) = -kT \log z \tag{8.86}$$

There is a value of z, denoted by z_0, at which (8.86) has two roots v_1 and v_2 for which $\Phi(v_1, z) = \Phi(v_2, z)$. The conditions for this to be so are that

$$-kT \log z_0 = \int_{v_0}^{v_1} dv'\, P_{\text{can}}(v') - v_1 P_{\text{can}}(v_1) = \int_{v_0}^{v_2} dv'\, P_{\text{can}}(v') - v_2 P_{\text{can}}(v_2)$$

$$\Phi(v_1, z_0) = \Phi(v_2, z_0) \tag{8.87}$$

The second condition is equivalent to $P_{\text{can}}(v_1) = P_{\text{can}}(v_2)$, by virtue of (8.84). Combining these conditions, we obtain

$$\int_{v_1}^{v_2} dv'\, P_{\text{can}}(v') = (v_2 - v_1) P_{\text{can}}(v_1) \tag{8.88}$$

which means that v_1 and v_2 are the end points of a Maxwell construction on P_{can}, as shown in Fig. 8.6.

In general we can find z as a function of $\bar{v}$ by solving (8.86) graphically, in a manner similar to that used in the last section for (8.68). The result

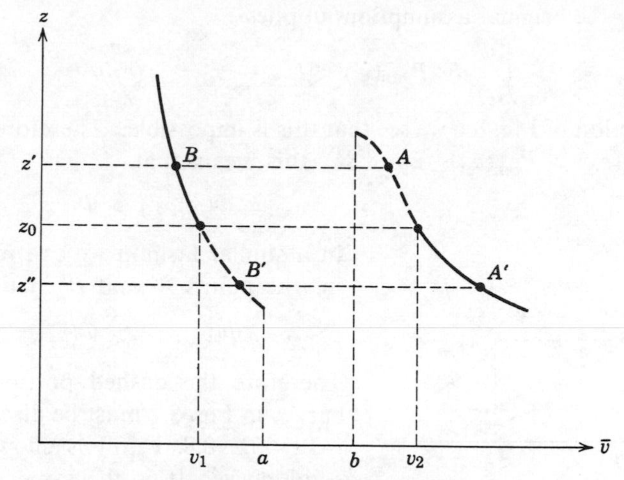

Fig. 8.7. z as a function of $\bar{v}$.

is qualitatively sketched in Fig. 8.7. As explained before the interval $a < \bar{v} < b$ must be excluded. By definition of the Maxwell construction the portions of the curves outside the interval $v_1 \leq \bar{v} \leq v_2$, shown in solid lines in Fig. 8.7, coincide with the corresponding portions in Fig. 8.3. We need to discuss further only the dashed portions of the curves.

Consider the points A and B in Fig. 8.7. Let their volumes be respectively v_A and v_B and let their common z value be z'. The fact that they are both solutions of (8.86) means that the function $\Phi(v, z')$ has two maxima, located respectively at $v = v_A$ and $v = v_B$. These maxima cannot be of the same height, because that would mean that v_A and v_B are respectively v_2 and v_1; which they are not. To determine which maximum is higher we note that by (8.75), (8.84), and the fact that z' is common to both,

$$\int_{v_B}^{v_A} dv' \, P_{\text{can}}(v') = v_A P_{\text{can}}(v_A) - v_B P_{\text{can}}(v_B) \tag{8.89}$$

Suppose $P_{\text{can}}(v_B) < P_{\text{can}}(v_A)$. Consider the point C indicated in Fig. 8.6. By inspection of Fig. 8.6 we see that

$$\int_{v_C}^{v_A} dv' \, P_{\text{can}}(v') < (v_A - v_C) P_{\text{can}}(v_A)$$

Subtracting (8.89) from this inequality, we obtain

$$\int_{v_C}^{v_B} dv' \, P_{\text{can}}(v') < v_B P_{\text{can}}(v_B) - v_C P_{\text{can}}(v_A)$$

which, by the original assumption, implies

$$\int_{v_C}^{v_B} dv' \, P_{\text{can}}(v') < (v_B - v_C)P_{\text{can}}(v_B)$$

By inspection of Fig. 8.6 we see that this is impossible. Therefore we must have $P_{\text{can}}(v_B) > P_{\text{can}}(v_A)$. By (8.84), this means that

$$\Phi(v_B, z') > \Phi(v_A, z')$$

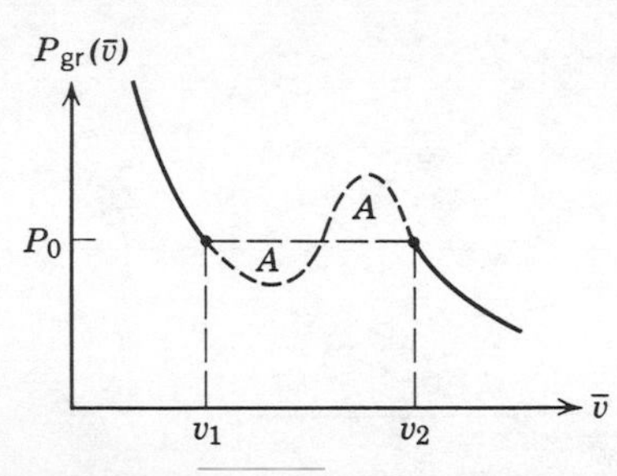

Fig. 8.8. The pressure in the grand canonical ensemble (solid lines).

In a similar fashion we can prove that, for the points A' and B' in Fig. 8.7,

$$\Phi(v_{A'}, z'') > \Phi(v_{B'}, z'')$$

Therefore the dashed portions of the curves in Fig. 8.7 must be discarded.

In Fig. 8.8, $P_{\text{gr}}(\bar{v})$ is shown as the solid curve. It is the same as $P_{\text{can}}(\bar{v})$ except that the portion between v_1 and v_2 is missing because there is no z that will give a $\bar{v}$ lying in that interval. In other words, in the grand canonical ensemble the system cannot have a volume in that interval. We can, however, fill in a horizontal line at P_0 by the usual arguments, namely, that since the systems at v_1 and v_2 have the same temperature, pressure, and chemical potential, a system at v_1 can coexist with a system at v_2 with any relative amount of each present.

It is an experimental fact that $\partial P/\partial v \le 0$. Thus it appears that the situation discussed in this section should never happen. The quantity P_{can} is the result of a calculation, however. Therefore whether or not $\partial P_{\text{can}}/\partial v \le 0$ is a mathematical question independent of experiments. In Chapter 15 we show that if P_{can} is calculated exactly with realistic intermolecular potentials then $\partial P_{\text{can}}/\partial v \le 0$. This statement is known as Van Hove's theorem, which shows that in principle the Maxwell construction need never be invoked. We rarely, however, make exact calculations (in fact, for physical problems, never). In an approximate calculation it is possible that we may obtain the result $\partial P_{\text{can}}/\partial v > 0$. In that event, if the same approximate calculation is to be done in the grand canonical ensemble, we can obtain the result simply by making the Maxwell construction.

PROBLEMS

8.1. (*a*) Obtain the pressure of a classical ideal gas as a function of N, V, and T, by calculating the partition function.
(*b*) Obtain the same by calculating the grand partition function.

8.2. Consider a classical system of N noninteracting diatomic molecules enclosed in a box of volume V at temperature T. The Hamiltonian for a *single* molecule is taken to be

$$H(\mathbf{p}_1, \mathbf{p}_2, \mathbf{r}_1, \mathbf{r}_2) = \frac{1}{2m}(p_1^2 + p_2^2) + \tfrac{1}{2}K\,|\mathbf{r}_1 - \mathbf{r}_2|^2$$

where $\mathbf{p}_1$, $\mathbf{p}_2$, $\mathbf{r}_1$, $\mathbf{r}_2$, are the momenta and coordinates of the two atoms in a molecule. Find

(*a*) the Helmholtz free energy of the system;
(*b*) the specific heat at constant volume;
(*c*) the mean square molecular diameter $\langle|\mathbf{r}_1 - \mathbf{r}_2|^2\rangle$.

8.3. Repeat the last problem, using the Hamiltonian

$$H(\mathbf{p}_1, \mathbf{p}_2, \mathbf{r}_1, \mathbf{r}_2) = \frac{1}{2m}(p_1^2 + p_2^2) + \epsilon\,|r_{12} - r_0|$$

where ϵ and r_0 are given positive constants and $r_{12} \equiv |\mathbf{r}_1 - \mathbf{r}_2|$.
Answer:

$$\frac{C_V}{Nk} = 6 - \frac{x^2[2(x^2 - 2) + (x + 2)^2 e^{-x}]}{(x^2 + 2 - e^{-x})^2} \qquad (x \equiv \epsilon r_0/kT)$$

8.4. Prove (8.27) with the help of the following hints.
(*a*)

$$H(p_1, q_1, N_1) + H(p_2, q_2, N_2) = H(p, q, N)$$

by the assumption that surface interactions can be neglected. Since the identities of the N_1 particles are not specified, there are $N!/N_1!\,(N - N_1)!$ ways to choose the set of N_1 particles.
(*b*)

$$Q_N(V, T) = \sum_{N_1=0}^{N} Q_{N_1}(V_1, T) Q_{N-N_1}(V - V_1, T)$$

8.5. From the equations (8.38) and (8.36), namely

$$\begin{cases} A = NkT \log z - kT \log \mathcal{Q}(z, V, T) \\ N = z\dfrac{\partial}{\partial z} \log \mathcal{Q}(z, V, T) \end{cases}$$

show that

$$\left(\frac{\partial A}{\partial N}\right)_{V,T} = kT \log z \equiv \mu \qquad \text{(chemical potential)}$$

8.6. (*a*) Show that in the grand canonical ensemble the mean square fluctuation of the number of particles can be expressed in the form

$$\langle N^2\rangle - \langle N\rangle^2 = z\frac{\partial}{\partial z} z \frac{\partial}{\partial z} \log \mathcal{Q}(z, V, T)$$

(*b*) Show that the statement in (*a*) is the same as (8.49), by first noticing that the right-hand side is $[\partial(\log z)/\partial N]^{-1}$, and then using the result of Problem 8.5.

8.7. Prove *Van Leeuwen's Theorem*: The phenomenon of diamagnetism does not exist in classical physics.

The following hints may be helpful:

(*a*) If $H(\mathbf{p}_1, \ldots, \mathbf{p}_N; \mathbf{q}_1, \ldots, \mathbf{q}_N)$ is the Hamiltonian of a system of charged particles in the absence of an external magnetic field, then $H[\mathbf{p}_1 - (e/c)\mathbf{A}_1, \ldots, \mathbf{p}_N - (e/c)\mathbf{A}_N; \mathbf{q}_1, \ldots, \mathbf{q}_N]$ is the Hamiltonian of the same system in the presence of an external magnetic field $\mathbf{B} = \nabla \times \mathbf{A}$, where $\mathbf{A}_i$ is the value of $\mathbf{A}$ at the position $\mathbf{q}_i$.

(*b*) The induced magnetization of the system along the direction of B is given by

$$M = \left\langle -\frac{\partial H}{\partial B} \right\rangle = kT \frac{\partial}{\partial B} \log Q_N$$

where H is the Hamiltonian in the presence of $\mathbf{B}$, $B = |\mathbf{B}|$, and Q_N is the partition function of the system in the presence of $\mathbf{B}$.

8.8. *Langevin's Theory of Paramagnetism:* Consider a system of N atoms, each of which has an intrinsic magnetic moment of magnitude μ. The Hamiltonian in the presence of an external magnetic field $\mathbf{B}$ is

$$H(p, q) - \mu B \sum_{i=1}^{N} \cos \alpha_i$$

where $H(p, q)$ is the Hamiltonian of the system in the absence of an external magnetic field, and α_i is the angle between $\mathbf{B}$ and the magnetic moment of the ith atom. Show that

(*a*) The induced magnetic moment is

$$M = N\mu\left(\coth \theta - \frac{1}{\theta}\right) \qquad (\theta \equiv \mu B/kT)$$

(*b*) The magnetic susceptibility per atom is

$$\chi = \frac{\mu^2}{kT}\left(\frac{1}{\theta^2} - \operatorname{csch}^2 \theta\right)$$

(*c*) At high temperatures χ satisfies Curie's law, namely $\chi \propto T^{-1}$. Find the proportionality constant, which is called Curie's constant.

8.9. *Imperfect Gas.* Consider a system of N molecules ($N \to \infty$) contained in a box of volume V ($V \to \infty$). The Hamiltonian of the system is

$$H = \sum_{i=1}^{N} \frac{p_i^{\,2}}{2m} + \sum_{i<j} v_{ij}$$

$$v_{ij} = v(|\mathbf{r}_i - \mathbf{r}_j|)$$

where $\mathbf{p}_i$ and $\mathbf{r}_i$ are respectively the momentum and the position of the ith molecule. The intermolecular potential $v(r)$ has the qualitative form shown in the accompanying figure. Let

$$f_{ij} \equiv f(|\mathbf{r}_i - \mathbf{r}_j|)$$

$$f(r) \equiv e^{-\beta v(r)} - 1$$

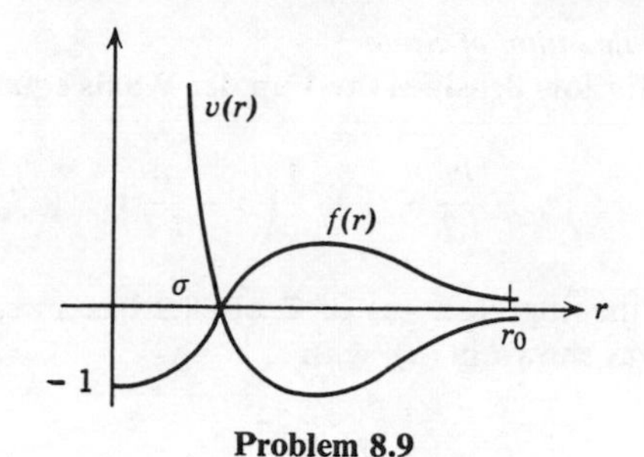

Problem 8.9

A sketch of $f(r)$ is also shown in the same figure.

(*a*) Show that the equation of state of the system is

$$\frac{Pv}{kT} = 1 + v\frac{\partial Z(v, T)}{\partial v}$$

where $v \equiv V/N$ and

$$Z(v, T) \equiv \frac{1}{N}\log\left[\frac{1}{V^N}\int d^3r_1 \cdots d^3r_N \prod_{i<j}(1 + f_{ij})\right]$$

(*b*) By expanding the product $\Pi(1 + f_{ij})$, show that

$$Z(v, T) = \frac{1}{N}\log\left[\frac{1}{V^N}\int d^3r_1 \cdots d^3r_N\left(1 + \sum_{i<j} f_{ij} + \cdots\right)\right]$$

$$= \log\left[1 + \frac{N}{2v}\int d^3r\, f(r) + \cdots\right]^{1/N}$$

(*c*) Show that at low densities, i.e.,

$$r_0^3/v \ll 1$$

it is a good approximation to retain only the first two terms in the series appearing in the expression $Z(v, T)$. Hence the equation of state is approximately given by

$$\frac{Pv}{kT} \approx 1 - \frac{1}{2v}\int_0^\infty dr\, 4\pi r^2 f(r)$$

The coefficient of $1/v$ is called the *second virial coefficient*.

Note: (i) Retaining the first two terms in the series appearing in $Z(v, T)$ is a good approximation because $Z(v, T)$ is the logarithm of the Nth root of the series. The approximation is certainly invalid for the series itself.

(ii) If all terms in the expansion of $\Pi(1 + f_{ij})$ were kept, we would have obtained a systematic expansion of Pv/kT in powers of $1/v$. Such an expansion is known as the *virial expansion*.

(iii) The complete virial expansion is difficult to obtain by the method described in this problem. It is obtained in Chapter 14 via the grand canonical ensemble. See (14.27) and (14.30).

8.10. *Van der Waals Equation of State*

(*a*) Show that for low densities the Van der Waals equation of state (2.28) reduces to

$$\frac{Pv}{kT} \approx 1 + \frac{1}{v}\left(b' - \frac{a'}{kT}\right)$$

(*b*) Show that the imperfect gas of Prob. 8.9 has an equation of state of the same form as shown in (*a*), with

$$b' = \frac{2\pi}{3}\sigma^3$$

$$a' = -2\pi kT \int_\sigma^\infty dr\, r^2(1 - e^{-\beta v(r)})$$

8.11. The equation of state for an N_2 gas can be written in the form

$$PV/NkT = 1 + a_2(T)(N/V)$$

for low densities. The second virial coefficient $a_2(T)$ has been measured as a function of temperature and is given in the accompanying table. Assume that the intermolecular potential $v(r)$ between N_2 molecules has the form shown in the accompanying sketch. From the data given, determine what you consider to be the best choices for the constants σ, r_0, and ϵ.

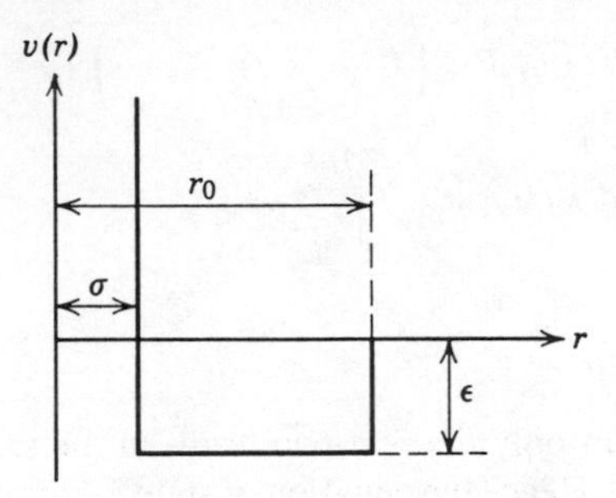

Problem 8.11

Temperature, °K	$a_2(T)$, °K/atm
100	-1.80
200	-4.26×10^{-1}
300	-5.49×10^{-2}
400	1.12×10^{-1}
500	2.05×10^{-1}

chapter 9

QUANTUM STATISTICAL MECHANICS

9.1 THE POSTULATES OF QUANTUM STATISTICAL MECHANICS

All systems in nature obey quantum mechanics. In quantum mechanics an observable of a system is associated with a hermitian operator, which operates on a Hilbert space. A state of the system is a vector $| \Psi \rangle$ in the same Hilbert space. If $| q \rangle$ is an eigenvector of the position operators of all the particles in the system, then $\langle q | \Psi \rangle \equiv \Psi(q)$ is the wave function of the system in the state $| \Psi \rangle$. The wave function furnishes a complete description of the state.

At any instant of time the wave function Ψ of a truly isolated system may be expressed as a linear superposition of a complete orthonormal set of stationary wave functions $\{\Phi_n\}$:

$$\Psi = \sum_n c_n \Phi_n \tag{9.1}$$

where c_n is a complex number and is generally a function of time. The index n stands for a set of quantum numbers, which are eigenvalues of certain chosen dynamical operators of the system. The square modulus $|c_n|^2$ is the probability that a measurement performed on the system will find it to have the quantum numbers n.

In statistical mechanics we always deal with systems that interact with the external world. Here we can regard the system plus the external world as a truly isolated system. The wave function Ψ for this whole system will depend on both the coordinates of the system under consideration and the coordinates of the external world. If $\{\Phi_n\}$ denotes a complete set of orthonormal stationary wave functions of the system, then Ψ is still formally given by (9.1), but c_n is to be interpreted as a wave function of the external world. It depends on the coordinates of the external world as well as on the time.

Suppose $\mathcal{O}$ is an operator corresponding to an observable of the system. According to the rules of quantum mechanics, the average result of a large number of measurements of this observable is instantaneously given by the expectation value

$$\frac{(\Psi, \mathcal{O}\Psi)}{(\Psi, \Psi)} = \frac{\sum_n \sum_m (c_n, c_m)(\Phi_n, \mathcal{O}\Phi_m)}{\sum_n (c_n, c_n)} \tag{9.2}$$

where (c_n, c_m), the scalar product of the nth and the mth wave function of the external world, is a function of time. The denominator in (9.2), being identical with (Ψ, Ψ), is independent of time, because the Hamiltonian of the system plus external world is hermitian. When we actually measure an observable in the laboratory, we measure not its instantaneous value but a time average. Thus the directly measurable quantity is not (9.2) but the following quantity:

$$\langle \mathcal{O} \rangle \equiv \frac{\overline{(\Psi, \mathcal{O}\Psi)}}{(\Psi, \Psi)} = \frac{\sum_n \sum_m \overline{(c_n, c_m)}(\Phi_n, \mathcal{O}\Phi_m)}{\sum_n \overline{(c_n, c_n)}} \tag{9.3}$$

where $\overline{(c_n, c_m)}$ is the average of (c_n, c_m) over a time interval that is short compared to the resolving time of the measuring apparatus but long compared to molecular times (e.g., collision times or periods of molecular motion). We note that $\sum_n \overline{(c_n, c_n)}$ is identical with $\sum_n (c_n, c_n)$, because the latter is independent of time.

The postulates of quantum statistical mechanics are postulates concerning the coefficients $\overline{(c_n, c_m)}$, when (9.3) refers to a macroscopic observable of a macroscopic system in thermodynamic equilibrium.

For definiteness, we consider a macroscopic system which, although not truly isolated, interacts so weakly with the external world that its energy is approximately constant. Let the number of particles in the system be N and the volume of the system be V, and let its energy lie between E and $E + \Delta$ ($\Delta \ll E$). Let H be the Hamiltonian of the system. For such a system it is convenient (but not necessary) to choose a standard set of complete orthonormal wave functions $\{\Phi_n\}$ such that every Φ_n is a wave function for N particles contained in the volume V and is an eigenfunction of H with the eigenvalue E_n:

$$H\Phi_n = E_n\Phi_n \tag{9.4}$$

The postulates of quantum statistical mechanics are the statements (9.5) and (9.6).

Postulate of Equal a Priori Probability

$$\overline{(c_n, c_n)} = \begin{cases} 1 & (E < E_n < E + \Delta) \\ 0 & \text{(otherwise)} \end{cases} \tag{9.5}$$

Postulate of Random Phases

$$\overline{(c_n, c_m)} = 0 \qquad (n \neq m) \tag{9.6}$$

As a consequence of these postulates we may *effectively* regard the wave function of the system as given by

$$\Psi = \sum_n b_n \Phi_n \tag{9.7}$$

where

$$|b_n|^2 = \begin{cases} 1 & (E < E_n < E + \Delta) \\ 0 & \text{(otherwise)} \end{cases} \tag{9.8}$$

and where the phases of the complex numbers $\{b_n\}$ are random numbers. In this manner the effect of the external world is taken into account in an average way. The observed value of an observable associated with the operator $\mathcal{O}$ is then given by

$$\langle \mathcal{O} \rangle = \frac{\sum_n |b_n|^2 (\Phi_n, \mathcal{O}\,\Phi_n)}{\sum_n |b_n|^2} \tag{9.9}$$

It should be emphasized that for (9.7) and (9.8) to be effectively valid the system must interact with the external world. Otherwise the postulate of random phases is false. By the randomness of the phases we mean no more and no less than the absence of interference of probability amplitudes, as expressed by (9.9). For a truly isolated system such a circumstance may be true at an instant, but it cannot be true for all times.

The postulate of random phases implies that the state of a system in equilibrium may be regarded as an *incoherent* superposition of eigenstates. It is possible to think of the system as one member of an infinite collection of systems, each of which is in an eigenstate whose wave function is Φ_n. Since these systems do not interfere with one another, it is possible to form a mental picture of each system *one at a time*. We call this mental picture an *ensemble* of systems. The ensemble defined by the previous postulates is called the *microcanonical ensemble*.

The postulates of quantum statistical mechanics are to be regarded as working hypotheses whose justification lies in the fact that they lead to results in agreement with experiments. Such a point of view is not entirely satisfactory, because these postulates cannot be independent of, and should be derivable from, the quantum mechanics of molecular systems. A rigorous derivation is at present lacking. We return to this subject very briefly at the end of this chapter.

We should recognize that the postulates of quantum statistical mechanics, even if regarded as phenomenological statements, are more fundamental than the laws of thermodynamics. The reason is twofold. First, the postulates of quantum statistical mechanics not only imply the laws of thermodynamics; they also lead to definite formulas for all the thermodynamic functions of a given substance. Second, they are more directly related to molecular dynamics than are the laws of thermodynamics.

The concept of an ensemble is a familiar one in quantum mechanics. A trivial example is the description of an incident beam of particles in the theory of scattering. The incident beam of particles in a scattering experiment is composed of many particles, but in the theory of scattering we consider the particles one at a time. That is, we calculate the scattering cross section for a single incident particle and then add the cross sections for all the particles to obtain the physical cross section. Inherent in this method is the assumption that the wave functions of the particles in the incident beam bear no definite phase with respect to one another. What we have described is in fact an ensemble of particles.

A less trivial example is the description of a beam of incident electrons whose spin can be polarized. If an electron has the wave function

$$\left[A\begin{pmatrix}1\\0\end{pmatrix} + B\begin{pmatrix}0\\1\end{pmatrix}\right] e^{i\mathbf{k}\cdot\mathbf{r}}$$

where A and B are definite complex numbers, the electron has a spin pointing in some definite direction. This corresponds to an incident beam of completely polarized electrons. In the cross section calculated with this wave function there will appear interference terms proportional to $A^*B + AB^*$. If we have an incident beam that is partially polarized, we first calculate the cross section with a wave function proportional to $\begin{pmatrix}1\\0\end{pmatrix}$ and then do the same thing for $\begin{pmatrix}0\\1\end{pmatrix}$, adding the two cross sections with appropriate weighting factors. This is equivalent to describing the incident beam by an ensemble of electrons in which the states $\begin{pmatrix}1\\0\end{pmatrix}$ and $\begin{pmatrix}0\\1\end{pmatrix}$ occur with certain relative probabilities.

9.2 DENSITY MATRIX

An ensemble is an incoherent superposition of states. Its relevance to physics has been postulated in the previous section. We note that only the square moduli $|b_n|^2$ appear in (9.9). Hence it should be possible to describe an ensemble in such a way that the random phases of the states never need to be mentioned. Such a goal is achieved by introducing the density matrix.

Before we define the density matrix let us note that an operator is defined when all its matrix elements with respect to a complete set of states are defined. Its matrix elements with respect to any other complete set of states can be found by the well-known rules of transformation theory in quantum mechanics. Therefore, if all the matrix elements of an operator are defined in one representation, the operator is thereby defined in any representation.

We define the density matrix ρ_{mn} corresponding to a given ensemble by

$$\rho_{mn} \equiv (\Phi_n, \rho\Phi_m) \equiv \delta_{mn}\, |b_n|^2 \tag{9.10}$$

where Φ_n and b_n have the same meaning as in (9.7). In this particular representation ρ_{mn} is a diagonal matrix, but in some other representation it need not be. Equation (9.10) also defines the density operator ρ whose matrix elements are ρ_{mn}. The operator ρ operates on state vectors in the Hilbert space of the system under consideration.

In terms of the density matrix, (9.9) can be rewritten in the form

$$\langle \mathcal{O} \rangle = \frac{\sum_n (\Phi_n, \mathcal{O}\, \rho\Phi_n)}{\sum_n (\Phi_n, \rho\Phi_n)} = \frac{Tr(\mathcal{O}\rho)}{Tr\rho} \tag{9.11}$$

The notation TrA denotes the trace of the operator A and is the sum of all the diagonal matrix elements of A in *any* representation. An elementary property of the trace is that

$$Tr(AB) = Tr(BA) \tag{9.12}$$

It follows immediately that TrA is independent of the representation; if TrA is calculated in one representation, its value in another representation is

$$Tr(SAS^{-1}) = Tr(S^{-1}SA) = TrA$$

The introduction of the density matrix merely introduces a notation. It does not introduce new physical content. The usefulness of the density matrix lies solely in the fact that with its help (9.11) is presented in a form that is manifestly independent of the choice of the basis $\{\Phi_n\}$, although this independence is a property that this expectation value always possesses.

The density operator ρ defined by (9.10) contains all the information about an ensemble. It is independent of time if it commutes with the Hamiltonian of the system and if the Hamiltonian is independent of time.

This statement is an immediate consequence of the equation of motion of ρ:

$$i\hbar \frac{\partial \rho}{\partial t} = [H, \rho] \tag{9.13}$$

which is the quantum mechanical version of Liouville's theorem.

Formally we can represent the density operator ρ as

$$\rho = \sum_n |\Phi_n\rangle |b_n|^2 \langle \Phi_n | \tag{9.14}$$

where $|\Phi_n\rangle$ is the state vector whose wave function is Φ_n. To prove (9.14), we verify that it has the matrix elements (9.10):

$$\rho_{mn} \equiv (\Phi_m, \rho\Phi_n) \equiv \langle \Phi_m | \rho | \Phi_n \rangle = \sum_k \langle \Phi_m | \Phi_k \rangle |b_k|^2 \langle \Phi_k | \Phi_n \rangle = \delta_{mn} |b_n|^2$$

(QED)

The time-averaging process through which we averaged out the effect of the external world on the system under consideration may be reformulated in terms of the density matrix.

Formula (9.2) is a general formula for the expectation value of any operator $\mathscr{O}$ with respect to an arbitrary wave function Ψ. It may be trivially rewritten in the form

$$\frac{(\Psi, \mathscr{O}\, \Psi)}{(\Psi, \Psi)} = \frac{\sum_n \sum_m R_{mn} \mathscr{O}_{nm}}{\sum_n R_{nn}} = \frac{Tr(R\mathscr{O})}{TrR}$$

where $R_{nm} \equiv (c_m, c_n) \equiv (\Phi_n, R\Phi_m)$, the last identity being a definition of the operator R, and $\mathscr{O}_{nm} \equiv (\Phi_n, \mathscr{O}\Phi_m)$. Although R may depend on the time, TrR is independent of time. The density operator is the time average of R:

$$\rho \equiv \overline{R}$$

9.3 ENSEMBLES IN QUANTUM STATISTICAL MECHANICS

Microcanonical Ensemble

The density matrix for the microcanonical ensemble in the representation in which the Hamiltonian is diagonal is

$$\rho_{mn} = \delta_{mn} |b_n|^2 \tag{9.15}$$

where

$$|b_n|^2 = \begin{cases} 1 & (E < E_n < E + \Delta) \\ 0 & (\text{otherwise}) \end{cases} \tag{9.16}$$

where $\{E_n\}$ are the eigenvalues of the Hamiltonian. The density operator is

$$\rho = \sum_{E<E_n<E+\Delta} |\Phi_n\rangle\langle\Phi_n| \tag{9.17}$$

The trace of ρ is equal to the number of states whose energy lies between E and $E + \Delta$:

$$Tr\rho = \sum_n \rho_{nn} \equiv \Gamma(E) \tag{9.18}$$

For macroscopic systems the spectrum $\{E_n\}$ almost forms a continuum. For $\Delta \ll E$, we may take

$$\Gamma(E) = \omega(E)\,\Delta \tag{9.19}$$

where $\omega(E)$ is the density of states at energy E. The connection between the microcanonical ensemble and thermodynamics is established by identifying the entropy as

$$S(E, V) = k \log \Gamma(E) \tag{9.20}$$

where k is Boltzmann's constant. This definition is the same as in classical statistical mechanics, except that $\Gamma(E)$ must be calculated in quantum mechanics. From this point on all further developments become exactly the same as in classical statistical mechanics and so they need not be repeated. No Gibbs paradox will result from (9.20) because the correct counting of states is automatically implied by the definition of $\Gamma(E)$ in (9.18).

The only new result following from (9.20) that is not obtainable in classical statistical mechanics is the third law of thermodynamics, which we discuss separately in Sec. 9.4.

Canonical Ensemble

The derivation of the canonical ensemble from the microcanonical ensemble given in Chapter 8 did not make essential use of classical mechanics. That derivation continues to be valid in quantum statistical mechanics, with the trivial change that the integration over Γ-space is replaced by a sum over all the states of the system:

$$\frac{1}{N!\,h^{3N}}\int dp\,dq \to \sum_n \tag{9.21}$$

Thus the canonical ensemble is defined by the density matrix

$$\rho_{mn} = \delta_{mn} e^{-\beta E_n} \tag{9.22}$$

where $\beta = 1/kT$. This result states that at the temperature T the relative probability for the system to have the energy eigenvalue E_n is $e^{-\beta E_n}$, which is called the *Boltzmann factor*. The partition function is given by

$$Q_N(V, T) = Tr\rho = \sum_n e^{-\beta E_n} \tag{9.23}$$

where it must be emphasized that *the sum on the right-hand side is a sum over states and not over energy eigenvalues*. The connection with thermodynamics is the same as in classical statistical mechanics.

The density operator ρ is

$$\rho = \sum_n |\Phi_n\rangle e^{-\beta E_n}\langle\Phi_n| = e^{-\beta H}\sum_n |\Phi_n\rangle\langle\Phi_n|$$

where H is the Hamiltonian operator. Now the operator $\sum_n |\Phi_n\rangle\langle\Phi_n|$ is the identity operator, by the completeness property of eigenstates. Therefore

$$\rho = e^{-\beta H} \tag{9.24}$$

The partition function can be written in the form

$$Q_N(V, T) = Tre^{-\beta H} \tag{9.25}$$

where the trace is to be taken over all states of the system that has N particles in the volume V. This form, which is explicitly independent of the representation, is sometimes convenient for calculations. The ensemble average of $\mathcal{O}$ in the canonical ensemble is

$$\langle\mathcal{O}\rangle = \frac{Tr(\mathcal{O}e^{-\beta H})}{Q_N} \tag{9.26}$$

Grand Canonical Ensemble

For the grand canonical ensemble the density operator ρ operates on a Hilbert space with an indefinite number of particles. We do not display it because we do not need it. It is sufficient to state that the grand partition function is

$$\mathcal{Q}(z, V, T) = \sum_{N=0}^{\infty} z^N Q_N(V, T) \tag{9.27}$$

where Q_N is the partition function for N particles. The connection between $\log \mathcal{Q}$ and thermodynamics is the same as in classical statistical mechanics. The ensemble average of $\mathcal{O}$ in the grand canonical ensemble is

$$\langle\mathcal{O}\rangle = \frac{1}{\mathcal{Q}}\sum_{N=0}^{\infty} z^N\langle\mathcal{O}\rangle_N \tag{9.28}$$

where $\langle \mathcal{O} \rangle_N$ is the ensemble average (9.26) in the canonical ensemble for N particles.

9.4 THIRD LAW OF THERMODYNAMICS

The definition of entropy is given by (9.20). At the absolute zero of temperature a system is in its ground state, i.e., a state of lowest energy. For a system whose energy eigenvalues are discrete, (9.20) implies that at absolute zero $S = k \log G$, where G is the degeneracy of the ground state. If the ground state is unique, then $S = 0$ at absolute zero. If the ground state is not unique, but $G \lessdot N$, where N is the total number of molecules in the system, then at absolute zero $S \lessdot k \log N$. In both of these cases the third law of thermodynamics holds, because the entropy per molecule is essentially zero at absolute zero.

The energy eigenvalues for most macroscopic systems, however, almost form a continuous spectrum. For these systems the previous argument only shows that the entropy per molecule approaches zero when the temperature T is so low that

$$kT \ll \Delta E$$

where ΔE is the energy difference between the first excited state and the ground state. As an estimate let us put

$$\Delta E \approx \frac{\hbar^2}{mV^{2/3}}$$

where m = mass of nucleon, $V = 1\ \text{cm}^3$. Then we find that $T \approx 5 \times 10^{-15}\,^\circ\text{K}$. Clearly this phenomenon has nothing to do with the third law of thermodynamics, which is a phenomenological statement based on experiments performed above 1°K.

To verify the third law of thermodynamics for systems having an almost continuous energy spectrum we must study the behavior of the density of states $\omega(E)$ near $E = 0$. Most of the substances known to us become crystalline solids near absolute zero. For these substances all thermodynamic functions near absolute zero may be obtained through Debye's theory, which is discussed in Sec. 12.2. It is shown there that the third law of thermodynamics is fulfilled.

The only known substance that remains a liquid at absolute zero is helium, which is discussed in Chapter 18. It is shown in Sec. 18.3 that near absolute zero the density of states for liquid helium is qualitatively the same as that for a crystalline solid. Therefore the third law of thermodynamics is also fulfilled for liquid helium.

Apart from these specific examples, which include all known substances, we cannot give a more universal proof of the third law of thermodynamics. But this is perhaps sufficient; after all, the third law of thermodynamics is a summary of empirical data gathered from known substances.

9.5 THE IDEAL GASES: MICROCANONICAL ENSEMBLE

The simplest system of N identical particles is that composed of N noninteracting members. The Hamiltonian is

$$H = \sum_{i=1}^{N} \frac{p_i^2}{2m} \tag{9.29}$$

where $p_i^2 = \mathbf{p}_i \cdot \mathbf{p}_i$, and $\mathbf{p}_i$ is the momentum operator of the ith particle. The Hamiltonian is independent of the positions of the particles or any other coordinates, e.g., spin, if any.

In nature a system of N identical particles is one of two types: A *Bose system* or a *Fermi system.** A complete set of eigenfunctions for a Bose system is the collection of those eigenfunctions of H that are symmetric under an interchange of any pair of particle coordinates. A complete set of eigenfunctions for a Fermi system is the collection of those eigenfunctions of H that are antisymmetric under an interchange of any pair of particle coordinates. Particles forming a Bose system are called *bosons*, and particles forming a Fermi system are called *fermions*.

In addition to these two types of systems we define, for mathematical comparison, what is called a Boltzmann system. It is defined as a system of particles whose eigenfunctions are *all* the eigenfunctions of H; but the rule for counting these eigenfunctions shall be the "correct Boltzmann counting." The set of eigenfunctions for a Boltzmann system includes those for a Bose system and those for a Fermi system, and *more*. There is no known system of this type in nature. It is a useful model, however, because at high temperatures the thermodynamic behavior of both the Bose system and the Fermi system approaches that of the Boltzmann system.

For noninteracting identical particles we have three cases: The ideal Bose gas, the ideal Fermi gas, and the ideal Boltzmann gas. We first work out the thermodynamics of these ideal gases in the formalism of the microcanonical ensemble. For this purpose it is necessary to find out, for each of the three cases, the number of states $\Gamma(E)$ of the system having an energy eigenvalue that lies between E and $E + \Delta$. That is, we must learn how to count.

* See Appendix A, Sec. A.1.

To avoid all unnecessary complications we confine our discussion to spinless particles. Any energy eigenvalue of an ideal system is a sum of single-particle energies, called *levels*. These are given by

$$\epsilon_{\mathbf{p}} = \frac{p^2}{2m} \tag{9.30}$$

where $p \equiv |\mathbf{p}|$ and $\mathbf{p}$ is the momentum eigenvalue of the single particle:

$$\mathbf{p} = \frac{2\pi\hbar}{L}\mathbf{n} \tag{9.31}$$

in which $\mathbf{n}$ is a vector whose components are 0 or $\pm$ integers and L is the cube root of the volume of the system:

$$L \equiv V^{1/3}$$

In the limit as $V \to \infty$ the possible values of $\mathbf{p}$ form a continuum. Then a sum over $\mathbf{p}$ can sometimes be replaced by an integration

$$\sum_{\mathbf{p}} \to \frac{V}{h^3}\int d^3p \tag{9.32}$$

where $h = 2\pi\hbar$ is Planck's constant.*

A state of an ideal system can be specified by specifying a set of occupation numbers $\{n_{\mathbf{p}}\}$ so defined that there are $n_{\mathbf{p}}$ particles having the momentum $\mathbf{p}$ in the state under consideration. Obviously the total energy E and the total number of particles N of the state are given by

$$\begin{aligned} E &= \sum_{\mathbf{p}} \epsilon_{\mathbf{p}} n_{\mathbf{p}} \\ N &= \sum_{\mathbf{p}} n_{\mathbf{p}} \end{aligned} \tag{9.33}$$

For spinless bosons and fermions $\{n_{\mathbf{p}}\}$ uniquely defines a state of the system. The allowed values for any $n_{\mathbf{p}}$ are

$$n_p = \begin{cases} 0, 1, 2, \ldots & \text{(for bosons)} \\ 0, 1 & \text{(for fermions)} \end{cases} \tag{9.34}$$

For a Boltzmann gas $n_{\mathbf{p}} = 0, 1, 2, \ldots$, but $\{n_{\mathbf{p}}\}$ specifies $N!/\prod_{\mathbf{p}}(n_{\mathbf{p}}!)$ states of the N-particle system. This is because an interchange of the momenta of two particles in the system in general leads to a new state but leaves $\{n_{\mathbf{p}}\}$ unchanged.

The total energy is a given number E to within a small uncertainty Δ, whose value is unimportant. Hence $\Gamma(E)$ may be found as follows. As

* For an explanation of (9.31) and (9.32), see Appendix A, Sec. A.2.

$V \to \infty$, the levels (9.30) form a continuum. Let us divide the spectrum of (9.30) into groups of levels containing respectively $g_1, g_2, \ldots$ levels, as shown in Fig. 9.1. Each group is called a cell and has an average energy ϵ_i. The occupation number of the ith cell, denoted by n_i, is the sum of $n_\mathbf{p}$ over all the levels in the ith cell. Each g_i is assumed to be very large, but its exact value is unimportant. Let

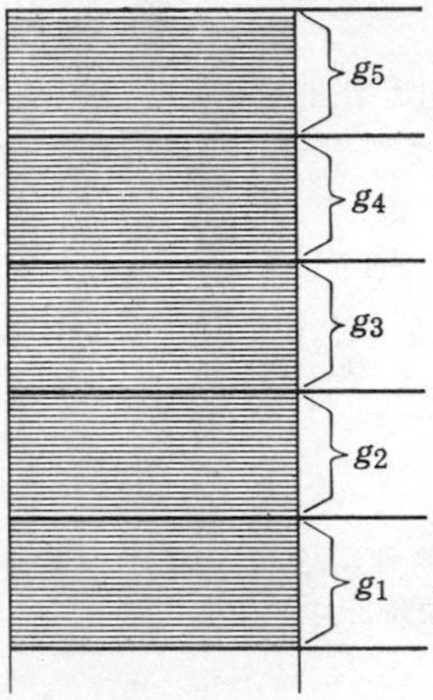

$$W\{n_i\} \equiv \text{no. of states of the system corresponding to the set of occupation numbers } \{n_i\} \tag{9.35}$$

Then

$$\Gamma(E) = \sum_{\{n_i\}} W\{n_i\} \tag{9.36}$$

where the sum extends over all sets of integers $\{n_i\}$ satisfying the conditions

$$E = \sum_i \epsilon_i n_i \tag{9.37}$$

$$N = \sum_i n_i \tag{9.38}$$

Fig. 9.1. Division of the single-particle energy spectrum into cells.

To find $W\{n_i\}$ for a Bose gas and a Fermi gas it is sufficient to find w_i, the number of ways in which n_i particles can be assigned to the ith cell (which contains g_i levels). Since interchanging particles in different cells does not lead to a new state of the system, we have $W\{n_i\} = \prod_j w_j$. For a Boltzmann gas interchanging particles in different cells leads to a new state of the system, and we consider all N particles together. The three cases are worked out as follows.

Bose Gas. Each level can be occupied by any number of particles. Picture the ith cell to have g_i subcells, with $g_i - 1$ partitions, as follows:

$$\cdot\cdot\,|\,\cdot\,|\,\cdot\cdot\,|\,\cdot\,| \quad |\,\cdot\cdot\,| \quad |\,\cdot$$

subcell 1 2 3 $g_i - 1$ g_i

The number w_i is the number of permutations of the n_i particles plus the $g_i - 1$ partitions that give rise to distinct arrangements:

$$w_i = \frac{(n_i + g_i - 1)!}{n_i!\,(g_i - 1)!}$$

Hence

$$W\{n_i\} = \prod_i w_i = \prod_i \frac{(n_i + g_i - 1)!}{n_i!\,(g_i - 1)!} \qquad \text{(Bose)} \tag{9.39}$$

Fermi Gas. The number of particles in each of the g_i subcells of the

ith cell is either 0 or 1. Therefore w_i is equal to the number of ways in which n_i things can be chosen out of g_i things:

$$w_i = \binom{g_i}{n_i} = \frac{g_i!}{n_i!\,(g_i - n_i)!} \tag{9.40}$$

Hence

$$W\{n_i\} = \prod_i w_i = \prod_i \frac{g_i!}{n_i!\,(g_i - n_i)!} \qquad \text{(Fermi)} \tag{9.41}$$

Boltzmann Gas. The N particles are first placed into cells, the ith cell having n_i particles. There are $N!/\prod_i(n_i!)$ ways to do this. Within the ith cell there are g_i levels. Among the n_i particles in the ith cell, the first one can occupy these levels g_i ways. The second and all subsequent ones also can occupy the levels g_i ways. Therefore there are $(g_i)^{n_i}$ ways in which n_i particles can occupy the g_i levels. The total number of ways to obtain $\{n_i\}$ is therefore

$$N!\prod_i \frac{g_i^{\,n_i}}{n_i!}$$

However, $W\{n_i\}$ is defined to be $1/N!$ of the last quantity:

$$W\{n_i\} = \prod_i \frac{g_i^{\,n_i}}{n_i!} \qquad \text{(Boltzmann)} \tag{9.42}$$

This definition corresponds to the rule of "correct Boltzmann counting" and does not correspond to any physical property of the particles in the system. It is just a rule that defines the mathematical model.

The fact that the rule for the counting of states is different for the three cases gives rise to the terminology *Bose statistics*, *Fermi statistics*, and *Boltzmann statistics*, which refer to the three rules of counting respectively.

To obtain the entropy $S = k \log \Gamma(E)$ we need to sum $W\{n_i\}$ over $\{n_i\}$ in accordance with (9.35). This is a formidable task. For the Boltzmann gas it was explicitly done in Sec. 7.5. As we might correctly guess, however, $\Gamma(E)$ is quite well approximated by $W\{\bar{n}_i\}$, where $\{\bar{n}_i\}$ is the set of occupation numbers that maximizes $W\{\bar{n}_i\}$ subject to (9.37) and (9.38). We adopt this approximation and verify its correctness by showing that the fluctuations are small. Accordingly the entropy is taken to be

$$S = k \log W\{\bar{n}_i\} \tag{9.43}$$

To find $\{\bar{n}_i\}$ we maximize $W\{n_i\}$ by varying the n_i subject to (9.37) and (9.38). The details of this calculation are similar to that in Sec. 4.3 and will not be reproduced. We merely give the answers:

$$\bar{n}_i = \begin{cases} \dfrac{g_i}{z^{-1}e^{\beta\epsilon_i} \mp 1} & \begin{pmatrix}\text{Bose}\\ \text{Fermi}\end{pmatrix} \\ g_i z e^{-\beta\epsilon_i} & \text{(Boltzmann)} \end{cases} \tag{9.44}$$

We deduce from this that

$$\bar{n}_{\mathbf{p}} = \begin{cases} \dfrac{1}{z^{-1}e^{\beta\epsilon_{\mathbf{p}}} \mp 1} & \begin{pmatrix}\text{Bose} \\ \text{Fermi}\end{pmatrix} \\ ze^{-\beta\epsilon_{\mathbf{p}}} & \text{(Boltzmann)} \end{cases} \tag{9.45}$$

The parameters z and β are two Lagrange multipliers to be determined from the conditions

$$\begin{aligned} \sum_{\mathbf{p}} \epsilon_{\mathbf{p}} \bar{n}_{\mathbf{p}} &= E \\ \sum_{\mathbf{p}} \bar{n}_{\mathbf{p}} &= N \end{aligned} \tag{9.46}$$

The first of these leads to the identification $\beta = 1/kT$, and the second identifies z as the fugacity.

Using Stirling's approximation and neglecting 1 compared to g_i we have from (9.42) and (9.43),

$$\frac{S}{k} = \log W\{\bar{n}_i\} = \begin{cases} \sum_i \left[\bar{n}_i \log\left(1 + \dfrac{g_i}{\bar{n}_i}\right) + g_i \log\left(1 + \dfrac{\bar{n}_i}{g_i}\right) \right] & \text{(Bose)} \\ \sum_i \left[\bar{n}_i \log\left(\dfrac{g_i}{\bar{n}_i} - 1\right) - g_i \log\left(1 - \dfrac{\bar{n}_i}{g_i}\right) \right] & \text{(Fermi)} \\ \sum_i \bar{n}_i \log (g_i/\bar{n}_i) & \text{(Boltzmann)} \end{cases} \tag{9.47}$$

More explicitly,

$$\frac{S}{k} = \begin{cases} \sum_i g_i \left[\dfrac{\beta\epsilon_i - \log z}{z^{-1}e^{\beta\epsilon_i} - 1} - \log(1 - ze^{-\beta\epsilon_i}) \right] & \text{(Bose)} \\ \sum_i g_i \left[\dfrac{\beta\epsilon_i - \log z}{z^{-1}e^{\beta\epsilon_i} + 1} + \log(1 + ze^{-\beta\epsilon_i}) \right] & \text{(Fermi)} \\ z \sum_i g_i e^{-\beta\epsilon_i} (\beta\epsilon_i - \log z) & \text{(Boltzmann)} \end{cases} \tag{9.48}$$

The validity of these equations depends on the assumption that

$$\frac{\overline{n_i^2} - \bar{n}_i^2}{\bar{n}_i^2} \ll 1 \tag{9.49}$$

This can be easily verified by a calculation similar to that in the derivation of (4.54). From (9.48) all other thermodynamic functions can be determined after z is determined in terms of N from (9.46).

The Boltzmann gas will be worked out explicitly. From (9.38) and (9.44) we have

$$N = z \sum_i g_i e^{-\beta\epsilon_i} = z \sum_{\mathbf{p}} e^{-\beta\epsilon_{\mathbf{p}}} = \frac{zV}{h^3} \int_0^\infty dp\, 4\pi p^2 e^{-\beta p^2/2m} = \frac{zV}{\lambda^3} \tag{9.50}$$

where

$$\lambda = \sqrt{\frac{2\pi\hbar^2}{mkT}} \tag{9.51}$$

This quantity is called the *thermal wavelength* because it is of the order of the de Broglie wavelength of a particle of mass m with the energy kT. Writing $v = V/N$ we obtain

$$z = \frac{\lambda^3}{v} \tag{9.52}$$

The condition $E = \sum n_i\epsilon_i$ requires that

$$E = z\sum_i g_i\epsilon_i e^{-\beta\epsilon_i} = z\sum_{\mathbf{p}} \epsilon_{\mathbf{p}} e^{-\beta\epsilon_{\mathbf{p}}} = \frac{zV}{h^3}\int_0^\infty dp\, 4\pi p^2\left(\frac{p^2}{2m}\right)e^{-\beta p^2/2m} = \tfrac{3}{2}NkT \tag{9.53}$$

Therefore T is the absolute temperature. The entropy is, by (9.48) and (9.46),

$$\begin{aligned} \frac{S}{k} &= z\sum_{\mathbf{p}} e^{-\beta\epsilon_{\mathbf{p}}}(\beta\epsilon_{\mathbf{p}} - \log z) = \beta E - N\log z \\ &= \tfrac{3}{2}N - N\log\left[\frac{N}{V}\left(\frac{2\pi\hbar^2}{mkT}\right)^{3/2}\right] \end{aligned} \tag{9.54}$$

This is the Sackur-Tetrode equation. The fact that the constant $h = 2\pi\hbar$ is Planck's constant follows from (9.31), where $\hbar$ first makes its appearance. The equation of state is deduced from the function $U(S, V)$, which is E expressed in terms of S and V. We straightforwardly find $PV = NkT$. It is to be noted that (9.54) does not satisfy the third law of thermodynamics. This should cause no concern, because a Boltzmann gas is not a physical system. It is only a model towards which the Bose and Fermi gases converge at high temperatures. This shows, however, that the third law of thermodynamics is not an automatic consequence of the general principles of quantum mechanics but depends on the nature of the density of states near the ground state.

The Bose and Fermi gases can be worked out along similar lines. They are more conveniently discussed, however, in the grand canonical ensemble, which we consider in the next section.

9.6 THE IDEAL GASES: GRAND CANONICAL ENSEMBLE

The partition functions for the ideal gases are

$$Q_N(V, T) = \sum_{\{n_{\mathbf{p}}\}} g\{n_{\mathbf{p}}\}e^{-\beta E\{n_{\mathbf{p}}\}} \tag{9.55}$$

where

$$E\{n_{\mathbf{p}}\} = \sum_{\mathbf{p}} \epsilon_{\mathbf{p}} n_{\mathbf{p}} \tag{9.56}$$

and the occupation numbers are subject to the condition

$$\sum_{\mathbf{p}} n_{\mathbf{p}} = N \tag{9.57}$$

For a Bose gas and a Boltzmann gas $n_{\mathbf{p}} = 0, 1, 2, \ldots$. For a Fermi gas $n_{\mathbf{p}} = 0, 1$. The number of states corresponding to $\{n_{\mathbf{p}}\}$ is

$$g\{n_{\mathbf{p}}\} = \begin{cases} 1 & \text{(Bose and Fermi)} \\ \dfrac{1}{N!}\left(\dfrac{N!}{\prod_{\mathbf{p}} n_{\mathbf{p}}!}\right) & \text{(Boltzmann)} \end{cases} \tag{9.58}$$

We first work out the Boltzmann gas:

$$Q_N = \sum_{\substack{n_0, n_1, \cdots \\ \Sigma n_i = N}} \left(\frac{e^{-\beta n_0 \epsilon_0}}{n_0!} \frac{e^{-\beta n_1 \epsilon_1}}{n_1!} \cdots\right) = \frac{1}{N!}(e^{-\beta\epsilon_0} + e^{-\beta\epsilon_1} + \cdots)^N$$

This equality is the multinomial theorem. In the limit as $V \to \infty$ we can write

$$\sum_{\mathbf{p}} e^{-\beta\epsilon_{\mathbf{p}}} = \frac{V}{h^3}\int_0^\infty dp\, 4\pi p^2 e^{-\beta p^2/2m} = V\left(\frac{mkT}{2\pi\hbar^2}\right)^{3/2} \tag{9.59}$$

Therefore

$$\frac{1}{N}\log Q_N = \log\left[\frac{V}{N}\left(\frac{mkT}{2\pi\hbar^2}\right)^{3/2}\right] \tag{9.60}$$

from which easily follows the Sackur-Tetrode equation for the entropy and the equation of state $PV = NkT$. The grand partition function is trivial and will not be considered.

For the Bose gas and the Fermi gas the partition function cannot be evaluated easily because of the condition (9.57). Instead of the partition function we consider the grand partition function

$$\begin{aligned} \mathcal{Q}(z, V, T) &= \sum_{N=0}^{\infty} z^N Q_N(V, T) = \sum_{N=0}^{\infty} \sum_{\substack{\{n_{\mathbf{p}}\} \\ \Sigma n_{\mathbf{p}} = N}} z^N e^{-\beta \Sigma \epsilon_{\mathbf{p}} n_{\mathbf{p}}} \\ &= \sum_{N=0}^{\infty} \sum_{\substack{\{n_{\mathbf{p}}\} \\ \Sigma n_{\mathbf{p}} = N}} \prod_{\mathbf{p}} (z e^{-\beta\epsilon_{\mathbf{p}}})^{n_{\mathbf{p}}} \end{aligned} \tag{9.61}$$

Now it is to be noted that the double summation just given is equivalent to summing each $n_{\mathbf{p}}$ independently. To prove this we must show that every term in one case appears once and only once in the other, and vice versa. This is easily done mentally. Therefore

$$\begin{aligned} \mathcal{Q}(z, V, T) &= \sum_{n_0}\sum_{n_1} \cdots [(ze^{-\beta\epsilon_0})^{n_0}(ze^{-\beta\epsilon_1})^{n_1} \cdots] \\ &= \left[\sum_{n_0} (ze^{-\beta\epsilon_0})^{n_0}\right]\left[\sum_{n_1} (ze^{-\beta\epsilon_1})^{n_1}\right] \cdots \\ &= \prod_{\mathbf{p}} \left[\sum_{n} (ze^{-\beta\epsilon_{\mathbf{p}}})^{n}\right] \end{aligned}$$

where the sum $\sum_n$ extends over the values $n = 0, 1, 2, \ldots$ for the Bose gas and the values $n = 0, 1$ for the Fermi gas. The results are

$$\mathcal{Q}(z, V, T) = \begin{cases} \prod_{\mathbf{p}} \dfrac{1}{1 - ze^{-\beta\epsilon_{\mathbf{p}}}} & \text{(Bose)} \\ \prod_{\mathbf{p}} (1 + ze^{-\beta\epsilon_{\mathbf{p}}}) & \text{(Fermi)} \end{cases} \tag{9.62}$$

The equations of state are

$$\frac{PV}{kT} = \log \mathcal{Q}(z, V, T) = \begin{cases} -\sum_{\mathbf{p}} \log (1 - ze^{-\beta\epsilon_{\mathbf{p}}}) & \text{(Bose)} \\ \sum_{\mathbf{p}} \log (1 + ze^{-\beta\epsilon_{\mathbf{p}}}) & \text{(Fermi)} \end{cases} \tag{9.63}$$

from which z is to be eliminated with the help of the equations

$$N = z \frac{\partial}{\partial z} \log \mathcal{Q}(z, V, T) = \begin{cases} \sum_{\mathbf{p}} \dfrac{ze^{-\beta\epsilon_{\mathbf{p}}}}{1 - ze^{-\beta\epsilon_{\mathbf{p}}}} & \text{(Bose)} \\ \sum_{\mathbf{p}} \dfrac{ze^{-\beta\epsilon_{\mathbf{p}}}}{1 + ze^{-\beta\epsilon_{\mathbf{p}}}} & \text{(Fermi)} \end{cases} \tag{9.64}$$

The average occupation numbers $\langle n_{\mathbf{p}} \rangle$ are given by

$$\begin{aligned} \langle n_{\mathbf{p}} \rangle &\equiv \frac{1}{\mathcal{Q}} \sum_{N=0}^{\infty} z^N \sum_{\substack{\{n_{\mathbf{p}}\} \\ \Sigma n_{\mathbf{p}} = N}} n_{\mathbf{p}} e^{-\beta \Sigma \epsilon_{\mathbf{p}} n_{\mathbf{p}}} = -\frac{1}{\beta} \frac{\partial}{\partial \epsilon_{\mathbf{p}}} \log \mathcal{Q} \\ &= \frac{ze^{-\beta\epsilon_{\mathbf{p}}}}{1 \mp ze^{-\beta\epsilon_{\mathbf{p}}}} \quad \begin{pmatrix} \text{Bose} \\ \text{Fermi} \end{pmatrix} \end{aligned} \tag{9.65}$$

which are the same as (9.45). The equations (9.64) are none other than the statement

$$N = \sum_{\mathbf{p}} \langle n_{\mathbf{p}} \rangle \tag{9.66}$$

The results here are completely equivalent to those in the microcanonical ensemble, as they should be.

Now we let $V \to \infty$, and replace sums over $\mathbf{p}$ by integrals over $\mathbf{p}$ in the manner indicated in (9.32), whenever possible. Such a replacement is clearly valid if the summand in question is finite for all $\mathbf{p}$. In (9.63) and (9.64), the fugacity z is non-negative for both the ideal Fermi gas and the ideal Bose gas. Because, if z were negative, then (9.64) cannot be satisfied for positive N. We see immediately that for the ideal Fermi gas it is permissible to replace the sums in (9.63) and (9.64) by integrals over $\mathbf{p}$. We then obtain the following equation of state.

Ideal Fermi Gas

$$\begin{cases} \dfrac{P}{kT} = \dfrac{4\pi}{h^3}\displaystyle\int_0^\infty dp\, p^2 \log\left(1 + ze^{-\beta p^2/2m}\right) \\ \dfrac{1}{v} = \dfrac{4\pi}{h^3}\displaystyle\int_0^\infty dp\, p^2 \dfrac{1}{z^{-1}e^{\beta p^2/2m} + 1} \end{cases} \tag{9.67}$$

where $v = V/N$. It can be verified in a straightforward fashion that (9.67) can also be written in the form

$$\begin{cases} \dfrac{P}{kT} = \dfrac{1}{\lambda^3} f_{5/2}(z) \\ \dfrac{1}{v} = \dfrac{1}{\lambda^3} f_{3/2}(z) \end{cases} \tag{9.68}$$

where $\lambda = \sqrt{2\pi\hbar^2/mkT}$ and

$$f_{5/2}(z) \equiv \frac{4}{\sqrt{\pi}}\int_0^\infty dx\, x^2 \log\left(1 + ze^{-x^2}\right) = \sum_{l=1}^{\infty} \frac{(-1)^{l+1} z^l}{l^{5/2}} \tag{9.69}$$

$$f_{3/2}(z) \equiv z\frac{\partial}{\partial z} f_{5/2}(z) = \sum_{l=1}^{\infty} \frac{(-1)^{l+1} z^l}{l^{3/2}} \tag{9.70}$$

For the ideal Bose gas the summands appearing in (9.63) and (9.64) diverge as $z \to 1$, because the single term corresponding to $\mathbf{p} = 0$ diverges. Thus the single term $\mathbf{p} = 0$ may be as important as the entire sum.* We split off the terms in (9.63) and (9.64) corresponding to $\mathbf{p} = 0$ and replace the rest of the sums by integrals. We then obtain the following equation of state.

Ideal Bose Gas

$$\begin{cases} \dfrac{P}{kT} = -\dfrac{4\pi}{h^3}\displaystyle\int_0^\infty dp\, p^2 \log\left(1 - ze^{-\beta p^2/2m}\right) - \dfrac{1}{V}\log(1 - z) \\ \dfrac{1}{v} = \dfrac{4\pi}{h^3}\displaystyle\int_0^\infty dp\, p^2 \dfrac{1}{z^{-1}e^{\beta p^2/2m} - 1} + \dfrac{1}{V}\dfrac{z}{1 - z} \end{cases} \tag{9.71}$$

where $v = V/N$. It can be verified in a straightforward fashion that (9.71) can also be written in the form

$$\begin{cases} \dfrac{P}{kT} = \dfrac{1}{\lambda^3} g_{5/2}(z) - \dfrac{1}{V}\log(1 - z) \\ \dfrac{1}{v} = \dfrac{1}{\lambda^3} g_{3/2}(z) + \dfrac{1}{V}\dfrac{z}{1 - z} \end{cases} \tag{9.72}$$

* That this is in fact the case is shown in Sec. 12.3 in connection with the Bose-Einstein condensation.

where $\lambda = \sqrt{2\pi\hbar^2/mkT}$, and

$$g_{5/2}(z) \equiv -\frac{4}{\sqrt{\pi}}\int_0^\infty dx\, x^2 \log(1 - ze^{-x^2}) = \sum_{l=1}^{\infty} \frac{z^l}{l^{5/2}} \tag{9.73}$$

$$g_{3/2}(z) \equiv z\frac{\partial}{\partial z} g_{5/2}(z) = \sum_{l=1}^{\infty} \frac{z^l}{l^{3/2}} \tag{9.74}$$

As (9.65) implies, the quantity $z/(1 - z)$ is the average occupation number $\langle n_0 \rangle$ for the single-particle level with $\mathbf{p} = 0$:

$$\frac{z}{1 - z} = \langle n_0 \rangle \tag{9.75}$$

This term contributes significantly to (9.72) if $\langle n_0 \rangle / V$ is a finite number, i.e., if a finite fraction of all the particles in the system occupy the single level with $\mathbf{p} = 0$. We see in Sec. 12.3 that such a circumstance gives rise to the phenomenon of Bose-Einstein condensation.

The internal energy for both the Fermi and the Bose gases may be found from the formula

$$U(z, V, T) = \frac{1}{\mathcal{Q}}\sum_{N=0}^{\infty} z^N \sum_{\substack{\{n_\mathbf{p}\} \\ \Sigma n_\mathbf{p} = N}} \left[e^{-\beta\Sigma\epsilon_\mathbf{p} n_\mathbf{p}} \sum_\mathbf{p} \epsilon_\mathbf{p} n_\mathbf{p}\right] = -\frac{\partial}{\partial\beta}[\log \mathcal{Q}(z, V, T)] \tag{9.76}$$

Since $\log \mathcal{Q} = PV/kT$, we obtain from (9.68) and (9.72) the results

$$\frac{1}{V}\, U(z, V, T) = \begin{cases} \dfrac{3}{2}\dfrac{kT}{\lambda^3} f_{5/2}(z) & \text{(Fermi)} \\[2ex] \dfrac{3}{2}\dfrac{kT}{\lambda^3}\, g_{5/2}(z) & \text{(Bose)} \end{cases} \tag{9.77}$$

To express U in terms of N, V, and T, we must eliminate z. The result would be a very complicated function. A comparison between (9.77), (9.68), and (9.72), however, shows that U is directly related to the pressure by*

$$U = \tfrac{3}{2}PV \qquad \text{(Bose and Fermi)} \tag{9.78}$$

This relation also holds for the ideal Boltzmann gas.

* It is assumed that the term $V^{-1}\log(1 - z)$ in (9.72) can be neglected. This is justified in Sec. 12.3.

The detailed study of the ideal gases together with their applications is taken up in Chapters 11 and 12.

9.7 FOUNDATIONS OF STATISTICAL MECHANICS

The present section contains no derivations. It merely furnishes an orientation on the subject of the derivation of statistical mechanics from molecular dynamics.*

It is recalled that a special case of statistical mechanics, the classical kinetic theory of gases, can be derived almost rigorously from molecular dynamics. The only *ad hoc* assumption in that derivation is the assumption of molecular chaos, which, however, plays a well-understood role; namely, the reduction of *reversible microscopic phenomena* to *irreversible macroscopic phenomena.* Since irreversibility is a necessary result of any successful derivation, an assumption of this kind is not only unavoidable but also desirable, because it serves to mark clearly the point at which irreversibility enters. An improvement on the existing derivation consists of replacing this assumption by one less *ad hoc*, but not of doing away with it altogether.

The derivation of the classical kinetic theory of gases may be considered largely satisfactory. When we consider the more general problem of the derivation of statistical mechanics, we may well keep this theory in mind as a model example. From this example, we learn that a satisfactory derivation of statistical mechanics must simultaneously fulfill two requirements:

(*a*) It must clearly display the connection between microscopic reversibility and macroscopic irreversibility.

(*b*) It must provide a detailed description of the approach to equilibrium.

Thus a satisfactory derivation of statistical mechanics must satisfy not only the philosophical desire of the physicist to base all natural phenomena on molecular dynamics, but also the practical desire of the physicist to calculate numbers with which to compare with experiments.

Logically speaking, it suffices to derive quantum statistical mechanics, of which classical statistical mechanics is a special case. If we want to understand nonequilibrium phenomena in the classical domain, however, it is expedient to use classical mechanics as a starting point. For this reason attempts to derive classical statistical mechanics from classical mechanics can be of great practical value.

* For a source of literature see *Fundamental Problems in Statistical Mechanics*, edited by E. G. D. Cohen (North-Holland Publishing Co., Amsterdam, 1962).

Attempts to derive statistical mechanics have so far been one of two types. Some take as their immediate goal the justification of the "ergodic hypothesis"; others aim at establishing the "master equation." Of the two, only the latter seems capable of fulfilling both the requirements set forth previously.

The ergodic hypothesis, first advanced by Boltzmann, states that a time average of a macroscopic quantity is, under equilibrium conditions, the same as an ensemble average. So far this hypothesis has been proved for a time average over an infinitely long time, both in classical and quantum mechanics, under certain assumptions that are too abstract to be easily stated and that remain to be justified. In physical experiments we do not average over an infinite time, but over a finite time interval that is in fact very short by macroscopic standards. It is plausible that this time interval can be effectively considered infinite because it is to be compared with characteristic molecular times, e.g., molecular collision time. But to show this would be the central problem of the approach. It is clear that we cannot show this unless we make full use of the details of molecular dynamics. The techniques used, however, in proving the ergodic hypothesis have been based on the avoidance of detailed molecular dynamics. Hence it appears that to make this approach relevant to physics drastically new techniques are needed.

The master equation is an equation governing the time development of the quantity $P_n(t)$, which is the probability that at the instant t the system is in the state n. If the word "state" is appropriately interpreted, $P_n(t)$ can be defined either in classical or quantum mechanics. To justify statistical mechanics, we have to show that $P_n(t)$ approaches the quantity $\overline{(c_n, c_n)}$ of (9.5) when t is much longer than a characteristic time of the system called the relaxation time, e.g., molecular collision time.

The master equation is

$$\frac{dP_n(t)}{dt} = \sum_m [W_{nm}P_m(t) - W_{mn}P_n(t)] \tag{9.79}$$

where W_{mn} is the transition probability per second from the state n to the state m. It was first derived by Pauli under the assumption that n refers to a single quantum state of the system and that the coefficients in the expansion (9.1) have random phases at all times. All subsequent work after Pauli's been has concerned with the improvement of these assumptions and with the solution of the master equation itself.

It can be shown that solutions to the master equation approach the desired limit as $t \to \infty$. Hence the task of deriving statistical mechanics reduces to the justification of the master equation and the calculation of the relaxation time.

The similarity between the master equation and the Boltzmann transport equation may be noted, although we should remember that the latter refers to μ-space whereas the former refers to Γ-space. The random-phase assumption here is similar to the assumption of molecular chaos in the Boltzmann transport equation. In both cases the solution for $t \to \infty$ is relatively easy to obtain, but the relaxation time is difficult to calculate.

The approach involving the master equation seems to hold greater promise for a satisfactory derivation of statistical mechanics and the concomitant understanding of general nonequilibrium phenomena. Further discussion of the master equation, however, is beyond the scope of this book.*

PROBLEMS

9.1. Find the density matrix for a partially polarized incident beam of electrons in a scattering experiment, in which a fraction f of the electrons are polarized along the direction of the beam and a fraction $1 - f$ is polarized opposite to the direction of the beam.

9.2. Verify (9.49).

9.3. Derive the equations of state (9.67) and (9.71), using the microcanonical ensemble.

9.4. Prove (8.14) in quantum statistical mechanics.

9.5. Calculate the grand partition function for a system of N noninteracting quantum mechanical harmonic oscillators, all of which have the same natural frequency ω_0. Do this for the following two cases:
(*a*) Boltzmann statistics
(*b*) Bose statistics.
Suggestion: Write down the energy levels of the N-oscillator system and determine the degeneracies of the energy levels for the two cases mentioned.

9.6. What is the equilibrium ratio of ortho- to para-hydrogen at a temperature of 30°K? What is this ratio in the limit of high temperatures? Assume that the distance between the protons in the molecule is 0.74 Å.
The following hints may be helpful.
(*a*) Boltzmann statistics is valid for H_2 molecules at the temperatures considered.
(*b*) The energy of a single H_2 molecule is a sum of terms corresponding to contributions from rotational motion, vibrational motion, translational motion, and excitation of the electronic cloud. Only the rotational energy need be taken into account.

* For a general discussion of the master equation, see N. G. Van Kampen, in Cohen, *op. cit.* An improvement on the random phase approximation is described by L. Van Hove, in Cohen, *op. cit.*

(*c*) The rotational energies are

$$E_{\text{para}} = \frac{\hbar^2}{2I} l(l+1) \qquad (l = 0, 2, 4, \ldots)$$

$$E_{\text{ortho}} = \frac{\hbar^2}{2I} l(l+1) \qquad (l = 1, 3, 5, \ldots)$$

where I is the moment of inertia of the H_2 molecule.
Answer: Let T = absolute temperature and $\beta = 1/kT$. Then

$$\frac{N_{\text{ortho}}}{N_{\text{para}}} = \frac{3 \sum\limits_{l\,\text{odd}} (2l+1)\, e^{-(\beta\hbar^2/2I)l(l+1)}}{\sum\limits_{l\,\text{even}} (2l+1)\, e^{-(\beta\hbar^2/2I)l(l+1)}}$$

chapter 10

THE PARTITION FUNCTION

10.1 DARWIN-FOWLER METHOD

Although the canonical ensemble may be derived from the microcanonical ensemble, as we have shown in Sec. 8.1, it may also be derived directly. Indeed, if we are not too concerned with rigor, the derivation is very simple. Consider an ensemble of M systems such that the energy averaged over all the systems is a given number U. We wish to find the most probable distribution of energies among these M systems in the limit as $M \to \infty$. By definition of an ensemble, the systems do not interact with one another; they may be considered one at a time; and they are consequently distinguishable from one another. Therefore our problem is mathematically identical with the problem of the most probable distribution for a classical ideal gas of particles. The answer as we know is the Maxwell-Boltzmann distribution, i.e., the energy value E_n occurs among the systems with relative probability $e^{-\beta E_n}$, where β is determined by the average energy U. This ensemble is the canonical ensemble. It is obvious that this derivation holds equally well in quantum and in classical statistical mechanics.

We want to present here a more rigorous derivation that avoids the use of Sterling's approximation, which is necessary in the usual derivation of the Maxwell-Boltzmann distribution. The purpose of this presentation is not only to derive the canonical ensemble directly but also to introduce the method of saddle point integration, which is a useful mathematical tool in statistical mechanics. The considerations that follow hold equally well for quantum and for classical statistical mechanics.

The problem has already been stated. We proceed to the derivation according to Darwin and Fowler. Assume that a system in the ensemble may have any one of the energy values E_k ($k = 0, 1, 2, \ldots$). By choosing the unit of energy to be sufficiently small, we can regard E_k as an integer. Among the systems in the ensemble let

$$
\begin{array}{ll}
m_0 & \text{systems have energy } E_0 \\
m_1 & \text{systems have energy } E_1 \\
\vdots & \\
m_k & \text{systems have energy } E_k \\
\vdots &
\end{array}
\tag{10.1}
$$

The set of integers $\{m_k\}$ describe an arbitrary distribution of energy among the systems. They must satisfy the conditions

$$
\begin{aligned}
\sum_{k=0}^{\infty} m_k &= M \\
\sum_{k=0}^{\infty} E_k m_k &= MU
\end{aligned}
\tag{10.2}
$$

where both M and U are integers. Our purpose is to find the most probable set $\{\bar{m}_k\}$.

Given an arbitrary set $\{m_k\}$ satisfying (10.2) there are generally more ways than one to construct an ensemble corresponding to (10.1), because the interchange of any two systems (which are distinguishable) leaves $\{m_k\}$ unchanged. Let $W\{m_k\}$ be the number of distinct ways in which we can assign energy values to systems so as to satisfy (10.1). Obviously

$$
W\{m_k\} = \frac{M!}{m_0!\, m_1!\, m_2! \cdots}
\tag{10.3}
$$

For the present case the postulate of equal *a priori* probability states that all distributions in energy among the systems are equally probable, subject to the conditions (10.2). Thus $\{\bar{m}_k\}$ is the set that maximizes (10.3). In anticipation of the fact that in the limit as $M \to \infty$ almost all possible sets $\{m_k\}$ are identical with $\{\bar{m}_k\}$, we can also find $\{\bar{m}_k\}$ by calculating the value of m_k averaged over all possible distributions in energy:

$$\langle m_k \rangle \equiv \frac{\sum'_{\{m_i\}} m_k W\{m_i\}}{\sum'_{\{m_i\}} W\{m_i\}} \tag{10.4}$$

where a prime over the sums indicate that they are sums over all sets $\{m_k\}$ subject to (10.2). We must also calculate the mean square fluctuation $\langle m_k^2 \rangle - \langle m_k \rangle^2$. If this vanishes as $M \to \infty$, then in that limit $\langle m_k \rangle \to \bar{m}_k$.

For convenience we modify the definition of $W\{m_k\}$ to

$$W\{m_k\} = \frac{M!\, g_0^{\ m_0} g_1^{\ m_1} \cdots}{m_0!\, m_1! \cdots} \tag{10.5}$$

where g_k is a number which at the end of the calculation will be set equal to unity. Let

$$\Gamma(M, U) \equiv \sum'_{\{m_i\}} W\{m_i\} \tag{10.6}$$

Then

$$\langle m_k \rangle = g_k \frac{\partial}{\partial g_k} \log \Gamma \tag{10.7}$$

The mean square fluctuation can be obtained as follows:

$$\begin{aligned}
\langle m_k^2 \rangle &= \frac{1}{\Gamma} \sum'_{\{m_i\}} m_k^2 W\{m_i\} = \frac{1}{\Gamma} g_k \frac{\partial}{\partial g_k}\left(g_k \frac{\partial \Gamma}{\partial g_k}\right) \\
&= g_k \frac{\partial}{\partial g_k}\left(\frac{1}{\Gamma} g_k \frac{\partial \Gamma}{\partial g_k}\right) - \left(\frac{\partial}{\partial g_k}\frac{1}{\Gamma}\right) g_k^{\ 2} \frac{\partial \Gamma}{\partial g_k} \\
&= g_k \frac{\partial}{\partial g_k}\left(g_k \frac{\partial}{\partial g_k} \log \Gamma\right) + \left(g_k \frac{\partial}{\partial g_k} \log \Gamma\right)^2
\end{aligned}$$

Therefore

$$\langle m_k^2 \rangle - \langle m_k \rangle^2 = g_k \frac{\partial}{\partial g_k}\left(g_k \frac{\partial}{\partial g_k} \log \Gamma\right) \tag{10.8}$$

Thus it is sufficient to calculate $\log \Gamma$.

By (10.6) and (10.5)

$$\Gamma = M! \sum'_{m_0,\, m_1,\, \ldots} \left(\frac{g_0^{\ m_0}}{m_0!} \cdot \frac{g_1^{\ m_1}}{m_1!} \cdots\right) \tag{10.9}$$

This cannot be explicitly evaluated because of the restriction (10.2). We are, however, only interested in this quantity in the limit as $M \to \infty$. To proceed, we define a generating function for Γ in the following manner. For any complex number z, let

$$G(M, z) \equiv \sum_{U=0}^{\infty} z^{MU}\Gamma(M, U) \tag{10.10}$$

Using (10.9) and (10.2) we obtain

$$G(M, z) = M! \sum_{U=0}^{\infty} \sum_{m_0, m_1, \ldots}' \left[\frac{(g_0 z^{E_0})^{m_0}}{m_0!} \cdot \frac{(g_0 z^{E_1})^{m_1}}{m_1!} \cdots\right] \tag{10.11}$$

It is easily seen that the double sum in (10.11) is equivalent to summing over all sets $\{m_k\}$ subject only to the condition $\Sigma m_k = M$. To show this we need only verify that every term in one sum appears once in the other and vice versa. Hence

$$\begin{aligned} G(M, z) &= \sum_{\substack{m_0, m_1, \ldots \\ \Sigma m_k = M}} \frac{M!}{m_0!\, m_1! \cdots} [(g_0 z^{E_0})^{m_0} (g_1 z^{E_1})^{m_1} \cdots] \\ &= (g_0 z^{E_0} + g_1 z^{E_1} + \cdots)^M \end{aligned} \tag{10.12}$$

The last step follows by use of the multinomial theorem. Let

$$f(z) \equiv \sum_{k=0}^{\infty} g_k z^{E_k} \tag{10.13}$$

Then

$$G(M, z) = [f(z)]^M \tag{10.14}$$

To obtain $\Gamma(M, U)$ from $G(M, z)$ we note that by definition $\Gamma(M, U)$ is the coefficient of z^{MU} in the expansion of $G(M, z)$ in powers of z. Therefore

$$\Gamma(M, U) = \frac{1}{2\pi i} \oint dz \frac{[f(z)]^M}{z^{MU+1}} \tag{10.15}$$

where the contour of integration is a closed path in the complex z plane about $z = 0$.

We may assume without loss of generality that the sequence $E_0, E_1, \ldots$ is a sequence of nondecreasing integers with no common divisor, because any common division τ can be removed by choosing the unit of energy τ times larger. Furthermore, we can set $E_0 = 0$, since this would only change the zero point of the energy. In so doing U would be changed to $U - E_0$, which we can again call U. The numbers $g_0, g_1, \ldots$ are as close to unity as we wish. For the immediate calculations we omit them temporarily. Hence

$$f(z) = 1 + z^{E_1} + z^{E_2} + \cdots \qquad (E_1 \le E_2 \le E_3 \le \cdots) \tag{10.16}$$

where $E_1, E_2, \ldots$ are integers with no common divisor. When z is a real positive number x, $f(x)$ is a monotonically increasing function of x with a radius of convergence at, say, $x = R$. The same is true for $[f(x)]^M$, as illustrated in Fig. 10.1. The function $1/z^{MU+1}$ on the other hand is a

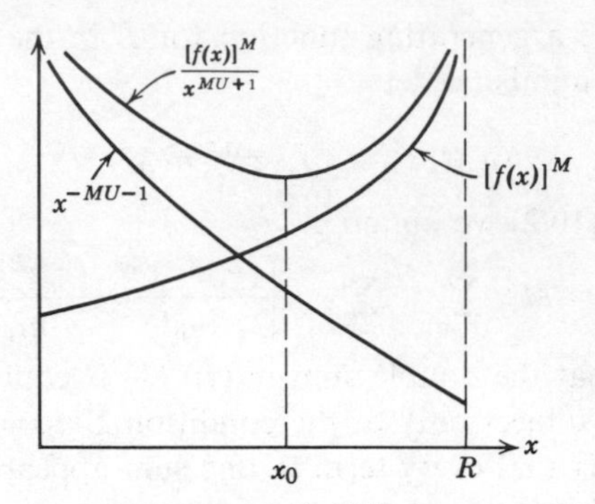

Fig. 10.1. The function $[f(x)]^M/x^{MU+1}$ for real positive x.

monotonically decreasing function along the real positive axis. The product of these two functions has a minimum at x_0 between 0 and R, as shown in Fig. 10.1. Now $f(z)$ is an analytic function for $|z| < R$, and z^{-MU-1} is analytic everywhere except at $z = 0$. Therefore the integrand of (10.18)

$$I(z) = \frac{[f(z)]^M}{z^{MU+1}} \tag{10.17}$$

is analytic at $z = x_0$. An analytic function has a unique derivative at a given point. Furthermore it satisfies the Cauchy-Riemann equation

$$\left(\frac{\partial^2}{\partial x^2} + \frac{\partial^2}{\partial y^2}\right) I(z) = 0 \qquad (z \equiv x + iy) \tag{10.18}$$

Hence

$$\left(\frac{\partial I}{\partial z}\right)_{z=x_0} = 0, \qquad \left(\frac{\partial^2 I}{\partial x^2}\right)_{z=x_0} > 0, \qquad \left(\frac{\partial^2 I}{\partial y^2}\right)_{z=x_0} < 0 \tag{10.19}$$

That is, in the complex z-plane, $I(z)$ has a minimum at $z = x_0$ along a path on the real axis but has a maximum at $z = x_0$ along a path parallel to the imaginary axis passing through x_0. The point x_0 is a saddle point, as illustrated in Fig. 10.2. Let $g(z)$ be defined by

$$\begin{aligned} I(z) &\equiv e^{Mg(z)} \\ g(z) &= \log f(z) - U \log z \end{aligned} \tag{10.20}$$

where we have neglected $1/M$ as compared to U. Then x_0 is the root of the equation

$$g'(x_0) = 0$$

or

$$\frac{\sum_k E_k x_0^{E_k}}{\sum_k x_0^{E_k}} = U \tag{10.21}$$

Furthermore

$$\left(\frac{\partial^2 I}{\partial x^2}\right)_{z=x_0} = Mg''(x_0) \exp\left[Mg(x_0)\right] \xrightarrow[M\to\infty]{} \infty \tag{10.22}$$

Hence the saddle point touches an infinitely sharp peak and an infinitely steep valley in the limit as $M \to \infty$. If we choose the contour of integration to be a circle about $z = 0$ with radius x_0, the contour will pass through x_0 in the imaginary direction. Thus along the contour the integrand has an extremely sharp maximum at the point $z = x_0$. If elsewhere along the contour there is no maximum comparable in height to this one, the contribution to the integral comes solely from the neighborhood of x_0. This is in fact true because for $z = x_0\, e^{i\theta}$

$$|I(z)| = \frac{1}{x_0^{\ MU}}\,|1 + (x_0 e^{i\theta})^{E_1} + (x_0 e^{i\theta})^{E_2} + \cdots|^M \tag{10.23}$$

The series (10.23) is maximum when all terms are real. This happens when and only when $\theta E_k = 2\pi n_k$, where n_k is 0 or an integer. If $\theta \neq 0$, then $2\pi/\theta$ must be a rational number, and this would mean that $E_k = (2\pi/\theta)n_k$, which is impossible unless $\theta = 2\pi$, because the E_k have no common divisor. Hence we conclude that the largest value of $I(z)$ occurs at $z = x_0$.

To do the integral (10.15) we expand the integrand about $z = x_0$:

$$g(z) = g(x_0) + \tfrac{1}{2}(z - x_0)^2\, g''(x_0) + \cdots$$

Hence

$$\Gamma(M, U) = \frac{1}{2\pi i}\oint dz e^{Mg(z)} \approx e^{Mg(x_0)}\,\frac{1}{2\pi i}\oint dz e^{\frac{1}{2}Mg''(x_0)(z-x_0)^2}$$

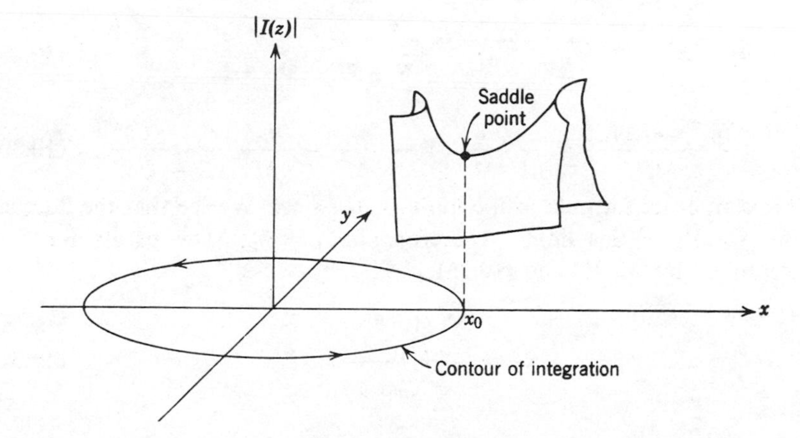

Fig. 10.2. The saddle point.

Putting $(z - x_0) = iy$, we obtained

$$\Gamma(M, U) \approx e^{Mg(x_0)} \frac{1}{2\pi} \int_{-\infty}^{+\infty} dy e^{-\frac{1}{2}Mg''(x_0)y^2} = \frac{e^{Mg(x_0)}}{\sqrt{2\pi M g''(x_0)}} \tag{10.24}$$

Hence

$$\frac{1}{M} \log \Gamma(M, U) \approx g(x_0) - \frac{1}{2M} \log [2\pi M g''(x_0)] \tag{10.25}$$

As $M \to \infty$ the first term gives the exact result. To evaluate $g(x_0)$ we first obtain from (10.20) the formulas

$$g(x_0) = \log f(x_0) - U \log x_0$$

$$g''(x_0) = \frac{f''(x_0)}{f(x_0)} - \frac{U(U-1)}{{x_0}^2}$$

Using $f(x_0)$ from (10.13) (restoring now the numbers g_k) and defining a parameter β by

$$x_0 \equiv e^{-\beta} \tag{10.26}$$

we obtain

$$g(x_0) = \log \left(\sum_{k=0}^{\infty} g_k e^{-\beta E_k} \right) + \beta U$$

$$g''(x_0) = \frac{e^{2\beta} \sum_{k=0}^{\infty} g_k({E_k}^2 - U^2) e^{-\beta E_k}}{\sum_{k=0}^{\infty} g_k e^{-\beta E_k}} \equiv e^{2\beta} \langle (E-U)^2 \rangle \tag{10.27}$$

Hence

$$\frac{1}{M} \log \Gamma(M, U) = \log \left(\sum_{k=0}^{\infty} g_k e^{-\beta E_k} \right) + \beta U - \frac{1}{2M} \log [2\pi M g''(x_0)] \tag{10.28}$$

from which we find, using (10.7) and (10.8),

$$\frac{\langle m_k \rangle}{M} = \frac{e^{-\beta E_k}}{\sum_{k=0}^{\infty} e^{-\beta E_k}} \tag{10.29}$$

$$\frac{\langle {m_k}^2 \rangle - \langle m_k \rangle^2}{M^2} = \frac{1}{M} \frac{\langle m_k \rangle}{M} \left[1 - \frac{\langle m_k \rangle}{M} - \frac{\langle m_k \rangle}{M} \frac{(E_k - U)^2}{\langle (E-U)^2 \rangle} \right] \tag{10.30}$$

This is an exact formula in the limit as $M \to \infty$. We see that the fluctuations vanish in this limit. Therefore $\langle m_k \rangle = \bar{m}_k$. The parameter β is determined by (10.21) and (10.26):

$$U = \frac{\sum_{k=0}^{\infty} E_k e^{-\beta E_k}}{\sum_{k=0}^{\infty} e^{-\beta E_k}} = \langle E \rangle \tag{10.31}$$

Hence β can be identified as $1/kT$, where T is the absolute temperature.

In the most probable distribution the probability of finding a system in the ensemble having the energy E_k is (10.29). The ensemble with such an energy distribution is the canonical ensemble.

10.2 CLASSICAL LIMIT OF THE PARTITION FUNCTION

Let H be the Hamiltonian operator of a system of N identical spinless particles.* Let H be the sum of two operators, the kinetic energy operator K, and the potential energy operator Ω:

$$H = K + \Omega \tag{10.32}$$

If $\Psi(\mathbf{r}_1, \ldots, \mathbf{r}_N)$ is a wave function of the system, then

$$K\Psi(\mathbf{r}_1, \ldots, \mathbf{r}_N) = -\frac{\hbar^2}{2m}\sum_{i=1}^{N} \nabla_i^2 \Psi(\mathbf{r}_1, \ldots, \mathbf{r}_N) \tag{10.33}$$

$$\Omega\Psi(\mathbf{r}_1, \ldots, \mathbf{r}_N) = \Omega(\mathbf{r}_1, \ldots, \mathbf{r}_N)\, \Psi(\mathbf{r}_1, \ldots, \mathbf{r}_N) \tag{10.34}$$

where m is the mass of a particle, ∇_i^2 is the Laplacian operator with respect to $\mathbf{r}_i$, and $\Omega(\mathbf{r}_1, \ldots, \mathbf{r}_N)$ is a sum of two-body potentials:

$$\Omega(\mathbf{r}_1, \ldots, \mathbf{r}_N) = \sum_{i<j} v_{ij} \tag{10.35}$$

where

$$v_{ij} \equiv v(|\mathbf{r}_i - \mathbf{r}_j|) \tag{10.36}$$

Whenever convenient, we use the abbreviation $(1, \ldots, N)$ for $(\mathbf{r}_1, \ldots, \mathbf{r}_N)$.

The partition function of the system is

$$Q_N(V, T) = Tr e^{-\beta H} = \sum_n (\Phi_n, e^{-\beta H}\Phi_n) \tag{10.37}$$

where $\Phi_n(1, \ldots, N)$ is a member of any complete orthonormal set of wave functions of the system and $\Phi_n^*(1, \ldots, N)$ is its complex conjugate. For any operator $\mathcal{O}$,

$$(\Phi_n, \mathcal{O}\,\Phi_n) \equiv \int d^{3N}r\, \Phi_n^*(1, \ldots, N)\, \mathcal{O}\Phi_n(1, \ldots, N) \tag{10.38}$$

Each Φ_n satisfies the boundary conditions imposed on the system and is normalized in the box of volume V containing the system. It is a symmetric (antisymmetric) function of $\mathbf{r}_1, \ldots, \mathbf{r}_N$, if the system is a system of bosons (fermions).

* It is straightforward to generalize the following considerations to the case of particles with spin and to the case of a mixed system of two or more different kinds of particles.

It will be shown that when the temperature is sufficiently high we can make the approximation

$$Q_N(V, T) \approx \frac{1}{N!\, h^{3N}} \int d^{3N}p\, d^{3N}r\, e^{-\beta H(p, r)} \tag{10.39}$$

where h is Planck's constant and $H(p, r)$ is the classical Hamiltonian

$$H(p, r) \equiv \sum_{i=1}^{N} \frac{p_i^{\,2}}{2m} + \Omega(\mathbf{r}_1, \ldots, \mathbf{r}_N) \tag{10.40}$$

This will prove that at sufficiently high temperatures the partition function approaches the classical partition function with "correct Boltzmann counting." In the course of proving (10.39) we obtain the criterion for a "sufficiently high temperature."

Let us first consider an ideal gas, for which $\Omega(1, \ldots, N) \equiv 0$. The eigenfunctions of the Hamiltonian are the free-particle wave functions $\Phi_p(1, \ldots, N)$ described in Appendix A, Sec. A.2. They are labeled by a set of N momenta

$$p \equiv \{\mathbf{p}_1, \ldots, \mathbf{p}_N\} \tag{10.41}$$

and satisfy the eigenvalue equation

$$K\Phi_p(1, \ldots, N) = K_p \Phi_p(1, \ldots, N) \tag{10.42}$$

where

$$K_p \equiv \frac{1}{2m}(p_1^{\,2} + \cdots + p_N^{\,2}) \tag{10.43}$$

For convenience we impose periodic boundary conditions with respect to the volume V. It follows that each $\mathbf{p}_i$ has the allowed values

$$\mathbf{p}_i = \frac{2\pi\hbar\mathbf{n}}{V^{1/3}} \tag{10.44}$$

where $\mathbf{n}$ is a vector whose components may be $0, \pm 1, \pm 2, \ldots$. More explicitly, $\Phi_p(1, \ldots, N)$ is given by

$$\begin{aligned} \Phi_p(1, \ldots, N) &= \frac{1}{\sqrt{N!}} \sum_P \delta_P [u_{\mathbf{p}_1}(P1) \cdots u_{\mathbf{p}_N}(PN)] \\ &= \frac{1}{\sqrt{N!}} \sum_P \delta_P [u_{P\mathbf{p}_1}(1) \cdots u_{P\mathbf{p}_N}(N)] \end{aligned} \tag{10.45}$$

$$u_{\mathbf{p}}(\mathbf{r}) = \frac{1}{\sqrt{V}} e^{i\mathbf{p}\cdot\mathbf{r}/\hbar} \tag{10.46}$$

The notation is explained in Appendix A, Sec. A.2. While Φ_p is labeled by $p \equiv \{\mathbf{p}_1, \ldots, \mathbf{p}_N\}$ it changes by at most a sign under any permutation of $\mathbf{p}_1, \ldots, \mathbf{p}_N$. Therefore a sum over all states is $1/N!$ times a sum over all the independent vectors $\mathbf{p}_1, \ldots, \mathbf{p}_N$. Accordingly

$$\begin{aligned} Tr e^{-\beta K} &= \frac{1}{N!} \sum_{\mathbf{p}_1 \cdots \mathbf{p}_N} (\Phi_p, e^{-\beta K} \Phi_p) \\ &= \frac{1}{N!} \sum_{\mathbf{p}_1 \cdots \mathbf{p}_N} e^{-\beta K_p} \int d^{3N}r \, |\Phi_p(1, \ldots, N)|^2 \end{aligned} \tag{10.47}$$

In the limit as $V \to \infty$ a sum over $\mathbf{p}_i$ may be replaced by an integration over $\mathbf{p}_i$:

$$\sum_{\mathbf{p}_i} \to \frac{V}{h^3} \int d^3 p_i \tag{10.48}$$

Therefore

$$Tr e^{-\beta K} = \frac{1}{N! \, h^{3N}} \int d^{3N}r \, J(1, \ldots, N) \tag{10.49}$$

where

$$J(1, \ldots, N) \equiv V^N \int d^{3N}p \, e^{-\beta K_p} |\Phi_p(1, \ldots, N)|^2 \tag{10.50}$$

By (10.45)

$$\begin{aligned} J(1, \ldots, N) = V^N \int d^{3N}p \, e^{\ \beta K_p} \sum_P \delta_P [u_{\mathbf{p}_1}{}^*(P1) \cdots u_{\mathbf{p}_N}{}^*(PN)] \\ \times \sum_{P'} \delta_{P'} [u_{P'\mathbf{p}_1}(1) \cdots u_{P'\mathbf{p}_N}(N)] \end{aligned}$$

Every one of the $N!$ terms in the sum $\sum_{P'}$ yields the same contribution after integration because they can all be reduced to the same integral by permuting the variables of integration. Hence

$$J(1, \ldots, N) = V^N \int d^{3N}p \, e^{-\beta K_p} \sum_P \delta_P [u_{\mathbf{p}_1}{}^*(P1) u_{\mathbf{p}_1}(1)] \cdots [u_{\mathbf{p}_N}{}^*(PN) u_{\mathbf{p}_N}(N)]$$

Using (10.43) and (10.46) we obtain

$$\begin{aligned} J(1, \ldots, N) &= \sum_P \delta_P \int d^{3N}p \prod_{j=1}^{N} e^{-\beta(p_j{}^2/2m) + i\mathbf{p}_j \cdot (\mathbf{r}_j - P\mathbf{r}_j)/\hbar} \\ &= \sum_P \delta_P \left[\int d^3p \, e^{-\beta(p^2/2m) + i\mathbf{p} \cdot (\mathbf{r}_1 - P\mathbf{r}_1)/\hbar} \right] \cdots \\ &\qquad \times \left[\int d^3p \, e^{-\beta(p^2/2m) + i\mathbf{p} \cdot (\mathbf{r}_N - P\mathbf{r}_N)/\hbar} \right] \\ &= \left[\int d^3p \, e^{-\beta p^2/2m} \right]^N \sum_P \delta_P [f(\mathbf{r}_1 - P\mathbf{r}_1) \cdots f(\mathbf{r}_N - P\mathbf{r}_N)] \end{aligned} \tag{10.51}$$

where

$$f(\mathbf{r}) \equiv \frac{\int d^3p\, e^{-\beta(p^2/2m)+i\mathbf{p}\cdot\mathbf{r}/\hbar}}{\int d^3p\, e^{-\beta p^2/2m}} = e^{-\pi r^2/\lambda^2} \tag{10.52}$$

where $r \equiv |\mathbf{r}|$ and $\lambda = \sqrt{2\pi\hbar^2/mkT}$, the thermal wave length. Substituting (10.51) into (10.49) we obtain

$$Tr e^{-\beta K} = \frac{1}{N!\, h^{3N}} \int d^{3N}p\, d^{3N}r\, e^{-\beta(p_1^2+\cdots+p_N^2)/2m} \times \sum_P \delta_P [f(\mathbf{r}_1 - P\mathbf{r}_1)\cdots f(\mathbf{r}_N - P\mathbf{r}_N)] \tag{10.53}$$

This is an exact identity. For high temperatures the integrand may be approximated as follows. The sum $\sum_P$ contains $N!$ terms. The term corresponding to the unit permutation $P = 1$ is $[f(0)]^N = 1$. The term corresponding to a permutation which only interchanges $\mathbf{r}_i$ and $\mathbf{r}_j$ is $[f(\mathbf{r}_i - \mathbf{r}_j)]^2$. Thus by enumerating the permutations in increasing order of the number of coordinates interchanged we arrive at the expansion

$$\sum_P \delta_P [f(\mathbf{r}_1 - P\mathbf{r}_1)\cdots f(\mathbf{r}_N - P\mathbf{r}_N)] = 1 \pm \sum_{i<j} f_{ij}^2 + \sum_{i,j,k} f_{ij} f_{ik} f_{kj} \pm \cdots \tag{10.54}$$

where $f_{ij} \equiv f(\mathbf{r}_i - \mathbf{r}_j)$ and where the $+$ sign applies to bosons and the $-$ sign to fermions. According to (10.52), f_{ij} vanishes extremely rapidly if $|\mathbf{r}_i - \mathbf{r}_j| \gg \lambda$. Therefore

$$Tr e^{-\beta K} \approx \frac{1}{N!\, h^{3N}} \int d^{3N}p\, d^{3N}r\, e^{-\beta(p_1^2+\cdots+p_N^2)/2m} \tag{10.55}$$

when the temperature is so high that

$$(\textit{thermal wavelength}) \ll (\textit{average interparticle distance}) \tag{10.56}$$

This proves (10.39) for an ideal gas.

It is of some interest to examine the first quantum correction to the classical partition function of an ideal gas. If $|\mathbf{r}_i - \mathbf{r}_j| \gg \lambda$, we may approximate the right-hand side of (10.54) by $1 \pm \Sigma f_{ij}^2$. To the same order of approximation, however, we can also write

$$1 \pm \sum_{i<j} f_{ij}^2 \approx \prod_{i<j} (1 \pm f_{ij}^2) = \exp\left(-\beta \sum_{j<i} \tilde{v}_{ij}\right) \tag{10.57}$$

where

$$\tilde{v}_{ij} \equiv -kT \log(1 \pm f_{ij}^2) = -kT \log\left[1 \pm \exp\left(-\frac{2\pi |\mathbf{r}_i - \mathbf{r}_j|^2}{\lambda^2}\right)\right] \tag{10.58}$$

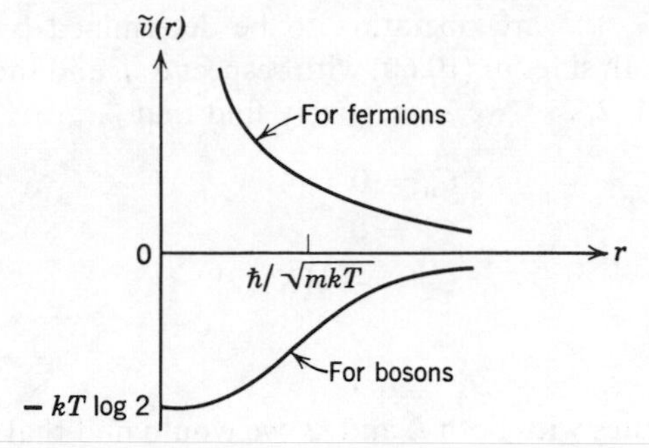

Fig. 10.3. The "statistical potential" between particles in an ideal gas arising from the symmetry properties of the N-particle wave function.

with the + sign for bosons and the − sign for fermions. Therefore an improvement over (10.55) is the formula

$$Tr e^{-\beta K} \approx \frac{1}{N!\, h^{3N}} \int d^{3N}p\, d^{3N}r \exp\left[-\beta\left(\sum_i \frac{p_i^2}{2m} + \sum_{i<j} \tilde{v}_{ij}\right)\right] \tag{10.59}$$

This shows that the first quantum correction to the partition function of an ideal gas has the same effect as that of endowing the particles with an interparticle potential* $\tilde{v}(r)$ and treating the problems classically. The potential $\tilde{v}(r)$ is attractive for bosons and repulsive for fermions, as illustrated in Fig. 10.3. In this sense we sometimes speak of the "statistical attraction" between bosons and the "statistical repulsion" between fermions. It must be emphasized, however, that $\tilde{v}(r)$ originates solely from the symmetry properties of the wave function. Furthermore, it depends on the temperature and thus cannot be regarded as a true interparticle potential.

We now turn our attention to the more general case in which the particles of the system interact with one another. For the calculation of traces we may continue to use the free-particle wave functions Φ_p, because any complete orthonormal set of wave functions will do.

First it is to be noted that in general K does not commute with Ω. Hence

$$e^{-\beta H} = e^{-\beta(K+\Omega)} \neq e^{-\beta K}\, e^{-\beta\Omega}$$

because the left-hand side is invariant under the exchange of K and Ω whereas the right-hand side is not. To find a suitable approximation for $e^{-\beta H}$ when $\beta \to 0$, let us assume that the following expansion is possible:

$$e^{-\beta(K+\Omega)} = e^{-\beta K}\, e^{-\beta\Omega}\, e^{C_0}\, e^{\beta C_1}\, e^{\beta^2 C_2} \cdots \tag{10.60}$$

* First discussed by G. E. Uhlenbeck and L. Gropper, *Phys. Rev.*, **41**, 79 (1932).

where $C_0, C_1, C_2, \ldots$ are operators to be determined by taking the nth derivatives of both sides of (10.60) with respect to β and then setting $\beta = 0$. Letting $n = 0, 1, 2, \ldots$, we successively find that

$$
\begin{aligned}
C_0 &= 0 \\
C_1 &= 0 \\
C_2 &= -\tfrac{1}{2}[K, \Omega] \\
&\vdots
\end{aligned}
\tag{10.61}
$$

If $[K, \Omega]$ commutes with both K and Ω, we would find that $C_n = 0\,(n > 2)$. In our case this is untrue but we shall assume that for β sufficiently small the following is a good approximation:

$$e^{-\beta(K+\Omega)} \approx e^{-\beta K} e^{-\beta\Omega} e^{-\frac{1}{2}\beta^2[K,\Omega]} \tag{10.62}$$

Consequently

$$Q_N(V, T) \approx Tr(e^{-\beta K} e^{-\beta\Omega}\, e^{-\frac{1}{2}\beta^2[K,\Omega]}) \tag{10.63}$$

From (10.33) and (10.34) it can be easily verified that

$$[K, \Omega] = -\frac{\hbar^2}{2m}\sum_{j=1}^{N} \nabla_j^2 \Omega - \frac{\hbar^2}{m}\sum_{j=1}^{N} (\nabla_j \Omega) \cdot \nabla_j \tag{10.64}$$

where it is assumed that $\nabla_j^2 \Omega$ and $\nabla_j \Omega$ both exist. There is no loss in generality owing to this assumption, because actual intermolecular potentials, to the extent that they exist, can always be represented as smooth functions of the coordinates. In order to simplify future calculations let us replace $-\nabla_j \Omega$ by a constant vector $\mathbf{F}$, the average force acting on a molecule. Accordingly, we write

$$-\tfrac{1}{2}\beta^2[K, \Omega] \approx -\beta\Omega' - \frac{i\beta\hbar}{2m}\mathbf{F} \cdot \mathbf{P}_{\text{op}} \tag{10.65}$$

where $\mathbf{P}_{\text{op}}$ is the total momentum operator and

$$\Omega' \equiv \frac{\beta\hbar^2}{4m}\sum_{j=1}^{N} \nabla_j^2 \Omega = \frac{\beta\hbar^2}{2m}\sum_{i<j} \phi_{ij} \tag{10.66}$$

with

$$
\begin{aligned}
\phi_{ij} &\equiv \phi(|\mathbf{r}_i - \mathbf{r}_j|) \\
\phi(r) &\equiv \nabla^2 v(r)
\end{aligned}
\tag{10.67}
$$

From (10.65) and (10.62) we have

$$Q_N(V, T) \approx \frac{1}{N!}\sum_{\mathbf{p}_1 \cdots \mathbf{p}_N} (\Phi_p,\, e^{-\beta(K_p+\Omega+\Omega') - i(\beta^2\hbar/2m)\mathbf{F} \cdot \mathbf{P}} \Phi_p) \tag{10.68}$$

where $\mathbf{P} = \mathbf{p}_1 + \cdots + \mathbf{p}_N$. Proceeding in a way similar to that for the ideal gas we find

$$Q_N(V, T) \approx \frac{1}{N!\, h^{3N}} \int d^{3N}r\, e^{-\beta(\Omega+\Omega')} J'(1, \ldots, N) \tag{10.69}$$

where

$$J'(1, \ldots, N) \equiv V^N \int d^{3N}p\, e^{-\beta K_p - i(\beta^2\hbar/2m)\mathbf{F}\cdot\mathbf{P}}\, |\Phi_p(1, \ldots, N)|^2 \tag{10.70}$$

The calculation of $J'(1, \ldots, N)$ is similar to that of $J(1, \ldots, N)$. Let

$$\mathbf{R} \equiv \frac{\beta^2\hbar^2}{2m}\mathbf{F} \tag{10.71}$$

Then

$$\begin{aligned} J'(1, \ldots, N) &= \sum_P \delta_P \int d^{3N}p \prod_{j=1}^{N} e^{-\beta(p_j^2/2m) + \mathbf{p}_j\cdot(\mathbf{r}_j - P\mathbf{r}_j - \mathbf{R})/\hbar} \\ &= \left[\int d^3p\, e^{-\beta p^2/2m}\right]^N \sum_P \delta_P [f(\mathbf{r}_1 - P\mathbf{r}_1 - \mathbf{R}) \cdots f(\mathbf{r}_N - P\mathbf{r}_N - \mathbf{R})] \end{aligned} \tag{10.72}$$

where $f(\mathbf{r})$ is defined by (10.52). The approximations (10.54) and (10.57) can again be introduced. Therefore

$$Q_N(V, T) \approx \frac{1}{N!\, h^{3N}} \int d^{3N}p\, d^{3N}r\, e^{-\beta H(p,r)}\, e^{-\beta(\Omega'+\Omega'')} \tag{10.73}$$

where Ω' is given by (10.66) and Ω'' is defined by

$$e^{-\beta\Omega''} \equiv \prod_{i<j} [1 \pm f(\mathbf{r}_i - P\mathbf{r}_j - \mathbf{R})] \qquad \begin{pmatrix}\text{Bose}\\ \text{Fermi}\end{pmatrix} \tag{10.74}$$

Thus (10.39) is obtained if $\Omega' + \Omega''$ can be neglected. By (10.66) and (10.74),

$$\Omega' + \Omega'' = \sum_{i<j} \tilde{v}_{ij}' \tag{10.75}$$

where

$$\tilde{v}_{ij}' \equiv \tilde{v}'(\mathbf{r}_i - \mathbf{r}_j) \tag{10.76}$$

with

$$\begin{aligned} \tilde{v}'(\mathbf{r}) &\equiv \frac{\lambda^2}{4\pi} \nabla^2 v(r) - kT \log(1 \pm e^{-\rho}) \qquad \begin{pmatrix}\text{Bose}\\ \text{Fermi}\end{pmatrix} \\ \rho &= \frac{\pi}{\lambda^2}\left|\mathbf{r} - \frac{\beta^2\hbar^2}{2m}\mathbf{F}\right|^2 \end{aligned} \tag{10.77}$$

Thus $\Omega' + \Omega''$ can be neglected if both the following conditions are fulfilled:

$$(\textit{thermal wavelength}) \ll (\textit{average interparticle distance}) \tag{10.78}$$

$$(\textit{thermal wavelength}) \ll (\textit{some characteristic length of interparticle potential}) \tag{10.79}$$

These are the qualitative conditions under which (10.39) is valid.

10.3 THE VARIATIONAL PRINCIPLE

In quantum mechanics we are familiar with the following variational principle. If the Hamiltonian of the system is a hermitian operator, the ground state energy of a system is the smallest possible value of the expectation value of the Hamiltonian with respect to a normalized wave function that is arbitrary, except that it satisfies the boundary conditions and the symmetries of the system. This principle can be used to find an upper bound of the ground state energy. We shall describe a similar principle for the Helmholtz free energy of a system. It is based on the following theorem of Peierls.*

THEOREM. Let H be the hermitian Hamiltonian operator of a system. Let $\{\Phi_n\}$ be an arbitrary orthonormal set of wave functions† for the system. Then the partition function Q satisfies the following inequality:

$$Q \geq \sum_n e^{-\beta(\Phi_n, H\Phi_n)} \tag{10.80}$$

The equality holds if $\{\Phi_n\}$ is the complete set of eigenfunctions of H.

According to this theorem we may in principle obtain Q by varying the set $\{\Phi_n\}$ until the right-hand side of (10.80) attains its maximum possible value. We recall that the Helmholtz free energy decreases if the partition function increases. Furthermore, $(\Phi_n, H\Phi_n)$ is the energy in first-order perturbation theory, with $\{\Phi_n\}$ as unperturbed wave functions. Thus the following variational principle for the calculation of the Helmholtz free energy may be stated.:

To find the Helmholtz free energy of a system, first find all the energy levels of the system in first-order perturbation theory, with a set of arbitrary unperturbed wave functions $\{\Phi_n\}$. These energy levels define a partition function and hence a Helmholtz free energy. Vary $\{\Phi_n\}$ until the latter attains its smallest possible value. This minimum value is the true Helmholtz free energy.

* R. E. Peierls, *Phys. Rev.*, **54**, 918 (1938). The proof we shall use is due to N. G. Van Kampen (unpublished).

† Note that $\{\Phi_n\}$ does not have to be a complete set.

To prove (10.80) we need only prove it under the assumption that $\{\Phi_n\}$ is a complete set. A set that is not complete may be regarded as a set obtained from a complete set by the omission of certain members. Since the right-hand side of (10.80) is a sum of positive terms, (10.80) holds *a fortiori* for a set that is not complete, if we have proved it for a complete set.

Let $\{\Phi_n\}$ be a complete set of orthonormal wave functions having the symmetry properties and boundary conditions of the problem. Then

$$Q \equiv \sum_n (\Phi_n, e^{-\beta H}\Phi_n) \tag{10.81}$$

Let

$$q \equiv \sum_n e^{-\beta(\Phi_n, H\Phi_n)} \tag{10.82}$$

As noted previously, to prove (10.80) it is sufficient to prove that

$$Q \geq q \tag{10.83}$$

It will be shown that (10.83) is a consequence of the following lemma.

Lemma. Let

(*a*) $\{x_n\}$ be a set of real numbers,

(*b*) $\{c_n\}$ be a set of real numbers such that

$$c_n \geq 0$$

$$\sum_n c_n = 1$$

(*c*) $\overline{f(x)} \equiv \sum_n c_n f(x_n)$, for any $f(x)$.

If $f''(x) \geq 0$, then

$$\overline{f(x)} \geq f(\bar{x})$$

Proof. By the mean value theorem

$$f(x) = f(\bar{x}) + (x - \bar{x})f'(\bar{x}) + \tfrac{1}{2}(x - \bar{x})^2 f''(x_1)$$

where x_1 is a fixed real number. Using the fact that $\sum_n c_n = 1$ we obtain

$$\overline{f(x)} = f(\bar{x}) + \overline{\tfrac{1}{2}(x - \bar{x})^2 f''(x_1)}$$

The second term here is non-negative, because $c_n \geq 0$ and $f''(x) \geq 0$. (QED)

The proof of (10.83) is as follows. Let $\{\Psi_n\}$ be a complete set of orthonormal eigenfunctions of H:

$$H\Psi_n = E_n\Psi_n \tag{10.84}$$

Since H is hermitian E_n is real. There exists a unitary transformation which brings $\{\Psi_n\}$ to $\{\Phi_n\}$:

$$\Phi_n = \sum_m S_{nm}\Psi_m \tag{10.85}$$

where $\{S_{nm}\}$ is a set of complex numbers satisfying

$$\sum_l S_{ln}{}^* S_{lm} = \sum_l S_{nl}{}^* S_{ml} = \delta_{nm} \tag{10.86}$$

Hence

$$Q = \sum_n (\Phi_n, e^{-\beta H} \Phi_n) = \sum_n (\Psi_n, e^{-\beta H} \Psi_n) = \sum_n e^{-\beta E_n} \tag{10.87}$$

We note that

$$(\Phi_n, H\Phi_n) = \sum_l |S_{nl}|^2 E_l \tag{10.88}$$

and

$$\sum_n |S_{nl}|^2 = 1 \tag{10.89}$$

The difference $Q - q$ can therefore be written in the form

$$Q - q = \sum_n \Big(\sum_l |S_{nl}|^2 e^{-\beta E_l} - e^{-\beta \Sigma_l |S_{nl}|^2 E_l} \Big) \tag{10.90}$$

For any n the following definitions satisfy the requirements of the previous lemma:

$$\begin{cases} \bar{E} \equiv \sum_l |S_{nl}|^2 E_l \\ \overline{f(E)} \equiv \sum_l |S_{nl}|^2 e^{-\beta E_l} \end{cases} \tag{10.91}$$

Each term in the sum of (10.90) is of the form $\overline{f(E)} - f(\bar{E})$, which, by the lemma, is non-negative. Therefore $Q - q \geq 0$, which was to be proved.

PROBLEMS

10.1. Derive with the help of the method of saddle point integration a formula for the partition function for an ideal Bose gas of N particles.

10.2. (*a*) Find the equations of state for an ideal Bose gas and an ideal Fermi gas in the limit of high temperatures. Include the first correction due to quantum effects. (Consultation with Problem 8.9 may be helpful.)
(*b*) Estimate, for each of the following ideal gases, the temperature below which quantum effects would become important: H_2, He, N_2.

10.3. *Pair Correlation Function*: The pair correlation function $D(\mathbf{r}_1, \mathbf{r}_2)$ of a system of particles is defined as follows:

$D(\mathbf{r}_1, \mathbf{r}_2)\, d^3r_1\, d^3r_2 \equiv$ probability of simultaneously finding a particle in the volume element $d^3\mathbf{r}_1$ about $\mathbf{r}_1$ and a particle in the volume element d^3r_2 and $\mathbf{r}_2$.

Calculate $D(\mathbf{r}_1, \mathbf{r}_2)$ for an ideal Bose gas and a ideal Fermi gas in the limit of high temperatures. Include quantum corrections only to the lowest approximation.

Solution: Classically we have

$$D(\mathbf{r}_1, \mathbf{r}_2) = \frac{N(N-1) \int d^{3N}p\, d^3r_3 \cdots d^3r_N\, e^{-\beta H(p,r)}}{\int d^{3N}p\, d^{3N}r\, e^{-\beta H(p,r)}}$$

For our problem we use this formula with

$$H(p, r) = \sum_{i=j}^{N} \frac{p_i^{\,2}}{2m} + \sum_{i<j} \tilde{v}_{ij}$$

To avoid complications assume that the density of the gas is almost zero. The limit $N \to \infty$, $V \to \infty$ should be so taken that $N/V \to 0$. Then

$$D(\mathbf{r}_1, \mathbf{r}_2) \approx \frac{N(N-1)V^{N-2}\left[1 \pm f_{12}^{\,2} \pm \dfrac{N(N-1)}{2V}\displaystyle\int d^3r\, f^2(r)\right]}{1 \pm \dfrac{N(N-1)}{2V}\displaystyle\int d^3r\, f^2(r)}$$

$$\approx \frac{1}{v^2}\left[1 \pm \exp\left(-\frac{2\pi}{\lambda^2}|\mathbf{r}_1 - \mathbf{r}_2|^2\right)\right]$$

This result continues to hold for finite v with $\lambda^3/v \ll 1$, although our derivation did not justify such a conclusion.

10.4. *Models for Ferromagnetism*: Consider a lattice of N fixed atoms of spin $\frac{1}{2}$. The quantum mechanical spin operators of the ith atom are the Pauli spin matrices $\boldsymbol{\sigma}_i$. Assuming that only nearest neighbors interact via a spin-spin interaction, we obtain the *Heisenberg model of ferromagnetism.* The Hamiltonian is

$$H_{\text{Heisenberg}} = -\lambda \sum_{\langle ij\rangle} \boldsymbol{\sigma}_i \cdot \boldsymbol{\sigma}_j - \mu \sum_{i=1}^{N} \boldsymbol{\sigma}_i \cdot \mathbf{B}$$

where $\langle ij\rangle$ denotes a nearest-neighbor pair, $\mathbf{B}$ is a uniform external magnetic field, and λ and μ are positive constants.

Another model, the *Ising model*, is constructed by associating with the ith atom a *number* s_i that is either $+1$ or -1 and taking the Hamiltonian to be

$$H_{\text{Ising}} = -\lambda \sum_{\langle ij\rangle} s_i s_j - \mu \sum_{i=1}^{N} s_i B$$

where B is the z component of $\mathbf{B}$.

Using the variational principle prove that, for the same temperature, the Helmholtz free energy of the Heisenberg model is not greater than that of the Ising model.

chapter 11

IDEAL FERMI GAS

11.1 EQUATION OF STATE OF AN IDEAL FERMI GAS

The equation of state of a spinless ideal Fermi gas is obtained by eliminating z from the equations (9.67). We first study the behavior of z as determined by the second equation of (9.67), namely

$$\frac{\lambda^3}{v} = f_{3/2}(z) \tag{11.1}$$

where $v = V/N$ and

$$f_{3/2}(z) = \frac{4}{\sqrt{\pi}} \int_0^\infty dx \, \frac{x^2}{z^{-1} e^{x^2} + 1} \tag{11.2}$$

It can be easily verified that $f_{3/2}(z)$ is a monotonically increasing function of z. For small z we have the power series expansion

$$f_{3/2}(z) = z - \frac{z^2}{2^{3/2}} + \frac{z^3}{3^{3/2}} - \frac{z^4}{4^{3/2}} + \cdots \tag{11.3}$$

For large z an asymptotic expansion may be obtained through a method due to Sommerfeld, as follows. Let $kT\nu$ be the chemical potential, which is defined by

$$kT\nu = \left(\frac{\partial A}{\partial N}\right)_{V,T} \tag{11.4}$$

and* is related to z by

$$\nu = \log z \tag{11.5}$$

Then

$$f_{3/2}(z) = \frac{4}{\sqrt{\pi}} \int_0^\infty dx \frac{x^2}{e^{x^2-\nu} + 1}$$
$$= \frac{2}{\sqrt{\pi}} \int_0^\infty dy \frac{\sqrt{y}}{e^{y-\nu} + 1} = \frac{4}{3\sqrt{\pi}} \int_0^\infty dy \frac{y^{3/2} e^{y-\nu}}{(e^{y-\nu} + 1)^2} \tag{11.6}$$

The last step is obtained through a partial integration. Expanding $y^{3/2}$ in a Taylor series about ν, we obtain

$$f_{3/2}(z) = \frac{4}{3\sqrt{\pi}} \int_0^\infty dy \frac{e^{y-\nu}}{(e^{y-\nu} + 1)^2} [\nu^{3/2} + \tfrac{3}{2}\nu^{1/2}(y-\nu) + \tfrac{3}{8}\nu^{-1/2}(y-\nu)^2 + \cdots]$$
$$= \frac{4}{3\sqrt{\pi}} \int_{-\nu}^\infty dt \frac{e^t}{(e^t + 1)^2} (\nu^{3/2} + \tfrac{3}{2}\nu^{1/2} t + \tfrac{3}{8}\nu^{-1/2} t^2 + \cdots)$$

Now we write

$$\int_{-\nu}^\infty = \int_{-\infty}^{+\infty} - \int_{-\infty}^{-\nu}$$

The second integral is of order $e^{-\nu}$. Therefore

$$f_{3/2}(z) = \frac{4}{3\sqrt{\pi}} \int_{-\infty}^{+\infty} dt \frac{e^t}{(e^t + 1)^2} (\nu^{3/2} + \tfrac{3}{2}\nu^{1/2} t + \tfrac{3}{8}\nu^{-1/2} t^2 + \cdots) + O(e^{-\nu})$$
$$= \frac{4}{3\sqrt{\pi}} (I_0 \nu^{3/2} + \tfrac{3}{2} I_1 \nu^{1/2} + \tfrac{3}{8} I_2 \nu^{-1/2} + \cdots) + O(e^{-\nu}) \tag{11.7}$$

where

$$I_n \equiv \int_{-\infty}^{+\infty} dt \frac{t^n e^t}{(e^t + 1)^2} \tag{11.8}$$

Apart from the factor t^n, the integrand is an even function of t. Hence $I_n = 0$ for odd n. For even n, we have

$$I_0 = -2 \int_0^\infty dt \frac{d}{dt} \frac{1}{(e^t + 1)} = 1 \tag{11.9}$$

and for even $n > 0$,

$$I_n = -2 \left[\frac{\partial}{\partial \lambda} \int_0^\infty dt \frac{t^{n-1}}{e^{\lambda t} + 1} \right]_{\lambda=1} = 2n \int_0^\infty du \frac{u^{n-1}}{e^u + 1}$$
$$= (n-1)!\,(2n)(1 - 2^{1-n})\zeta(n) \tag{11.10}$$

* See Problem 8.5.

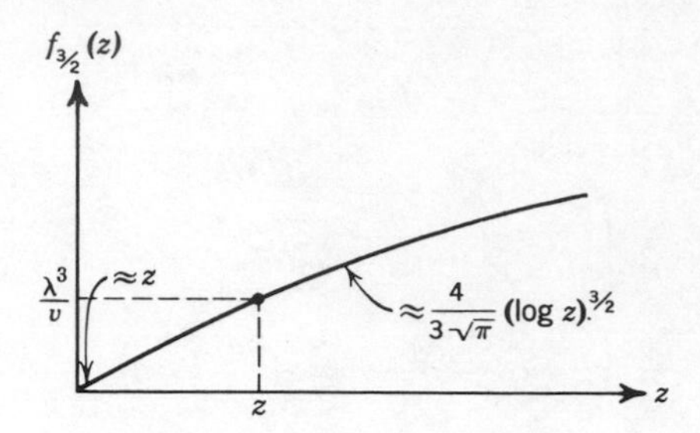

Fig. 11.1. The function $f_{\frac{3}{2}}(z)$.

where $\zeta(n)$ is the Riemann zeta function of n and is a tabulated function.* Some values of $\zeta(n)$ are:

$$\zeta(2) = \frac{\pi^2}{6}, \qquad \zeta(4) = \frac{\pi^4}{90}, \qquad \zeta(6) = \frac{\pi^6}{945}$$

Hence

$$f_{3/2}(z) = \frac{4}{3\sqrt{\pi}}\left[(\log z)^{3/2} + \frac{\pi^2}{8}(\log z)^{-1/2} + \cdots\right] + O(z^{-1}) \tag{11.11}$$

A graph of $f_{3/2}(z)$ is shown in Fig. 11.1. For any given positive value of λ^3/v, the value of z determined by (11.1) can be read off such a graph. It is seen that z increases monotonically as λ^3/v increases. For fixed v, z increases monotonically as the temperature decreases.

High Temperatures and Low Densities ($\lambda^3/v \ll 1$)

For $\lambda^3/v \ll 1$ the average interparticle separation $v^{1/3}$ is much larger than the thermal wavelength λ. We would expect quantum effects to be negligible. From (11.1) and (11.3) we obtain

$$\frac{\lambda^3}{v} = z - \frac{z^2}{2^{3/2}} + \cdots$$

which may be solved to give

$$z = \frac{\lambda^3}{v} + \frac{1}{2\sqrt{2}}\left(\frac{\lambda^3}{v}\right)^2 + \cdots \tag{11.12}$$

Thus z reduces to that of the Boltzmann gas [equation (9.52)] when $\lambda^3 \to 0$ ($T \to \infty$). The average occupation number (9.65) reduces to Maxwell-Boltzmann form:

$$\langle n_{\mathbf{p}} \rangle \approx \frac{\lambda^3}{v} e^{-\beta\epsilon_{\mathbf{p}}} \tag{11.13}$$

* See Jahnke and Emde, *Tables of Functions* (Dover, New York, 1945), p. 269.

The equation of state (9.67) becomes

$$\frac{Pv}{kT} = \frac{v}{\lambda^3}\left(z - \frac{z^2}{2^{5/2}} + \cdots\right) = 1 + \frac{1}{2^{5/2}}\frac{\lambda^3}{v} + \cdots \tag{11.14}$$

This is in the form of a virial expansion. The corrections to the classical ideal gas law, however, are not due to molecular interactions, but to quantum effects. The second virial coefficient in this case is

$$\frac{\lambda^3}{2^{5/2}} = \frac{1}{2^{5/2}}\left(\frac{2\pi\hbar^2}{mkT}\right)^{3/2} \tag{11.15}$$

All other thermodynamic functions reduce to those for a classical ideal gas plus small corrections.

Low Temperatures and High Densities ($\lambda^3/v \gg 1$)

For $\lambda^3/v \gg 1$ the average de Broglie wavelength of a particle is much greater than the average interparticle separation. Thus quantum effects, in particular the effects of the Pauli exclusion principle, become all important.

In the neighborhood of absolute zero we have, from (11.1) and (11.11),

$$\frac{1}{v}\left(\frac{2\pi\hbar^2}{mkT}\right)^{3/2} \approx \frac{4}{3\sqrt{\pi}}(\log z)^{3/2} \tag{11.16}$$

Hence

$$z \approx e^{\beta\epsilon_F} \tag{11.17}$$

where

$$\epsilon_F \equiv \frac{\hbar^2}{2m}\left(\frac{6\pi^2}{v}\right)^{2/3} \tag{11.18}$$

This is called the *Fermi energy*. To study its physical significance, let us examine $\langle n_{\mathbf{p}}\rangle$ near absolute zero:

$$\langle n_{\mathbf{p}}\rangle \approx \frac{1}{e^{\beta(\epsilon_{\mathbf{p}}-\epsilon_F)} + 1} \tag{11.19}$$

If $\epsilon_{\mathbf{p}} < \epsilon_F$, then the exponential in the denominator vanishes as $T \to 0$ ($\beta \to \infty$). Hence $\langle n_{\mathbf{p}}\rangle = 1$. Otherwise, $\langle n_{\mathbf{p}}\rangle = 0$. Thus

$$\langle n_{\mathbf{p}}\rangle_{T=0} = \begin{cases} 1 & (\epsilon_{\mathbf{p}} < \epsilon_F) \\ 0 & (\epsilon_{\mathbf{p}} > \epsilon_F) \end{cases} \tag{11.20}$$

The physical meaning of this formula is clear. On account of the Pauli exclusion principle no two particles can be in the same level. Therefore, in the ground state of the system, the particles occupy the lowest possible levels and fill the levels up to the finite energy level ϵ_F. In momentum space the particles fill a sphere of radius p_F. This sphere is sometimes called the Fermi sphere. Thus ϵ_F is simply the single-particle energy level below which there are exactly N states. With this interpretation, let us calculate ϵ_F independently. The condition determining ϵ_F is

$$\sum_{\mathbf{p}} \langle n_{\mathbf{p}} \rangle_{T=0} = N$$

which, by (11.20), is equivalent to

$$\frac{V}{h^3} \int_{\epsilon_{\mathbf{p}} < \epsilon_F} d^3p = N$$

Putting $\epsilon_F = p_F^2/2m$, we find that p_F must satisfy the condition

$$\frac{V}{(2\pi\hbar)^3} \frac{4\pi}{3} p_F^3 = N \tag{11.21}$$

which leads to (11.18).

The last calculation also shows how ϵ_F is to be modified if the particles have spin. If the spin of a particle is $\hbar s$, then for a given momentum $\mathbf{p}$ there are $2s + 1$ single-particle states, all having the same energy $\epsilon_{\mathbf{p}}$. Therefore (11.21) must be modified to read

$$(2s + 1) \frac{V}{(2\pi\hbar)^3} \frac{4\pi}{3} p_F^3 = N \tag{11.22}$$

which leads to

$$\epsilon_F = \frac{\hbar^2}{2m} \left(\frac{6\pi^2}{2s + 1} \frac{1}{v} \right)^{2/3} \tag{11.23}$$

We can also interpret (11.22) in the following way. A particle of spin $\hbar s$ can have $2s + 1$ different spin orientations. Particles that have different spin orientations can have any symmetry relative to the interchange of their *position* coordinates. Hence we can consider a system of N fermions of spin $\hbar s$ to be made up of $2s + 1$ independent Fermi gases each having $N/(2s + 1)$ spinless fermions, and (11.22) immediately follows.

To obtain the thermodynamic functions for low temperatures and high densities we first obtain the expansion for the chemical potential from (11.1) and (11.11):

$$kT\nu = kT \log z = \epsilon_F \left[1 - \frac{\pi^2}{12} \left(\frac{kT}{\epsilon_F} \right)^2 + \cdots \right] \tag{11.24}$$

The expansion parameter is kT/ϵ_F. If we define the *Fermi temperature* T_F, which is a function of density, by

$$kT_F \equiv \epsilon_F \tag{11.25}$$

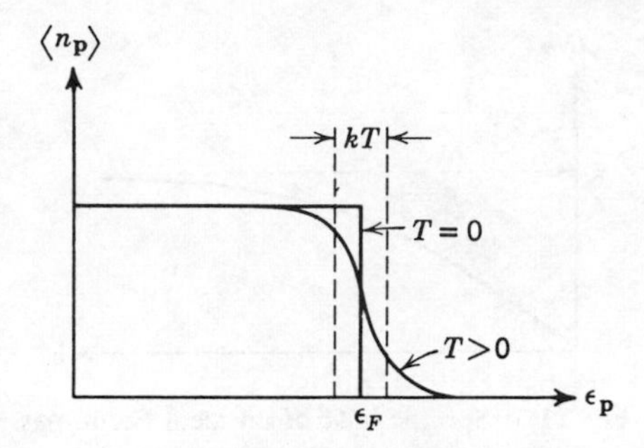

Fig. 11.2. Average occupation number in an ideal Fermi gas.

then low temperature and high density means $T \ll T_F$. In this domain the gas is said to be *degenerate* because the particles tend to go to the lowest energy levels possible. For this reason T_F is also called the *degeneracy temperature*.

The average occupation number is

$$\langle n_{\mathbf{p}} \rangle = \frac{1}{e^{\beta \epsilon_{\mathbf{p}} - \nu} + 1} \tag{11.26}$$

where ν is given by (11.24). Since $\epsilon_{\mathbf{p}} = p^2/2m$, $n_{\mathbf{p}}$ depends on $\mathbf{p}$ only through p^2. A sketch of $n_{\mathbf{p}}$ is shown in Fig. 11.2.

The internal energy is

$$U = \sum_{\mathbf{p}} \epsilon_{\mathbf{p}} \langle n_{\mathbf{p}} \rangle = \frac{V}{h^3} \frac{4\pi}{2m} \int_0^\infty dp\, p^4 \langle n_{\mathbf{p}} \rangle$$

After a partial integration we obtain

$$U = \frac{V}{4\pi^2 m \hbar^3} \int_0^\infty dp \frac{p^5}{5} \left(-\frac{\partial}{\partial p} \langle n_{\mathbf{p}} \rangle \right) = \frac{\beta V}{20\pi^2 m^2 \hbar^3} \int_0^\infty dp \frac{p^6 e^{\beta \epsilon_{\mathbf{p}} - \nu}}{(e^{\beta \epsilon_{\mathbf{p}} - \nu} + 1)^2} \tag{11.27}$$

It is apparent from Fig. 11.2 that $\partial \langle n_{\mathbf{p}} \rangle / \partial p$ is sharply peaked at $p = p_F$. In fact, at absolute zero it is a δ-function at $p = p_F$. Therefore the integral in (11.27) can be evaluated by expanding the factor p^6 about $p = p_F$. The procedure is similar to that used in obtaining (11.11). After inserting ν from (11.24) we obtain the asymptotic expansion

$$U = \tfrac{3}{5} N \epsilon_F \left[1 + \tfrac{5}{12}\pi^2 \left(\frac{kT}{\epsilon_F} \right)^2 + \cdots \right] \tag{11.28}$$

The first term is the ground state energy of the Fermi gas at the given density, as we can verify by showing the following:

$$\sum_{|\mathbf{p}| < p_F} \frac{p^2}{2m} = \tfrac{3}{5} N \epsilon_F \tag{11.29}$$

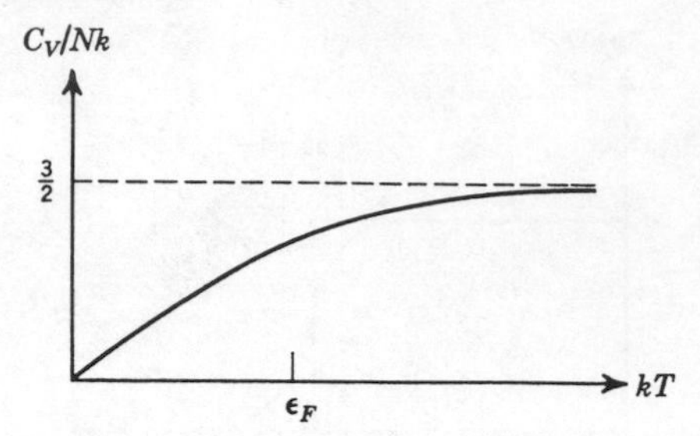

Fig. 11.3. Specific heat of an ideal Fermi gas.

The specific heat at constant volume can be immediately obtained from (11.28);

$$\frac{C_V}{Nk} \approx \frac{\pi^2}{2} \frac{kT}{\epsilon_F} \tag{11.30}$$

It vanishes linearly as $T \to 0$, thus verifying the third law of thermodynamics. We know that C_V/Nk approaches $\frac{3}{2}$ as $T \to \infty$. Thus a rough sketch of C_V/Nk can be made, as shown in Fig. 11.3. The fact that it is proportional to T at these low temperatures can be understood as follows. At a temperature $T > 0$, $\langle n_\mathbf{p} \rangle$ differs from that at $T = 0$ because a certain number of particles are excited to energy levels $\epsilon_\mathbf{p} > \epsilon_F$. Roughly speaking, particles with energies of order kT below ϵ_F are excited to energies of order kT above ϵ_F (see Fig. 11.2). The number of particles excited is therefore of the order of $(kT/\epsilon_F)N$. Therefore the total excitation energy above the ground state is $\Delta U \approx (kT/\epsilon_F)NkT$, from which follows $C_V \approx (kT/\epsilon_F)Nk$.

From (9.78) and (11.28) follows the equation of state

$$P = \frac{2}{3}\frac{U}{V} = \frac{2}{5}\frac{\epsilon_F}{v}\left[1 + \frac{5\pi^2}{12}\left(\frac{kT}{\epsilon_F}\right)^2 + \cdots\right] \tag{11.31}$$

This shows that even at absolute zero it is necessary to contain the ideal Fermi gas with externally fixed walls because the pressure does not vanish. This is a manifestation of the Pauli exclusion principle, which allows only one particle to have zero momentum. All other particles must have finite momentum and give rise to the zero-point pressure.

To obtain the thermodynamic function for arbitrary values of λ^3/v numerical methods must be employed to calculate the functions $f_{3/2}(z)$ and $f_{3/2}(z)$.

11.2 THEORY OF WHITE DWARF STARS

It is an empirical rule that the brightness of a star is proportional to its color (i.e., predominant wavelength emitted). The proportionality constant is roughly the same for all stars. Thus if we make a plot of brightness against color, we obtain what is known as the Hertzprung-Russell diagram, in which most stars fall within a linear strip called the main sequence, as shown in Fig. 11.4. There are, however, stars that are exceptions to this rule. There are the red giant stars, huge stars which are abnormally bright for their red color; and there are the white dwarf stars, small stars which are abnormally faint for their white color. The white dwarf star makes an interesting subject for our study, because to a good approximation it is a degenerate Fermi gas.

A detailed study of the constitution of white dwarf stars leads to the conclusion that they lack brightness because the hydrogen supply, which is the main energy source of stars, has been used up, and they are composed mainly of helium. What little brightness they have is derived from the gravitational energy released through a slow contraction of the star. Probably these stars have reached the end point of stellar evolution. One of the nearest stars to the solar system, the companion of Sirius, 8 light years from us, is a white dwarf. Being so faint that it escapes the naked eye, it was first predicted by the calculations of Bessel, who tried to explain why Sirius apparently moves about a point in empty space.

An idealized model of a white dwarf may be constructed from some typical data for such a star:

$$\text{Content: mostly helium}$$

$$\text{Density} \approx 10^7 \text{ gram/cc} \approx 10^7 \rho_\odot$$

$$\text{Mass} \approx 10^{33} \text{ grams} \approx M_\odot$$

$$\text{Central temperature} \approx 10^7\ {}^\circ\text{K} \approx \text{T}_\odot$$

where the subscript $\odot$ denotes quantities referring to the sun. Thus a white dwarf star is a mass of helium at an extremely high temperature and under extreme compression. The temperature $10^7\ {}^\circ$K corresponds to a

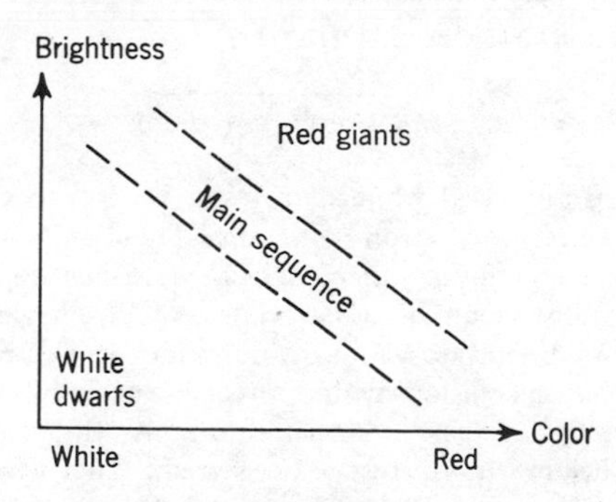

Fig. 11.4. Russell-Hertzprung diagram.

thermal energy of 1000 ev. Hence the helium atoms are expected to be completely ionized, and the star may be regarded as a gas composed of helium nuclei and electrons. We regard the gas of electrons as an ideal Fermi gas, with a density of approximately 10^{30} electrons/cc. This corresponds to a Fermi energy of

$$\epsilon_F \approx \frac{\hbar^2}{2m} \frac{1}{v^{2/3}} \approx 20 \text{ Mev}$$

and a Fermi temperature of

$$T_F \approx 10^{11} \,^\circ\text{K}$$

Since the Fermi temperature is much greater than the temperature of the star, the electron gas is a highly degenerate Fermi gas, which behaves no differently from an electron gas at absolute zero. In fact we regard the electron gas to be an ideal Fermi gas in its ground state. The enormous zero-point pressure exerted by the electron gas is counteracted by the gravitational attraction that binds the star. This gravitational binding is due almost entirely to the helium nuclei in the star. The pressure due to kinetic motion of the helium nuclei, and to any radiation that may be present, will be neglected.

Thus we arrive at the following idealized model: A white dwarf is taken to be a system of N electrons in its ground state, at such a density that the electrons must be treated by relativistic dynamics. The electrons move in a background of $N/2$ motionless helium nuclei which provide the gravitational attraction to hold the entire system together.* This model must then exhibit properties that are the combined effects of the Pauli principle, relativistic dynamics, and the gravitational law.

First let us work out the pressure exerted by a Fermi gas of relativistic electrons in the ground state. The states for a single electron are specified by the momentum $\mathbf{p}$ and the spin quantum number $s = \pm\frac{1}{2}$. The single-particle energy levels are independent of s:

$$\epsilon_{\mathbf{p}s} = \sqrt{(pc)^2 + (m_e c^2)^2}$$

* The temperature in an actual white dwarf star is so high that electron-positron pairs can be created in electron-electron collisions. These pairs in turn annihilate into radiation. Therefore in equilibrium there should be a certain number of electron-positron pairs and a certain amount of radiation present. We neglect the effects of these. It has been speculated that neutrinos can also be created in electron-electron, electron-positron and photon-photon collisions with appreciable probability. This leads to some interesting phenomena, for neutrinos interact so weakly with matter that they do not come to thermal equilibrium with the rest of the system. They simply leave the star and cause a constant drain of energy. [H. Y. Chiu and P. Morrison, *Phys. Rev. Lett.* **5,** 573 (1960).] Our model is based on the neglect of these effects.

where m_e is the mass of an electron. The ground state energy of the Fermi gas is

$$E_0 = 2 \sum_{|\mathbf{p}|<p_F} \sqrt{(pc)^2 + (m_e c^2)^2} = \frac{2V}{h^3}\int_0^{p_F} dp\, 4\pi p^2 \sqrt{(pc)^2 + (m_e c^2)^2} \qquad (11.32)$$

where p_F, the Fermi momentum, is defined by

$$\frac{V}{h^3}\left(\frac{4}{3}\pi p_F^{\,3}\right) = \frac{N}{2}$$

or

$$p_F = \hbar\left(\frac{3\pi^2}{v}\right)^{1/3} \qquad (11.33)$$

Changing the variable of integration in (11.32) to $x = p/m_e c$ we obtain

$$\frac{E_0}{N} = \frac{m_e^{\,4} c^5}{\pi^2 \hbar^3} v f(x_F) \qquad (11.34)$$

where

$$f(x_F) = \int_0^{x_F} dx\, x^2\sqrt{1 + x^2} = \begin{cases} \frac{1}{3}x_F^{\,3}(1 + \frac{3}{10}x_F^{\,2} + \cdots) & (x_F \ll 1) \\ \frac{1}{4}x_F^{\,4}\left(1 + \dfrac{1}{x_F^{\,2}} + \cdots\right) & (x_F \gg 1) \end{cases} \qquad (11.35)$$

and

$$x_F \equiv \frac{p_F}{m_e c} = \frac{\hbar}{m_e c}\left(\frac{3\pi^2}{v}\right)^{1/3} \qquad (11.36)$$

If the total mass of the star is M and the radius of the star is R, then

$$M = (m_e + 2m_P)N \approx 2m_P N$$

$$R = \left(\frac{3V}{4\pi}\right)^{1/3} \qquad (11.37)$$

where m_P is the mass of a proton. In terms of M and R we have

$$v = \frac{8\pi}{3}\frac{m_P R^3}{M} \qquad (11.38)$$

and

$$x_F = \frac{\hbar}{m_e c}\frac{1}{R}\left(\frac{9\pi}{8}\frac{M}{m_P}\right)^{1/3} \equiv \frac{\bar{M}^{1/3}}{\bar{R}} \qquad (11.39)$$

where

$$\bar{M} = \frac{9\pi}{8}\frac{M}{m_P}$$

$$\bar{R} = \frac{R}{(\hbar/m_e c)} \qquad (11.40)$$

The pressure exerted by the Fermi gas is

$$P_0 = -\frac{\partial E_0}{\partial V} = \frac{m_e^4 c^5}{\pi^2 \hbar^3}\left[-f(x_F) - \frac{\partial f(x_F)}{\partial x_F} v \frac{\partial x_F}{\partial v}\right]$$

$$= \frac{m_e^4 c^5}{\pi^2 \hbar^3}\left[\frac{1}{3} x_F^3 \sqrt{1 + x_F^2} - f(x_F)\right] \tag{11.41}$$

The nonrelativistic and extreme relativistic limits of P_0 are given by

$$P_0 \approx \left(\frac{m_e^4 c^5}{15\pi^2 \hbar^3}\right) x_F^5 = \tfrac{4}{5} K \frac{\bar{M}^{5/3}}{\bar{R}^5} \qquad \text{(nonrel.: } x_F \ll 1) \tag{11.42}$$

$$P_0 \approx \left(\frac{m_e^4 c^5}{12\pi^2 \hbar^3}\right)(x_F^4 - x_F^2) = K\left(\frac{\bar{M}^{4/3}}{\bar{R}^4} - \frac{\bar{M}^{2/3}}{\bar{R}^2}\right) \qquad \text{(extreme rel.: } x_F \gg 1) \tag{11.43}$$

where

$$K = \frac{m_e c^2}{12\pi^2}\left(\frac{m_e c}{\hbar}\right)^3 \tag{11.44}$$

A qualitative plot of P_0 against R for fixed M is shown in Fig. 11.5. It is seen that, for small R, P_0 becomes smaller than what is expected on the basis of nonrelativistic dynamics.

The condition for equilibrium of the star may be obtained through the following argument. Let us first imagine that there is no gravitational interaction. Then the density of the system will be uniform, and external walls will be needed to keep the Fermi gas at a given density. The amount of work that an external agent has to do in order to compress the star of given mass from a state of infinite diluteness to a state of finite density would be given by

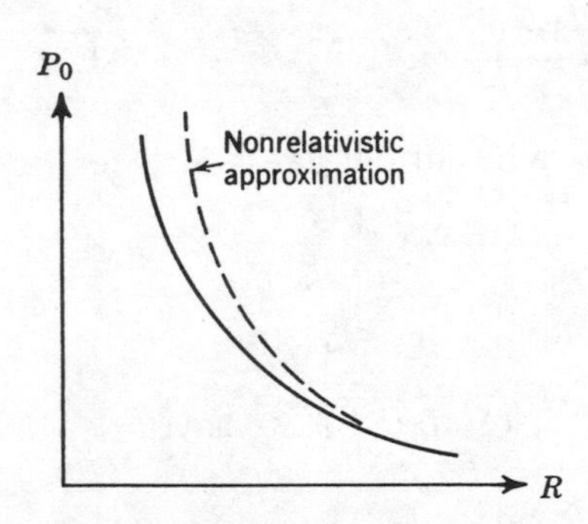

Fig. 11.5. Pressure of an ideal Fermi gas at absolute zero.

$$-\int_\infty^R P_0 4\pi r^2\, dr \tag{11.45}$$

where P_0 is the pressure of a uniform Fermi gas and R is the radius of the star. Now imagine that the gravitational interaction is "switched on." Different parts of the star will now attract one another, resulting in a decrease of the energy of the star by an amount that is called the gravitational self-energy. On dimensional grounds the gravitational self-energy must have the form

$$-\frac{\alpha\gamma M^2}{R} \tag{11.46}$$

where γ is the gravitational constant and α is a pure number of the order of unity. The exact value of α depends on the functional form of the density as a function of spatial distance and cannot be determined by our argument. If R is the equilibrium radius of the star, the gravitational self-energy must exactly compensate the work done in bringing the star together. Hence

$$\int_{\infty}^{R} P_0 4\pi r^2 \, dr = -\frac{\alpha \gamma M^2}{R} \tag{11.47}$$

Differentiating (11.47) with respect to R we obtain the condition for equilibrium:

$$P_0 = \frac{\alpha}{4\pi} \frac{\gamma M^2}{R^4} = \frac{\alpha}{4\pi} \gamma \left(\frac{8m_P}{9\pi}\right)^2 \left(\frac{m_e c}{\hbar}\right)^4 \frac{\bar{M}^2}{\bar{R}^4} \tag{11.48}$$

Strictly speaking, (11.47) merely defines α. Its physical content is furnished by the assumption that α is of the order of unity. We now determine the relation between M and R by inserting an appropriate expression for P_0 into (11.48). This will be done for the following three different cases:

(*a*) Suppose the temperature of the electron gas is much higher than the Fermi temperature. Then the electron gas may be considered as an ideal Boltzmann gas, with

$$P_0 = \frac{kT}{v} = \frac{3kT}{8\pi m_P} \frac{M}{R^3}$$

Substitution of this into (11.48) yields the linear relation

$$R = \tfrac{2}{3} \alpha M \frac{m_P \gamma}{kT} \tag{11.49}$$

This case, however, is never applicable for a white dwarf star.

(*b*) Suppose the electron gas is at such a low density that nonrelativistic dynamics may be used ($x_F \ll 1$). Then P_0 is given by (11.42), and (11.48) leads to the equilibrium condition

$$\tfrac{4}{5} K \frac{\bar{M}^{5/3}}{\bar{R}^5} = K' \frac{\bar{M}^2}{\bar{R}^4}$$

where

$$K' = \frac{\alpha}{4\pi} \gamma \left(\frac{8m_P}{9\pi}\right)^2 \left(\frac{m_e c}{\hbar}\right)^4 \tag{11.50}$$

Thus the radius of the star decreases as the mass of the star increases:

$$\bar{M}^{1/3} \bar{R} = \frac{4}{5} \frac{K}{K'} \tag{11.51}$$

This condition is valid when the density is low. Hence it is valid for small M and large R.

(*c*) Suppose the electron gas is at such a high density that relativistic effects are important ($x_F \gg 1$). Then P_0 is given by (11.43). The equilibrium condition becomes

$$K\left(\frac{\bar{M}^{4/3}}{\bar{R}^4} - \frac{\bar{M}^{2/3}}{\bar{R}^2}\right) = K'\frac{\bar{M}^2}{\bar{R}^4} \tag{11.52}$$

or

$$\bar{R} = \bar{M}^{2/3}\sqrt{1 - (\bar{M}/\bar{M}_0)^{2/3}} \tag{11.53}$$

where

$$\bar{M}_0 = \left(\frac{K}{K'}\right)^{3/2} = \left(\frac{27\pi}{64\alpha}\right)^{3/4}\left(\frac{\hbar c}{\gamma m_P{}^2}\right)^{3/2} \tag{11.54}$$

Numerically,

$$\frac{\hbar c}{\gamma m_P{}^2} \approx 10^{39} \tag{11.55}$$

This interesting pure number is the rest energy of X divided by the gravitational attraction of two protons separated by the Compton wavelength of X, where X is anything. The mass M_0 corresponding to the reduced quantity $\bar{M}_0$ is (taking $\alpha \approx 1$):

$$M_0 = \frac{8}{9\pi}\, m_P \bar{M}_0 \approx 10^{33}\ \text{g} \approx M_\odot \tag{11.56}$$

the mass of the sun. The formula (11.53) is valid for high densities or for $R \to 0$. Hence it is valid for M near M_0. Our model yields the remarkable predication that no white dwarf star can have a mass larger than M_0, because otherwise (11.53) would give an imaginary radius. The physical reason underlying this result is that if the mass is greater than a certain amount, the pressure coming from the Pauli exclusion principle is not sufficient to support the gas against gravitational collapse.

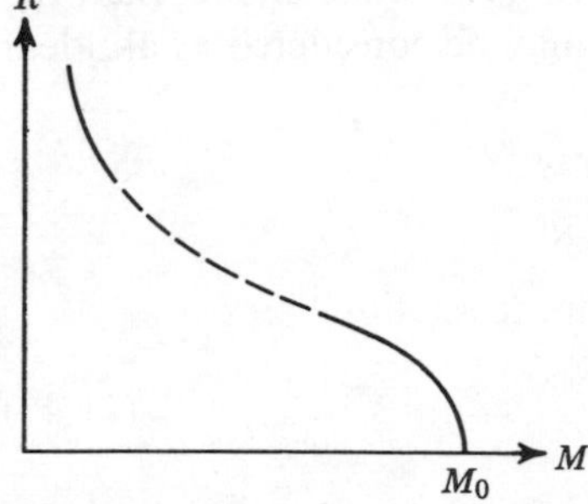

Fig. 11.6. Radius-mass relationship of a white dwarf star.

The radius-mass relationship of a white dwarf star, according to our model, has the form shown in Fig. 11.6, where the solid lines indicate the regions covered by formulas (11.51) and (11.53). We have not been able to calculate α, so that an exact value of M_0 cannot be obtained. More refined considerations* give the result

$$M_0 = 1.4 M_\odot \tag{11.57}$$

* S. Chandrasekhar, *Stellar Structure* (Dover, New York, 1957), Chapter XI.

This mass is known as the Chandrasekhar limit. Thus according to our model no star can become a white dwarf unless its mass is less than $1.4M_{\odot}$. This conclusion has so far been verified by astronomical observations. The answer to the question of how a star evolves if its mass is greater than $1.4M_{\odot}$ is not known. It is possible that these stars end their lives by becoming supernovae.

11.3 LANDAU DIAMAGNETISM

Van Leeuwen's theorem* states that the phenomenon of diamagnetism is absent in classical statistical mechanics. Landau† first showed how diamagnetism arises from the quantization of the orbits of charged particles in a magnetic field.

The magnetic susceptibility per unit volume of a system is defined to be

$$\chi \equiv \frac{\partial \mathscr{M}}{\partial B} \tag{11.58}$$

where $\mathscr{M}$ is the average induced magnetic moment per unit volume of the system along the direction of an external magnetic field B:

$$\mathscr{M} \equiv \frac{1}{V}\left\langle -\frac{\partial H}{\partial B}\right\rangle \tag{11.59}$$

where H is the Hamiltonian of the system in the presence of an external magnetic field B. For weak fields the Hamiltonian H depends on B linearly. In the canonical ensemble we have

$$\mathscr{M} = kT\frac{\partial}{\partial B}\frac{\log Q_N}{V} \tag{11.60}$$

and in the grand canonical ensemble we have

$$\mathscr{M} = kT\frac{\partial}{\partial B}\left(\frac{\log \mathscr{Q}}{V}\right)_{T,V,z} \tag{11.61}$$

where z is to be eliminated in terms of N by the usual procedure.

A system is said to be diamagnetic if $\chi < 0$; paramagnetic if $\chi > 0$. In order to understand diamagnetism in the simplest possible terms, we construct an idealized model of a physical substance that exhibits diamagnetism. The magnetic properties of a physical substance are mainly due to the electrons in the substance. These electrons are either bound to atoms or nearly free. In the presence of an external magnetic field two

* See Problem 8.7.
† L. Landau, *Z. Phys.*, **64**, 629 (1930).

effects are important for the magnetic properties of the substance: (*a*) The electrons, free or bound, move in quantized orbits in the magnetic field. (*b*) The spins of the electrons tend to be aligned parallel to the magnetic field. The atomic nuclei contribute little to the magnetic properties except through their influence on the wave functions of the electrons. They are too massive to have significant orbital magnetic moments, and their intrinsic magnetic moments are about 10^{-3} times smaller than the electron's. The alignment of the electron spin with the external magnetic field gives rise to paramagnetism, whereas the orbital motions of the electrons give rise to diamagnetism. In a physical substance these two effects compete. We completely ignore paramagnetism for the present, however. The effect of atomic binding on the electrons is also ignored. Thus we consider the idealized problem of a free spinless electron gas in an external magnetic field. This model will illustrate the important fact that diamagnetism arises from the quantization of orbits, but it is too simplified a model to be of use in physical applications.

Consider a system of N spinless electrons contained in a volume V. The electrons are free except for their interactions with an uniform external magnetic field $\mathbf{B}$. To calculate the partition function we first calculate the energy levels of a single particle. This we do via a short cut, the "old" quantum theory.* According to the "old" quantum theory the "allowed" orbits of a charged particle in an external field are the classical orbits that satisfy the quantum conditions

$$\oint \mathbf{p} \cdot d\mathbf{r} = (j + \tfrac{1}{2})h \qquad (j = 0, 1, 2, \ldots) \tag{11.62}$$

where $\mathbf{p}$, $\mathbf{r}$ are the *classical* canonical variables of a particle. The allowed energies of a particle are those values of the classical Hamiltonian $H(\mathbf{p}, \mathbf{r})$ consistent with (11.62). In this way the "old" quantum theory furnishes us with the correct energy levels and their degeneracies, which are sufficient for the calculation of the partition function.

The Hamiltonian for a single electron is taken to be

$$H(\mathbf{p}, \mathbf{r}) = \frac{1}{2m}\left(\mathbf{p} + \frac{e}{c}\mathbf{A}\right)^2 \tag{11.63}$$

where m is the mass and $-e$ is the charge of an electron and the vector potential $\mathbf{A}$ is related to the uniform external magnetic field $\mathbf{B}$ by

$$\mathbf{B} = \nabla \times \mathbf{A} \tag{11.64}$$

* An alternative procedure consists of solving the Schrödinger equation, which would yield not only the energy levels but also the wave functions.

If there is no motion along the direction of **B**, a classical orbit is a circle of radius a normal to **B**, as shown in Fig. 11.7. The velocity is tangent to the circle, with a constant magnitude given by the familiar expression

$$|\mathbf{v}| = \frac{ea}{mc} B \tag{11.65}$$

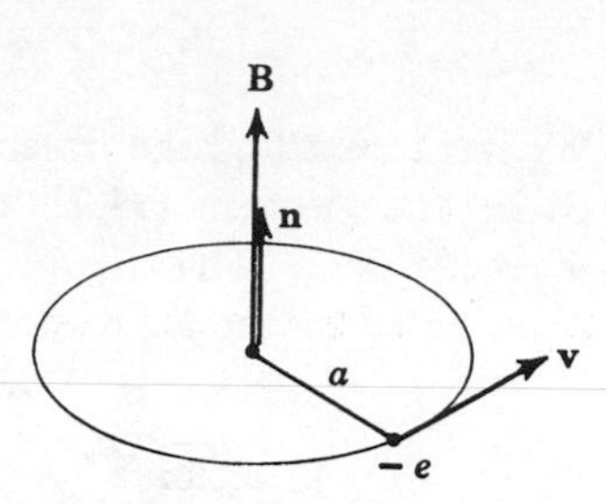

Fig. 11.7. Charged particle moving in a uniform magnetic field.

The canonical momentum $\mathbf{p}$ is not equal to $m\mathbf{v}$ but is to be found from the Hamiltonian equation

$$\mathbf{v} = \nabla_{\mathbf{p}} H = \frac{1}{m}\left(\mathbf{p} + \frac{e}{c}\mathbf{A}\right)$$

which leads to

$$\mathbf{p} = m\mathbf{v} - \frac{e}{c}\mathbf{A} \tag{11.66}$$

That is, $\mathbf{p}$ is the sum of electron and field momenta.

Now, only those orbits that satisfy (11.62) are "allowed." Hence we must have

$$\oint \left(m\mathbf{v} - \frac{e}{c}\mathbf{A}\right) \cdot d\mathbf{r} = (j + \tfrac{1}{2})h \tag{11.67}$$

where the line integral extends over one revolution of the orbit along the direction of motion of the electron. From (11.64) we have

$$\oint \mathbf{A} \cdot d\mathbf{r} = \iint \nabla \times \mathbf{A} \cdot \mathbf{n}\, dS = \iint \mathbf{B} \cdot \mathbf{n}\, dS = \pi a^2 B \tag{11.68}$$

Hence the quantum conditions (11.67) become

$$2\pi a m v - \pi a^2 \frac{eB}{c} = (j + \tfrac{1}{2})h$$

Accordingly the "allowed" orbits have radii satisfying the condition

$$a^2 = \frac{2c}{eB}(j + \tfrac{1}{2})\hbar \qquad (j = 0, 1, 2, \ldots) \tag{11.69}$$

For given B the minimum size of an "allowed" orbit is $c\hbar/eB$. The energy corresponding to the jth "allowed" orbit is

$$\frac{1}{2m}(\mathbf{p} - e\mathbf{A})^2 = \tfrac{1}{2}m\,|\mathbf{v}|^2 = \frac{e\hbar}{mc}B(j + \tfrac{1}{2}) \quad (j = 0, 1, 2, \ldots) \tag{11.70}$$

The allowed energies of an electron are (11.70) plus the kinetic energy of motion along the direction of B:

$$\epsilon(p_z, j) = \frac{p_z^{\,2}}{2m} + \frac{e\hbar}{mc}B(j + \tfrac{1}{2}) \qquad (j = 0, 1, 2, \ldots) \tag{11.71}$$

where p_z is the momentum associated with the motion along the field direction. Its allowed values are taken to be

$$p_z = \frac{2\pi\hbar l}{V^{1/3}} \qquad (l = 0, \pm 1, \pm 2, \ldots) \tag{11.72}$$

We must next find the degeneracy of $\epsilon(p_z, j)$. This can be done by comparing the spectrum (11.71) with that for a free electron in the absence of magnetic field. That is, for a fixed value of p_z, we compare the spectra represented by the expressions

$$\frac{1}{2m}(p_x^2 + p_y^2) \quad \text{and} \quad \frac{e\hbar}{mc} B(j + \tfrac{1}{2})$$

In the two-dimensional space spanned by p_x and p_y the energy surfaces are circles. Divide this space into concentric annular rings all having the same area, as shown in Fig. 11.8. Each annular region contains the same number of states. When the field is turned on, the energy of a state changes by only a finite amount. For the energy spectrum to result in an equally spaced discrete spectrum, all states originally in an annular ring (of appropriate area) must, when the field is turned on, assume the same energy. Therefore the degeneracy of $\epsilon(p_z, j)$, for given p_z, is the same for all j and is given by

$$g = \text{no. of free particle states satisfying } (p_x^2 + p_y^2)/2m < (e\hbar B/mc) \tag{11.73}$$

Let p be defined by $(p^2/2m) = (e\hbar/mc)B$. Then

$$g = \frac{V^{2/3}}{h^2}\pi p^2 = \frac{V^{2/3}}{2\pi}\frac{eB}{\hbar c} \tag{11.74}$$

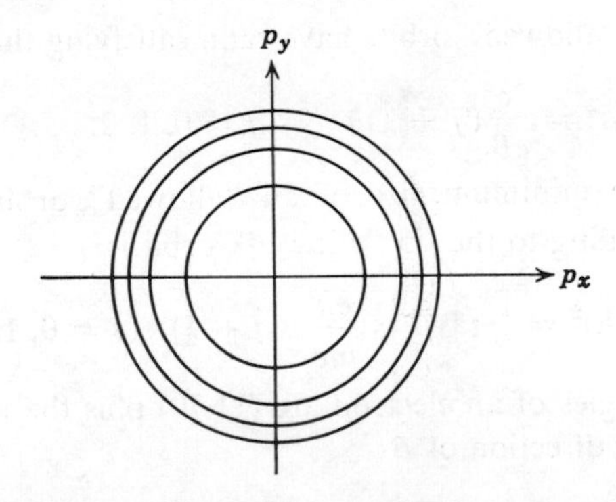

Fig. 11.8. Momentum space of a particle. The annular rings have the same area and hence contain the same number of states.

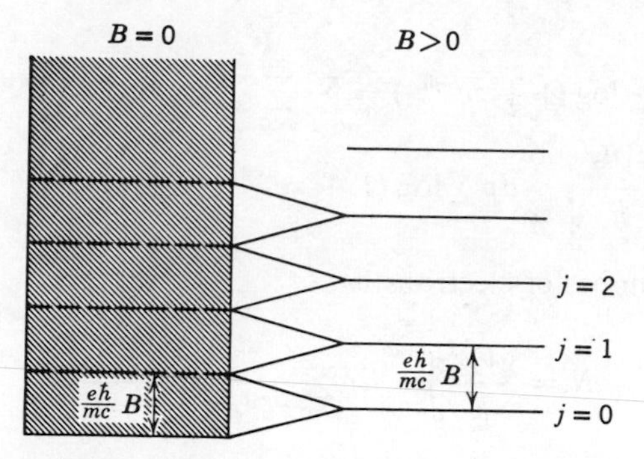

Fig. 11.9. Comparison of the energy spectra of a charged particle with and without magnetic field.

The presence of a magnetic field B causes g free-electron energy levels to degenerate, as illustrated in Fig. 11.9. Physically this degeneracy arises from the fact that an orbit in the magnetic field has the same energy, wherever its center. A convenient way to incorporate (11.74) and (11.71) is to label each state of a single electron by the quantum numbers (p_z, j, α) where α ranges from 1 to g:

$$\epsilon(p_z, j, \alpha) = \frac{p_z^2}{2m} + \frac{e\hbar B}{mc}(j + \tfrac{1}{2}) \tag{11.75}$$

where

$$\begin{aligned} p_z &= \frac{2\pi\hbar}{V^{1/3}}\, l \qquad (l = 0, \pm 1, \pm 2, \ldots) \\ j &= 0, 1, 2, \ldots \\ \alpha &= 1, 2, \ldots, g \end{aligned} \tag{11.76}$$

The partition function is

$$Q_N = \sum_{\{n_\lambda\}}{}' \exp\left(-\beta \sum_\lambda \epsilon_\lambda n_\lambda\right) \tag{11.77}$$

where λ denotes the set of quantum numbers $\{p_z, j, \alpha\}$, and the prime over the sum implies the restrictions

$$\begin{aligned} n_\lambda &= 0, 1 \\ \sum_\lambda n_\lambda &= N \end{aligned} \tag{11.78}$$

The grand partition function is

$$\mathcal{Q} = \prod_\lambda (1 + ze^{-\beta\epsilon_\lambda}) \tag{11.79}$$

and

$$\log \mathcal{Q} = \sum_{\lambda} \log(1 + ze^{-\beta\epsilon_\lambda}) = \sum_{\alpha=1}^{g} \sum_{j=0}^{\infty} \sum_{p_z} \log(1 + ze^{-\beta\epsilon(p_z, j, \alpha)})$$
$$= \frac{gV^{1/3}}{h} \int_{-\infty}^{+\infty} dp \sum_{j=0}^{\infty} \log(1 + ze^{-\beta\epsilon(p, j, \alpha)}) \tag{11.80}$$

The average number of electrons is

$$N = \frac{gV^{1/3}}{h} \int_{-\infty}^{+\infty} dp \sum_{j=0}^{\infty} \frac{1}{z^{-1}e^{\beta\epsilon(p,j,\alpha)} + 1} \tag{11.81}$$

We shall study the high-temperature and low-temperature properties of the system.

For high temperatures (11.81) requires that $z \to 0$, in order that the average number of electrons remain finite. It is then a good approximation to retain only the first term in the expansion of (11.80) in powers of z:

$$\begin{cases} \log \mathcal{Q} \approx \dfrac{zgV^{1/3}}{h} \displaystyle\int_{-\infty}^{+\infty} dp \sum_{j=0}^{\infty} e^{-\beta[(p^2/2m) + (e\hbar B/mc)(j+1/2)]} \\ N \approx \log \mathcal{Q} \end{cases} \tag{11.82}$$

Explicit evaluation yields

$$\frac{1}{V} \log \mathcal{Q} \approx \frac{z}{2\pi^2} \frac{eB}{\hbar^2 c} \sqrt{\frac{\pi m}{2\beta}} \frac{e^{-x}}{1 - e^{-2x}} \tag{11.83}$$

where

$$x \equiv \frac{e\hbar}{2mc} \frac{B}{kT} \tag{11.84}$$

For weak fields we expand (11.83) about $x = 0$:

$$\frac{1}{V} \log \mathcal{Q} \approx \frac{z}{4\pi^2} \frac{eB}{\hbar^2 c} \sqrt{\frac{\pi m}{2\beta}} \frac{1}{x}\left(1 - \frac{x^2}{6}\right) = \frac{z}{\lambda^3}\left[1 - \frac{1}{6}\left(\frac{e\hbar}{2mc}\right)^2 \left(\frac{B}{kT}\right)^2\right] \tag{11.85}$$

where λ is the thermal wavelength. From (11.58) and (11.61) we find that

$$\chi \approx -\frac{z}{kT\lambda^3}\left(\frac{e\hbar}{2mc}\right)^2 \tag{11.86}$$

To eliminate z we use (11.82) and (11.85) and obtain, in the limit $B \to 0$,

$$\frac{z}{\lambda^3} = \frac{N}{V} \equiv \frac{1}{v} \tag{11.87}$$

Therefore at high temperatures the magnetic susceptibility per unit volume as a function of temperature T and specific volume v is

$$\chi \approx -\frac{1}{3kTv}\left(\frac{e\hbar}{2mc}\right)^2 \tag{11.88}$$

This exhibits the $1/T$ dependence in conformity with Curie's law. The Curie constant is

$$\frac{1}{3v}\left(\frac{e\hbar}{2mc}\right)^2$$

which owes its existence to the quantization of orbits. In fact, $e\hbar/2mc$ is none other than the magnetic moment of the smallest "allowed" orbit.

As $T \to 0$, χ approaches a value that depends on the field strength B and changes discontinuously as B changes. We study this effect next.

11.4 DE HAAS-VAN ALPHEN EFFECT

We now study the orbital magnetism of the ideal Fermi gas, discussed in the last section, at absolute zero. If E_0 is the ground state energy of the system, the induced magnetic moment per unit volume at absolute zero is simply given by

$$\mathscr{M} = -\frac{1}{V}\frac{\partial E_0}{\partial B} \tag{11.89}$$

The magnetic susceptibility per unit volume at absolute zero is given by

$$\chi = \frac{\partial \mathscr{M}}{\partial B} = -\frac{1}{V}\frac{\partial^2 E_0}{\partial B^2} \tag{11.90}$$

Thus it is sufficient to obtain E_0 as a function of B. To simplify the mathematics, we ignore the motion of the electrons along the direction of B and set $p_z = 0$. Thus we consider a two-dimensional Fermi gas. The single-particle energy levels are then given by

$$\epsilon_j = \frac{e\hbar B}{mc}(j + \tfrac{1}{2}) \qquad (j = 0, 1, 2, \ldots) \tag{11.91}$$

each of which is g-fold degenerate, with

$$g = \frac{L^2}{2\pi}\frac{eB}{\hbar c} \tag{11.92}$$

where L^2 is the total area occupied by the two-dimensional Fermi gas.

The ground state energy E_0 is the sum of ϵ_j over the lowest N single-particle *states*. Since g depends on B, the maximum number of particles that can have the energy ϵ_j depends on B. If B is such that $g \geq N$, then all particles can occupy the lowest energy level, and $E_0 = N(e\hbar/2mc)B$. If B is such that $g < N$, some particles will have to occupy higher energy levels. To find a general expression for E_0 as a function of B, let us consult the single-particle energy spectrum sketched below.

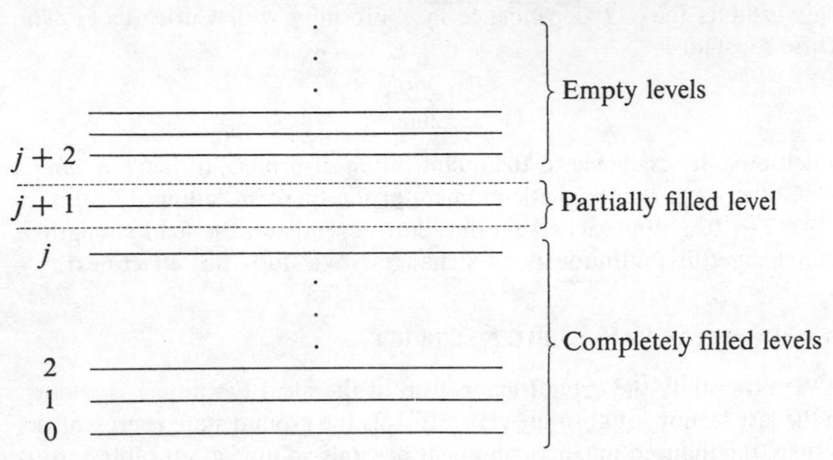

Each level holds g particles. Suppose B is such that the j lowest levels are completely filled with g particles in each level, the $(j + 1)$st level is only partially filled, and the higher levels are empty. Then

$$(j + 1)g < N < (j + 2)g$$

or

$$\frac{1}{j + 1} > \frac{B}{B_0} > \frac{1}{j + 2} \tag{11.93}$$

where

$$B_0 \equiv 2 \frac{\hbar c}{e} \frac{\pi N}{L^2} \tag{11.94}$$

If B lies in the interval (11.93), the ground state energy is

$$\begin{aligned}
E_0(B) &= g \sum_{i=0}^{j} \epsilon_i + [N - (j + 1)g]\epsilon_{j+1} \\
&= \frac{e\hbar}{mc} B\left\{g \sum_{i=0}^{j} (i + \tfrac{1}{2}) + [N - (j + 1)g](j + \tfrac{3}{2})\right\} \\
&= \frac{e\hbar}{mc} B\{g[\tfrac{1}{2}j(j + 1) + \tfrac{1}{2}(j + 1)] + [N - (j + 1)g](j + \tfrac{3}{2})\} \\
&= NB_0 \frac{e\hbar}{mc} \frac{B}{B_0}\left[(j + \tfrac{3}{2}) - \tfrac{1}{2}(j + 1)(j + 2) \frac{B}{B_0}\right]
\end{aligned} \tag{11.95}$$

In summary, we have

$$\frac{1}{N}E_0(B) = \begin{cases} \dfrac{e\hbar}{2mc} B_0 x & (x > 1) \\ \dfrac{e\hbar}{2mc} B_0 x[(2j+3) - (j+1)(j+2)x] & \\ \qquad \left(\dfrac{1}{j+2} < x < \dfrac{1}{j+1}, \quad j = 0, 1, 2, \ldots\right) \end{cases} \tag{11.96}$$

where

$$x \equiv \frac{B}{B_0} \tag{11.97}$$

The magnetization per unit volume and the magnetic susceptibility per unit volume are respectively given by

$$\mathscr{M} = \begin{cases} -\dfrac{1}{v}\dfrac{e\hbar}{2mc} & (x > 1) \\ \dfrac{1}{v}\dfrac{e\hbar}{2mc}[2(j+1)(j+2)x - (2j+3)] & \\ \qquad \left(\dfrac{1}{j+2} < x < \dfrac{1}{j+1}, \quad j = 0, 1, 2, \ldots\right) \end{cases} \tag{11.98}$$

$$\chi = \begin{cases} 0 & (x > 1) \\ \dfrac{1}{vB_0}\dfrac{e\hbar}{mc}(j+1)(j+2) & \left(\dfrac{1}{j+2} < x < \dfrac{1}{j+1}, \quad j = 0, 1, 2, \ldots\right) \end{cases} \tag{11.99}$$

These are shown in Fig. 11.10.

We have considered only particles moving in two dimensions. Taking p_z into account would smooth out the discontinuities in $\mathscr{M}$ and χ, but the oscillatory behavior of $\mathscr{M}$ would persist. This interesting effect, first

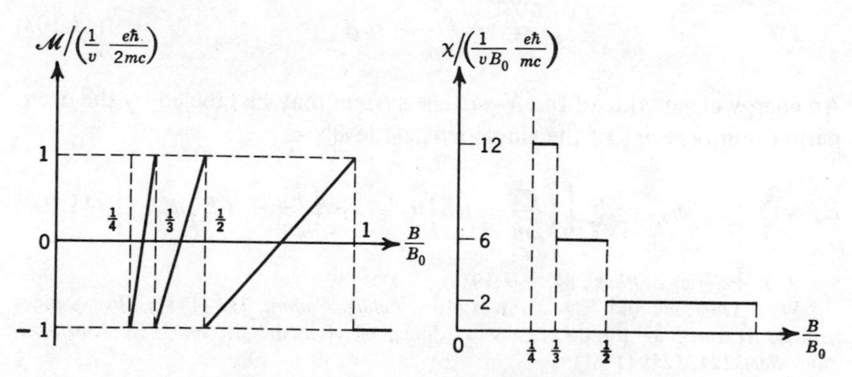

Fig. 11.10. DeHaas-Van Alphen effect.

discussed by Peierls,* should occur for metals at low temperatures, since it is known that charge carriers in metals are to a good approximation free electrons. A qualitatively similar effect is indeed found experimentally—the de Haas-Van Alphen effect.†

11.5 PAULI PARAMAGNETISM

The Hamiltonian of a nonrelativistic free electron in an external magnetic field B is given by

$$\frac{1}{2m}\left(\mathbf{p} + \frac{e}{c}\mathbf{A}\right)^2 - \boldsymbol{\mu}\cdot\mathbf{B} \tag{11.100}$$

where $\boldsymbol{\mu}$ is the intrinsic magnetic moment operator of the electron:

$$\boldsymbol{\mu} = \mu\boldsymbol{\sigma}$$
$$\mu = \frac{e\hbar}{2mc} \tag{11.101}$$

where $\boldsymbol{\sigma}$ is the spin operator. The value of μ follows from Dirac's theory of the electron. The first term in (11.100) gives rise to diamagnetism, as we have studied. The second term gives rise to paramagnetism. We now consider its effect alone.‡

We consider a system of N free fermions of spin $\hbar/2$, each of which is described by the single-particle Hamiltonian

$$H = \frac{p^2}{2m} - \boldsymbol{\mu}\cdot\mathbf{B} \tag{11.102}$$

The eigenvalues of $\boldsymbol{\sigma}\cdot\mathbf{B}$ are sB, where $s = \pm 1$. Hence the single-particle energy levels are

$$\epsilon_{\mathbf{p},s} = \frac{p^2}{2m} - s\mu B \tag{11.103}$$

An energy eigenvalue of the N-particle system may be labeled by the occupation numbers $n_{\mathbf{p},s}$ of the single-particle levels $\epsilon_{\mathbf{p},s}$:

$$E_n = \sum_{\mathbf{p}}\sum_{s} \epsilon_{\mathbf{p},s} n_{\mathbf{p},s} = \sum_{\mathbf{p}}\left[\left(\frac{p^2}{2m} - \mu B\right)n_{\mathbf{p},+1} + \left(\frac{p^2}{2m} + \mu B\right)n_{\mathbf{p},-1}\right] \tag{11.104}$$

* R. E. Peierls, *Z. Phys.*, **81,** 186 (1933).

† W. J. De Haas and P. M. Van Alphen, *Leiden Comm.*, **212** (1931). For a more realistic treatment of the de Hass-Van Alphen effect in metals, see J. M. Luttinger, *Phys. Rev.*, **121,** 1251 (1961).

‡ Following W. Pauli, *Z. Physik.*, **41,** 81 (1927).

where

$$\begin{aligned} n_{\mathbf{p},s} &= 0, 1 \\ \sum_s \sum_{\mathbf{p}} n_{\mathbf{p},s} &= N \end{aligned} \tag{11.105}$$

Let

$$\begin{aligned} n_{\mathbf{p},+1} &\equiv n_{\mathbf{p}}^+ \\ n_{\mathbf{p},-1} &\equiv n_{\mathbf{p}}^- \\ \sum_{\mathbf{p}} n_{\mathbf{p},+1} &\equiv N_+ \\ \sum_{\mathbf{p}} n_{\mathbf{p},-1} &\equiv N_- = N - N_+ \end{aligned} \tag{11.106}$$

Then an energy eigenvalue of the system can also be written in the form

$$E_n = \sum_{\mathbf{p}} (n_{\mathbf{p}}^+ + n_{\mathbf{p}}^-) \frac{p^2}{2m} - \mu B(N_+ - N_-) \tag{11.107}$$

The partition function is

$$Q_N = \sum_{\{n_{\mathbf{p}}^+\},\{n_{\mathbf{p}}^-\}}' \exp\left[-\beta \sum_{\mathbf{p}} (n_{\mathbf{p}}^+ + n_{\mathbf{p}}^-) \frac{p^2}{2m} + \beta\mu B(N_+ - N_-)\right] \tag{11.108}$$

where the prime over the sum denotes the restrictions (11.105). The sum can be evaluated as follows. First we choose an arbitrary integer N_+ and sum over all sets $\{n_{\mathbf{p}}^+\}$, $\{n_{\mathbf{p}}^-\}$ such that $\sum_{\mathbf{p}} n_{\mathbf{p}}^+ = N_+$, and $\sum_{\mathbf{p}} n_{\mathbf{p}}^- = N - N_+$. Then we sum over all integers N_+ from 0 to N. In this manner we arrive at the formula

$$Q_N = \sum_{N_+=0}^{N} e^{\beta\mu B(2N_+ - N)} \sum_{\{n_{\mathbf{p}}^+\}}'' \exp\left(-\beta \sum_{\mathbf{p}} \frac{p^2}{2m} n_{\mathbf{p}}^+\right) \sum_{\{n_{\mathbf{p}}^-\}}''' \exp\left(-\beta \sum_{\mathbf{p}} \frac{p^2}{2m} n_{\mathbf{p}}^-\right) \tag{11.109}$$

where $\sum''$ is subject to the restriction $\sum_{\mathbf{p}} n_{\mathbf{p}}^+ = N_+$, and $\sum'''$ is subject to the restriction $\sum_{\mathbf{p}} n_{\mathbf{p}}^- = N_- = N - N_+$. Let $Q_N^{(0)}$ denote the partition function of the ideal Fermi gas of N *spinless* particles of mass m:

$$Q_N^{(0)} \equiv \sum_{\Sigma n_{\mathbf{p}}=N} \exp\left(-\beta \sum_{\mathbf{p}} \frac{p^2}{2m} n_{\mathbf{p}}\right) \equiv e^{-\beta A(N)} \tag{11.110}$$

Then

$$\begin{aligned} Q_N &= e^{-\beta\mu BN} \sum_{N_+=0}^{N} e^{2\beta\mu BN_+} Q_{N_+}^{(0)} Q_{N-N_+}^{(0)} \\ \frac{1}{N} \log Q_N &= -\beta\mu B + \frac{1}{N} \log \sum_{N_+=0}^{N} e^{2\beta\mu BN_+ - \beta A(N_+) - \beta A(N_-)} \end{aligned} \tag{11.111}$$

There are $N + 1$ positive terms in the sum just given. The logarithm of this sum is equal to the logarithm of the largest term in the sum plus a contribution of the order of $\log N$. Therefore, neglecting a term of order $N^{-1} \log N$, we have

$$\frac{1}{N} \log Q_N = \beta f(\bar{N}_+) \tag{11.112}$$

where

$$f(\bar{N}_+) = \text{Max}\,[f(N_+)]$$
$$f(N_+) \equiv \mu B\left(\frac{2N_+}{N} - 1\right) - \frac{1}{N}[A(N_+) + A(N - N_+)] \tag{11.113}$$

Obviously we can interpret $\bar{N}_+$ as the average number of particles with spin up. If $\bar{N}_+$ is known, the magnetization per unit volume can be obtained through the formula

$$\mathscr{M} = \frac{\mu(2\bar{N}_+ - N)}{V} \tag{11.114}$$

We now explicitly find $\bar{N}_+$. The condition (11.113) is equivalent to the condition*

$$\left[\frac{\partial f(N_+)}{\partial N_+}\right]_{N_+ = \bar{N}_+} = 0$$

or

$$2\mu B - \left[\frac{\partial A(N')}{\partial N'}\right]_{N' = \bar{N}_+} - \left[\frac{\partial A(N - N')}{\partial N'}\right]_{N' = \bar{N}_+} = 0 \tag{11.115}$$

Let $kT\nu(N)$ be the chemical potential of an ideal Fermi gas of N *spinless* particles:

$$kT\nu(N) = \frac{\partial A(N)}{\partial N} \tag{11.116}$$

Then

$$\left[\frac{\partial A(N')}{\partial N'}\right]_{N' = \bar{N}_+} = kT\nu(\bar{N}_+)$$

$$\left[\frac{\partial A(N - N')}{\partial N'}\right]_{N' = \bar{N}_+} = -\left[\frac{\partial A(N - N')}{\partial (N - N')}\right]_{N - N' = N - \bar{N}_+} = -kT\nu(N - \bar{N}_+)$$

Thus (11.115) becomes

$$kT[\nu(\bar{N}_+) - \nu(N - \bar{N}_+)] = 2\mu B \tag{11.117}$$

* To be careful, we should make sure that (11.115) determines a maximum and not a minimum and that $\bar{N}_+$ lies between 0 and N. It can be verified that (11.115) has only one real root which automatically satisfies these requirements.

This condition states that at a given temperature the average number of particles with spin up is such that the chemical potential of the particles with spin up is greater than that of the particles with spin down by $2\mu B$. We solve (11.117) in the low-temperature and high-temperature limits.

Let the Fermi energy for the present system be

$$\epsilon_F(N) \equiv \left(\frac{3\pi^2 N}{V}\right)^{2/3} \frac{\hbar^2}{2m} \tag{11.118}$$

In the low-temperature region ($kT \ll \epsilon_F$), we can use the expansion (11.24)* for $kT\nu(N)$:

$$kT\nu(N) = \epsilon_F(2N)\left\{1 - \frac{\pi^2}{12}\left[\frac{kT}{\epsilon_F(2N)}\right]^2 + \cdots\right\}$$

Thus (11.117) becomes

$$\epsilon_F(2\bar{N}_+) - \epsilon_F(2N - 2\bar{N}_+) - \frac{\pi^2(kT)^2}{12} \times \left[\frac{1}{\epsilon_F(2\bar{N}_+)} - \frac{1}{\epsilon_F(2N - 2\bar{N}_+)}\right] + \cdots = 2\mu B \tag{11.119}$$

Let

$$r \equiv \frac{2\bar{N}_+}{N} - 1 \qquad (-1 \le r \le +1) \tag{11.120}$$

Then (11.119) becomes

$$(1 + r)^{2/3} - (1 - r)^{2/3} - \frac{\pi^2}{12}\left(\frac{kT}{\epsilon_F}\right)^2 [(1 + r)^{-2/3} - (1 - r)^{-2/3}] + \cdots = \frac{2\mu B}{\epsilon_F} \tag{11.121}$$

At absolute zero, r satisfies the equation

$$(1 + r)^{2/3} - (1 - r)^{2/3} = \frac{2\mu B}{\epsilon_F} \tag{11.122}$$

This may be solved graphically, as shown in Fig. 11.11. For $B \ll \epsilon_F/2\mu$ an approximate solution is

$$\begin{aligned} r &\approx \frac{3\mu B}{2\epsilon_F} \\ \bar{N}_+ &\approx \frac{N}{2}\left(1 + \frac{3\mu B}{2\epsilon_F}\right) \end{aligned} \tag{11.123}$$

* Note that in (11.24) the symbol ϵ_F stands for the Fermi energy of N spinless particles and does not have the same meaning as ϵ_F here.

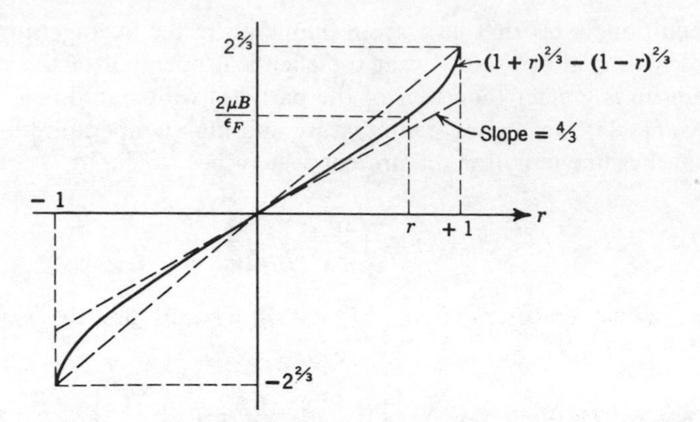

Fig. 11.11. Graphical solution of (11.122).

Thus when $B = 0$ half the particles have spin up, and the other half spin down. When $B > 0$ the balance shifts in favor of spin up. From (11.114) and (11.123) we obtain, for absolute zero,

$$
\begin{aligned}
\mathscr{M} &= \frac{\mu r}{v} \approx \frac{3\mu^2 B}{2\epsilon_F v} \\
\chi &\approx \frac{3\mu^2}{2\epsilon_F v}
\end{aligned}
\tag{11.124}
$$

For $0 < kT \ll \epsilon_F$ and $\mu B \ll \epsilon_F$ we can solve (11.121) by expanding the left-hand side in powers of r, and we obtain

$$
\begin{aligned}
r &\approx \frac{3\mu B}{2\epsilon_F}\left[1 - \frac{\pi^2}{12}\left(\frac{kT}{\epsilon_F}\right)^2\right] \\
\chi &\approx \frac{3\mu^2}{2\epsilon_F v}\left[1 - \frac{\pi^2}{12}\left(\frac{kT}{\epsilon_F}\right)^2\right]
\end{aligned}
\tag{11.125}
$$

For high temperatures ($kT \gg \epsilon_F$) we use (11.12):

$$
\nu(N) \approx \log\left(\frac{N\lambda^3}{V}\right)
$$

Hence (11.117) gives

$$
\log\left[\frac{\lambda^3(1+r)}{v}\right] - \log\left[\frac{\lambda^3(1-r)}{v}\right] = \frac{2\mu B}{kT}
$$

or

$$
r = \tanh\frac{\mu B}{kT} \approx \frac{\mu B}{kT} \tag{11.126}
$$

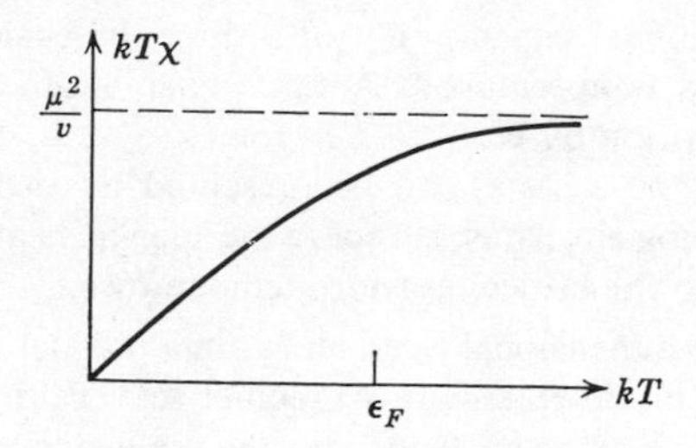

Fig. 11.12. Pauli paramagnetism.

The magnetic susceptibility per unit volume is then given by

$$\chi \approx \frac{\mu^2}{kTv} \tag{11.127}$$

A qualitative plot of $kT\chi$ is shown in Fig. 11.12.

PROBLEMS

11.1. Give numerical estimates for the Fermi energy of
(*a*) electrons in a typical metal;
(*b*) nucleons in a heavy nucleus;
(*c*) He^3 atoms in liquid He^3 (atomic volume = 46.2 Å^3/atom). Treat all the mentioned particles as free particles.

11.2. Show that for the ideal Fermi gas of N particles the Helmholtz free energy at low temperatures is given by

$$A = \tfrac{3}{5}\,\epsilon_F\left[1 - \frac{5\pi^2}{12}\left(\frac{kT}{\epsilon_F}\right)^2 + \cdots\right]$$

11.3. A collection of free nucleons is enclosed in a box of volume V. The energy of a single nucleon of momentum $\mathbf{p}$ is

$$\epsilon_{\mathbf{p}} = \frac{p^2}{2m} + mc^2$$

where $mc^2 = 1000$ Mev.
(*a*) Pretending that there is no conservation law for the number of nucleons, calculate the partition function of a system of nucleons (which obey Fermi statistics) at temperature T.
(*b*) Calculate the average energy density.
(*c*) Calculate the average particle density.
(*d*) Discuss the necessity for a conservation law for the number of nucleons, in the light of the foregoing calculations.

11.4. (*a*) What is the heat capacity C_V of a three-dimensional cubic lattice of atoms at room temperature? Assume each atom to be bound to its equilibrium position by Hooke's law forces.
(*b*) Assuming that a metal can be presented by such a lattice of atoms plus freely moving electrons, compare the specific heat due to the electrons with that due to the lattice, at room temperature.

11.5. Consider a two-dimensional electron gas in a magnetic field strong enough for there to be no de Haas-Van Alphen effect. Taking into account both orbital and spin paramagnetism, find the magnetization at absolute zero.

chapter 12

IDEAL BOSE GAS

12.1 PHOTONS

We consider the equilibrium properties of electromagnetic radiation enclosed in a cubical volume V at temperature T. Such a system is sometimes known as a "black-body cavity." It can be experimentally produced by making a cavity in any material, evacuating the cavity completely, and then heating the material to a given temperature. The atoms in the walls of this cavity will constantly emit and absorb electromagnetic radiation, so that in equilibrium there will be a certain amount of electromagnetic radiation in the cavity, and nothing else. If the cavity is sufficiently large, the thermodynamic properties of the radiation in the cavity should be independent of the nature of the wall. Accordingly we can impose on the radiation field any boundary condition that is convenient.

It is well known that the Hamiltonian for a free electromagnetic field can be written as a sum of terms, each having the form of a Hamiltonian for a harmonic oscillator of some frequency. This possibility corresponds to the possibility of regarding any radiation field as a linear superposition of plane waves of various frequencies. In quantum theory each harmonic oscillator of frequency ω can only have the energies $(n + \frac{1}{2})\hbar\omega$, where $n = 0, 1, 2, \ldots$. This fact leads to the concept of photons as quanta of the electromagnetic field. A state of the free electromagnetic field is specified by the number n for each of the oscillators. In other words, it is specified by enumerating the number of photons present for each frequency.

According to the quantum theory of the electromagnetic field, a photon may be regarded as a massless particle of spin $\hbar$ and of definite momentum and energy. It always moves with the velocity of light c. The fact that a photon has no rest mass implies that its spin can have only two independent orientations: Parallel or antiparallel to the momentum. A photon in a definite spin state corresponds to a plane electromagnetic wave that is either right- or left-circularly polarized. We may, however, superimpose two photon states with definite spins and obtain a photon state which is linearly polarized but which is not an eigenstate of spin. In the following we consider linearly polarized photons.

For our purpose it is sufficient to know that a photon of frequency ω has the following properties:

$$\begin{aligned} \text{Energy} &= \hbar\omega \\ \text{Momentum} &= \hbar\mathbf{k}, \qquad |\mathbf{k}| = \frac{\omega}{c} \\ \text{Polarization vector} &= \boldsymbol{\epsilon}, \qquad |\boldsymbol{\epsilon}| = 1, \qquad \mathbf{k}\cdot\boldsymbol{\epsilon} = 0 \end{aligned} \tag{12.1}$$

Such a photon corresponds* to a plane wave of electromagnetic radiation whose electric field vector is

$$\mathbf{E}(\mathbf{r}, t) = \boldsymbol{\epsilon} e^{i(\mathbf{k}\cdot\mathbf{r} - \omega t)} \tag{12.2}$$

The direction of $\boldsymbol{\epsilon}$ is the direction of the electric field. The condition $\boldsymbol{\epsilon}\cdot\mathbf{k} = 0$ is a consequence of the transversality of the electric field, i.e., $\nabla\cdot\mathbf{E} = 0$. Thus for given $\mathbf{k}$ there are two and only two independent polarization vectors $\boldsymbol{\epsilon}$. If we impose periodic boundary conditions on $\mathbf{E}(\mathbf{r}, t)$ in a cube of volume $V = L^3$, we obtain the following allowed values of $\mathbf{k}$:

$$\begin{aligned} \mathbf{k} &= \frac{2\pi\mathbf{n}}{L} \\ \mathbf{n} &= \text{a vector whose components are } 0, \pm 1, \pm 2, \ldots \end{aligned} \tag{12.3}$$

Thus the number of allowed momentum values between k and $k + dk$ is

$$\frac{V}{(2\pi)^3} 4\pi k^2\, dk \tag{12.4}$$

Photons obey Bose statistics, for they are indistinguishable and there can be any number of photons with the same $\mathbf{k}$ and $\boldsymbol{\epsilon}$. Since atoms can emit and absorb photons the number of photons in a black-body cavity is not definite.

* What this means precisely is explained in any book on the quantum theory of the electromagnetic field.

The total energy of the state of the electromagnetic field in which there are $n_{\mathbf{k},\boldsymbol{\epsilon}}$ photons of momentum $\mathbf{k}$ and polarization $\boldsymbol{\epsilon}$ is given by

$$E\{n_{\mathbf{k},\boldsymbol{\epsilon}}\} = \sum_{\mathbf{k},\boldsymbol{\epsilon}} \hbar\omega n_{\mathbf{k},\boldsymbol{\epsilon}} \tag{12.5}$$

where

$$\omega = c\,|\mathbf{k}|$$

$$n_{\mathbf{k},\boldsymbol{\epsilon}} = 0, 1, 2, \ldots \tag{12.6}$$

Since the number of photons is indefinite, the partition function is

$$Q = \sum_{\{n_{\mathbf{k},\boldsymbol{\epsilon}}\}} e^{-\beta E\{n_{\mathbf{k},\boldsymbol{\epsilon}}\}} \tag{12.7}$$

with no restriction on $\{n_{\mathbf{k},\boldsymbol{\epsilon}}\}$. The calculation of Q is trivial:

$$Q = \sum_{\{n_{\mathbf{k},\boldsymbol{\epsilon}}\}} \exp\left(-\beta \sum_{\mathbf{k},\boldsymbol{\epsilon}} \hbar\omega n_{\mathbf{k},\boldsymbol{\epsilon}}\right) = \prod_{\mathbf{k},\boldsymbol{\epsilon}} \sum_{n=0}^{\infty} e^{-\beta\hbar\omega n} = \prod_{\mathbf{k},\boldsymbol{\epsilon}} \frac{1}{1 - e^{-\beta\hbar\omega}}$$

$$\log Q = -\sum_{\mathbf{k},\boldsymbol{\epsilon}} \log\left(1 - e^{-\beta\hbar\omega}\right) = -2\sum_{\mathbf{k}} \log\left(1 - e^{-\beta\hbar\omega}\right) \tag{12.8}$$

The average occupation number for photons of momentum $\mathbf{k}$, regardless of polarization, is

$$\langle n_{\mathbf{k}}\rangle = -\frac{1}{\beta}\frac{\partial}{\partial(\hbar\omega)} \log Q = \frac{2}{e^{\beta\hbar\omega} - 1} \tag{12.9}$$

and the internal energy is

$$U = -\frac{\partial}{\partial\beta} \log Q = \sum_{\mathbf{k}} \hbar\omega \langle n_{\mathbf{k}}\rangle \tag{12.10}$$

To find the pressure, we express Q in the form

$$\log Q = -2\sum_{\mathbf{n}} \log\left(1 - e^{\beta\hbar c 2\pi|\mathbf{n}|V^{-1/3}}\right) \tag{12.11}$$

from which we obtain

$$P = \frac{1}{\beta}\frac{\partial}{\partial V} \log Q = \frac{1}{3V}\sum_{\mathbf{k}} \hbar\omega\langle n_k\rangle$$

Comparison between this equation and (12.10) leads to the equation of state

$$PV = \tfrac{1}{3}U \tag{12.12}$$

We now calculate U in the limit as $V \to \infty$. From (12.10), (12.9), and (12.3) we have

$$U = \frac{2V}{(2\pi)^3}\int_0^\infty dk 4\pi k^2 \frac{\hbar ck}{e^{\beta\hbar ck} - 1} = \frac{V\hbar}{\pi^2 c^3}\int_0^\infty d\omega \frac{\omega^3}{e^{\beta\hbar\omega} - 1}$$

Hence the internal energy per unit volume is

$$\frac{U}{V} = \int_0^\infty d\omega\, u(\omega, T) \tag{12.13}$$

where

$$u(\omega, T) = \frac{\hbar}{\pi^2 c^3} \frac{\omega^3}{e^{\beta\hbar\omega} - 1} \tag{12.14}$$

This is the energy density due to photons of frequency ω and is the well-known radiation formula of Planck. The integral (12.13) can be explicitly evaluated and leads to

$$\frac{U}{V} = \frac{\pi^2}{15} \frac{(kT)^4}{(\hbar c)^3} \tag{12.15}$$

It follows that the specific heat per unit volume is

$$c_V = \frac{4\pi^2 k^4 T^3}{15(\hbar c)^3} \tag{12.16}$$

The specific heat is not bounded as $T \to \infty$, because the number of photons in the cavity is not bounded.

Both (12.14) and (12.15) can be verified experimentally by opening the black-body cavity to the external world through an infinitesimal tunnel. Radiation would then escape from the cavity with the velocity c, so that the amount of total energy radiated per second per unit area of the opening is

$$I(T) = \frac{cU}{V} = \frac{\pi^2(kT)^4}{15\hbar^3 c^2} \tag{12.17}$$

This is Stefan's law. The universal constant $\pi^2 k^4/15\hbar^3 c^3$ is known as Stefan's constant. If we filter the escaping radiation through frequency filters, we can measure the energy flux of radiation of frequency ω:

$$I(\omega, T) = cu(\omega, T) \tag{12.18}$$

This function is shown in Fig. 12.1. Thus the radiation is most intense at a frequency that is an increasing function of T. The area under the curves shown in Fig. 12.1 increases like T^4. All these conclusions are in excellent agreement with experiments.

It should be noted that although the form of $u(\omega, T)$ can be arrived at only through quantum theory the equation of state $PV = U/3$ and the fact that $U \propto T^4$ can be derived in classical physics.

The equation of state may be derived as follows. Consider first a plane wave whose electric and magnetic field vectors are **E** and **B**. The average energy density is

$$\tfrac{1}{2}\overline{(E^2 + B^2)} = \overline{E^2}$$

The radiation pressure, which is equal to the average momentum flux, is

$$\overline{|\mathbf{E} \times \mathbf{B}|} = \overline{E^2}$$

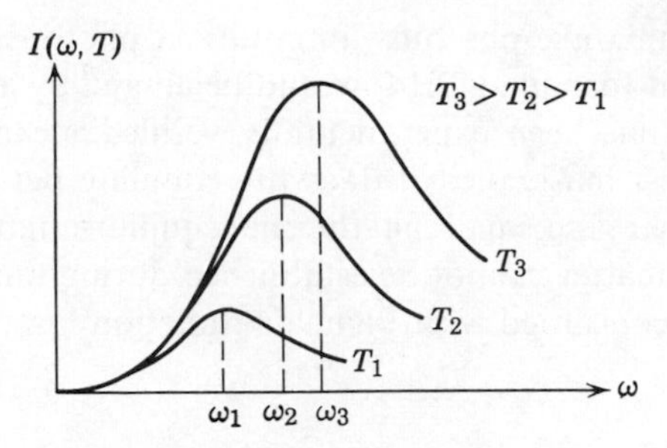

Fig. 12.1. Planck's radiation law.

Thus the energy density is numerically equal to the radiation pressure. Now consider an amount of isotropic radiation contained in a cubical box. The radiation field in the box may be considered an *incoherent* superposition of plane waves propagating in all directions. The relative intensities of the plane waves depend only on the temperature as determined by the walls of the box. The radiation pressure on any wall of the box is one-third of the energy density in the box, because, whereas all the plane waves contribute to the energy density, only one-third of the plane waves contribute to the radiation pressure on any wall of the box.

To derive $U \propto T^4$, recall that the second law of thermodynamics implies the following relation, which holds for all systems:

$$\left(\frac{\partial U}{\partial V}\right)_T = T\left(\frac{\partial P}{\partial T}\right)_V - P \tag{12.19}$$

From $PV = U/3$ and the fact that P depends on temperature alone we have

$$\left(\frac{\partial U}{\partial V}\right)_T = 3P = \frac{U}{V} \equiv u(T) \tag{12.20}$$

Using (12.19) we have

$$u = \frac{T}{3}\frac{du}{dT} - \tfrac{1}{3}u$$

$$\frac{du}{u} = 4\frac{dT}{T}$$

Hence

$$u = CT^4 \tag{12.21}$$

The constant C cannot be obtained through classical considerations.

If the photon had a finite rest mass, no matter how small, then it would have three independent polarizations instead of two.* There would be,

* If the photon had a finite rest mass, it could be transformed to rest by a Lorentz transformation. We could then make a second Lorentz transformation in an arbitrary direction, so that the spin would lie neither parallel nor antiparallel to the momentum.

in addition to transverse photons, longitudinal photons. If this were so, Planck's radiation formula (12.14) would be altered by a factor of $\frac{3}{2}$. The fact that (12.14) has been experimentally verified means that either the photon has no rest mass, or if it does the coupling between longitudinal photons and matter is so small that thermal equilibrium between longitudinal photons and matter cannot be established during the course of any of our experiments concerned with Planck's radiation law.

12.2 PHONONS

Phonons are quanta of the field of sound waves in a macroscopic body. Mathematically they emerge in the same way that photons arise from the quantization of the electromagnetic field. As we have mentioned earlier, the electromagnetic field in a cavity can be Fourier-analyzed into plane waves. This analysis decomposes the Hamiltonian of the electromagnetic field into a sum of terms each of which represents a harmonic oscillator. The quanta of these harmonic oscillators are photons. On the other hand the Hamiltonian for a solid, which is made up of atoms arranged in a crystal lattice, may be approximated by a sum of terms, each representing a harmonic oscillator, corresponding to a normal mode of lattice oscillation.* Each normal mode is classically a wave of distortion of the lattice planes—a sound wave. In quantum theory these normal modes give rise to quanta called phonons.

From what has been said we see that a quantum state of a crystal lattice near its ground state may be specified by enumerating all the phonons present. Therefore at a very low temperature a solid can be regarded as a volume containing a gas of noninteracting phonons.

Since a phonon is a quantum of a certain harmonic oscillator, it has a characteristic frequency ω_i and an energy $\hbar\omega_i$. The state of the lattice in which one phonon is present corresponds to a sound wave of the form

$$\boldsymbol{\epsilon} e^{i(\mathbf{k}\cdot\mathbf{r}-\omega t)} \tag{12.22}$$

where the propagation vector $\mathbf{k}$ has the magnitude

$$|\mathbf{k}| = \frac{\omega}{c} \tag{12.23}$$

in which c is the velocity of sound.† The polarization vector $\boldsymbol{\epsilon}$ is not necessarily perpendicular to $\mathbf{k}$. Thus it can have three independent directions, corresponding to one longitudinal mode of compression wave

* In as much as anharmonic forces between atoms, which at high temperatures allow the lattice to melt, can be neglected.

† We assume, for simplicity, that c is independent of the polarization vector $\boldsymbol{\epsilon}$.

and two transverse modes of shear wave. Since an excited state of a harmonic oscillator may contain any number of quanta, the phonons obey Bose statistics, with no conservation of their total number.

If a solid has N atoms, it has $3N$ normal modes. Therefore there will be $3N$ different types of phonon with the characteristic frequencies

$$\omega_1, \omega_2, \ldots, \omega_{3N} \tag{12.24}$$

The values of these frequencies depend on the nature of the lattice. In the Einstein model of a lattice they are taken to be equal to one another. An improved model is that of Debye, who assumed that for the purpose of finding the frequencies (12.24), and only for this purpose, the solid is approximately an elastic continuum of volume V. The frequencies (12.24) are then taken to be the lowest $3N$ normal frequencies of such a system. Since an elastic continuum has a continuous distribution of normal frequencies we shall be interested in the number of normal modes whose frequency lies between ω and $\omega + d\omega$. To find this number we must know the boundary conditions on a sound wave in the elastic medium. Taking periodic boundary conditions, we find as usual that $\mathbf{k} = (2\pi/L)\mathbf{n}$, where $L = V^{1/3}$ and $\mathbf{n}$ has the components $0, \pm 1, \pm 2, \ldots$. The number we seek is

$$f(\omega)\, d\omega \equiv \begin{array}{c}\text{no. of normal modes with}\\ \text{frequency between } \omega \text{ and } \omega + d\omega\end{array} = \frac{3V}{(2\pi)^3} 4\pi k^2\, dk \tag{12.25}$$

where the factor 3 comes from the three possible polarizations. Since $k = \omega/c$ we have

$$f(\omega)\, d\omega = V \frac{3\omega^2}{2\pi^2 c^3}\, d\omega \tag{12.26}$$

The maximum frequency ω_m is obtained by the requirement that

$$\int_0^{\omega_m} f(\omega)\, d\omega = 3N \tag{12.27}$$

which gives, with $v = V/N$,

$$\omega_m = c\left(\frac{6\pi^2}{v}\right)^{1/3} \tag{12.28}$$

The wavelength corresponding to ω_m is

$$\lambda_m = \frac{2\pi c}{\omega_m} = \left(\tfrac{4}{3}\pi v\right)^{1/3} \approx \text{interparticle distance} \tag{12.29}$$

This is a reasonable criterion because for wavelengths shorter than λ_m a wave of displacements of atoms becomes meaningless.

We now calculate the equilibrium properties of a solid at low temperatures by calculating the partition function for an appropriate gas of phonons. The energy of the state in which there are n_i phonons of the ith type is*

$$E\{n_i\} = \sum_{i=1}^{3N} n_i \hbar\omega_i \tag{12.30}$$

The partition function is

$$Q = \sum_{\{n_i\}} e^{-\beta E\{n_i\}} = \prod_{i=1}^{3N} \frac{1}{1 - e^{-\beta\hbar\omega_i}}$$

Hence

$$\log Q = -\sum_{i=1}^{3N} \log\left(1 - e^{-\beta\hbar\omega_i}\right) \tag{12.31}$$

The average occupation number is

$$\langle n_i \rangle = -\frac{1}{\beta}\frac{\partial}{\partial(\hbar\omega_i)} \log Q = \frac{1}{e^{\beta\hbar\omega_i} - 1} \tag{12.32}$$

The internal energy is

$$U = -\frac{\partial}{\partial\beta} \log Q = \sum_{i=1}^{3N} \hbar\omega_i \langle n_i \rangle = \sum_{i=1}^{3N} \frac{\hbar\omega_i}{e^{\beta\hbar\omega_i} - 1} \tag{12.33}$$

Passing to the limit $V \to \infty$ we obtain, with the help of (12.26),

$$U = \frac{3V}{2\pi^2 c^3} \int_0^{\omega_m} d\omega\, \omega^2 \frac{\hbar\omega}{e^{\beta\hbar\omega} - 1} \tag{12.34}$$

or

$$\frac{U}{N} = \frac{9(kT)^4}{(\hbar\omega_m)^3} \int_0^{\beta\hbar\omega_m} dt\, \frac{t^3}{e^t - 1} \tag{12.35}$$

We define the Debye function $D(x)$ by

$$D(x) \equiv \frac{3}{x^3} \int_0^x dt\, \frac{t^3}{e^t - 1} = \begin{cases} 1 - \frac{3}{8}x + \frac{1}{20}x^2 + \cdots & (x \ll 1) \\ \dfrac{\pi^4}{5x^3} + O(e^{-x}) & (x \gg 1) \end{cases} \tag{12.36}$$

and the Debye temperature T_D by

$$kT_D \equiv \hbar\omega_m = \hbar c \left(\frac{6\pi^2}{v}\right)^{1/3} \tag{12.37}$$

Then

$$\frac{U}{N} = 3kT\, D(\lambda) = \begin{cases} 3kT\left(1 - \dfrac{3}{8}\dfrac{T_D}{T} + \cdots\right) & (T \gg T_D) \\ 3kT\left[\dfrac{\pi^4}{5}\left(\dfrac{T}{T_D}\right)^3 + O(e^{-T_D/T})\right] & (T \ll T_D) \end{cases} \tag{12.38}$$

* We should add to (12.30) an unknown constant representing the ground state energy of the solid, but this constant does not affect any subsequent results and hence can be ignored.

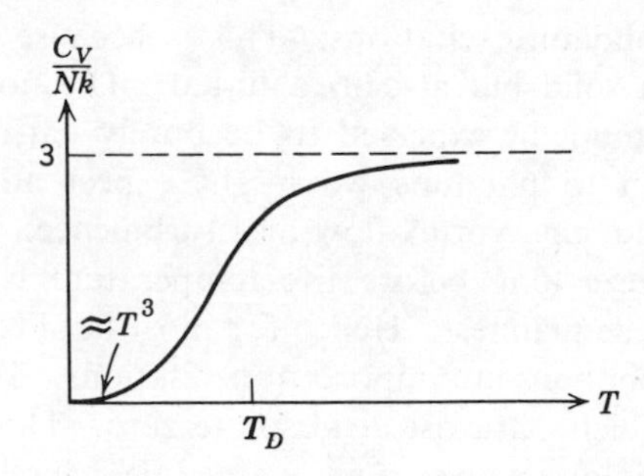

Fig. 12.2. Specific heat of a crystal lattice in Debye's theory.

where $\lambda \equiv T_D/T$. Then the specific heat is given by

$$\frac{C_V}{Nk} = 3D(\lambda) + 3T\frac{dD(\lambda)}{dT} = 3\left[4D(\lambda) - \frac{3\lambda}{e^\lambda - 1}\right] \tag{12.39}$$

The high- and low-temperature behaviors of C_V are as follows:

$$\frac{C_V}{Nk} = \begin{cases} 3\left[1 - \dfrac{1}{20}\left(\dfrac{T_D}{T}\right)^2 + \cdots\right] & (T \gg T_D) \\ \dfrac{12\pi^4}{5}\left(\dfrac{T}{T_D}\right)^3 + O(e^{-T_D/T}) & (T \ll T_D) \end{cases} \tag{12.40}$$

A plot of the specific heat is shown in Fig. 12.2, which agrees quite well with experimental findings.

At low temperatures C_V vanishes like T^3, verifying the third law of thermodynamics. When the temperature is much greater than the Debye temperature the lattice behaves classically, as indicated by the fact that $C_V \approx 3Nk$. For most solids the Debye temperature is of the order of 200°K. This is why the Dulong-Petit law $C_V \approx 3Nk$ holds at room temperatures. At extremely high temperatures the model of noninteracting phonons breaks down because the lattice eventually melts. The melting of the lattice is made possible by the fact that the forces between the atoms in the lattice are not strictly harmonic forces. In the phonon language the phonons are not strictly free. They must interact with each other, and this interaction becomes strong at very high temperatures.

The discussion of Debye's model of a solid furnishes a good illustration of the concept of phonons. The essential point in this model is *not* that at low temperature the excitations of a solid are phonons, but that the excitations are solely accounted for by phonons. The reason for this, as explained before, is that the normal modes of the system are purely harmonic oscillations. Intuitively we expect that at temperatures low enough for quantum effects to be manifest not only solids, but liquids, too, should have phonon excitations. This is because sound waves can exist not only in a solid but also in a liquid. The normal modes of a liquid, however, cannot be expected to be purely harmonic oscillations, so that in addition to phonons we might expect additional types of excitation in a liquid, e.g., vortex flow and turbulence.

Most liquids freeze long before the temperature is low enough for quantum effects to be manifest. Hence for most liquids it is an academic question whether phonons are important excitations. The only exception is liquid helium, which can exist at absolute zero. Thus it is relevant to ask whether at very low temperatures we can describe liquid helium as a gas of phonons *and nothing else*. Experimentally this is answered in the affirmative for liquid He^4 but not for liquid He^3. The theoretical reason* has to do with the Bose statistics obeyed by He^4 atoms.

12.3 BOSE-EINSTEIN CONDENSATION

Equation (9.71) gives the equation of state for the ideal Bose gas of N particles of mass m contained in a volume V. To study in detail the properties of the equation of state we must find the fugacity z as a function of temperature and specific volume by solving the second equation of (9.71), namely

$$\frac{1}{v} = \frac{1}{\lambda^3} g_{3/2}(z) + \frac{1}{V} \frac{z}{1-z} \tag{12.41}$$

where $v = V/N$, and $\lambda = \sqrt{2\pi\hbar^2/mkT}$, the thermal wavelength. To do this, we must first study the properties of the function $g_{3/2}(z)$, which is a special case of a more general class of functions

$$g_n(z) \equiv \sum_{l=1}^{\infty} \frac{z^l}{l^n} \tag{12.42}$$

These functions have been studied† and tabulated‡ in the literature.

It is obvious that for real values of z between 0 and 1, $g_{3/2}(z)$ is a bounded, positive, monotonically increasing function of z. In order to satisfy (12.41) it is necessary that

$$0 \le z \le 1$$

For comparison we recall that $0 \le z < \infty$ in the case of Fermi statistics. For small z, the power series (12.42) furnishes a practical way to calculate $g_{3/2}(z)$:

$$g_{3/2}(z) = z + \frac{z^2}{2^{3/2}} + \frac{z^3}{3^{3/2}} + \cdots \tag{12.43}$$

* See Sec. 18.3.

† J. E. Robinson, *Phys. Rev.*, **83,** 678 (1951).

‡ F. London, *Superfluids*, Vol. II (John Wiley and Sons, New York, 1954), Appendix.

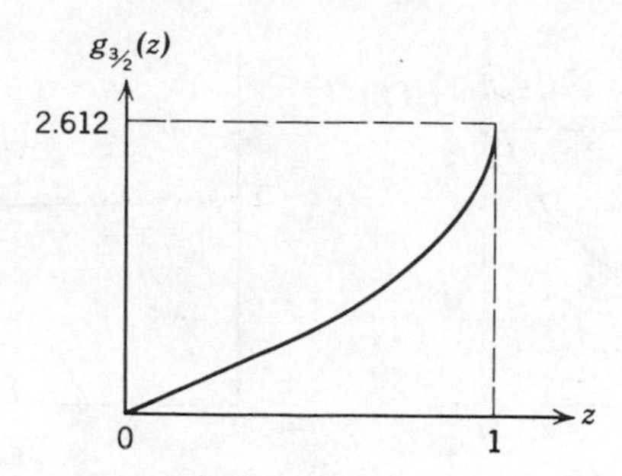

Fig. 12.3. The function $g_{3/2}(z)$.

At $z = 1$ its derivative diverges, but its value is finite:

$$g_{3/2}(1) = \sum_{l=1}^{\infty} \frac{1}{l^{3/2}} = \zeta(\tfrac{3}{2}) = 2.612 \cdots \tag{12.44}$$

where $\zeta(x)$ is the Riemann zeta function of x. Thus for all z between 0 and 1,

$$g_{3/2}(z) \leq 2.612 \cdots \tag{12.45}$$

A graph of $g_{3/2}(z)$ is shown in Fig. 12.3.

Let us rewrite (12.41) in the form

$$\lambda^3 \frac{\langle n_0 \rangle}{V} = \frac{\lambda^3}{v} - g_{3/2}(z) \tag{12.46}$$

This implies that $\langle n_0 \rangle / V > 0$ when the temperature and the specific volume are such that

$$\frac{\lambda^3}{v} > g_{3/2}(1) \tag{12.47}$$

This means that a finite fraction of the particles occupies the level with $\mathbf{p} = 0$. This phenomenon is known as the *Bose-Einstein condensation.* The condition (12.47) defines a subspace of the thermodynamic P-v-T space of the ideal Bose gas, which corresponds to the transition region of the Bose-Einstein condensation. As we see later, in this region the system can be considered to be a mixture of two thermodynamic phases, one phase being composed of particles with $\mathbf{p} = 0$ and the other with $\mathbf{p} \neq 0$. We refer to the region (12.47) as the condensation region. It is separated from the rest of the P-v-T space by the two-dimensional surface

$$\frac{\lambda^3}{v} = g_{3/2}(1) \tag{12.48}$$

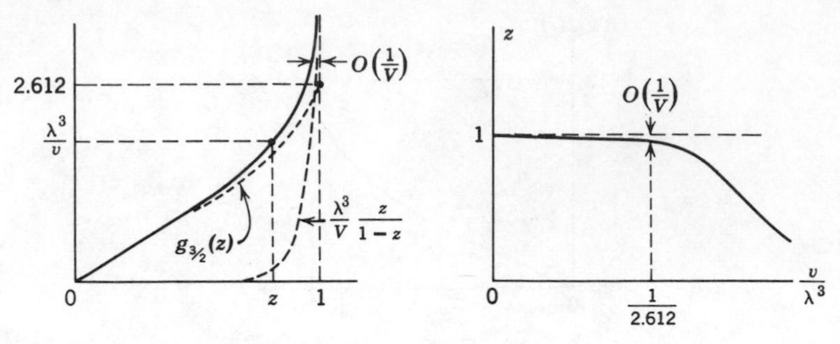

Fig. 12.4. (*a*) Graphical solution of (12.41); (*b*) the fugacity for an ideal Bose gas contained in a finite volume V.

For a given specific volume v, (12.48) defines a critical temperature T_c:

$$\lambda_c^3 = v g_{3/2}(1) \tag{12.49}$$

or

$$kT_c = \frac{2\pi\hbar^2/m}{[v g_{3/2}(1)]^{2/3}} \tag{12.50}$$

As indicated by (12.49), T_c is the temperature at which the thermal wavelength is of the same order of magnitude as the average interparticle separation. For a given temperature T, (12.48) defines a critical volume v_c:

$$v_c = \frac{\lambda^3}{g_{3/2}(1)} \tag{12.51}$$

In terms of T_c and v_c the region of condensation is the region in which $T < T_c$ or $v < v_c$.

To find z as a function of T and v we solve (12.41) graphically. For a large but finite value of the total volume V the graphical construction in Fig. 12.4*a* yields the curve for z shown in Fig. 12.4*b*. In the limit as $V \to \infty$ we obtain

$$z = \begin{cases} 1 & \left(\dfrac{\lambda^3}{v} \geq g_{3/2}(1)\right) \\ \text{the root of } g_{3/2}(z) = \lambda^3/v & \left(\dfrac{\lambda^3}{v} \leq g_{3/2}(1)\right) \end{cases} \tag{12.52}$$

For $(\lambda^3/v) \leq g_{3/2}(1)$, the value of z must be found by numerical methods. A graph of z is given in Fig. 12.5.

To make these considerations more rigorous the following point must be noted. It is recalled that (12.41) is derived from the condition

$$\frac{N}{V} = \frac{1}{V}\sum_{\mathbf{p}\neq 0} \langle n_\mathbf{p}\rangle + \frac{\langle n_0\rangle}{V}$$

by replacing the sum on the right side by an integral. It is clear that this integral is unchanged if we subtract from the sum any *finite* number of terms. More generally, (12.41) should be replaced by the equation

$$\frac{N}{V} = \frac{1}{\lambda^3} g_{3/2}(z) + \frac{\langle n_0 \rangle}{V} + \left(\frac{\langle n_1 \rangle}{V} + \frac{\langle n_2 \rangle}{V} + \cdots \right)$$

where, in the parentheses, there appear any finite number of terms. Every term in the parentheses, however, approaches zero as $V \to \infty$. For example,

$$\frac{\langle n_1 \rangle}{V} = \frac{1}{V} \frac{1}{z^{-1} e^{\beta \epsilon_1} - 1} \leq \frac{1}{V} \frac{1}{e^{\beta \epsilon_1} - 1}$$

where

$$2m \, \epsilon_1 = (2\pi\hbar)^2 \frac{l_1}{V^{2/3}}$$

$$l_1 = \text{sum of the squares of three integers not all zero}$$

Hence

$$\frac{\langle n_1 \rangle}{V} \leq \frac{1}{V} \frac{2m\beta V^{2/3}}{(2\pi\hbar)^2 \beta^2 l_1{}^2} \xrightarrow[V \to \infty]{} 0 \tag{12.53}$$

This shows that (12.41) is valid.

By (12.52) and the fact that $\langle n_0 \rangle = z/(1 - z)$ we can write

$$\frac{\langle n_0 \rangle}{N} = \begin{cases} 0 & \left(\dfrac{\lambda^3}{v} \leq g_{3/2}(1) \right) \\ 1 - \left(\dfrac{T}{T_c} \right)^{3/2} = 1 - \dfrac{v}{v_c} & \left(\dfrac{\lambda^3}{v} \geq g_{3/2}(1) \right) \end{cases} \tag{12.54}$$

A plot of $\langle n_0 \rangle / N$ is shown in Fig. 12.6. It is seen that when $T < T_c$, a finite fraction of the particles in the system occupy the single level with $\mathbf{p} = 0$. On the other hand (12.53) shows that $\langle n_\mathbf{p} \rangle / N$ is always zero for $\mathbf{p} \neq 0$. Therefore we have the following situation: For $T > T_c$ no single

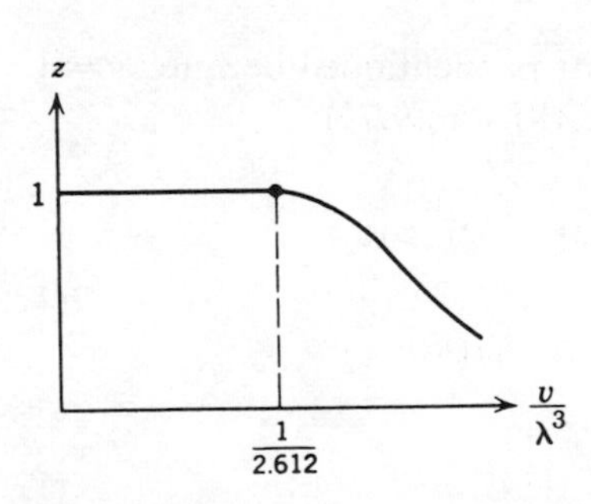

Fig. 12.5. The fugacity for an ideal Bose gas of infinite volume.

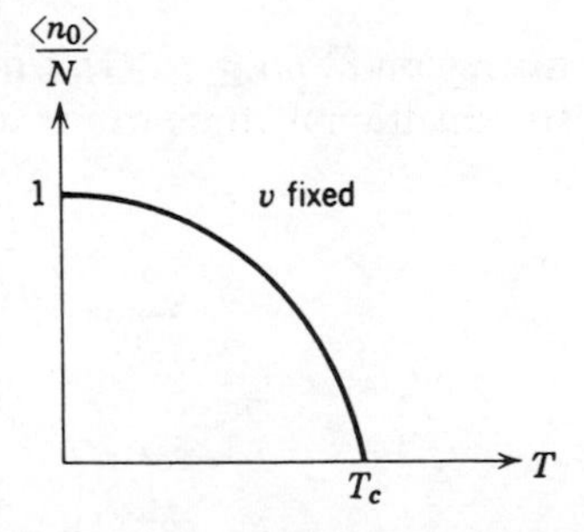

Fig. 12.6. Average occupation number of the level with $\mathbf{p} = 0$.

level is occupied by a finite fraction of all the particles. The particles "spread thinly" over all levels. For $T < T_c$ a finite fraction $1 - (T/T_c)^{3/2}$ occupies the level with $\mathbf{p} = 0$ while the rest of the particles "spread thinly" over the levels with $\mathbf{p} \neq 0$. At absolute zero all particles occupy the level with $\mathbf{p} = 0$.

The Bose-Einstein condensation is sometimes described as a "condensation in momentum space." We shall see, however, that its thermodynamic manifestations are those of a first-order phase transition. If we examine the equation of state alone, we discern no difference between the Bose-Einstein condensation and an ordinary gas-liquid condensation. If the particles of the ideal Bose gas are placed in a gravitational field, then in the condensation region there will be a spatial separation of the two phases, just as in a gas-liquid condensation.* The term "momentum-space condensation" merely serves to emphasize the fact that the cause of the Bose-Einstein condensation lies in the symmetry of the wave function and not in any interparticle interaction.

By virtue of (12.52) all thermodynamic functions of the ideal Bose gas will be given by different analytical expressions for the region of condensation and for the complement of that region. Only in the condensation region will these analytical expressions be simple. In the other region numerical computations would be necessary to obtain explicit formulas.

Throughout the remainder of this section let z be defined only for the region $(\lambda^3/v) \leq g_{3/2}(1)$. Some equivalent definitions of z are

$$\begin{aligned} g_{3/2}(z) &= \frac{\lambda^3}{v} \\ \frac{g_{3/2}(z)}{g_{3/2}(1)} &= \frac{v_c}{v} \\ \frac{g_{3/2}(z)}{g_{3/2}(1)} &= \left(\frac{T_c}{T}\right)^{3/2} \end{aligned} \tag{12.55}$$

In the region $(\lambda^3/v) \geq g_{3/2}(1)$, z need not be mentioned because $z = 1$.

The equation of state can be obtained from (9.71):

$$\frac{P}{kT} = \begin{cases} \dfrac{1}{\lambda^3} g_{5/2}(z) & (v > v_c) \\[2ex] \dfrac{1}{\lambda^3} g_{5/2}(1) & (v < v_c) \end{cases} \tag{12.56}$$

where

$$g_{5/2}(1) = \zeta(\tfrac{5}{2}) = 1.342 \cdots \tag{12.57}$$

* W. Lamb and A. Nordsieck, *Phys. Rev.*, **59**, 677 (1941).

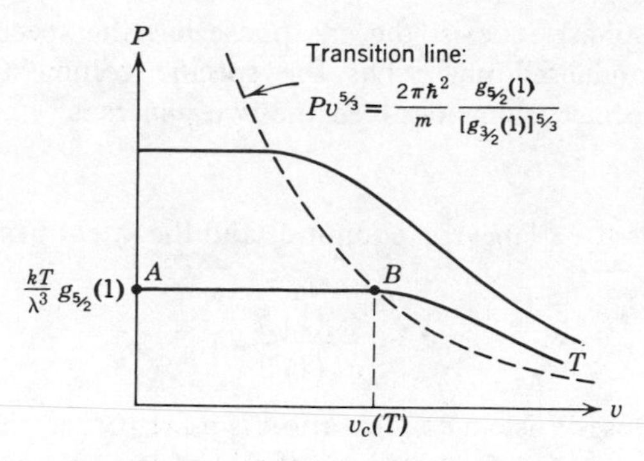

Fig. 12.7. Isotherms of the ideal Bose gas.

The term $V^{-1} \log(1 - z)$ in (9.71) is zero as $V \to \infty$. For $v > v_c$ this is obvious. Foı $v < v_c$, it is also true, because $(1 - z) \propto V^{-1}$. It is immediately seen that for $v < v_c$, P is independent of v. The isotherms are shown in Fig. 12.7, and the *P-T* diagram is shown in Fig. 12.8. We may, as in the case of a gas-liquid condensation, interpret the horizontal portion of an isotherm to mean that in that region the system is a mixture of two phases. In the present case these two phases correspond to the two points labeled *A* and *B* in Fig. 12.7. We refer to these respectively as the condensed phase and the gas phase. The horizontal portion of the isotherm is the region of phase transition between the two phases. The vapor pressure is

$$P_0(T) = \frac{kT}{\lambda^3} g_{5/2}(1) \tag{12.58}$$

Differentiation of this equation leads to

$$\frac{dP_0(T)}{dT} = \frac{5}{2} \frac{kg_{5/2}(1)}{\lambda^3} = \frac{1}{Tv_c}\left[\frac{5}{2} kT \frac{g_{5/2}(1)}{g_{3/2}(1)}\right] \tag{12.59}$$

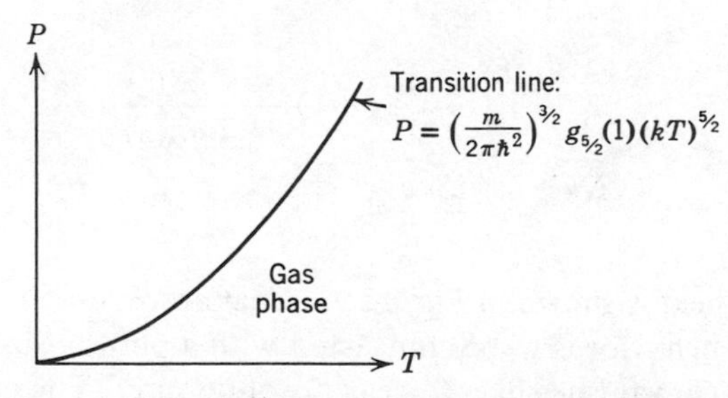

Fig. 12.8. *P–T* diagram of the ideal Bose gas. Note that the space above the transition curve does not correspond to anything. The condensed phase lies on the transition line itself.

When the two phases coexist the gas phase has the specific volume v_c, whereas the condensed phase has the specific volume 0. Hence the difference in specific volume between the two phases is

$$\Delta v = v_c \tag{12.60}$$

In fact (12.59) is the Clapeyron equation, and the latent heat of transition per particle is

$$L = \frac{g_{5/2}(1)}{g_{3/2}(1)} \frac{5}{2} kT \tag{12.61}$$

Therefore the Bose-Einstein condensation is a first-order phase transition.

Other thermodynamic functions for the ideal Bose gas are given in the following. For each thermodynamic function the upper equation refers to the region $v > v_c$ (or $T > T_c$) and the lower equation refers to the region $v < v_c$ (or $T < T_c$):

$$\frac{U}{N} = \tfrac{3}{2}Pv = \begin{cases} \dfrac{3}{2}\dfrac{kTv}{\lambda^3} g_{5/2}(z) \\[2ex] \dfrac{3}{2}\dfrac{kTv}{\lambda^3} g_{5/2}(1) \end{cases} \tag{12.62}$$

$$-\frac{A}{NkT} = \begin{cases} \dfrac{v}{\lambda^3} g_{5/2}(z) - \log z \\[2ex] \dfrac{v}{\lambda^3} g_{5/2}(1) \end{cases} \tag{12.63}$$

$$\frac{G}{NkT} = \begin{cases} \log z \\ 0 \end{cases} \tag{12.64}$$

$$\frac{S}{Nk} = \begin{cases} \dfrac{5}{2}\dfrac{v}{\lambda^3} g_{5/2}(z) - \log z \\[2ex] \dfrac{5}{2}\dfrac{v}{\lambda^3} g_{5/2}(1) \end{cases} \tag{12.65}$$

$$\frac{C_V}{Nk} = \begin{cases} \dfrac{15}{4}\dfrac{v}{\lambda^3} g_{5/2}(z) - \dfrac{9}{4}\dfrac{g_{3/2}(z)}{g_{1/2}(z)} \\[2ex] \dfrac{15}{4}\dfrac{v}{\lambda^3} g_{5/2}(1) \end{cases} \tag{12.66}$$

The specific heat is shown in Fig. 12.9. Near absolute zero, C_V vanishes like $T^{3/2}$. This behavior is to be contrasted with a photon gas or a phonon gas, for which C_V vanishes like T^3 near absolute zero. The reason for this difference lies in the difference between the particle spectrum $\epsilon_p = p^2/2m$ and the photon or phonon spectrum $\epsilon_p = cp$. At the same energy the particle spectrum has a higher density of states than the photon or phonon spectrum. Consequently there are more modes of excitation available for a particle, and the specific heat is greater.

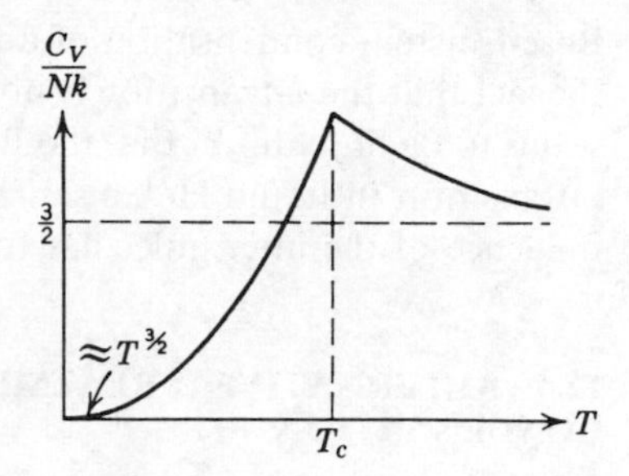

Fig. 12.9. Specific heat of the ideal Bose gas.

From (12.65) we see that $S = 0$ at $T = 0$, in accordance with the third law of thermodynamics. This means that the condensed phase (which exists at $T = 0$) has no entropy. At any finite temperature the total entropy is entirely due to the gas phase. The fraction of particles in the gas phase in the transition region is v/v_c, or $(T/T_c)^{3/2}$. If we rewrite S in the transition region in the form

$$\frac{S}{N} = \left(\frac{T}{T_c}\right)^{3/2} s = \left(\frac{v}{v_c}\right) s \tag{12.67}$$

we find that

$$s = \frac{g_{5/2}(1)}{g_{3/2}(1)} \frac{5}{2} k \tag{12.68}$$

which is the entropy per particle of the gas phase. The difference in specific entropy between the gas phase and the condensed phase is

$$\Delta s = s = \frac{g_{5/2}(1)}{g_{3/2}(1)} \frac{5}{2} k \tag{12.69}$$

Comparing this with (12.61), we find that

$$L = T \Delta s \tag{12.70}$$

This shows that the interpretation of the Bose-Einstein condensation as a first-order phase transition is self-consistent.

The only Bose system known to exist at low temperatures is liquid He^4. At a temperature of 2.18°K, He^4 exhibits the remarkable λ-transition, at which the specific heat becomes logarithmically infinite. Since He^4 atoms obey Bose statistics, we cannot help suspecting that this transition is the Bose-Einstein condensation modified by intermolecular interactions. This suspicion is supported by the fact that no such transition occurs in liquid He^3, whose atoms obey Fermi statistics. Furthermore, substituting the mass of He^4 and the density of liquid helium into (12.50) leads to the transition temperature $T_c = 3.14$°K, which is of the right order of magnitude. The difference between the λ-transition in liquid He^4 and the

Bose-Einstein condensation of an ideal Bose gas lies most significantly in the fact that the λ-transition is not a first-order transition. Although there seems to be little doubt that the Bose statistics plays a dominant role in the λ-transition in liquid He^4, a satisfactory theory that takes into account the influence of the intermolecular forces is yet to be devised.

12.4 ALTERNATIVE TREATMENT OF BOSE-EINSTEIN CONDENSATION

The partition function of the ideal Bose gas is

$$Q_N(V, T) = \sum_{\{n_\mathbf{p}\}} \exp\left(-\beta \sum_\mathbf{p} n_\mathbf{p}\epsilon_\mathbf{p}\right) \tag{12.71}$$

where $\epsilon_\mathbf{p} = p^2/2m$ and the summation extends over all sets of occupation numbers $\{n_\mathbf{p}\}$ satisfying the condition

$$\sum_\mathbf{p} n_\mathbf{p} = N \tag{12.72}$$

Since $\epsilon_0 = 0$, n_0 does not appear in the exponent of the summand in (12.71). We may perform the partition sum as follows. First choose an integer n_0 and sum over all $n_\mathbf{p}$ with $\mathbf{p} \neq 0$, subject to the condition

$$\sum_{\mathbf{p}\neq 0} n_\mathbf{p} = N - n_0 \tag{12.73}$$

After this is done, sum over n_0 from 0 to N. Thus,

$$Q_N(V, T) = \sum_{n_0=0}^{N} {\sum_{\{n_\mathbf{p}\}}}' \exp\left(-\beta \sum_{\mathbf{p}\neq 0} n_\mathbf{p}\epsilon_\mathbf{p}\right) \tag{12.74}$$

Let

$$\xi \equiv \frac{n_0}{N} \qquad \left(\xi = 0, \frac{1}{N}, \frac{2}{N}, \ldots, 1\right) \tag{12.75}$$

Let $Q(\xi)$ be the partition function of a fictitious ideal Bose gas of $N(1 - \xi)$ particles in which the level with $\mathbf{p} = 0$ is artificially removed:

$$Q(\xi) \equiv {\sum_{\{n_\mathbf{p}\}}}' \exp\left(-\beta \sum_{\mathbf{p}\neq 0} n_\mathbf{p}\epsilon_\mathbf{p}\right) \tag{12.76}$$

In the limit as $V \to \infty$ an expression for $Q(\xi)$ may be obtained directly from (12.63):

$$\frac{1}{N} \log Q(\xi) = \frac{v}{\lambda^3} g_{5/2}(z) - (1 - \xi) \log z \tag{12.77}$$

where z is the root of the equation

$$\frac{\lambda^3}{v}(1 - \xi) = g_{3/2}(z) \tag{12.78}$$

With these definitions we can rewrite (12.74) in the form

$$\frac{1}{N}\log Q_N(V, T) = \frac{1}{N}\log\sum_{\xi=0}^{1} Q(\xi) \tag{12.79}$$

The sum (12.79) contains $N + 1$ terms, all positive. Let $Q(\bar{\xi})$ be the largest possible value of $Q(\xi)$, Then

$$Q(\bar{\xi}) \leq Q_N(V, T) \leq (N + 1)Q(\bar{\xi})$$

or

$$\frac{1}{N}\log Q(\bar{\xi}) \leq \frac{1}{N}\log Q_N(V, T) \leq \frac{1}{N}\log Q(\bar{\xi}) + \frac{1}{N}\log (N + 1)$$

In the limit as $N \to \infty$ we have

$$\frac{1}{N}\log Q_N(V, T) = \frac{1}{N}\log Q(\bar{\xi}) \tag{12.80}$$

where

$$\frac{1}{N}\log Q(\bar{\xi}) \equiv \text{Max}\left[\frac{1}{N}\log Q(\xi)\right] \tag{12.81}$$

Obviously $\bar{\xi}$ is the average fraction of particles occupying the level with $\mathbf{p} = 0$. We show that

$$\bar{\xi} = \frac{\langle n_0 \rangle}{N} \tag{12.82}$$

where $\langle n_0 \rangle$ is given by (12.54). When this is achieved the Bose-Einstein condensation follows.

To maximize $\log Q(\xi)$ we proceed as follows.

Step 1. From (12.78) we can obtain ξ as a function of z for given λ^3/v. We thus obtain a family of ξ–z curves, one for each value of λ^3/v.

Step 2. For fixed λ^3/v we find $\bar{\xi}$ by maximizing $\log Q(\xi)$ along the appropriate ξ–z curve.

From the fact that $g_{3/2}(z)$ is a monotonically increasing function of z, we immediately see that, for given λ^3/v, $1 - \xi$ is a monotonically decreasing function of z. Furthermore, since $g_{3/2}(0) = 0$, all ξ–z curves pass through the point $\xi = 1$, $z = 0$. The family of ξ–z curves is qualitatively sketched in Fig. 12.10. Along a ξ–z curve we have

$$\frac{d}{d\xi}\log Q(\xi) = \frac{\partial}{\partial\xi}(\log Q)_z + \left(\frac{1}{z}\frac{\partial z}{\partial\xi}\right)z\frac{\partial}{\partial z}(\log Q)_\xi$$

The second term is zero by (12.78). Hence

$$\frac{d}{d\xi}\log Q(\xi) = \frac{\partial}{\partial\xi}(\log Q)_z = N\log z \leq 0 \tag{12.83}$$

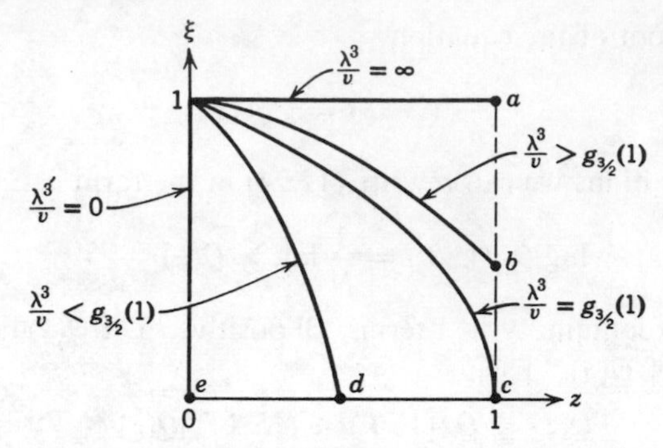

Fig. 12.10. Family of ξ–z curves. The equilibrium values $\bar{\xi}$ of ξ lie on the curve *abcde*.

Therefore the largest value of $\log Q(\xi)$ occurs at the smallest possible value of ξ on the particular ξ–z curve. It is obvious by inspection of Fig. 12.10 that

$$\bar{\xi} = \begin{cases} 0 & [\lambda^3/v \leq g_{3/2}(1)] \\ 1 - \dfrac{v g_{3/2}(1)}{\lambda^3} & [\lambda^3/v \geq g_{3/2}(1)] \end{cases} \tag{12.84}$$

This proves (12.82). The remaining discussion reduces to that of the last section.

PROBLEMS

12.1. Some experimental values* for the specific heat of liquid He^4 are given in the accompanying table. The values are obtained along the vapor pressure curve of liquid He^4, but we may assume that they are not very different from the values of c_V at the same temperatures.

Temperature, deg K	Specific Heat, joule/g-deg
0.60	0.0051
0.65	0.0068
0.70	0.0098
0.75	0.0146
0.80	0.0222
0.85	0.0343
0.90	0.0510
0.95	0.0743
1.00	0.1042

* Taken from H. C. Kramers, "Some Properties of Liquid Helium below 1°K," Dissertation, Leiden (1955).

(*a*) Show that the behavior of the specific heat is characteristic of that of a gas of phonons.
(*b*) Find the velocity of sound in liquid He^4 at low temperature.

12.2. Equation (12.64) states that $G = 0$ for $v < v_c$. Using the formula $S = -(\partial G/\partial T)_P$, we would obtain $S = 0$ for $v < v_c$, in contradiction to (12.65). What is wrong with the previous statement?

12.3. In the neighborhood of $z = 1$ the following expansion may be obtained (F. London, *loc. cit.*):

$$g_{5/2}(z) = 2.363\nu^{3/2} + 1.342 - 2.612\nu - 0.730\nu^2 + \cdots$$

where $\nu \equiv -\log z$. From this the corresponding expansions for $g_{3/2}$, $g_{1/2}$, and $g_{-1/2}$ may be obtained by the recursion formula $g_{n-1} = -\partial g_n/\partial \nu$. Using this expansion show that for the ideal Bose gas the discontinuity of $\partial C_V/\partial T$ at $T = T_c$ is given by

$$\left(\frac{\partial}{\partial T}\frac{C_V}{Nk}\right)_{T\to T_c^+} - \left(\frac{\partial}{\partial T}\frac{C_V}{Nk}\right)_{T\to T_c^-} = \frac{3.66}{T_c}$$

12.4. Show that the equation of state of the ideal Bose gas in the gas phase can be written in the form of a virial expansion, i.e.,

$$\frac{Pv}{kT} = 1 - \frac{1}{4\sqrt{2}}\left(\frac{\lambda^3}{v}\right) + \left(\frac{1}{8} - \frac{2}{9\sqrt{3}}\right)\left(\frac{\lambda^3}{v}\right)^2 - \cdots$$

12.5. (*a*) Calculate the grand partition function $\mathcal{Q}(z, V, T)$ for a two-dimensional ideal Bose gas and obtain the limit

$$\lim_{V\to\infty} \frac{1}{V} \log \mathcal{Q}(z, V, T)$$

where $V = L^2$ is the area available to the system.
(*b*) Find the average number of particles per unit area as a function of z and T.
(*c*) Show that there is no Bose-Einstein condensation for a two-dimensional ideal Bose gas.

chapter 13

IMPERFECT GASES AT LOW TEMPERATURES

13.1 DEFINITION OF THE PROBLEM

An imperfect gas is an extremely dilute system of particles which interact among themselves through an interparticle potential of finite range and of such a nature that there exists no two-particle bound state. The diluteness of the gas enables us to treat the interparticle interaction as a small perturbation on the ideal gas. An imperfect gas, therefore, is the first improvement on the ideal gas as a model for a physical gas. We shall consider an imperfect gas at extremely low temperatures. For such a system there are two important parameters of the dimension of length: The thermal wavelength λ and the average interparticle separation $v^{1/3}$. These two lengths may be of comparable magnitude, but they must be much larger than the range of the interparticle potential, or any other length in the problem, except the size of the container.

In quantum mechanics a particle cannot be localized within its de Broglie wavelength, which in the present case may be replaced by the thermal wavelength. Thus in the present case a particle "spreads" over a distance much larger than the range of the interaction potential. Within the range of interaction of any given particle, the probability of finding another particle is small. Therefore

(*a*) the effective interaction experienced by a particle is small, even though the interparticle potential may have large values;

(*b*) the details of the interparticle potential are unimportant, because a particle that is spread out in space sees only an averaged effect of the potential.

In the quantum theory of scattering it is known that at low energies the scattering of a particle by a potential does not depend on the shape of the potential, but depends only on a single parameter obtainable from the potential—the scattering length a. The total scattering cross section at low energies is $4\pi a^2$. Hence roughly speaking a is the effective diameter of the potential. We may also say that at low energies the scattering from a potential looks like that from a hard sphere of diameter a.

The previous discussion makes it plausible that at extremely low temperatures it is possible to describe an imperfect gas solely in terms of the three parameters λ, $v^{1/3}$, and a. Our problem is to formulate a method by which all the thermodynamic functions of the imperfect gas can be obtained to lowest order in the small parameters a/λ and $a/v^{1/3}$.

The more general problem of finding a systematic method for the calculation of the partition function is discussed in Chapter 14. The systematic method, although exact, is difficult to apply in actual calculations. The method we describe, although limited in validity, can be easily used.

To attain the limited goal desired, we show that for the purpose of calculating the low-lying energy levels of an imperfect gas the Hamiltonian of the system may be replaced by an effective Hamiltonian in which only scattering parameters, such as the scattering length, appear explicitly. The partition function of the imperfect gas can then be calculated with the help of the effective Hamiltonian. This method, first introduced by Fermi,* is known as the method of pseudopotentials.

13.2 METHOD OF PSEUDOPOTENTIALS IN TWO-BODY PROBLEMS

We consider a system of two particles interacting through a finite-ranged potential which has no bound state. The object of the method of pseudopotentials is to obtain all the energy levels of the system in terms of the scattering phase shifts of the potential. For the sake of concreteness we first assume that the potential is the hard-sphere potential with diameter a. The wave function for the two particles may be written in the form†

$$\Psi(\mathbf{r}_1, \mathbf{r}_2) = e^{i\mathbf{P}\cdot\mathbf{R}}\psi(\mathbf{r}) \tag{13.1}$$

* E. Fermi, *Ricerca Sci.*, **7**, 13 (1936). Our presentation follows that of K. Huang and C. N. Yang, *Phys. Rev.*, **105**, 767 (1957).

† We suppress spin coordinates, if any.

where

$$\begin{aligned} \mathbf{R} &= \tfrac{1}{2}(\mathbf{r}_1 + \mathbf{r}_2) \\ \mathbf{r} &= \mathbf{r}_2 - \mathbf{r}_1 \end{aligned} \tag{13.2}$$

and $\mathbf{P}$ is the total momentum vector. The Schrödinger equation in the center-of-mass system is

$$\begin{aligned} (\nabla^2 + k^2)\psi(\mathbf{r}) &= 0 \qquad (r > a) \\ \psi(\mathbf{r}) &= 0 \qquad (r \leq a) \end{aligned} \tag{13.3}$$

The hard-sphere potential is no more than a boundary condition for the relative wave function $\psi(\mathbf{r})$. It is understood that some boundary condition for $r \to \infty$ is specified, but what it is is irrelevant to our considerations. The number k is the relative wave number, and (13.3) presents an eigenvalue problem for k. When the allowed values of k are known, the energy eigenvalues of the system are given by

$$E(\mathbf{P}, \mathbf{k}) = \frac{P^2}{2M} + \frac{\hbar^2 k^2}{2\mu}$$

where M is the total mass and μ the reduced mass of the system.

The aim of the method of pseudopotentials is to replace the hard-sphere boundary condition by an inhomogeneous term for the wave equation. Such an idea is familiar in electrostatics, where to find the electrostatic potential in the presence of a metallic sphere (with some given boundary condition at infinity) we may replace the sphere by a distribution of charges on the surface of the sphere and find the potential set up by the fictitious charges. We can further replace the surface charges by a collection of multipoles at the center of the sphere with appropriate strengths. If we solve the Poisson equation with these multipole sources, we obtain the exact electrostatic potential *outside the sphere.* In an analogous way, the method of pseudopotentials replaces the boundary condition on $\psi(\mathbf{r})$ by a collection of sources at the point $\mathbf{r} = 0$. Instead of producing electrostatic multipole potentials, however, these sources will produce scattered S-waves, P-waves, D-waves, etc.

Let us first consider spherically symmetric (S-wave) solutions of (13.3) at very low energies ($k \to 0$). The equations (13.3) become

$$\begin{aligned} \frac{1}{r^2}\frac{d}{dr}\left(r^2 \frac{d\psi}{dr}\right) &= 0 \qquad (r > a) \\ \psi(r) &= 0 \qquad (r \leq a) \end{aligned} \tag{13.4}$$

The solution is obviously

$$\psi(r) = \begin{cases} \text{const.}\left(1 - \dfrac{a}{r}\right) & (r > a) \\ 0 & (r \leq a) \end{cases} \tag{13.5}$$

Now define an extended wave function $\psi_{ex}(r)$ such that

$$(\nabla^2 + k^2)\psi_{ex}(r) = 0 \qquad \text{(everywhere except at } r = 0) \tag{13.6}$$

with the boundary condition

$$\psi_{ex}(a) = 0 \tag{13.7}$$

For $k \to 0$ we have

$$\psi_{ex}(r) \xrightarrow[r\to 0]{} \left(1 - \frac{a}{r}\right)\chi \tag{13.8}$$

where χ is a constant that depends on the boundary condition at $r = \infty$. We can avoid explicit use of this boundary condition by writing

$$\chi = \left[\frac{\partial}{\partial r}(r\psi_{ex})\right]_{r=0} \tag{13.9}$$

which is an immediate consequence of (13.8). To eliminate the explicit requirement (13.7), we generalize the equation (13.6) to include the point $r = 0$. This can be easily done by finding the behavior of $(\nabla^2 + k^2)\psi_{ex}$ near $r = 0$, as required by (13.8). Since $k \to 0$, it is sufficient to note that according to (13.8)

$$\nabla^2\psi_{ex}(r) \xrightarrow[r\to 0]{} 4\pi a\ \delta(\mathbf{r})\chi = 4\pi a\ \delta(\mathbf{r})\frac{\partial}{\partial r}(r\psi_{ex}) \tag{13.10}$$

Therefore as $k \to 0$ the function $\psi_{ex}(r)$ everywhere satisfies the equation

$$(\nabla^2 + k^2)\psi_{ex}(r) = 4\pi a\ \delta(\mathbf{r})\frac{\partial}{\partial r}(r\psi_{ex}) \tag{13.11}$$

The operator $\delta(\mathbf{r})(\partial/\partial r)r$ is the pseudopotential.* For small k and for $r \geq a$, $\psi_{ex}(r)$ satisfies the same equation and the same boundary condition as $\psi(r)$. Therefore $\psi_{ex}(r) = \psi(r)$ for $r \geq a$, and the eigenvalues of k are the same in both cases.

The equation (13.11) is not the exact equation we desire, because only the S-wave solutions with small k coincide with the actual solutions of the physical problem. To obtain an equation for an extended wave function that rigorously coincides with $\psi(r)$ for $r \geq a$ it is necessary to generalize (13.11) to arbitrary values of k and to nonspherically symmetric solutions. The generalization is given in Appendix B. It suffices for the present to state that the result of the generalization consists of the following modifications of (13.11):

(*a*) The exact S-wave pseudopotential is

$$-\frac{4\pi}{k\cot\eta_0}\,\delta(\mathbf{r})\frac{\partial}{\partial r}r \tag{13.12}$$

* The foregoing derivation is due to J. M. Blatt and V. F. Weisskopf, *Theoretical Nuclear Physics* (John Wiley and Sons, New York, 1952), p. 74.

where η_0 is the S-wave phase shift for the hard-sphere potential:

$$-\frac{1}{k\cot\eta_0} = \frac{\tan ka}{k} = a[1 + \tfrac{1}{3}(ka)^2 + \cdots] \tag{13.13}$$

(*b*) An infinite series of pseudopotentials is added to the right-hand side of (13.11), representing the effects of P-wave scattering, D-wave scattering, etc. The lth-wave pseudopotential is proportional to a^{2l+1}.

From these results it is seen that (13.11) is correct up to the order a^2. That is, if the wave function $\psi(r)$ and the eigenvalue k are expanded in a power series in a, then (13.11) correctly gives the coefficients of a and a^2.

The differential operator $(\partial/\partial r)r$ in the pseudopotential (13.11) may be replaced by unity if $\psi_{ex}(\mathbf{r})$ is well behaved at $r = 0$, for then

$$\left[\frac{\partial}{\partial r}(r\psi_{ex})\right]_{r=0} = \psi_{ex}(0) + \left[r\frac{\partial}{\partial r}\psi_{ex}\right]_{r=0} = \psi_{ex}(0) \tag{13.14}$$

If $\psi_{ex}(\mathbf{r}) \xrightarrow[r\to 0]{} Ar^{-1} + B$, however, then

$$\left[\frac{\partial}{\partial r}(r\psi_{ex})\right]_{r=0} = B \tag{13.15}$$

An illustration of the effect of $(\partial/\partial r)r$ is given in Problem 13.1.

We now turn to the method of pseudopotentials for the case of two particles interacting through a general finite-ranged potential which has no bound state. Here (13.3) is replaced by the equation

$$\frac{\hbar^2}{\mu}(\nabla^2 + k^2)\psi(\mathbf{r}) = v(r)\psi(\mathbf{r}) \tag{13.16}$$

with some given boundary condition for $r \to \infty$. At low energies only S-wave scattering is important. Therefore let us consider only spherically symmetric solutions. Then (13.16) reduces to

$$u''(r) + k^2u(r) = \frac{\mu}{\hbar^2}v(r)u(r) \tag{13.17}$$

where

$$u(r) \equiv r\psi(r) \tag{13.18}$$

By assumption $v(r)$ is finite-ranged and has no bound state. Therefore, as $r \to \infty$, $u(r)$ approaches a sinusoidal function:

$$u(r) \xrightarrow[r\to\infty]{} u_\infty(r) \tag{13.19}$$

where

$$u_\infty(r) \equiv r\psi_\infty(r) = \text{const.}\,(\sin kr + \tan\eta_0 \cos kr) \tag{13.20}$$

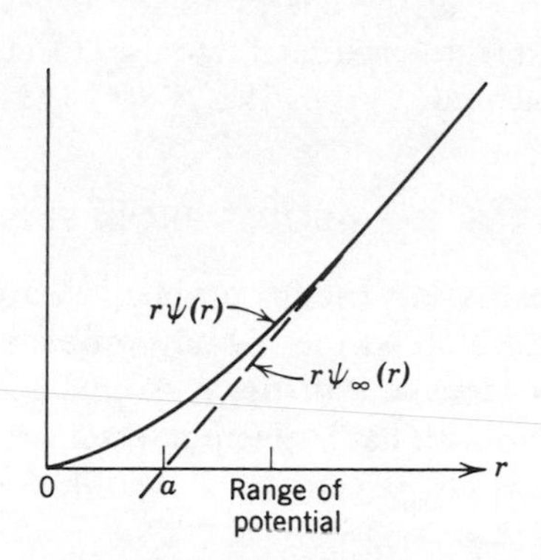

Fig. 13.1. Wave function in a repulsive potential with positive scattering length.

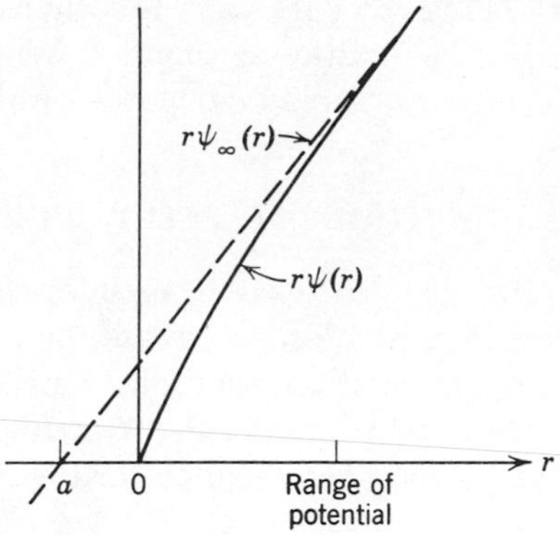

Fig. 13.2. Wave function in an attractive potential with negative scattering length.

where η_0 is by definition the *S*-wave phase shift. For $k \to 0$,

$$\psi_\infty(r) \xrightarrow[r \to 0]{} \text{const.} \left(1 + \frac{\tan \eta_0}{kr}\right) \tag{13.21}$$

In general η_0 is a function of k. For small k there is a well-known expansion analogous to (13.13), known as the effective range expansion:

$$k \cot \eta_0 = -\frac{1}{a} + \frac{1}{2} k^2 r_0 + \cdots \tag{13.22}$$

where a is called the scattering length and r_0 the effective range. The meaning of the scattering length can be seen by substituting (13.22) into (13.20). For $k \to 0$ we obtain (13.8). As illustrated in Figs. 13.1 and 13.2, the scattering length is the intercept of the asymptotic wave function $r\psi_\infty(r)$ with the coordinate axis. For the hard-sphere potential the scattering length is the hard-sphere diameter. In general a may be either positive or negative. It is positive for a predominantly repulsive potential (Fig. 13.1) and negative for a predominantly attractive potential (Fig. 13.2).

At low energies we may neglect all terms in (13.22) except $-1/a$ and obtain

$$-\frac{1}{k \cot \eta_0} \approx a$$

This approximation, known as the "shape-independent approximation," states that at low energies the potential acts as if it were a hard-sphere potential of diameter a. Therefore (13.11) can be taken over.* In general

* The derivation of (13.11) remains valid if a is negative.

(13.11) is certainly valid for the calculation of the energy to the lowest order in the scattering length a. Whether it is still meaningful to use (13.11) for higher orders in a depends on the potential.

13.3 METHOD OF PSEUDOPOTENTIALS IN N-BODY PROBLEMS

Having introduced the pseudopotentials in the two-body problem we are now in a position to discuss the generalization to the N-body problem. The considerations that follow are independent of statistics.

Let us first consider the N-body problem with hard-sphere interactions. The Schrödinger equation for the system is

$$-\frac{\hbar^2}{2m}(\nabla_1{}^2 + \cdots + \nabla_N{}^2)\Psi(\mathbf{r}_1, \ldots, \mathbf{r}_N) = E\Psi(\mathbf{r}_1, \ldots, \mathbf{r}_N)$$

$$(|\mathbf{r}_i - \mathbf{r}_j| > a, \text{ all } i \neq j) \quad (13.23)$$

$$\Psi(\mathbf{r}_1, \ldots, \mathbf{r}_N) = 0 \qquad \text{(otherwise)}$$

We also require the Ψ satisfies some boundary condition on the surface of a large cube, e.g., that Ψ satisfies periodic boundary conditions. The hard-sphere interactions are equivalent to a boundary condition that requires Ψ to vanish whenever $|\mathbf{r}_i - \mathbf{r}_j| = a$, for all $i \neq j$. In the $3N$-dimensional configuration space the collection of all points for which $|\mathbf{r}_i - \mathbf{r}_j| = a$ represents a tree-like hypersurface, a portion of which we schematically represent by Fig. 13.3. Thus we draw a cylinder, labeled 12, to represent the surface in which $|\mathbf{r}_1 - \mathbf{r}_2| = a$, whereas $\mathbf{r}_3, \ldots, \mathbf{r}_N$ may have arbitrary values. The whole "tree" is the totality of all such cylinders, $\frac{1}{2}N(N-1)$ in number, which mutually intersect in a complicated way. If the hard-sphere diameter a is small, these cylinders have a small radius. To find the wave function outside the "tree," it is natural to replace the "tree" by a series of "multipoles" at the "axes," i.e., at the lines $|\mathbf{r}_i - \mathbf{r}_j| = 0$.

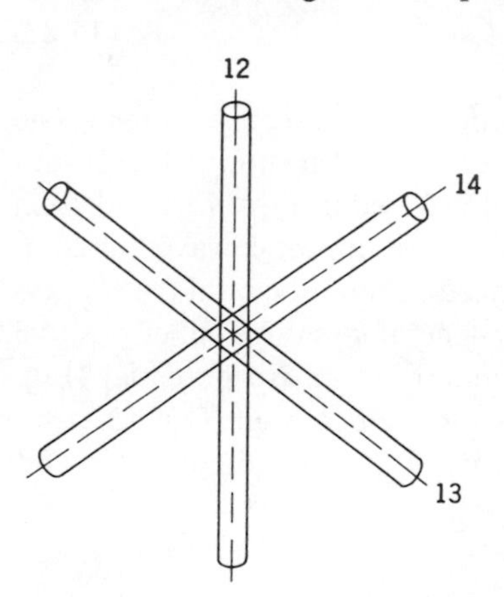

Fig. 13.3. The tree-like hypersurface in the 3N-dimensional configuration space. The hard-sphere interactions are equivalent to the boundary condition that the wave function vanishes on the surface of the "tree."

It can be easily shown that replacing the effect of each cylinder by multipoles along its axis amounts to introducing the two-body pseudopotentials described in the previous section. Our extended wave function would then satisfy a Schrödinger equation containing the sum of $\frac{1}{2}N(N-1)$ two-body pseudopotentials. These two-body pseudopotentials, however, do not exactly replace the effect of the "tree." Although they correctly give the behavior of Ψ near a cylinder and far away from any intersection of cylinders, they do not necessarily give the correct behavior of Ψ near an intersection of two or more cylinders. For example, the intersection corresponding to $|\mathbf{r}_1 - \mathbf{r}_2| = a$ and $|\mathbf{r}_1 - \mathbf{r}_3| = a$ represents a configuration in which particles 1, 2, and 3 collide simultaneously–an intrinsically three-body effect which has not been taken into account in the two-body pseudopotentials. The sum of two-body pseudopotentials accounts only for the effects of binary collisions.

Using our geometrical picture, we see that in addition to the two-body pseudopotentials it may be necessary to place additional multipoles (pseudopotentials) at each intersection of two or more cylinders. To find the exact magnitude of these three- and more-body pseudopotentials we would have to solve three- and more-body problems. Their dependence on the hard-sphere diameter a, however, can be found by a dimensional argument.

As an example, the three-body pseudopotential needed at the intersection of the lines $|\mathbf{r}_1 - \mathbf{r}_2| = 0$ and $|\mathbf{r}_1 - \mathbf{r}_3| = 0$ must appear in the three-body Schrödinger equation in the form

$$(\nabla_1^2 + \nabla_2^2 + \nabla_3^2 + k^2)\psi$$
$$= (\text{sum of two-body pseudopotentials}) + \delta(\mathbf{r}_1 - \mathbf{r}_2)\,\delta(\mathbf{r}_1 - \mathbf{r}_3)K\psi$$

The quantity K must be of the dimension (length)^4. At low energies ($k \to 0$) the only length in the problem is a. Therefore K must be of the order a^4. In a similar way we deduce that four-body pseudopotentials are of the order a^7, and so forth. These pseudopotentials may be ignored, if we are only interested in an accuracy up to the order a^2. The necessity for such n-body pseudopotentials shows that the pseudopotentials are not additive. This is analogous to the well-known situation in electrostatics that image charges are not additive. For example, the images of a point charge in front of two mutually orthogonal plane conductors are not simply the two images produced by each plane conductor taken separately.

If the interparticle potential is not the hard-sphere potential but a finite-ranged potential that has no bound state, the considerations just given can be taken over. The effective Hamiltonian for an imperfect gas of N identical particles of mass m may be taken to be

$$H = -\frac{\hbar^2}{2m}(\nabla_1^2 + \cdots + \nabla_N^2) + \frac{4\pi a\hbar^2}{m}\sum_{i<j}\delta(\mathbf{r}_i - \mathbf{r}_j)\frac{\partial}{\partial r_{ij}}\,r_{ij} \quad (13.24)$$

where a is the scattering length. This is valid for both fermions and bosons. The eigenvalues of this Hamiltonian will be the correct eigenvalues for an imperfect hard-sphere gas up to order a^2. For a general imperfect gas they will be correct to the lowest order in a.

We note that (13.24) is not a hermitian operator because $(\partial/\partial r)r$ is not a hermitian operator. This need not cause concern because, by its derivation, (13.24) has been shown to have real eigenvalues that are the approximate eigenvalues of the real problem. The nonhermiticity reflects the fact that the eigenfunctions of (13.24) do not everywhere coincide with the eigenfunctions of the real problem, but do so only in the asymptotic region. The fact, however, that (13.24) is not hermitian means that we cannot find its eigenvalues by variational methods.

For the remaining part of this chapter we diagonalize (13.24) only to the lowest order in a. That is, the pseudopotentials in (13.24) are regarded as small perturbations to be treated only to the first order in perturbation theory. Thus the operators $(\partial/\partial r)r$ will always act on unperturbed free-particle wave functions, which are well-behaved. Hence the operators $(\partial/\partial r)r$ can be set equal to unity, and we work with the Hamiltonian

$$H' = -\frac{\hbar^2}{2m}(\nabla_1^{\,2} + \cdots + \nabla_N^{\,2}) + \frac{4\pi a\hbar^2}{m}\sum_{i<j}\delta(\mathbf{r}_i - \mathbf{r}_j) \quad (13.25)$$

It is to be emphasized that *this Hamiltonian is valid only for the purpose of applying first-order perturbation theory.* We must not diagonalize (13.25) exactly, because the exact eigenvalues are the same as those for a free-particle system—it being well known that a three-dimensional δ-function potential produces no scattering.

13.4 AN IMPERFECT FERMI GAS*

The Energy Levels

We consider a dilute system of N identical fermions of mass m and spin $\hbar/2$, contained in a box of volume V, at very low temperatures. The fermions interact with one another through a two-body interaction characterized by the scattering length a. The energy levels to the first order in a may be obtained from (13.25) through the use of first-order perturbation theory.

Let the unperturbed wave functions be free-particle wave functions Φ_n, labeled by the occupation numbers $\{\ldots, n_{\mathbf{p},s}, \ldots\}$, where $n_{\mathbf{p},s}$ is the number

* K. Huang, *Tsing Hua Journal of Chinese Studies*, Special Number 1, 185 (1959).

of fermions with momentum $\mathbf{p}$ and spin quantum number s. The energy levels to the first order in a are

$$E_n \equiv (\Phi_n, H'\Phi_n) = \sum_{\mathbf{p},s} n_{\mathbf{p},s} \frac{p^2}{2m} + \frac{4\pi a\hbar^2}{m}\left(\Phi_n, \sum_{i<j} \delta(\mathbf{r}_i - \mathbf{r}_j)\Phi_n\right)$$

The second term is calculated in (A. 41) of Appendix A. With that, we have

$$E_n = \sum_{\mathbf{p}} \frac{p^2}{2m}(n_{\mathbf{p}}^+ + n_{\mathbf{p}}^-) + \frac{4\pi a\hbar^2}{mV} N_+N_- \tag{13.26}$$

where $n_{\mathbf{p}}^{\pm}$ and $N_{\pm}$ are defined in (11.106).

The pseudopotential employed in the derivation of (13.26) is valid only for $k\,|a| \ll 1$, where k is the relative wave number of any pair of particles in the system. Hence (13.26) is valid only for low-lying energy levels if

$$k_F\,|a| \ll 1 \tag{13.27}$$

where k_F is the wave number of a particle at the Fermi level:

$$k_F{}^2 = \frac{2m}{\hbar^2}\epsilon_F = \left(\frac{3\pi^2}{v}\right)^{2/3} \tag{13.28}$$

with $v = V/N$. Thus (13.27) is equivalent to the requirement of low density, namely, $a/v^{1/3} \ll 1$.

The physical meaning of (13.26) can be easily understood. It suffices to consider the case of a repulsive interparticle potential, so that $a > 0$. For a given distribution in momentum the energy is minimum when $N_+N_- = 0$, which means that all spins are aligned. This is a consequence of the Pauli exclusion principle, which requires that the relative wave function of two particles with parallel spin be spatially antisymmetric. Hence they are not likely to be found near each other, and the repulsive interaction energy is minimized. The case $a < 0$ can be discussed in a similar way.

Let us assume that $a > 0$ and that each fermion has an intrinsic magnetic moment of magnitude μ. In the presence of a uniform external magnetic field B, the energy levels of the system are given by

$$E_n = \sum_{\mathbf{p}} (n_{\mathbf{p}}^+ + n_{\mathbf{p}}^-)\frac{p^2}{2m} + \frac{4\pi a\hbar^2}{mV} N_+N_- - (N_+ - N_-)\mu B \tag{13.29}$$

We note that this differs from the energy levels (11.107) only by the presence of the second term, which has the tendency to align the spin of the particles. Thus we expect that it will make a positive contribution to the magnetic susceptibility of the system.

The Partition Function

The partition function* is

$$Q_N = \sum_{\{n_{\mathbf{p}}^+\},\{n_{\mathbf{p}}^-\}}' \exp\left\{-\beta\left[\sum_{\mathbf{p}} (n_{\mathbf{p}}^+ + n_{\mathbf{p}}^-)\frac{p^2}{2m} - \mu B(N_+ - N_-) + \frac{4\pi a\hbar^2 N_+ N_-}{mV}\right]\right\} \tag{13.30}$$

The notation is the same as that of (11.108). Proceeding in the same way as in the evaluation of (11.108), we obtain

$$\frac{1}{N}\log Q_N = \beta g(\bar{N}_+) \tag{13.31}$$

where

$$\begin{aligned} g(\bar{N}_+) &= \text{Max}\,[g(N_+)] \\ g(N_+) &\equiv \mu B\left(\frac{2N_+}{N} - 1\right) - \frac{4\pi a\hbar^2}{mV} N_+(N - N_+) - \frac{1}{N}[A(N_+) + A(N - N_+)] \end{aligned} \tag{13.32}$$

Thus $\bar{N}_+$ is the root of the equations

$$\begin{aligned} \left[\frac{\partial g(N_+)}{\partial N_+}\right]_{N_+=\bar{N}_+} &= 0 \\ \left[\frac{\partial^2 g(N_+)}{\partial N_+^{\,2}}\right]_{N_+=\bar{N}_+} &< 0 \end{aligned} \tag{13.33}$$

It must be noted that (13.33) locates the point at which the curve $g(N_+)$ passes through a maximum. It is conceivable (and in fact true) that $\bar{N}_+$ may occur not at a maximum of the curve $g(N_+)$ but at the boundary of the range of N_+, i.e., at $\bar{N}_+ = 0$ or $\bar{N}_+ = N$. We keep this in mind as we proceed. Let $kT\nu(N)$ be the chemical potential of the ideal Fermi gas of N *spinless* particles of mass m: $kT\nu(N) = \partial A(N)/\partial N$. We can write (13.33) in the manner

$$\begin{aligned} kT[\nu(\bar{N}_+) - \nu(N - \bar{N}_+)] &= 2\mu B + \frac{4\pi a\hbar^2}{mV}(2\bar{N}_+ - N) \\ kT[\nu'(\bar{N}_+) + \nu'(N - \bar{N}_+)] - \frac{8\pi a\hbar^2}{mV} &> 0 \end{aligned} \tag{13.34}$$

* Since the partition function (13.30) is calculated with the energy levels in first-order perturbation theory, the variational principle of Sec. 10.3 implies that it is smaller than $Tr e^{-\beta H'}$, where H' is (13.25). This is trivially true, because as we have pointed out the exact eigenvalues of (13.25) are free-particle energies. It is not true, however, that (13.30) is smaller than $Tr e^{-\beta H}$ where H is (13.24), because (13.24) is not a hermitian operator, and the variational principle does not apply to it.

where $\nu'(N) \equiv \partial\nu(N)/\partial N$. Let

$$r \equiv \frac{2\bar{N}_+}{N} - 1 \qquad (-1 \leq r \leq +1) \tag{13.35}$$

Then (13.34) becomes

$$\begin{aligned} kT\left\{\nu\left[\frac{N}{2}(1+r)\right] - \nu\left[\frac{N}{2}(1-r)\right]\right\} &= 2\mu B + \frac{a\lambda^2}{v}\, 2kTr \\ \frac{\partial}{\partial r}\left\{\nu\left[\frac{N}{2}(1+r)\right] - \nu\left[\frac{N}{2}(1-r)\right]\right\} - \frac{2a\lambda^2}{v} &> 0 \end{aligned} \tag{13.36}$$

where $\nu[x] \equiv \nu(x)$ and $\lambda = \sqrt{2\pi\hbar^2/mkT}$, the thermal wavelength. The low-temperature and high-temperature approximations for $\nu(Nx/2)$ are obtainable from (11.24) and (11.12) respectively. They are

$$kT\nu\left(\frac{N}{2}\,x\right) \approx x^{2/3}\epsilon_F\left[1 - \frac{\pi^2}{12}\left(\frac{kT}{\epsilon_F}\right)^2\frac{1}{x^{4/3}}\right] \qquad \left(\frac{kT}{\epsilon_F} \ll 1\right) \tag{13.37}$$

$$kT\nu\left(\frac{N}{2}\,x\right) \approx \log\frac{\lambda^2}{2v} \qquad \left(\frac{kT}{\epsilon_F} \gg 1\right) \tag{13.38}$$

where ϵ_F is the Fermi energy given by (13.28).

If r is known, the magnetization per unit volume and the magnetic susceptibility per unit volume are respectively given by

$$\begin{aligned} \mathscr{M} &= \frac{\mu r}{v} \\ \chi &= \frac{\partial \mathscr{M}}{\partial B} \end{aligned} \tag{13.39}$$

Spontaneous Magnetization

We first consider the case $B = 0$. At absolute zero, (13.36) reduces to

$$\begin{aligned} (1+r)^{2/3} - (1-r)^{2/3} &= \zeta r \\ \frac{1}{2}\left[\frac{1}{(1+r)^{1/3}} + \frac{1}{(1-r)^{1/3}}\right] &> \frac{3}{4}\zeta \end{aligned} \tag{13.40}$$

where

$$\zeta \equiv \frac{8}{3\pi}\, k_F a \tag{13.41}$$

Equation (13.40) is invariant under a change of sign of r. This is to be expected; in the absence of field, no absolute meaning can be attached to "up" or "down." Thus it is sufficient to consider $r \geq 0$. We may solve

(13.40) graphically by referring to Fig. 11.11, where $(1 + r)^{2/3} - (1 - r)^{2/3}$ is plotted against r. We need only obtain the intersection between the curve in Fig. 11.11 and the straight line ζr. It is seen that for $\zeta < \frac{4}{3}$, $r = 0$ is the only intersection. If ζ is such that

$$\tfrac{4}{3} < \zeta < 2^{2/3} \tag{13.42}$$

then there is an additional intersection $r > 0$, and the value $r > 0$ corresponds to a maximum, whereas the value $r = 0$ corresponds to a minimum. If $\zeta > 2^{2/3}$, then (13.40) has no solution. In this case the maximum of $g(N_+)$ must occur either at $N_+ = 0$ or at $N_+ = N$, unless $g(N_+)$ is a constant. Since $g(N_+)$ is not a constant, and since there is no distinction between $N_+ = 0$ and $N_+ = N$, we can choose to let $\bar{N}_+ = N$, or $r = 1$. The value of r at absolute zero as a function of the repulsive strength ζ, is summarized as follows:

$$\begin{array}{lll} r = 0 & (\zeta < \frac{4}{3}) & \text{(no spontaneous magnetization)} \\ 0 < r < 1 & (\frac{4}{3} < \zeta < 2^{2/3}) & \text{(partial spontaneous magnetization)} \\ r = 1 & (\zeta > 2^{2/3}) & \text{(saturated spontaneous magnetization)} \end{array} \tag{13.43}$$

That is, if the repulsive strength is sufficiently strong, the system becomes ferromagnetic. The critical value of a at which ferromagnetism first sets in ($\zeta = \frac{4}{3}$) corresponds to

$$k_F a = \frac{\pi}{2} \tag{13.44}$$

The foregoing results hold at absolute zero. At a finite but small temperature we also have to take into account the second term of (13.37), obtaining instead of (13.40) the following:

$$\begin{gathered} (1 + r)^{2/3} - (1 - r)^{2/3} - \frac{\pi^2}{12}\left(\frac{kT}{\epsilon_F}\right)^2 \left[\frac{1}{(1 + r)^{2/3}} - \frac{1}{(1 - r)^{2/3}}\right] = \zeta r \\ \frac{1}{2}\left\{\frac{1}{(1 + r)^{1/3}} + \frac{1}{(1 - r)^{1/3}} + \frac{\pi^2}{12}\left(\frac{kT}{\epsilon_F}\right)^2 \left[\frac{1}{(1 + r)^{5/3}} + \frac{1}{(1 - r)^{5/3}}\right]\right\} > \frac{3}{4}\zeta \end{gathered} \tag{13.45}$$

Let $r(T)$ be the solution at absolute temperature T. It is easily seen that if $r(0) = 0$, then $r(T) = 0$; if $r(0) > 0$, then $r(T) < r(0)$. Thus, if there is spontaneous magnetization at absolute zero, the magnetization decreases with temperature. The spontaneous magnetization vanishes above a critical temperature T_c (the Curie temperature), which is the value of T at which both equations in (13.45) are satisfied for $\zeta > \frac{4}{3}$ and $r = 0$. We find that

$$\frac{kT_c}{\epsilon_F} = \frac{3}{\pi}\sqrt{\zeta - \tfrac{4}{3}} = \frac{2}{\pi\sqrt{3}}\sqrt{\frac{2}{\pi} k_F a - 1} \tag{13.46}$$

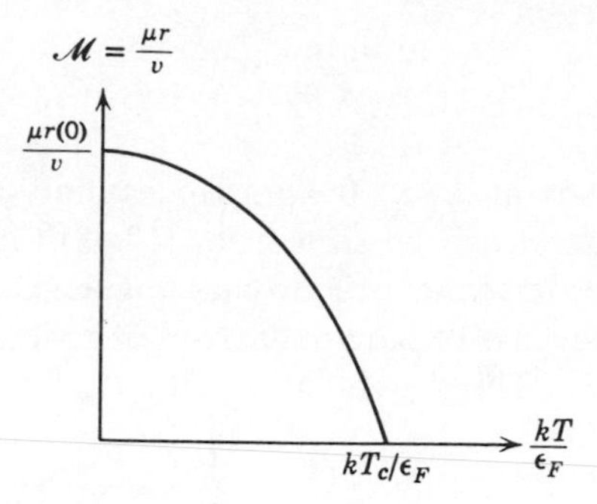

Fig. 13.4. Spontaneous magnetization of an imperfect Fermi gas with repulsive interactions.

Formula (13.46) is valid only if $\zeta - \frac{4}{3} \ll 1$, because we have made use of (13.37) in its derivation. For high temperatures substitution of (13.38) into (13.36) yields $r = 0$ as the only solution, as we expect. A qualitative plot of the magnetization as a function of temperature for the ferromagnetic case is shown in Fig. 13.4.

It must be pointed out that the model we have used is a physical model only if $k_F a \ll 1$. Therefore the case of ferromagnetism, which requires $k_F a > \pi/2$, is beyond the domain of validity of the model. It is instructive, however, to see how the spatial repulsion between the fermions can enhance the spin alignment to such an extent that, if we are willing to extrapolate the results of a weak interaction model, ferromagnetism results.

Paramagnetic Susceptibility

We now consider the case of $B > 0$. Let $r_0(T)$ be the value of r for $B = 0$, but for an arbitrary temperature. Putting

$$r = r_0(T) + \frac{\chi v}{\mu} B \tag{13.47}$$

and treating $\chi v B/\mu$ as a small quantity, we can solve (13.36) and obtain

$$\chi = \frac{2\mu^2/\epsilon_F v}{\dfrac{NkT}{2\epsilon_F}\left\{\nu'\left[\dfrac{N}{2}(1 + r_0)\right] + \nu'\left[\dfrac{N}{2}(1 - r_0)\right]\right\} - \dfrac{8}{3\pi} k_F a} \tag{13.48}$$

The low- and high-temperature limits are

$$\chi \xrightarrow[T\to 0]{} \frac{3\mu^2/\epsilon_F v}{(1 + r_0)^{-1/3} + (1 - r_0)^{-1/3} - (4/\pi)k_F a} \tag{13.49}$$

$$\chi \xrightarrow[T\to\infty]{} \frac{\mu^2}{kTv} \tag{13.50}$$

Hence Curie's constant is

$$C = \frac{\mu^2}{kv} \tag{13.51}$$

Note that r_0 depends on $k_F a$. It approaches unity when $k_F a$ exceeds a certain value. Thus it can be seen from (13.48) that in general $\chi > 0$. The system is either ferromagnetic or paramagnetic, never diamagnetic.

Consider now the case of paramagnetism, for which we require $r_0 = 0$ for all temperatures. This means that

$$k_F a < \frac{\pi}{2} \tag{13.52}$$

Here (13.48) becomes

$$\frac{T\chi}{C} = \frac{3kT}{2\epsilon_F} \frac{1}{f - (2/\pi)k_F a} \tag{13.53}$$

where

$$f \equiv \frac{3kT}{2\epsilon_F} \frac{N}{2} \nu'\left(\frac{N}{2}\right) \tag{13.54}$$

The function $(T\chi/C)$ rises linearly at $T = 0$, with a slope given by

$$\frac{\partial}{\partial T}\left(\frac{T\chi}{C}\right)_{T=0} = \frac{3}{2} \frac{1}{1 - (2/\pi)k_F a} \tag{13.55}$$

It reaches a maximum value, which is greater than unity, at $kT/\epsilon_F \approx 1$. Then it approaches unity as $T \to \infty$. A qualitative plot of $T\chi/C$ is shown in Fig. 13.5. If we calculate χ for an ideal Fermi gas endowed with the same magnetic moment, we find the slope

$$\frac{\partial}{\partial T}\left(\frac{T\chi}{C}\right)_{T=0} = \frac{3}{2} \quad \text{(ideal Fermi gas)} \tag{13.56}$$

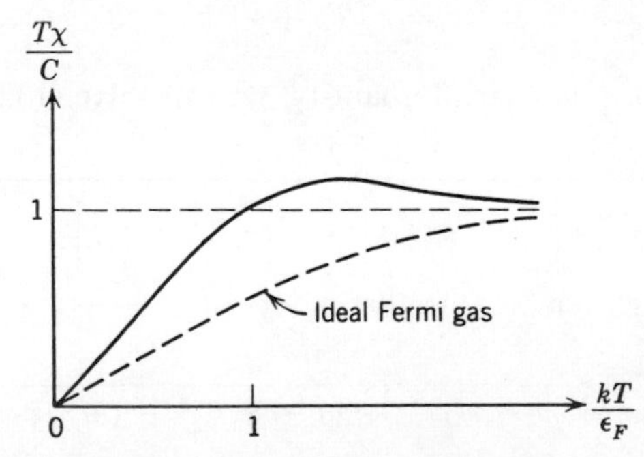

Fig. 13.5. Paramagnetic susceptibility of an imperfect Fermi gas with repulsive interactions. The model used is well founded only for $kT/\epsilon_F \ll 1$.

The imperfect gas has a steeper slope, as (13.55) shows, which is again a reflection of the enhancement of spin alignment by the repulsive interaction. The result is sometimes described by saying that imperfect gas behaves like an ideal gas with a higher Fermi energy.*

13.5 AN IMPERFECT BOSE GAS†

The Energy Levels

We consider a dilute system of N identical spinless bosons of mass m, contained in a box of volume V, at very low temperatures. The bosons interact with one another through binary collisions characterized by the scattering length a which is assumed to be positive. The energy levels to the first order in a may be obtained from (13.25) through the use of first-order perturbation theory.

Let the unperturbed wave functions be free-particle wave functions Φ_n, labeled by the occupation numbers $\{\ldots, n_\mathbf{p}, \ldots\}$, where $n_\mathbf{p}$ is the number of bosons with momentum $\mathbf{p}$. The energy levels to the first order in a are

$$E_n' \equiv (\Phi_n, H'\Phi_n) = \sum_\mathbf{p} \frac{p^2}{2m} n_\mathbf{p} + \frac{4\pi a\hbar^2}{m}\left(\Phi_n, \sum_{i<j} \delta(\mathbf{r}_i - \mathbf{r}_j)\Phi_n\right)$$

The second term is calculated in (A.36) of Appendix A. With that, we have

$$E_n' = \sum_\mathbf{p} \frac{p^2}{2m} n_\mathbf{p} + \frac{4\pi a\hbar^2}{mV}\left(N^2 - \tfrac{1}{2}\sum_\mathbf{p} n_\mathbf{p}^2\right) \tag{13.57}$$

This formula is valid only under the conditions

$$\begin{aligned} \frac{a}{v^{1/3}} &\ll 1 \\ ka &\ll 1 \end{aligned} \tag{13.58}$$

where k is the relative wave number of any pair of particles. Thus (13.57) becomes invalid if there are excited particles of high momentum.

Let us first study the implications of (13.57). The ground state energy per particle is obtained from (13.57) by setting all $n_\mathbf{p} = 0$ for $\mathbf{p} \neq 0$, and $n_0 = N$:

$$\frac{E_0}{N} = \frac{2\pi a\hbar^2}{mv} = \left(\frac{\hbar}{m}\right)^2 2\pi a\rho \tag{13.59}$$

* See, however, Problem 13.2.

† K. Huang, C. N. Yang, and J. M. Luttinger, *Phys. Rev.* **105**, 776 (1957).

where ρ is the mass density. It is proportional to the scattering length a and to the mass density, and it may be interpreted to be the energy shift of an average particle in the "optical approximation," whereby the effect of the rest of the system is replaced by a medium having an index of refraction. This interpretation can be justified as follows. In the shape-independent approximation we may replace a scattering potential by one of any shape, provided it gives the same scattering length. Let us replace the interparticle potential by a very shallow but very long-ranged square well such that the scattering length is still a. Now a particle moving through the system essentially "sees" a uniform potential of an appropriate depth. This gives (13.59).

For an excited state in which the particles have vanishingly small momenta the energy per particle is

$$\frac{E_n}{N} = \left(\frac{\hbar}{m}\right)^2 4\pi a \rho \left[1 - \tfrac{1}{2} \sum_{\mathbf{p}} \left(\frac{n_{\mathbf{p}}}{N}\right)^2\right]$$

The second term is most negative when all the excited particles are in the same momentum state. Thus we may say that "spatial repulsion leads to momentum space attraction." This is a consequence of the symmetry of the wave function.*

The "momentum space attraction" just mentioned also leads to an "energy gap" in the spectrum (13.57). This may be seen as follows. The energies of the very low excited states of the system are approximately given by

$$E_n' \approx \sum_{\mathbf{p}} \frac{p^2}{2m} n_{\mathbf{p}} + N \left(\frac{\hbar}{m}\right)^2 4\pi a \rho \left[1 - \frac{1}{2} \left(\frac{n_0}{N}\right)^2\right]$$

According to this formula, the excitation of one particle from the momentum state $\mathbf{p} = 0$ to a state of infinitesimal momentum changes the energy by the *finite* amount

$$\Delta = \left(\frac{\hbar}{m}\right)^2 2\pi a \rho$$

Thus the single-particle energy spectrum is separated from the zero-point of energy by the amount Δ. This "energy gap," however, is a feature only of the lowest-order formula. When the energy levels are calculated to higher orders in perturbation theory,† the energy gap disappears. Instead, there is only a decrease of level density just above the ground state, changing the single-particle spectrum $p^2/2m$ into a phonon spectrum $\hbar cp/2m$, where c is a constant. The "energy gap," which implies that the level density is

* See Problem 13.3.

† See Sec. 19.2.

strictly zero just above the ground state, is a crude approximation to the actual state of affairs.

The foregoing discussions make it clear that the energy levels (13.57), although not exact, possess many qualitative features of the effect of a repulsive interaction among bosons. We use them to calculate the partition function. The validity of this calculation is discussed as we proceed. We introduce a further simplification, namely, we take the energy levels to be

$$E_n = \sum_{\mathbf{p}} \frac{p^2}{2m} n_{\mathbf{p}} + \frac{4\pi a \hbar^2}{mV} (N^2 - \tfrac{1}{2} n_0^2) \tag{13.60}$$

The behavior of the model defined by (13.60) should be qualitatively the same as that by (13.57) when the temperature is so low that few particles are excited.

The Partition Function

The partition function is

$$Q_N = \sum_{\{n_{\mathbf{p}}\}} \exp \left\{ -\beta \left[\sum_{\mathbf{p}} \frac{p^2}{2m} n_{\mathbf{p}} + \frac{4\pi a \hbar^2}{mV} (N^2 - \tfrac{1}{2} n_0^2) \right] \right\} \tag{13.61}$$

Following the method and the notation of Sec. 12.4, we write

$$\begin{aligned} Q_N &= \exp\left(-N \frac{2a\lambda^2}{v}\right) \sum_{n_0=0}^{N} \exp\left(\frac{a\lambda^2 n_0^2}{V}\right) \sum_{\{n_{\mathbf{p}}\}}' \exp\left(-\beta \sum_{p\neq 0} \frac{p^2}{2m} n_{\mathbf{p}}\right) \\ &= e^{-N(2a\lambda^2/v)} \sum_{\xi=0}^{1} e^{(a\lambda^2/v)N\xi^2} \, Q(\xi) \end{aligned}$$

Where $Q(\xi)$ is defined by (12.76). Proceeding further, as in the derivation of (12.80), we obtain

$$\frac{1}{N} \log Q_N = -\frac{2a\lambda^2}{v} + f(\bar{\xi}) \tag{13.62}$$

where

$$f(\bar{\xi}) = \text{Max}\, f(\xi) \tag{13.63}$$

$$f(\xi) \equiv \frac{a\lambda^2}{v} \xi^2 + \frac{1}{N} \log Q(\xi) \tag{13.64}$$

By (12.77) and (12.78) we can determine $f(\xi)$ through the parametric equations

$$f(\xi) = \frac{a\lambda^2}{v} \xi^2 + \frac{v}{\lambda^3} g_{5/2}(z) - (1 - \xi) \log z \tag{13.65}$$

$$\frac{\lambda^3}{v} (1 - \xi) = g_{3/2}(z) \tag{13.66}$$

To find $f(\bar{\xi})$ we may proceed as follows:

Step 1. From (13.66) we can obtain ξ as a function of z for given λ^3/v. We thus obtain a family of ξ–z curves, one for each value of λ^3/v.

Step 2. For fixed λ^3/v we find $\bar{\xi}$ by maximizing $f(\xi)$ along the appropriate ξ–z curve.

The ξ–z curves are the same as those shown in Fig. 12.10. They are shown again in Fig. 13.6. Along a ξ–z curve we have

$$\frac{df(\xi)}{d\xi} = \log z + \frac{a\lambda^2}{v} 2\xi \tag{13.67}$$

$$\frac{d^2f(\xi)}{d\xi^2} = -\frac{\lambda^3}{v}\left[\frac{1}{g_{1/2}(z)} - \frac{2a}{\lambda}\right] < 0 \tag{13.68}$$

The last statement holds if $(2a/\lambda) < z$, which may be considered always fulfilled since $(a/\lambda) \ll 1$. Thus $f(\xi)$ is a decreasing function of ξ along a ξ–z curve only if $df/d\xi \leq 0$ at the smallest possible value of ξ on that curve.

For $(\lambda^3/v) < g_{3/2}(1)$ this is the case. Hence $\bar{\xi}$ is the smallest possible value of ξ, namely zero. (See Fig. 13.6) If $(\lambda^3/v) > g_{3/2}(1)$, then $f(\xi)$ passes through a maximum at $z = z_0$, which is determined by the condition $df/d\xi = 0$. Using (13.66) and (13.67) we find that

$$\bar{\xi} = \begin{cases} 0 & \left[\dfrac{\lambda^3}{v} < g_{3/2}(1)\right] \\ 1 - \dfrac{v g_{3/2}(z_0)}{\lambda^3} & \left[\dfrac{\lambda^3}{v} > g_{3/2}(1)\right] \end{cases} \tag{13.69}$$

where z_0 is the root of

$$-\log z_0 = \frac{2a\lambda^2}{v}\left[1 - \frac{v}{\lambda^3} g_{3/2}(z_0)\right] \tag{13.70}$$

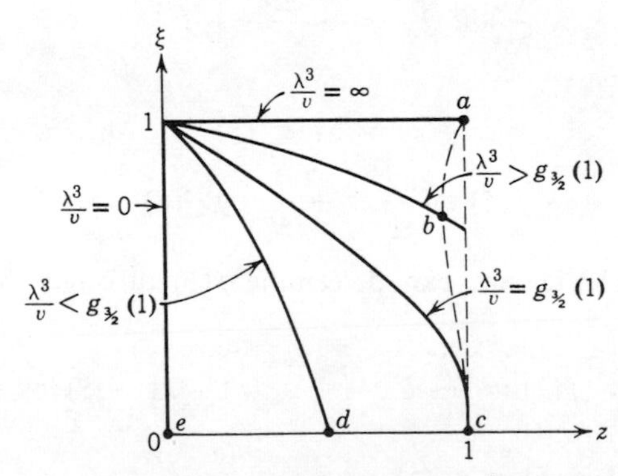

Fig. 13.6. Family of ξ–z curves. The equilibrium values $\bar{\xi}$ of ξ lie on the curve *abcde*.

On the thermodynamic v-T plane of the system the line

$$\frac{\lambda^3}{v} = g_{3/2}(1) \tag{13.71}$$

is the transition line of the Bose-Einstein condensation.

The transition curve is therefore the same as that of the ideal Bose gas, as are the transition temperature T_c and the transition volume v_c:

$$kT_c = \frac{2\pi\hbar^2}{m} \frac{1}{[vg_{3/2}(1)]^{2/3}} \tag{13.72}$$

$$v_c = \frac{\lambda^3}{g_{3/2}(1)} \tag{13.73}$$

An explicit solution of (13.70) can be obtained for two limiting cases:

$$z_0 \approx 1 - \frac{2a\lambda^2}{v}\left(1 - \frac{v}{v_c}\right) \qquad \left(\frac{a\lambda^2}{v} \ll 1\right) \tag{13.74}$$

$$z_0 \approx e^{-2a\lambda^2/v} \qquad \left(\frac{a\lambda^2}{v} \gg 1\right) \tag{13.75}$$

We shall consider further the first case only, because the second case corresponds to $kT \ll \Delta$, where our model is not reliable.

To the first order in a/λ and in $a\lambda^2/v$ we have

$$\xi = \begin{cases} 0 & (v > v_c,\ T > T_c) \\ \left(1 - \dfrac{v}{v_c}\right) = \left[1 + \dfrac{2a}{\lambda} g_{5/2}(1)\right] & (v < v_c,\ T < T_c) \end{cases} \tag{13.76}$$

and

$$\frac{1}{N}\log Q_N = \begin{cases} \dfrac{v}{\lambda^3} g_{5/2}(z) - \log z - \dfrac{2a\lambda^2}{v} & (v > v_c,\ T > T_c) \\ \dfrac{v}{\lambda^3} g_{5/2}(1) - 2\dfrac{a\lambda^2}{v}\left[1 - \frac{1}{2}\left(1 - \dfrac{v}{v_c}\right)^2\right] & (v < v_c,\ T < T_c) \end{cases} \tag{13.77}$$

where z is the root of the equation

$$\frac{\lambda^3}{v} = g_{3/2}(z) \tag{13.78}$$

The form of (13.77) is very simple. It shows that the free energy is equal to the free energy of the ideal Bose gas, plus the second term in (13.60) with n_0 replaced by $\langle n_0 \rangle$ of the ideal Bose gas.

The Equation of State

From (13.77) we obtain the equation of state

$$P = \begin{cases} P^{(0)} + \dfrac{4\pi a\hbar^2}{mv^2} & (v > v_c,\ T > T_c) \\ P^{(0)} + \dfrac{2\pi a\hbar^2}{m}\left(\dfrac{1}{v^2} + \dfrac{1}{{v_c}^2}\right) & (v < v_c,\ T < T_c) \end{cases} \tag{13.79}$$

where $P^{(0)}$ is the pressure of the ideal Bose gas. An isotherm is shown in Fig. 13.7, and the P-T diagram is shown in Fig. 13.8. The Bose-Einstein condensation is here a second-order transition. The specific heat decreases across the transition point by the amount

$$\frac{\Delta C_V}{Nk} = \frac{9a}{2\lambda_c} g_{3/2}(1) \tag{13.80}$$

We cannot deduce from these results that an imperfect Bose gas with repulsive interactions generally exhibits a second-order transition. The present model merely shows that the transition appears to be a second-order transition if higher-order effects in a/λ and $a\lambda^2/v$ are neglected.

If we consider a fictitious system whose partition function is exactly (13.61), we find that the exact equation of state in the grand canonical ensemble exhibits a first-order transition.* This result, however, has only academic interest, because the derivation of the model shows that it corresponds to a physical system only to the first order in a.

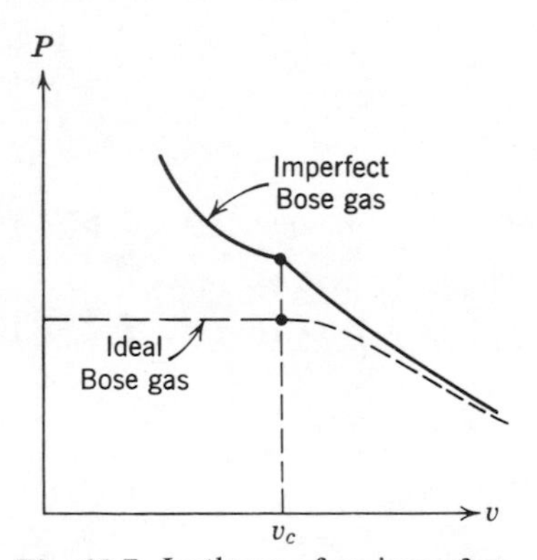

Fig. 13.7. Isotherm of an imperfect Bose gas with repulsive interactions.

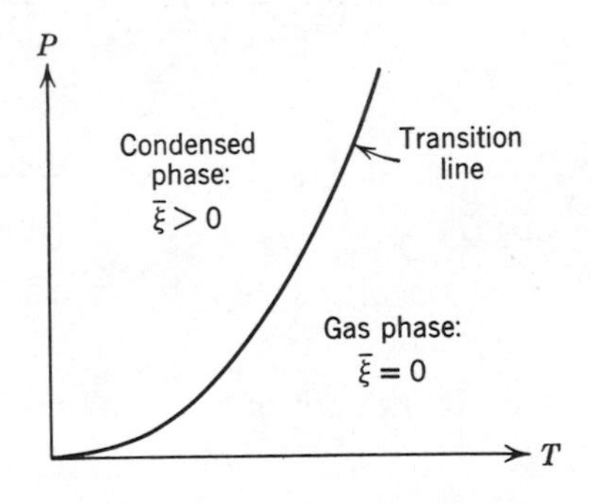

Fig. 13.8. *P-T* diagram of the imperfect Bose gas. In contradistinction to that of the diagram of the ideal Bose gas, Fig. 12.8, the space above the transition now corresponds to the condensed phase.

* K. Huang, C. N. Yang, and J. M. Luttinger, *loc. cit.*

PROBLEMS

13.1. (*a*) Find all spherically symmetric solutions and corresponding eigenvalues of the equation

$$(\nabla^2 + k_n^2)\psi_n(r) = 0$$

in the region between two concentric spheres of radii R and a $(R > a)$, with the boundary conditions

$$\psi(R) = \psi(a) = 0$$

(*b*) Expand the eigenvalues k_n^2 in powers of a, keeping terms up to order a^2.

(*c*) Using the method of pseudopotentials, calculate the eigenvalue k_n^2 up to order a^2 and show that it agrees with the answer to (*b*).

Reference. K. Huang and C. N. Yang, *Phys. Rev.* **105,** 767 (1957), §2(*b*).

13.2. (*a*) Show that for the imperfect Fermi gas discussed in Sec. 13.4 the specific heat at constant volume is given by

$$\frac{C_V}{N} = -2k\frac{\partial}{\partial T}\left[I(r)T^2\frac{\partial r}{\partial T}\right] + \frac{32\pi a\hbar^2}{mv}T\left[\left(\frac{\partial r}{\partial T}\right)^2 + r\frac{\partial^2 r}{\partial T^2}\right]$$

where

$$I(r) \equiv v\left[\frac{N}{2}(1+r)\right] - v\left[\frac{N}{2}(1-r)\right]$$

(*b*) Show that when there is no spontaneous magnetization

$$C_V = (C_V)_{\text{ideal gas}}$$

and hence the interpretation that the imperfect gas behaves like an ideal gas with a higher Fermi energy cannot be consistently maintained.

13.3. Consider two free bosons contained in a box of volume V with periodic boundary conditions. Let the momenta of the two particles be $\mathbf{p}$ and $\mathbf{q}$.

(*a*) Write down the normalized wave function $\psi_{\mathbf{pq}}(\mathbf{r}_1, \mathbf{r}_2)$ for both $\mathbf{p} \neq \mathbf{q}$ and $\mathbf{p} = \mathbf{q}$.

(*b*) Show that for $\mathbf{p} \neq \mathbf{q}$

$$|\psi_{\mathbf{pq}}(\mathbf{r}, \mathbf{r})|^2 > |\psi_{\mathbf{pp}}(\mathbf{r}, \mathbf{r})|^2$$

(*c*) Explain the meaning of the statement "spatial repulsion leads to momentum space attraction."

13.4. For the imperfect Bose gas discussed in Sec. 13.5, find the specific heat at constant volume near absolute zero.

Solution. The formula (13.77) is not valid for this calculation because (13.74) does not hold near absolute zero. We must consider (13.75). Near absolute zero the excited particles are not important because of the "energy gap." From (13.60) we may take the internal energy to be

$$U \approx \frac{4\pi a\hbar^2 N^2}{mV}(1 - \tfrac{1}{2}\xi^2)$$

Using (13.69) and (13.75), and remembering that $a\lambda^2/v \gg 1$, we obtain

$$\frac{C_V}{kN} = \frac{4a}{\lambda}\frac{\Delta}{kT}e^{-2\Delta/kT}$$

where Δ is the "energy gap". It vanishes exponentially as $T \to 0$, as is characteristic of a model with an "energy gap."

13.5. For the imperfect Bose gas discussed in Sec. 13.5, show that in the gas phase

$$\frac{Pv}{kT} = 1 + \left(-\frac{1}{4\sqrt{2}} + \frac{2a}{\lambda}\right)\frac{\lambda^3}{v} + \left(\frac{1}{8} - \frac{1}{3\sqrt{3}}\right)\left(\frac{\lambda^3}{v}\right)^2 - \cdots$$

Thus we can conclude that the third and higher virial coefficients, if they depend on a, must involve orders of a^2 or higher.

chapter 14

CLUSTER EXPANSIONS

14.1 CLASSICAL CLUSTER EXPANSION

Many systems of physical interest can be treated classically. A large class of such systems is described by a classical Hamiltonian for N particles of the form

$$H = \sum_{i=1}^{N} \frac{p_i^2}{2m} + \sum_{i<j} v_{ij} \tag{14.1}$$

where $\mathbf{p}_i$ is the momentum of the ith particle and $v_{ij} = v(|\mathbf{r}_i - \mathbf{r}_j|)$ is the potential energy of interaction between the ith and the jth particle. If the system occupies a volume V, the partition function is

$$Q_N(V, T) = \frac{1}{N!\, h^{3N}} \int d^{3N}p\, d^{3N}r \exp\left(-\beta \sum_i \frac{p_i^2}{2m} - \beta \sum_{i<j} v_{ij}\right) \tag{14.2}$$

where each coordinate $\mathbf{r}_i$ is integrated over the volume V. The integrations over momenta can be immediately effected, leading to

$$Q_N(V, T) = \frac{1}{\lambda^{3N} N!} \int d^{3N}r \exp\left(-\beta \sum_{i<j} v_{ij}\right) \tag{14.3}$$

where $\lambda = \sqrt{2\pi\hbar^2/mkT}$ is the thermal wavelength. The integral in (14.3) is called the *configuration integral*. For potentials v_{ij} of the usual type between molecules, a systematic method for the calculation of the configuration integral consists of expanding the integrand in powers of

$\exp(-\beta v_{ij}) - 1$. This leads to the cluster expansion of Ursell and Mayer.* As we shall see, this expansion is of practical use if the system is a dilute gas.

Let the configuration integral be denoted by $Z_N(V, T)$:

$$Z_N(V, T) \equiv \int d^3r_1 \cdots d^3r_N \exp\left(-\beta \sum_{i<j} v_{ij}\right) \tag{14.4}$$

in terms of which the partition function may be written as

$$Q_N(V, T) = \frac{1}{N!\, \lambda^{3N}} Z_N(V, T) \tag{14.5}$$

and the grand partition function as

$$\mathcal{Q}(z, V, T) = \sum_{N=0}^{\infty} \left(\frac{z}{\lambda^3}\right)^N \frac{Z_N(V, T)}{N!} \tag{14.6}$$

Let f_{ij} be defined by

$$e^{-\beta v_{ij}} \equiv 1 + f_{ij} \tag{14.7}$$

For the usual type of intermolecular potentials, v_{ij} and f_{ij} have the qualitative forms shown in Fig. 14.1. Thus f_{ij} is everywhere bounded and is negligibly small when $|\mathbf{r}_i - \mathbf{r}_j|$ is larger than the range of the intermolecular potential. In terms of f_{ij} the configuration integral may be represented by

$$Z_N(V, T) = \int d^3r_1 \cdots d^3r_N \prod_{i<j} (1 + f_{ij}) \tag{14.8}$$

in which the integrand is a product of $\frac{1}{2}N(N-1)$ terms, one for each distinct pair of particles. Expanding this product we obtain

$$Z_N(V, T) = \int d^3r_1 \cdots d^3r_N [1 + (f_{12} + f_{13} + \cdots) + (f_{12}f_{13} + f_{12}f_{14} + \cdots) + \cdots] \tag{14.9}$$

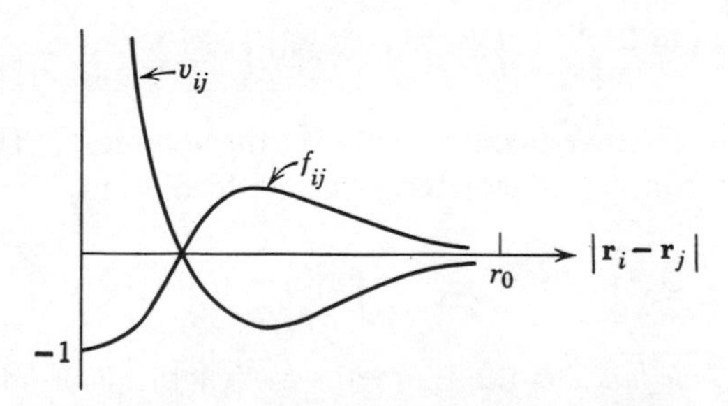

Fig. 14.1. Intermolecular potential v_{ij} and the function f_{ij}.

* For original literature, see J. E. Mayer and M. G. Mayer, *Statistical Mechanics* (John Wiley and Sons, New York, 1940), Chapter 13.

A convenient way to enumerate all the terms in the expansion (14.9) is to associate each term with a graph, defined as follows:

An N-particle graph is a collection of N distinct circles numbered 1, 2, . . . , *N, with any number of lines joining the same number of distinct pairs of circles. If the distinct pairs joined by lines are the pairs* $\alpha, \beta, \ldots, \lambda$, *then the graph represents the term*

$$\int d^3r_1 \cdots d^3r_N \, f_\alpha f_\beta \cdots f_\gamma \tag{14.10}$$

appearing in the expansion (14.9).

If the set of distinct pairs $\{\alpha, \beta, \ldots, \gamma\}$ is joined by lines in a given graph, replacing this set by a set $\{\alpha', \beta', \ldots, \gamma'\}$ that is not identical with $\{\alpha, \beta, \ldots, \gamma\}$ gives rise to a graph that is counted as distinct from the original one (although the integrals represented by the respective graphs have the same numerical value). For example, for $N = 3$, the following graphs are distinct:

① joined to ② and ③; ② joined to ③ and ①

but the following graphs are identical:

①—② ③ ②—① ③

We may regard a graph as a picturesque way of writing the integral (14.10). For example, we may write, for $N = 10$,

[①—② ③—⑨ ④ ⑤ ⑥—⑧, ⑥—⑦, ⑥—⑩, ⑧—⑩, ⑦—⑧] $= \int d^3r_1 \cdots d^3r_{10} \, f_{12} f_{39} f_{67} f_{68} f_{8,10} f_{6,10} f_{78}$ (14.11)

With such a convention, we can state that

$$Z_N = (\text{sum of all distinct } N\text{-particle graphs}) \tag{14.12}$$

The proof is obvious.

Any graph can in general be decomposed into smaller units. For example, the graph (14.11) is a product of five factors, namely

[①—② ③—⑨ ④ ⑤ ⑥—⑧, ⑥—⑦, ⑥—⑩, ⑧—⑩, ⑦—⑧] = [④] · [⑤] · [①—②] · [③—⑨] · [⑥—⑧, ⑥—⑦, ⑥—⑩, ⑧—⑩, ⑦—⑧]

It would therefore facilitate the analysis of Z_N if we first defined the basic units out of which an arbitrary graph can be composed.

We define an l-cluster to be an l-particle graph in which every circle is attached to at least one line and every circle is joined directly or indirectly to all other $l - 1$ circles. For example, the following is a 6-cluster:

[①—②, ②—③, ①—④, ④—⑥, ⑤—⑥] $= \int d^3r_1 \cdots d^3r_6 \, f_{12} f_{23} f_{14} f_{46} f_{56}$ (14.13)

We define a *cluster integral* $b_l(V, T)$ by

$$b_l(V, T) \equiv \frac{1}{l!\,\lambda^{3l-3}V}\,(\text{sum of all possible } l\text{-clusters}) \tag{14.14}$$

The normalization factor is so chosen that

(*a*) $b_l(V, T)$ is dimensionless;

(*b*) $\bar{b}_l(T) \equiv \lim_{V\to\infty} b_l(V, T)$ is a finite number.

The property (*b*) follows from the fact that f_{ij} has a finite range, so that in an l-cluster the only integration that gives rise to a factor V is the integration over the "center of gravity" of the l particles.* Some of the cluster integrals are

$$b_1 = \frac{1}{V}[①] = \frac{1}{V}\int d^3r_1 = 1 \tag{14.15}$$

$$b_2 = \frac{1}{2!\,\lambda^3 V}[①\!-\!②] = \frac{1}{\lambda^3 2V}\int d^3r_1\, d^3r_2\, f_{12} = \frac{1}{2\lambda^3}\int d^3r_{12}\, f_{12} \tag{14.16}$$

$$b_3 = \frac{1}{3!\,\lambda^6 V}\Big[\;①②③ + ①②③ + ①②③ + ①②③\;\Big] \tag{14.17}$$

Any N-particle graph is a product of a number of clusters, of which m_l are l-clusters, with

$$\sum_{l=1}^{N} l m_l = N \tag{14.18}$$

A given set of integers $\{m_l\}$ satisfying (14.18), however, does not uniquely specify a graph, because

(*a*) there are in general many ways to form an l-cluster, e.g.,

① ② ③   ① ② ③

(*b*) there are in general many ways to assign which particle belongs to which cluster, e.g.,

① ② ③   ① ② ④

Thus a set of integers $\{m_l\}$ specifies a collection of graphs. Let the sum of all the graphs corresponding to $\{m_l\}$ be denoted by $S\{m_l\}$. Then

$$Z_N = \sum_{\{m_l\}} S\{m_l\} \tag{14.19}$$

where the summation extends over all sets $\{m_l\}$ satisfying (14.18).

* A rigorous proof of the existence of b_l is implied in the results of Appendix C and is referred to in Sec. 15.3.

By definition, $S\{m_l\}$ can be obtained as follows. First write down an arbitrary N-particle graph that contains m_1 1-clusters, m_2 2-clusters, etc.; e.g.,

$$\underbrace{\{[\bigcirc]\cdots[\bigcirc]\}}_{m_1 \text{ factors}}\underbrace{\{[\bigcirc\!\!-\!\!\bigcirc]\cdots[\bigcirc\!\!-\!\!\bigcirc]\}}_{m_2 \text{ factors}}$$

$$\times \underbrace{\left\{[\ldots][\ldots][\ldots]\cdots[\ldots]\right\}}_{m_3 \text{ factors}}\cdots \tag{14.20}$$

There are exactly N circles appearing in (14.20), and these N circles are to be filled in by the numbers 1, 2, . . . , N in an arbitrary order. We can write down many more examples like (14.20); e.g., we may change the choice of some of the 3-clusters (there being four distinct topological shapes for a 3-cluster). Again we may permute the numbering of the circles in (14.20), and that would lead to a distinct graph. If we add up all these possibilities, we obtain $S\{m_l\}$. Thus we may write

$$S\{m_l\} = \sum_P [\bigcirc]^{m_1} [\bigcirc\!\!-\!\!\bigcirc]^{m_2}$$

$$\times \left[\ldots + \ldots + \ldots + \ldots\right]^{m_3} \left[\cdots\right]^{m_4} \cdots \tag{14.21}$$

The meaning of this formula is as follows. Each bracket contains the sum over all l-clusters. If all the brackets $[\cdots]^{m_l}$ are expanded in multinomial expansions, the summand of Σ_P will itself be a sum of a large number of terms in which every term contains exactly N circles. The sum Σ_P extends over all distinct ways of numbering these circles from 1 to N.

Now each graph is an integral whose value is independent of the way its circles are numbered. Therefore $S\{m_l\}$ is equal to the number of terms in the sum Σ_P times the value of any term in the sum. The number of terms in the sum Σ_P can be found by observing that

(*a*) there are m_l l-clusters, and a permutation of these m_l things does not lead to a new graph;

(*b*) in the sum over all l-clusters, such as (14.17), a permutation of the l particles within it does not lead to a new graph. Hence the number of terms in the sum Σ_P is*

$$\frac{N!}{[(1!)^{m_1}(2!)^{m_2}\cdots][m_1!\, m_2!\cdots]} \tag{14.22}$$

* To understand the method of counting the reader is advised to work out some simple examples.

and the value of any term is

$$(1!\ Vb_1)^{m_1}(2!\ \lambda^3 Vb_2)^{m_2}(3!\ \lambda^6 Vb_3)^{m_3}\cdots$$

Therefore

$$S\{m_l\} = N!\prod_{l=1}^{N}\frac{(V\lambda^{3l-3}b_l)^{m_l}}{m_l!} = N!\,\lambda^{3N}\prod_{l=1}^{N}\frac{1}{m_l!}\left(\frac{V}{\lambda^3}b_l\right)^{m_l} \tag{14.24}$$

From (14.5), (14.9), and (14.24) we obtain

$$Q_N(V, T) = \sum_{\{m_l\}}\prod_{l=1}^{N}\frac{1}{m_l!}\left(\frac{V}{\lambda^3}b_l\right)^{m_l} \tag{14.25}$$

This formula is complicated by the restriction (14.18). The grand partition function is simpler in appearance:

$$\mathscr{Q}(z, V, T) = \sum_{m_1=0}^{\infty}\sum_{m_2=0}^{\infty}\cdots\left[\frac{1}{m_1!}\left(\frac{V}{\lambda^3}zb_1\right)^{m_1}\frac{1}{m_2!}\left(\frac{V}{\lambda^3}z^2b_2\right)^{m_2}\cdots\right]$$

or

$$\frac{1}{V}\log\mathscr{Q}(z, V, T) = \frac{1}{\lambda^3}\sum_{l=1}^{\infty}b_lz^l \tag{14.26}$$

from which we obtain the equation of state in parametic form:

$$\begin{cases}\dfrac{P}{kT} = \dfrac{1}{\lambda^3}\displaystyle\sum_{l=1}^{\infty}b_lz^l\\[2ex]\dfrac{1}{v} = \dfrac{1}{\lambda^3}\displaystyle\sum_{l=1}^{\infty}lb_lz^l\end{cases} \tag{14.27}$$

This is known as the cluster expansion for the equation of state.*

If the system under consideration is a dilute gas, we may expand the pressure in powers of $1/v$ and obtain the virial expansion. For this purpose we may take the equation of state to be†

$$\begin{cases}\dfrac{P}{kT} = \dfrac{1}{\lambda^3}\displaystyle\sum_{l=1}^{\infty}\bar{b}_lz^l\\[2ex]\dfrac{1}{v} = \dfrac{1}{\lambda^3}\displaystyle\sum_{l=1}^{\infty}l\bar{b}_lz^l\end{cases} \tag{14.28}$$

where

$$\bar{b}_l(T) \equiv \lim_{V\to\infty} b_l(V, T) \tag{14.29}$$

The virial expansion of the equation of state is defined to be

$$\frac{Pv}{kT} = \sum_{l=1}^{\infty}a_l(T)\left(\frac{\lambda^3}{v}\right)^{l-1} \tag{14.30}$$

* Compare this derivation with that outlined in Problem 8.9.

† The justification of (14.28) is given in Sec. 15.3.

where $a_l(T)$ is called the lth virial coefficient. We can find the relationship between the virial coefficients a_l and the cluster integrals $\bar{b}_l$ by substituting (14.30) into (14.28) and requiring that the resulting equation be satisfied for every z:

$$\sum_{l=1}^{\infty} a_l \left(\sum_{n=1}^{\infty} n\bar{b}_n z^n \right)^{l-1} = \frac{\sum_{l=1}^{\infty} \bar{b}_l z^l}{\sum_{l=1}^{\infty} l\bar{b}_l z^l} \tag{14.31}$$

This is equivalent to the condition

$$(\bar{b}_1 z + 2\bar{b}_2 z^2 + 3\bar{b}_3 z^3 + \cdots)\left[a_1 + a_2 \left(\sum_{n=1}^{\infty} n\bar{b}_n z^n \right) + a_3 \left(\sum_{n=1}^{\infty} n\bar{b}_n z^n \right)^2 + \cdots \right]$$
$$= \bar{b}_1 z + \bar{b}_2 z^2 + \bar{b}_3 z^3 + \cdots \tag{14.32}$$

By equating the coefficient of each power of z we obtain

$$\begin{aligned} a_1 &= \bar{b}_1 = 1 \\ a_2 &= -\bar{b}_2 \\ a_3 &= 4\bar{b}_2^2 - 2\bar{b}_3 \\ a_4 &= -20\bar{b}_2^3 + 18\,\bar{b}_2\bar{b}_3 - 3\bar{b}_4 \\ &\cdots \end{aligned} \tag{14.33}$$

Each virial coefficient therefore involves only a straightforward computation of a number of integrals. The classical problem of the imperfect gas is hereby reduced to quadruture.

14.2 QUANTUM CLUSTER EXPANSION

Kahn and Uhlenbeck* develop a cluster expansion in quantum statistical mechanics. The method they introduce applies equally well to classical statistical mechanics.

Consider N identical particles enclosed in a volume V. Let the Hamiltonian H of the system have the same form as (14.1) but be an operator instead of a number. In the coordinate representation, $\mathbf{p}_j = -i\hbar\nabla_j$, and v_{ij} is the same function of the number $|\mathbf{r}_i - \mathbf{r}_j|$ as that shown in Fig. 14.1. The partition function is

$$Q_N(V, T) = Tr e^{-\beta H} = \int d^{3N}r \sum_{\alpha} \Psi_\alpha^*(1, \ldots, N) e^{-\beta H} \Psi_\alpha(1, \ldots, N) \tag{14.34}$$

where $\{\Psi_\alpha\}$ is a complete set of orthonormal wave functions appropriate

* B. Kahn and G. E. Uhlenbeck, *Physica*, **5**, 399 (1938).

to the system considered, and the set of coordinates $\{\mathbf{r}_1, \ldots, \mathbf{r}_N\}$ is denoted in abbreviation by $\{1, \ldots, N\}$. Let

$$W_N(1, \ldots, N) \equiv N!\, \lambda^{3N} \sum_\alpha \Psi_\alpha^*(1, \ldots, N) e^{-\beta H} \Psi_\alpha(1, \ldots, N) \tag{14.35}$$

The partition function can be written in the form

$$Q_N(V, T) = \frac{1}{N!\, \lambda^{3N}} \int d^{3N}r\, W_N(1, \ldots, N) \tag{14.36}$$

The integral appearing in (14.36) approaches the classical configuration integral in the limit of high temperatures. Some properties of the function $W_N(1, \ldots, N)$ are:

(*a*) $W_1(1) = 1$

Proof:

$$W_1(1) = W_1(\mathbf{r}_1) = \frac{\lambda^3}{V} \sum_{\mathbf{p}} e^{-i\mathbf{p}\cdot\mathbf{r}_1/\hbar} e^{(\beta\hbar^2/2m)\nabla^2} e^{i\mathbf{p}\cdot\mathbf{r}_1/\hbar}$$

$$= \left(\frac{\lambda}{h}\right)^3 \int d^3p\, e^{-\beta p^2/2m} = 1 \qquad \text{(QED)}$$

(*b*) $W_N(1, \ldots, N)$ is a symmetric function of its arguments.

(*c*) $W_N(1, \ldots, N)$ is invariant under a unitary transformation of the complete set of wave functions $\{\Psi_\alpha\}$ appearing in (14.35).

Proof: Suppose $\Psi_\alpha = \sum_\lambda S_{\alpha\lambda} \Phi_\lambda$, where $S_{\alpha\lambda}$ is a unitary matrix:

$$\sum_\alpha S_{\alpha\lambda}^* S_{\alpha\gamma} = \delta_{\lambda\gamma}$$

Then

$$\sum_\alpha (\Psi_\alpha, e^{-\beta H}\Psi_\alpha) = \sum_{\alpha,\lambda} S_{\alpha\lambda}^* S_{\alpha\gamma} (\Phi_\lambda, e^{-\beta H}\Phi_\gamma) = \sum_\lambda (\Phi_\lambda, e^{-\beta H}\Phi_\lambda) \qquad \text{(QED)}$$

The following property appears to be intuitively obvious, but it is difficult to establish quantitatively. Suppose the coordinates $\mathbf{r}_1, \ldots, \mathbf{r}_N$ have such values that they can be divided into two groups containing respectively A and B coordinates, with the property that any two coordinates $\mathbf{r}_i$ and $\mathbf{r}_j$ belonging to different groups must satisfy the condition

$$\begin{aligned} |\mathbf{r}_i - \mathbf{r}_j| &\gg r_0 \\ |\mathbf{r}_i - \mathbf{r}_j| &\gg \lambda \end{aligned} \tag{14.37}$$

Then

$$W_N(\mathbf{r}_1, \ldots, \mathbf{r}_N) \approx W_A(r_A) W_B(r_B) \tag{14.38}$$

where r_A and r_B denote collectively the respective coordinates in the two groups.

The property (14.38) can be sharpened for the ideal gases, for which W_N is proportional to the function $J(1, \ldots, N)$ introduced in (10.50). The result (10.51) shows that

$$W_N = W_A W_B + O(e^{-\pi r^2/\lambda^2}) \qquad \text{(ideal gases)} \tag{14.39}$$

where r is the minimum distance between the two groups.

We do not enter into a detailed discussion of the property (14.38) for a system of interacting particles because we do not use it except as a heuristic motivation for the next step.

To indicate what would be involved in a proof of (14.38), let us consider the case $N = 2$. If we assume that for $|\mathbf{r}_1 - \mathbf{r}_2| \gg r_0$ we can write

$$\exp\left\{-\beta\left[\frac{\hbar^2}{2m}(\nabla_1{}^2 + \nabla_2{}^2) + v_{12}\right]\right\} \approx \exp\left[\frac{-\beta\hbar^2}{2m}(\nabla_1{}^2 + \nabla_2{}^2)\right] \tag{14.40}$$

then the problem reduces to that of the noninteracting system, and we obtain (14.39). The main task of the proof would be to establish a quantitative criterion for (14.40). This is nontrivial, because v_{12} does not commute with $\nabla_1{}^2 + \nabla_2{}^2$.

To proceed with the development of Kahn and Uhlenbeck, let us first consider the case $N = 2$. According to (14.38) we should expect that as $|\mathbf{r}_1 - \mathbf{r}_2| \to \infty$,

$$W_2(1, 2) \to W_1(1)W_2(2)$$

If we define a function $U_2(1, 2)$ by $W_2(1, 2) = W_1(1)W_2(2) + U_2(1, 2)$, we should expect that, as $|\mathbf{r}_1 - \mathbf{r}_2| \to \infty$,

$$U_2(1, 2) \to 0$$

Hence the integral of $U_2(1, 2)$ over $\mathbf{r}_1$ and $\mathbf{r}_2$ should be the analog of the 2-cluster in classical statistical mechanics.

We proceed systematically in the following manner. Let a sequence of cluster functions $U_l(1, \ldots, l)$ be successively defined by the following scheme, in which the lth equation is a definition of $U_l(1, \ldots, l)$:

$$W_1(1) = U_1(1) = 1 \tag{14.41}$$

$$W_2(1, 2) = U_1(1)U_1(2) + U_2(1, 2) \tag{14.42}$$

$$\begin{aligned} W_3(1, 2, 3) = {} & U_1(1)U_1(2)U_1(3) + U_1(1)U_2(2, 3) \\ & + U_1(2)U_2(3, 1) + U_1(3)U_2(1, 2) + U_3(1, 2, 3) \end{aligned} \tag{14.43}$$

$$\vdots$$

The last equation in this scheme, defining $U_N(1, \ldots, N)$, is

$$W_N(1, \ldots, N) = \sum_{\{m_l\}} \sum_{P} \underbrace{[U_1(\,) \cdots U_1(\,)]}_{m_1 \text{ factors}} \underbrace{[U_2(\,,\,) \cdots U_2(\,,\,)]}_{m_2 \text{ factors}} \cdots \underbrace{[U_N(\,,\,,\cdots,\,)]}_{m_N \text{ factor}} \tag{14.44}$$

where m_l is zero or a positive integer and the set of integers $\{m_l\}$ satisfies the condition

$$\sum_{l=1}^{N} l m_l = N \tag{14.45}$$

The sum over $\{m_l\}$ in (14.44) extends over all sets $\{m_l\}$ satisfying (14.45). The arguments of the U_l are left blank in (14.44). There are exactly N

such blanks, and they are to be filled by the N coordinates $\mathbf{r}_1, \ldots, \mathbf{r}_N$ in any order. The sum Σ_P is a sum over all *distinct* ways of filling these blanks.

We can solve the equations (14.41)–(14.44) successively for U_1, U_2, etc., and obtain

$$U_1(1) = W_1(1) = 1 \tag{14.46}$$

$$U_2(1, 2) = W_2(1, 2) - W_1(1)W_1(2) \tag{14.47}$$

$$\begin{aligned} U_3(1, 2, 3) = W_3(1, 2, 3) &- W_2(1, 2)W_1(3) - W_2(2, 3)W_1(1) \\ &- W_2(3, 1)W_1(2) + 2W_1(1)W_1(2)W_1(3) \end{aligned} \tag{14.48}$$

$$\vdots$$

We see that $U_l(1, \ldots, l)$ is a symmetric function of its arguments and is determined by all the $W_{N'}$ with $N' \leq l$. By the property (14.38) we expect that $U_l \to 0$ as $|\mathbf{r}_i - \mathbf{r}_j| \to \infty$, where $\mathbf{r}_i$ and $\mathbf{r}_j$ are any two of the arguments of U_l.

The lth-cluster integral $b_l(V, T)$ is defined by

$$b_l(V, T) \equiv \frac{1}{l!\, \lambda^{3l-3} V} \int d^3r_1 \cdots d^3r_l\, U_l(1, \ldots, l) \tag{14.49}$$

It is clear that b_l is dimensionless. If U_l vanishes sufficiently rapidly whenever any two of its auguments are far apart from each other, the integral appearing in (14.49) is proportional to V as $V \to \infty$, and the limit $b_l(\infty, T)$ may be expected to exist. Whether this is true depends on the nature of the interparticle potential. We assume that it is.

We now show that the partition function is expressible directly in terms of the cluster integrals. According to (14.36) we need to integrate W_N over all the coordinates. Let us make use of the formula (14.44). An integration over all the coordinates will yield the same result for every term in the sum Σ_P. Thus the result of the integration is the number of terms in the sum Σ_P times the integral of any term in the sum Σ_P. The number of terms in the sum Σ_P is given by (14.22). Hence

$$\begin{aligned} &\int d^{3N}r\, W(1, \ldots, N) \\ &= \sum_{\{m_l\}} \frac{N!}{[(1!)^{m_1}(2!)^{m_2} \cdots](m_1!\, m_2! \cdots)} \int d^{3N}r[(U_1 \cdots U_1)(U_2 \cdots U_2) \cdots] \\ &= N! \sum_{\{m_l\}} \frac{1}{m_1!}\left[\frac{1}{1!} \int d^3r_1\, U_1(1)\right]^{m_1} \frac{1}{m_2!}\left[\frac{1}{2!} \int d^3r_1\, d^3r_2\, U_2(1, 2)\right]^{m_2} \cdots \\ &= N! \sum_{\{m_l\}} \prod_{l=1}^{N} \frac{(V\lambda^{3l-3}b_l)^{m_l}}{m_l!} = N!\, \lambda^{3N} \sum_{\{m_l\}} \prod_{l=1}^{N} \frac{1}{m_l!}\left(\frac{V}{\lambda^3} b_l\right)^{m_l} \end{aligned} \tag{14.50}$$

Therefore the partition function is given by

$$Q_N(V, T) = \sum_{\{m_l\}} \prod_{l=1}^{N} \frac{1}{m_l!} \left(\frac{V}{\lambda^3} b_l \right)^{m_l} \tag{14.51}$$

This is of precisely the same form as (14.25) for the classical partition function. The discussion following (14.25) therefore applies equally well to the present case and will not be repeated. We only point out the main differences between the quantum cluster integrals and the classical ones.

For an ideal gas we have seen in earlier chapters that

$$\bar{b}_l^{(0)} = \begin{cases} l^{-5/2} & \text{(ideal Bose gas)} \\ (-1)^{l+1} l^{-5/2} & \text{(ideal Fermi gas)} \end{cases} \tag{14.52}$$

Thus for a Bose and a Fermi gas $\bar{b}_l$ does not vanish for $l > 1$, even in the absence of interparticle interactions, in contradistinction to the classical ideal gas.

The calculation of $\bar{b}_l$ in the classical case only involves the calculation of a number of integrals—a finite task. In the quantum case, however, the calculation of $\bar{b}_l$ necessitates a knowledge of U_l, which in turn necessitates a knowledge of $W_{N'}$ for $N' \leq l$. Thus to find b_l for $l > 1$ we would have to solve an l-body problem. There is no finite prescription for doing this except for the case $l = 2$, which is the subject of the next section.*

14.3 THE SECOND VIRIAL COEFFICIENT

To calculate the second virial coefficient a_2 for any system it is sufficient to calculate $\bar{b}_2$, since $a_2 = -\bar{b}_2$. A general formula for $\bar{b}_2$ (in fact, for all b_l) has already been given for the classical case. Only the quantum case is considered here.†

To find $\bar{b}_2$ we need to know $W_2(1, 2)$, which is a property of the two-body system. Let the Hamiltonian for the two-body system in question be

$$H = -\frac{\hbar^2}{2m}(\nabla_1^{\,2} + \nabla_2^{\,2}) + v(|\mathbf{r}_1 - \mathbf{r}_2|) \tag{14.53}$$

and let its normalized eigenfunctions be $\Psi_\alpha(1, 2)$, with eigenvalues E_α:

$$H\Psi_\alpha(1, 2) = E_\alpha \Psi_\alpha(1, 2) \tag{14.54}$$

* There exist formal methods for the calculation of b_l. For a source of literature, see *Proceedings of the International Congress on Many-Particle Problems*, Supplement to *Physica* (December, 1960).

† The following development is due to E. Beth and G. E. Uhlenbeck, *Physica*, **4**, 915 (1937).

Let

$$\begin{aligned} \mathbf{R} &= \tfrac{1}{2}(\mathbf{r}_1 + \mathbf{r}_2) \\ \mathbf{r} &= \mathbf{r}_2 - \mathbf{r}_1 \end{aligned} \tag{14.55}$$

Then

$$\begin{aligned} \Psi_\alpha(1, 2) &= \frac{1}{\sqrt{V}} e^{i\mathbf{P}\cdot\mathbf{R}} \psi_n(\mathbf{r}) \\ E_\alpha &= \frac{P^2}{4m} + \epsilon_n \end{aligned} \tag{14.56}$$

where the quantum number α refers to the set of quantum numbers $(\mathbf{P}, n)$. The relative wave function $\psi_n(\mathbf{r})$ satisfies the eigenvalue equation

$$\left[-\frac{\hbar^2}{m} \nabla^2 + v(r) \right] \psi_n(\mathbf{r}) = \epsilon_n \psi_n(\mathbf{r}) \tag{14.57}$$

with the normalization condition

$$\int d^3r \, |\psi_n(\mathbf{r})|^2 = 1 \tag{14.58}$$

Using (14.56) to be the wave functions for the calculation of $W_2(1, 2)$, we find from (14.35) that

$$W_2(1, 2) = 2\lambda^6 \sum_\alpha |\Psi_\alpha(1, 2)|^2 e^{-\beta E_\alpha} = \frac{2\lambda^6}{V} \sum_{\mathbf{P}} \sum_n |\psi_n(\mathbf{r})|^2 e^{-\beta P^2/4m} e^{-\beta \epsilon_n} \tag{14.59}$$

In the limit as $V \to \infty$ the sum over $\mathbf{P}$ can be effected immediately:

$$\frac{1}{V} \sum_{\mathbf{P}} e^{-\beta P^2/4m} = \frac{4\pi}{h^3} \int_0^\infty dP \, P^2 e^{-\beta P^2/4m} = \frac{2\sqrt{2}}{\lambda^3} \tag{14.60}$$

where $\lambda = \sqrt{2\pi\hbar^2/mkT}$, the thermal wavelength. Therefore

$$W_2(1, 2) = 4\sqrt{2}\lambda^3 \sum_n |\psi_n(\mathbf{r})|^2 e^{-\beta\epsilon_n} \tag{14.61}$$

If we repeat all the calculations so far for a two-body system of noninteracting particles, we obtain

$$W_2^{(0)}(1, 2) = 4\sqrt{2}\lambda^3 \sum_n |\psi_n^{(0)}(\mathbf{r})|^2 e^{-\beta\epsilon_n^{(0)}} \tag{14.62}$$

where the superscript $^{(0)}$ refers to quantities of the noninteracting system. From (14.49) and (14.47) we have

$$b_2 = \frac{1}{2\lambda^3 V} \int d^3r_1 \, d^3r_2 \, U_2(1, 2) = \frac{1}{2\lambda^3 V} \int d^3R \, d^3r [W_2(1, 2) - 1]$$

Hence

$$b_2 - b_2^{(0)} = \frac{1}{2\lambda^3 V}\int d^3R\, d^3r[W_2(1,2) - W_2^{(0)}(1,2)]$$

$$= 2\sqrt{2}\int d^3r \sum_n [|\psi_n(\mathbf{r})|^2\, e^{-\beta\epsilon_n} - |\psi_n^{(0)}(\mathbf{r})|^2\, e^{-\beta\epsilon_n^{(0)}}]$$

$$= 2\sqrt{2}\sum_n (e^{-\beta\epsilon_n} - e^{-\beta\epsilon_n^{(0)}}) \tag{14.63}$$

where

$$b_2^{(0)} = \begin{cases} 2^{-5/2} & \text{(ideal Bose gas)} \\ -2^{-5/2} & \text{(ideal Fermi gas)} \end{cases} \tag{14.64}$$

To analyze (14.63) further we must study the energy spectra $\epsilon_n^{(0)}$ and ϵ_n. For the noninteracting system, $\epsilon_n^{(0)}$ forms a continuum. We write

$$\epsilon_n^{(0)} = \frac{\hbar^2 k^2}{m} \tag{14.65}$$

which defines the relative wave number k. For the interacting system the spectrum of ϵ_n in general contains a discrete set of values ϵ_B, corresponding to two-body bound states, and a continuum. In the continuum, we define the wave number k for the interacting system by putting

$$\epsilon_n = \frac{\hbar^2 k^2}{m} \tag{14.66}$$

Let $g(k)\, dk$ be the number of states with wave number lying between k and $k + dk$, and let $g^{(0)}(k)\, dk$ denote the corresponding quantity for the noninteracting system. Then (14.63) can be written in the form

$$b_2 - b_2^{(0)} = 2\sqrt{2}\sum_B e^{-\beta\epsilon_B} + 2\sqrt{2}\int_0^\infty dk[g(k) - g^{(0)}(k)]e^{-\beta\hbar^2k^2/m} \tag{14.67}$$

where ϵ_B denotes the energy of a bound state of the interacting two-body system.

Let $\eta_l(k)$ be the scattering phase shift of the potential $v(r)$ for the lth partial wave of wave number k. It will be shown that

$$g(k) - g^{(0)}(k) = \frac{1}{\pi}\sum_l{}' (2l+1)\frac{\partial\eta_l(k)}{\partial k} \tag{14.68}$$

where the sum Σ' extends over the values

$$l = \begin{cases} 0, 2, 4, 6, \ldots & \text{(bosons)} \\ 1, 3, 5, 7, \ldots & \text{(fermions)} \end{cases} \tag{14.69}$$

Therefore

$$\mathfrak{b}_2 - \mathfrak{b}_2^{(0)} = 2\sqrt{2}\sum_B e^{-\beta\epsilon_B} + \frac{2\sqrt{2}}{\pi}\int_0^\infty dk \sum_l{}' (2l+1)\frac{\partial\eta_l(k)}{\partial k} e^{-\beta\hbar^2 k^2/m} \tag{14.70}$$

A partial integration leads finally to the formula

$$\mathfrak{b}_2 - \mathfrak{b}_2^{(0)} = 2\sqrt{2}\sum_B e^{-\beta\epsilon_B} + \frac{2\sqrt{2}\lambda^2}{\pi^2}\sum_l{}' (2l+1)\int_0^\infty dk k\eta_l(k) e^{-\beta\hbar^2 k^2/m} \tag{14.71}$$

The second virial coefficient of any system is hereby reduced to quadrature.

It remains to prove (14.68). We may choose both $\psi_n(\mathbf{r})$ and $\psi_n^{(0)}(\mathbf{r})$ to be pure spherical harmonics, because $v(r)$ does not depend on the angles of $\mathbf{r}$ with respect to any fixed axis. Thus we write

$$\begin{aligned} \psi_{klm}(\mathbf{r}) &= A_{klm} Y_l^m(\theta, \phi)\frac{u_{kl}(r)}{r} \\ \psi_{klm}^{(0)}(\mathbf{r}) &= A_{klm}^{(0)} Y_l^m(\theta, \phi)\frac{u_{kl}^{(0)}(r)}{r} \end{aligned} \tag{14.72}$$

For bosons $\psi(\mathbf{r}) = \psi(-\mathbf{r})$, and for fermions $\psi(\mathbf{r}) = -\psi(-\mathbf{r})$. Therefore

$$l = \begin{cases} 0, 2, 4, 6, \ldots & \text{(bosons)} \\ 1, 3, 5, 7, \ldots & \text{(fermions)} \end{cases} \tag{14.73}$$

Let the boundary conditions be

$$u_{kl}(R) = u_{kl}^{(0)}(R) = 0 \tag{14.74}$$

where R is a very large radius which approaches infinity at the end of the calculation. The asymptotic forms of u_{kl} and $u_{kl}^{(0)}$ are

$$\begin{aligned} u_{kl}(r) &\xrightarrow[r\to\infty]{} \sin\left[kr + \frac{l\pi}{2} + \eta_l(k)\right] \\ u_{kl}^{(0)}(r) &\xrightarrow[r\to\infty]{} \sin\left(kr + \frac{l\pi}{2}\right) \end{aligned} \tag{14.75}$$

This defines $\eta_l(k)$. The eigenvalues k are determined by the boundary conditions (14.74):

$$\begin{aligned} kR + \frac{l\pi}{2} + \eta_l(k) &= \pi n \qquad \text{(interacting system)} \\ kR + \frac{l\pi}{2} &= \pi n \qquad \text{(noninteracting system)} \end{aligned} \tag{14.76}$$

where $n = 0, 1, 2, \ldots$. It is seen that the eigenvalues k depends on n and l but not on m. Since there are $2l + 1$ spherical harmonics Y_l^m for a given l, each eigenvalue k is $(2l + 1)$-fold degenerate.

For a given l, changing n by one unit causes k to change by the respective amounts Δk, $\Delta k^{(0)}$:

$$\begin{aligned} \Delta k &= \frac{\pi}{R + [\partial \eta_l(k)/\partial k]} \\ \Delta k^{(0)} &= \frac{\pi}{R} \end{aligned} \tag{14.77}$$

These are the spacings of eigenvalues for a given l. Let the number of states of a given l with wave number lying between k and $k + dk$ be denoted by $g_l(k)\,dk$ and $g_l^{(0)}(k)\,dk$ for the two cases. We must have

$$\begin{aligned} \frac{g_l(k)\,\Delta k}{2l+1} &= 1 \\ \frac{g_l^{(0)}(k)\,\Delta k^{(0)}}{2l+1} &= 1 \end{aligned} \tag{14.78}$$

or

$$\begin{aligned} g_l(k) &= \frac{2l+1}{\pi}\left[R + \frac{\partial \eta_l(k)}{\partial k}\right] \\ g_l^{(0)}(k) &= \frac{2l+1}{\pi} R \end{aligned} \tag{14.79}$$

Therefore

$$g_l(k) - g_l^{(0)}(k) = \frac{2l+1}{\pi}\frac{\partial \eta_l(k)}{\partial k} \tag{14.80}$$

Summing (14.80) over all l consistent with (14.73) we obtain (14.68).

For $l > 2$ there is no known formula for b_l comparable in simplicity to (14.71), because there is no known treatment of the l-body problem for $l > 2$ comparable to the phase shift analysis of the two-body problem.*

PROBLEMS

14.1. (*a*) Calculate b_2 and b_3 for a classical hard-sphere gas with hard-sphere diameter a.
(*b*) Express the equation of state of a classical hard-sphere gas in the form of a virial expansion. Include terms up to the third virial coefficient.

14.2. Find b_2 for an ideal Bose gas and compare it with b_2. Is the difference significant?

14.3. Calculate the second virial coefficients for a spinless hard-sphere Bose gas and a spinless hard-sphere Fermi gas to the *two* lowest nonvanishing orders in a/λ, where a = hard sphere diameter and λ = thermal wavelength.

* An analysis of b_3 is given by A. Pais and G. E. Uhlenbeck, *Phys. Rev.*, **116**, 250 (1959).

Answers:

$$b_2 = 2^{-5/2} - \frac{2a}{\lambda} - \frac{10\pi^2}{3}\left(\frac{a}{\lambda}\right)^5 + \cdots \qquad \text{(Bose)}$$

$$b_2 = -2^{-5/2} - 6\pi\left(\frac{a}{\lambda}\right)^3 + 18\pi^2\left(\frac{a}{\lambda}\right)^5 + \cdots \qquad \text{(Fermi)}$$

14.4. Let the free-particle wave functions for a system of distinguishable particles be

$$\Theta_p(1, \ldots, N) \equiv \frac{1}{V^{N/2}} e^{i(\mathbf{p}_1 \cdot \mathbf{r}_1 + \cdots + \mathbf{p}_N \cdot \mathbf{r}_N)}$$

Let

$$\langle 1, \ldots, N | e^{-\beta H} | 1', \ldots, N' \rangle = \sum_{\mathbf{p}_1} \cdots \sum_{\mathbf{p}_N} \Theta_p^*(1, \ldots, N)\, e^{-\beta H} \Theta_p(1', \ldots, N')$$

The symbol $| 1, \ldots, N\rangle$ may be regarded as an eigenvector of the position operators of N distinguishable particles. Show that with the help of this quantity (14.35) may be expressed in the form

$$\frac{1}{\lambda^{3N}} W_N(1, \ldots, N) = \sum_P \delta_P \langle 1, \ldots, N | e^{-\beta H} | P1, \ldots, PN \rangle$$

14.5. Consider a system of particles obeying Boltzmann statistics. Show that the partition function can be written in the form

$$Q_N^{\text{Bo}} = \frac{1}{N!\,\lambda^{3N}} \int d^{3N}r\, W^{\text{Bo}}(1, \ldots, N)$$

where $$\frac{1}{\lambda^{3N}} W^{\text{Bo}}(1, \ldots, N) = \langle 1, \ldots, N | e^{-\beta H} | 1, \ldots, N \rangle$$

chapter 15

PHASE TRANSITIONS

15.1 FORMULATION OF THE PROBLEM

Phase transitions are common occurrences in matter. From experience we find it simplest to characterize a phase transition as the manifestation of a certain singularity or discontinuity in the equation of state. The purpose of the present chapter is to show that in the general formalism of statistical mechanics the phenomenon of phase transition is a *possible* consequence of molecular interactions.

To seek the right question to ask, let us be very specific and consider a classical system of N molecules contained in a volume V, with the Hamiltonian

$$H = K + \Omega(1, \ldots, N) \tag{15.1}$$

where

$$\begin{aligned} K &= \sum_{i=1}^{N} \frac{p_i^{\,2}}{2m} \\ \Omega(1, \ldots, N) &= \sum_{i<j} v_{ij} \end{aligned} \tag{15.2}$$

with $v_{ij} = v(|\mathbf{r}_i - \mathbf{r}_j|)$. The intermolecular potential $v(r)$ shall have the general property that

$$\begin{aligned} v(r) &= \infty && (r \leq a) \\ 0 < v(r) &< -\epsilon && (a < r < r_0) \\ v(r) &= 0 && (r \geq r_0) \end{aligned} \tag{15.3}$$

Thus each molecule may be pictured as a hard sphere of diameter a surrounded by an attractive potential of range r_0 and maximum depth ϵ. To a good approximation $v(r)$ is the intermolecular potential in ordinary matter. A qualitative sketch of $v(r)$ is shown in Fig. 15.1. Throughout the present chapter we deal only with the system just described. The classical partition function of the system is

$$Q_N(V) = \frac{1}{N!\,\lambda^{3N}} \int d^{3N}r\, e^{-\beta\,\Omega(1,\ldots,N)} \tag{15.4}$$

where $\lambda = \sqrt{2\pi\hbar^2/mkT}$, the thermal wavelength. The temperature, being fixed, will not be displayed unless necessary. The grand partition function is

$$\mathcal{Q}(z, V) = \sum_{N=0}^{\infty} z^N Q_N(V) \tag{15.5}$$

The equation of state is to be obtained by eliminating z between the parametric equations

$$\begin{cases} \beta P = \dfrac{1}{V} \log \mathcal{Q}(z, V) \\[2ex] \dfrac{1}{v} = \dfrac{1}{V} z \dfrac{\partial}{\partial z} \log \mathcal{Q}(z, V) \end{cases} \tag{15.6}$$

where v is the specific volume, a parameter independent of the total volume V. We consider the purely mathematical question of the most general behavior of P as a function of v for a given fixed value of the total volume V.

For a given value of the total volume V there is a maximum number of molecules $N_m(V)$ that can be accommodated in V. If $N > N_m(V)$, then at least two molecules will "touch" each other and

$$Q_N(v) = 0 \qquad [N > N_m(v)] \tag{15.7}$$

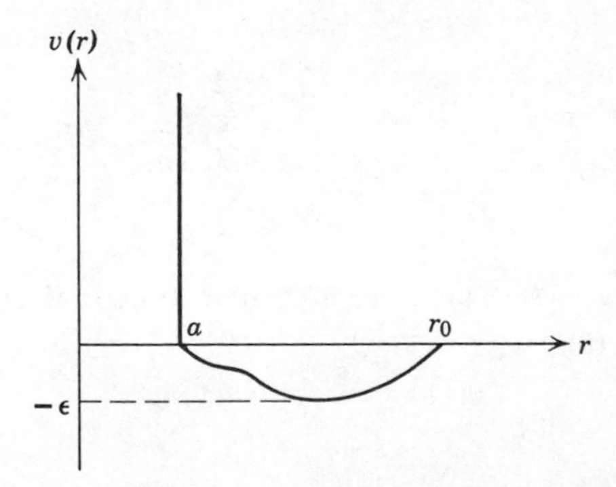

Fig. 15.1. The interparticle potential assumed throughout this chapter.

Thus the grand partition function is a polynomial in z of degree $N_m(V)$:

$$\mathscr{Q}(z, V) = 1 + zQ_1(v) + z^2Q_2(v) + \cdots + z^{N_m}Q_{N_m}(V) \tag{15.8}$$

All the coefficients in this polynomial are by definition positive. Hence it has no real positive root. The equation of state now takes the parametric form

$$\begin{aligned} \beta P &= \frac{1}{V}\log\,[1 + zQ_1(V) + z^2Q_2(V) + \cdots + z^{N_m}Q_{N_m}(V)] \\ \frac{1}{v} &= \frac{1}{V}\frac{zQ_1(V) + 2z^2Q_2(V) + \cdots + N_m z^{N_m}Q_{N_m}(V)}{1 + zQ_1(V) + z^2Q_2(V) + \cdots + z^{N_m}Q_{N_m}(V)} \end{aligned} \tag{15.9}$$

Clearly both P and v are analytic functions of z in a region of the complex z plane that includes the real positive axis. Therefore P is an analytic function of v in a region of the complex v plane that includes the real positive axis. The following statements hold:

(*a*)
$$P \geq 0 \tag{15.10}$$

Proof. This is obvious, because $Q_N \geq 0$ for each N.

(*b*)
$$\frac{V}{N_m} \leq v < \infty \tag{15.11}$$

Proof. This follows from the definition of N_m.

(*c*)
$$\frac{\partial P}{\partial v} \leq 0 \tag{15.12}$$

Proof.

$$\frac{\partial P}{\partial v} = \frac{\partial P}{\partial z}\frac{\partial z}{\partial v} = \frac{1}{vz}\frac{\partial z}{\partial v} = \frac{1}{vz(\partial v/\partial z)} = -\frac{1}{Vv^4\langle[N - \langle N\rangle]^2\rangle} \leq 0$$

The foregoing conclusions remain valid for a quantum mechanical system because (15.7) remains valid. To see this, we note that (15.1) continues to be the Hamiltonian provided $\mathbf{p}_i$ is taken to be the momentum operator for the ith particle. The hard-sphere potential contained in Ω now requires that any eigenfunction of H vanish whenever two particles "touch." For evaluation of Q_N we can use the formulas (14.35) and (14.36) and choose as a complete set of wave functions the eigenfunctions of H. It is then obvious that (15.7) holds.

For any finite value of the total volume V, no matter how large, we cannot easily recognize a phase transition unless the equation of state can be explicitly calculated. Suppose, for example, that the system under consideration actually exhibits a first-order transition. For a finite value of the total volume, no matter how large, the pressure cannot be strictly constant in the transition region because it is an analytic function of v. Nevertheless, $\partial P/\partial v$ might be extremely small in some range of v, e.g., of the order of 10^{-23} atm/cc. On a macroscopic scale we would regard the pressure as constant, but there is no way to see this from the partition function unless the function $P(v)$ has been explicitly calculated.

In order that we may recognize a phase transition without explicitly calculating the partition function, it is necessary for us to consider the system in the limit as $V \to \infty$. In this limit, the equation of state is given by

$$\begin{cases} \beta P = \lim\limits_{V\to\infty} \left[\dfrac{1}{V} \log \mathcal{Q}(z, V)\right] \\[2ex] \dfrac{1}{v} = \lim\limits_{V\to\infty} \left[\dfrac{1}{V}\, z \dfrac{\partial}{\partial z} \log \mathcal{Q}(z, V)\right] \end{cases} \tag{15.13}$$

The limit as $V \to \infty$ must be understood in the strict mathematical sense. In particular, the order of the operations $\lim\limits_{V\to\infty}$ and $z(\partial/\partial z)$ may not be freely interchanged. We may now hope to find possible singularities in the equation of state that can be identified as phase transitions.

15.2 THE THEORY OF YANG AND LEE

It has been shown that the equation of state of a system whose Hamiltonian is (15.1), classically or quantum mechanically, exhibits no unusual behavior when the volume of the system is finite, no matter how large. The mathematical reason is that there are no real positive roots of the equation $\mathcal{Q}(z, V) = 0$ for any finite value of V. This means that regarded as a function of the complex variable z the zeros of the function $\mathcal{Q}(z, V)$ are distributed in the complex z plane but are never on the positive real axis. As V increases the number of zeros increases [since the degree of the polynomial (15.8) is a function of V], and their positions may move about in the complex z plane. In the limit as $V \to \infty$ some of the roots may converge towards the positive real axis.

Suppose that in the complex z plane there is a region R which contains a segment of the real positive z axis and which is free of zeros of $\mathcal{Q}(z, V)$ for all V. It is reasonable to expect that as $V \to \infty$ the properties (15.10)–(15.12) continue to hold. Thus such a region R may correspond to a single homogeneous phase of the system. If there are many nonoverlapping regions R, each region may be expected to correspond to a phase of the system. To study the possible phase transitions, we only have to study the most general behavior of the equation of state (15.13) as z goes from one R region to another.

This idea forms the basis of the theory of Yang and Lee,* which is contained in the following two theorems.

THEOREM 1. $\lim_{V\to\infty} [V^{-1} \log \mathcal{Q}(z, V)]$ exists for all $z > 0$. This limit is independent of the shape of the volume V and is a continuous, non-decreasing function of z.

It is assumed that as $V \to \infty$ the surface area of V increases no faster than $V^{2/3}$.

THEOREM 2. Let R be a region in the complex z plane that contains a segment of the positive real axis and contains no root of the equation $\mathcal{Q}(z, V) = 0$ for any V. Then for all z in R the quantity $V^{-1} \log \mathcal{Q}(z, V)$ converges *uniformly* to a limit as $V \to \infty$. This limit is an analytic function of z for all z in R.

These theorems are proved in Appendix C for classical statistical mechanics. We discuss their consequences here.

Let a phase of the system be defined as the collection of thermodynamic states corresponding to values of z lying in any single region R as defined by theorem 2. Let

$$F_\infty(z) \equiv \lim_{V\to\infty} \frac{1}{V} \log \mathcal{Q}(z, V) \tag{15.14}$$

Then in any single phase $V^{-1} \log \mathcal{Q}(z, V)$ uniformly converges to $F_\infty(z)$ as $V \to \infty$. This implies that in the equation of state (15.13) the order of the operations $\lim_{V\to\infty}$ and $z(\partial/\partial z)$ can be interchanged. Therefore in any single phase the equation of state is

$$\begin{cases} \beta P(z) = F_\infty(z) \\ \dfrac{1}{v(z)} = z \dfrac{\partial F_\infty(z)}{\partial z} \end{cases} \tag{15.15}$$

It is obvious that the properties (15.10)–(15.12) remain valid in the limit as $V \to \infty$. In particular

$$\frac{\partial P}{\partial v} \leq 0 \tag{15.16}$$

Some possible behaviors of the equation of state consistent with theorems 1 and 2 are illustrated by the following examples.

Suppose the region R includes the entire positive z axis, as shown in Fig. 15.2. Then the system is always in a single phase. The equation of state may be obtained by graphically eliminating z in (15.15), as shown in Fig. 15.3.

* C. N. Yang and T. D. Lee, *Phys. Rev.*, **87,** 404 (1952).

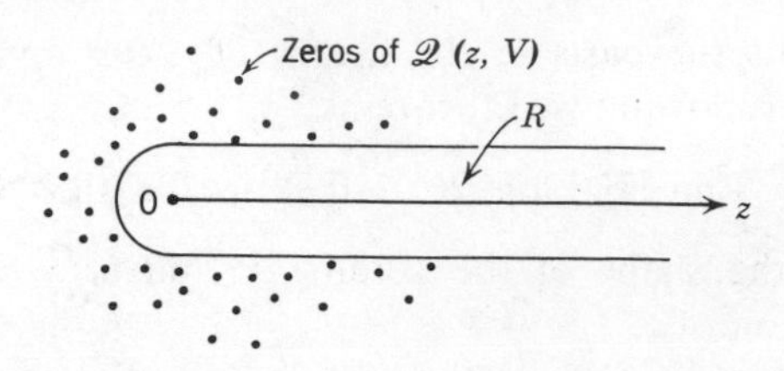

Fig. 15.2. Region R that is free of zeros of $\mathscr{Q}(z, V)$.

If on the other hand a zero of $\mathscr{Q}(z, V)$ approaches the point z_0 on the real-position z axis as $V \to \infty$, then there will be two regions R_1 and R_2 in which theorem 2 holds separately, as shown in Fig. 15.4. At $z = z_0$, $P(z)$ must be continuous, as required by theorem 1. Its derivative, however, may be discontinuous. An example of this behavior is shown in Fig. 15.5. The system possesses two phases, corresponding respectively to the regions $z < z_0$ and $z > z_0$. At $z = z_0$, $1/v(z)$ is discontinuous. We can show that $1/v(z)$ must increase when z passes from $z < z_0$ to $z > z_0$ in the following way: For any finite V,

$$z \frac{\partial}{\partial z}\left[\frac{1}{v(z)}\right] = z \frac{\partial}{\partial z}\, z \frac{\partial}{\partial z}\left[\frac{1}{V} \log \mathscr{Q}(z, V)\right] = \left\langle \left(\frac{N}{V}\right)^2 \right\rangle - \left\langle \frac{N}{V} \right\rangle^2 \geq 0$$

This suffices to show that if z_2 lies in R_2 and z_1 lies in R_1 with $z_2 > z_1$ then $v^{-1}(z_2) \geq v^{-1}(z_1)$. Thus we obtain a first-order transition between two phases. This is shown in the P-v diagram in Fig. 15.5, in which the horizontal portion of the curve between v_b and v_a originates from the

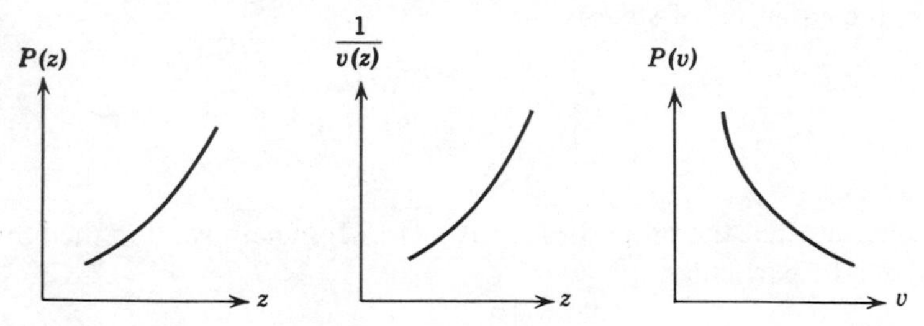

Fig. 15.3. Equation of state of a system that has only a single phase.

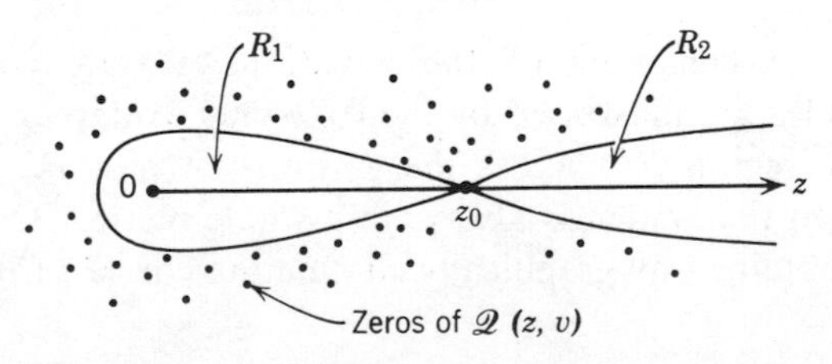

Fig. 15.4. Two regions R_1, R_2 each free of zeros of $\mathscr{Q}(z, V)$.

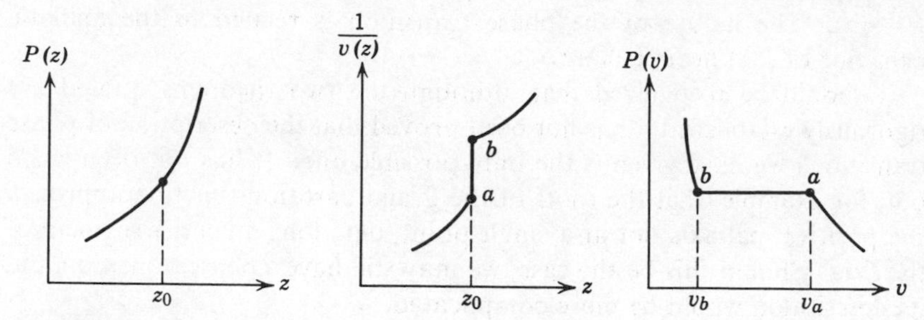

Fig. 15.5. Equation of state of a system with two phases connected by a first-order transition.

vertical portion between the points a and b of $1/v(z)$. The fact that $1/v(z)$ must actually assume all the values between the points a and b follows from the observation that the curve of $1/v(z)$ is a limiting curve as $V \to \infty$, and that for any finite value of V, $1/v(z)$ is a continuous curve.

If in the same example $dP(z)/dz$ is continuous at $z = z_0$ but $\partial^2 P(z)/\partial z^2$ is discontinuous, we would have instead a second-order phase transition, as shown in Fig. 15.6. It is clear that a phase transition of the nth order will result if $d^{n+1}P(z)/z^{n+1}$ is discontinuous at z_0 while all lower derivatives are continuous. It is also consistent with theorems 1 and 2 that some derivative of $P(z)$ may diverge at z_0, so that the resulting phase transition could not be described in terms of the order but would be similar to the λ-transition in liquid He^4.

To sum up, we have shown that it is *possible* that the equation of state (15.13) exhibits the phenomenon of phase transition. That is, the existence of phase transitions is not in contradiction to the Hamiltonian (15.1) and the general formalism of statistical mechanics. Furthermore, it is possible to relate the occurrence of a phase transition to the circumstance that a root z of the equation $\mathscr{Q}(z, V) = 0$ approaches the real positive z axis as

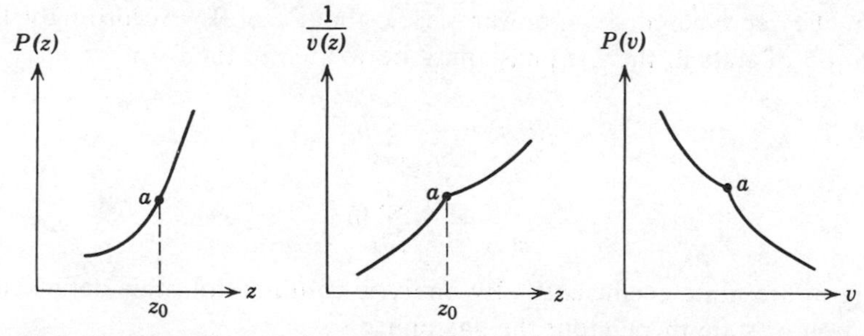

Fig. 15.6. Equation of state of system with two phases connected by a second-order transition.

$V \to \infty$. The nature of the phase transition is related to the analytic behavior of $P(z)$ near such a root.

It should be recognized that although the two theorems quoted are rigorously established it has not been proved that the description of phase transitions we have given is the only possible one. It has not been ruled out, for example, that the roots of the grand partition function approach the positive real axis not at a single point, but along an entire segment of the axis. Should this be the case, we may still have a phase transition, but its description would be more complicated.

In principle the uniqueness of our description can be proved or disproved, because the partition function of the system under discussion is well defined. All that is necessary is to calculate the partition function explicitly. Unfortunately the calculation is beyond our mathematical power.

In view of this deficiency it is interesting to test the validity of our description in simpler models. A notable model is the two-dimensional Ising model, whose partition function can be calculated exactly. (See Chapter 17.) The Ising model exhibits a phase transition which falls into one of the categories discussed previously. It can be shown* that in this case the roots of the grand partition function always lie on the unit circle in the complex z plane. As the size of the system approaches infinity the roots merge into a continuous distribution which extends at one point to the positive real axis. The phase transition is completely characterized by the distribution function of the roots. This example shows that the description of phase transitions given previously is valid in at least one problem.

15.3 THE GAS PHASE

The gas phase of a system, if it exists, is defined by the phase that corresponds to the region R that includes the origin $z = 0$ of the complex z plane. It follows from theorem 2 that $P(z)$ is analytic near $z = 0$, and it can be represented by a power series about $z = 0$. Accordingly the equation of state in the gas phase may be written in the form

$$\begin{aligned} \beta P(z) &= \frac{1}{\lambda^3} \sum_{l=1}^{\infty} b_l z^l \\ \frac{1}{v(z)} &= \frac{1}{\lambda^3} \sum_{l=1}^{\infty} l b_l z^l \end{aligned} \tag{15.17}$$

where b_l are finite coefficients. By analytic continuation, this defines the equation of state throughout the gas phase.

* T. D. Lee and C. N. Yang, *Phys. Rev.*, **87**, 410 (1952).

For a finite volume V the developments in Chapter 14 show that the equation of state can generally be reduced to the form (15.17), but with $\bar{b}_l$ replaced by $b_l(V)$, the lth cluster integral. The uniform convergence stated in theorem 2 then implies that in the gas phase

$$\lim_{V \to \infty} b_l(V) = \bar{b}_l \tag{15.18}$$

This shows that for the system whose Hamiltonian is (15.1) all the cluster integrals approach finite limits as $V \to \infty$ and that the virial expansion (14.30) is valid.

The equation (15.17) by itself, however, does not tell us the extent of the region R. Thus not only does it become invalid beyond the point of phase transition (if there is such a transition), it is also incapable of locating the transition point.

Mayer* has worked out the explicit equation of state following from (15.17), assuming that it is correct for all $z > 0$. He found that in general an isotherm has the form represented in Fig. 15.7. This result does not agree with the isotherm of any physical system, because the portion B indicated in Fig. 15.7 never rises again. From our general discussion we see that (a) the portion B is invalid; (b) the portion A of the isotherm is valid for $v > v_0$, but the value of v_0 cannot be determined from (15.17). Thus the virial expansion of the equation of state does not contain all the information about the equation of state.†

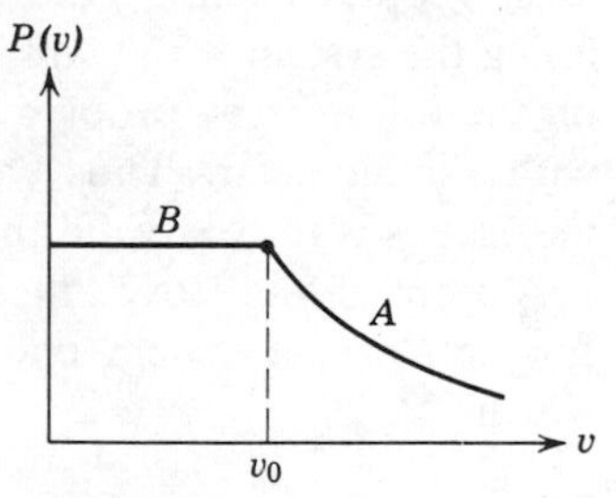

Fig. 15.7. Equation of state obtained by taking the virial expansion to be exact.

15.4 VAN HOVE'S THEOREM

We continue to consider a classical system whose Hamiltonian is (15.1). We prove the following theorem.‡

VAN HOVE'S THEOREM. The equation of state in the canonical ensemble is the same as that in the grand canonical ensemble.

* See Mayer and Mayer, *loc. cit.*

† Another example of the failure of the virial expansion, that for the ideal Bose gas, is discussed by W. H. J. Fuchs, *J. Rational Mech. and Anal.*, **4**, 647 (1955). As a trivial mathematical example consider $\mathcal{Q}(z, V) = (1 + z)^V(1 + z^{\alpha V})$ where α is a given positive number.

‡ L. Van Hove, *Physica*, **15**, 951 (1949). The original statement of Van Hove's theorem is (15.19). Our proof is somewhat different from the original one.

Van Hove's theorem implies that if $P_{\text{can}}(v)$ is the pressure in the canonical ensemble and if $\partial P_{\text{can}}(v)/\partial v$ exists, then

$$\frac{\partial}{\partial v} P_{\text{can}}(v) \leq 0 \tag{15.19}$$

for, if this were not so, the equation of state in the canonical ensemble would disagree with that in the grand canonical ensemble, as we have shown in Sec. 8.5.

In the last section we have considered the various types of possible phase transitions in the grand canonical ensemble. By Van Hove's theorem the results of that discussion can be taken over in the canonical ensemble.

Van Hove's theorem has a direct physical meaning. The pressure of a macroscopic system should be the same, whether we measure it by confining the system with walls and measuring the force per unit area acting on the walls, or by probing a small volume in the interior of the system with a manometer. Thus Van Hove's theorem must be true, if statistical mechanics is to be a valid theory of matter.

A proof of Van Hove's theorem is as follows. Let $Q_N(V)$ be the partition function of the system under consideration. We consider the limit in which

$$\begin{aligned} N &\to \infty \\ V &\to \infty \\ \frac{V}{N} &= v \end{aligned} \tag{15.20}$$

where v is a given finite number greater than a fixed multiple of the close-packing specific volume. Let

$$f_N(v) \equiv \frac{1}{N} \log Q_N(V) \tag{15.21}$$

$$f_\infty(v) \equiv \operatorname*{Lim}_{N\to\infty} \left[\frac{1}{N} \log Q_N(vN)\right] \tag{15.22}$$

The pressure $P_{\text{can}}(v)$ in the canonical ensemble is defined by

$$\beta P_{\text{can}}(v) \equiv \frac{\partial}{\partial v} f_\infty(v) \tag{15.23}$$

It is our object to prove that $P_{\text{can}}(v) = P(v)$ where $P(v)$ is defined by (15.13).

We first define a comparison partition function $\tilde{Q}_N(V, V_0)$ in the following manner. Suppose the volume V is a union of cubes. Then V can be covered with a number γ of elementary cubes of equal size. Within each elementary cube we construct a smaller cube, called a *cell*, of which each face is a distance $r_0/2$ from the nearest face of the elementary cube that contains it, where r_0 is the range of the interparticle potential. Thus the volume V contains γ cells, each separated by a distance r_0 from its nearest neighbors. Since r_0 is the range of the interparticle potential, two particles lying in different cells do not interact with each other. The construction is schematically shown in Fig. 15.8. Let the space between the cells be called the *corridor*. Let the volume of each cell be V_0 and let the volume of the corridor be V_c. Then

$$V_c = V - \gamma V_0 \tag{15.24}$$

The surface area of a cell is $6V_0^{2/3}$. The volume of the part of the corridor that surrounds a cell is less than $cr_0V_0^{2/3}$, where c is a numerical constant. Hence

$$V_c < c\gamma r_0 V_0^{2/3} \tag{15.25}$$

Since $V > \gamma V_0$ we have $(V_c/V) < c(r_0/V_0^{1/3})$. We shall eventually consider V_0 to be sufficiently large for the volume of the corridor to be negligible compared to the total volume, i.e., $(V_c/V) \approx 0$.

We define the comparison partition function $\tilde{Q}_N(V, V_0)$ to be the partition function of the system under the restriction that *no particle shall be in the corridor.*

In Appendix C it is established in classical statistical mechanics that

$$\tilde{Q}_N(V, V_0) \leq Q_N(V) \leq \tilde{Q}_N(V, V_0)e^{\sigma M} \tag{15.26}$$

where σ is a finite constant given by

$$\sigma = e^{\beta\epsilon(r_0/a)^3} \tag{15.27}$$

and M is the maximum number of particles in the corridor:

$$M = \frac{V_c}{\frac{4}{3}\pi a^3} < \text{const.}\ \frac{r_0}{a^3}\,\gamma V_0^{2/3} \tag{15.28}$$

Thus as $V_0 \to \infty$, $Q_N(V) \to \tilde{Q}_N(V, V_0)$.

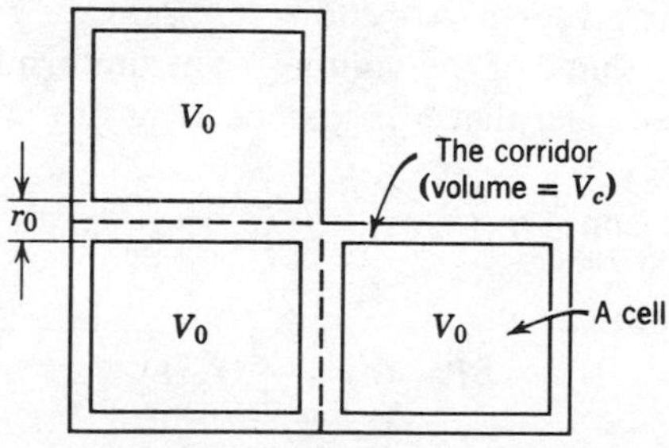

Fig. 15.8. Construction of cells within the volume of the system.

As $V_0 \to \infty$ it is necessary that $V \to \infty$, because $V > \gamma V_0$. The detailed manner in which $Q_N(V)$ approaches $\tilde{Q}_N(V, V_0)$ depends on the ratio V_0/V. We consider the limit (15.20), taken in conjunction with $V_0 \to \infty$, in such a manner that each cell always contains an enormous number of particles. The manner in which this can be done is not unique. For the sake of definiteness we choose to take the limit by fixing v and then letting

$$
\begin{aligned}
&N \to \infty \\
&V = vN \\
&V_0 = \frac{v^3}{a^6}\sqrt{N}
\end{aligned}
\tag{15.29}
$$

Thus V_0 tends to infinity much slower than V. The factor v^3 is introduced to insure that V_0 always contains a large number of particles.

Let

$$\tilde{f}_N(v) \equiv \frac{1}{N} \log \tilde{Q}_N(V, V_0) \tag{15.30}$$

$$\tilde{f}_\infty(v) \equiv \operatorname*{Lim}_{N\to\infty} \tilde{f}_N(v) \tag{15.31}$$

where in (15.30) the numbers V and V_0 are given by (15.29). It follows from (15.26)–(15.28) that

$$|\tilde{f}_N(v) - f_N(v)| < \frac{\sigma'}{N^{1/6}} \tag{15.32}$$

where σ' is a finite number independent of v and N:

$$\sigma' = \text{const.}\ \frac{r_0}{a}\, e^{\beta\epsilon(r_0/a)^3} \tag{15.33}$$

By a procedure similar to that described in Appendix C, Sec. C.2, we can establish the following facts:

(*a*) $\tilde{f}_\infty(v)$ exists.

(*b*) $f_\infty(v)$ exists and is equal to $\tilde{f}_\infty(v)$.

(*c*) $\tilde{f}_N(v)$ and $f_N(v)$ converge *uniformly* to their common limit as $N \to \infty$.

By an argument given in Appendix C, Sec. C.2, $f_\infty(v)$ and $\tilde{f}_\infty(v)$ are independent of the shape of the volume. The uniformity of convergence, a consequence of the fact that σ' is independent of v, implies

$$\operatorname*{Lim}_{N\to\infty} \frac{\partial}{\partial v} \tilde{f}_N(v) = \frac{\partial}{\partial v} \tilde{f}_\infty(v) = \frac{\partial}{\partial v} f_\infty(v) \tag{15.34}$$

Using (15.23) we obtain

$$\beta P_{\text{can}}(v) = \frac{\partial}{\partial v} \tilde{f}_\infty(v) \tag{15.35}$$

Thus to study $P_{\text{can}}(v)$ it is sufficient to study $\tilde{f}_N(v)$.

We calculate $\tilde{f}_N(v)$ by first calculating $\tilde{Q}_N(V, V_0)$, where the numbers V and V_0 are functions of N as given in (15.29). The system whose partition is $\tilde{Q}_N(V, V_0)$ is composed of noninteracting cells each of which contains an indefinite number of particles. If there are N_i particles in the ith cell, the partition function will be a product of the individual partition functions for each cell. To obtain $\tilde{Q}_N(V, V_0)$ we only have to add up all such products, with all possible assignments of particles into the individual cells. Thus

$$\tilde{Q}_N(V, V_0) = \sum_{\{N_i\}} Q_{N_1}(V_0) Q_{N_2}(V_0) \cdots Q_{N_\gamma}(V_0) \tag{15.36}$$

where the sum over the set of integers $\{N_i\}$ is subject to the restriction

$$\sum_{i=1}^{\gamma} N_i = N \tag{15.37}$$

We may rewrite (15.36) as

$$\tilde{Q}_N(V, V_0) = \sum_{n=0}^{N} Q_n(V_0) \left[\sum_{\{N_i\}}' Q_{N_2}(V_0) \cdots Q_{N_\gamma}(V_0) \right] \tag{15.38}$$

where the sum Σ' is subject to the condition

$$\sum_{i=2}^{\gamma} N_i = N - n \tag{15.39}$$

It is then easily recognized by comparing the expression in brackets on the right-hand side of (15.38) with the right-hand side of (15.36) that

$$\tilde{Q}_N(V, V_0) = \sum_{n=0}^{N} Q_n(V_0) \tilde{Q}_{N-n}(V - V_0, V_0) \tag{15.40}$$

Because of the hard-sphere repulsion between particles there exists a finite number κ such that

$$Q_n(v_0) = 0 \qquad (n > \kappa V_0) \tag{15.41}$$

Hence in (15.40) the index n is such that for every term in the sum

$$\frac{n}{N} < \frac{\kappa V_0}{N} \xrightarrow[N \to \infty]{} 0 \tag{15.42}$$

Thus as $N \to \infty$ we may expand $\tilde{Q}_{N-n}(V - V_0, V_0)$ in powers of n and keep only terms to the first order in n:

$$\begin{aligned} \log [\tilde{Q}_{N-n}(V - V_0, V_0)] &\equiv (N - n) \tilde{f}\left(\frac{V - V_0}{N - n}\right) \\ &\approx N\tilde{f}_N(v) - V_0 \frac{\partial}{\partial v} \tilde{f}_N(v) + n\left(v \frac{\partial}{\partial v} - 1\right) \tilde{f}_N(v) \end{aligned} \tag{15.43}$$

Substituting this into the right-hand side of (15.40) we obtain the following approximate equation, which becomes exact as $N \to \infty$:

$$\exp [N\tilde{f}_N(v)]$$
$$\approx \exp [N\tilde{f}_N(v)] \exp \left[-V_0 \frac{\partial}{\partial v} \tilde{f}_N(v) \right] \sum_{n=0}^{\infty} Q_n(V_0) \exp \left[n \left(v \frac{\partial}{\partial v} - 1 \right) \tilde{f}_N(v) \right]$$

or

$$\exp \left[V_0 \frac{\partial}{\partial v} \tilde{f}(v) \right] \approx \sum_{n=0}^{\infty} Q_n(V_0) \exp \left[n \left(v \frac{\partial}{\partial v} - 1 \right) \tilde{f}_N(v) \right] \tag{15.44}$$

where, because of (15.41), it is permissible to let the sum over n extend to ∞. Taking the logarithms of both sides of (15.44) we obtain an implicit equation for $\tilde{f}_N(v)$:

$$\left\{ \begin{aligned} \frac{\partial}{\partial v} \tilde{f}_N(v) &\approx \frac{1}{V_0} \log \sum_{n=0}^{\infty} z^n Q_n(V_0) \\ \log z &\approx \left(v \frac{\partial}{\partial v} - 1 \right) \tilde{f}_N(v) \end{aligned} \right. \tag{15.45}$$

where V_0 is defined in (15.29). Differentiating both sides of the equations (15.45) we obtain

$$\left\{ \begin{aligned} \frac{\partial^2}{\partial v^2} \tilde{f}_N(v) &\approx \frac{1}{z} \frac{\partial z}{\partial v} z \frac{\partial}{\partial z} \left[\frac{1}{V_0} \log \sum_{n=0}^{\infty} z^n Q_n(V_0) \right] \\ \frac{\partial^2}{\partial v^2} \tilde{f}_N(v) &\approx \frac{1}{vz} \frac{\partial z}{\partial v} \end{aligned} \right. \tag{15.46}$$

This shows that z can also be defined by the equation

$$\frac{1}{v} = z \frac{\partial}{\partial z} \left[\frac{1}{V_0} \log \sum_{n=0}^{\infty} z^n Q_n(V_0) \right] \tag{15.47}$$

Finally, letting $N \to \infty$ and using (15.23), we obtain

$$\left\{ \begin{aligned} \beta P_{\text{can}}(v) &= \lim_{V \to \infty} \left[\frac{1}{V} \log \sum_{n=0}^{\infty} z^n Q_n(V) \right] \\ \frac{1}{v} &= \lim_{V \to \infty} z \frac{\partial}{\partial z} \left[\frac{1}{V} \log \sum_{n=0}^{\infty} z^n Q_n(V) \right] \end{aligned} \right. \tag{15.48}$$

This is Van Hove's theorem.

C

SPECIAL TOPICS IN STATISTICAL MECHANICS

chapter 16

THE ISING MODEL

16.1 DEFINITION OF THE ISING MODEL

One of the most interesting phenomena in the physics of the solid state is ferromagnetism. In some metals, e.g., Fe and Ni, a finite fraction of the spins of the atoms becomes spontaneously polarized in the same direction, giving rise to a macroscopic magnetic field. This happens, however, only when the temperature is lower than a characteristic temperature known as the Curie temperature. Above the Curie temperature the spins are oriented at random, producing no net magnetic field. As the Curie temperature is approached from both sides the specific heat of the metal approaches infinity. The transition from the nonferromagnetic state to the ferromagnetic state is a phase transition of a kind not included in the Ehrenfest classification of phase transitions.

The Ising model is a crude attempt to simulate the structure of a physical ferromagnetic substance.* Its main virtue lies in the fact that a two-dimensional Ising model yields to an exact treatment in statistical mechanics. It is the only nontrivial example of a phase transition that can be worked out with mathematical rigor.

* More accurately, the Ising model simulates a "domain" in a ferromagnetic substance. A discussion of the physical properties and the atomic structure of ferromagnets (and antiferromagnets) is beyond the scope of the present discussion. For information in these aspects consult a book on solid-state physics.

In the Ising model* the system considered is an array of N fixed points called lattice sites that form an n-dimensional periodic lattice ($n = 1, 2, 3$). The geometrical structure of the lattice may (for example) be cubic or hexagonal. Associated with each lattice site is a spin variable s_i ($i = 1, \ldots, N$) which is a *number* that is either $+1$ or -1. There are no other variables. If $s_i = +1$, the ith site is said to have spin up, and if $s_i = -1$, it is said to have spin down. A given set of numbers $\{s_i\}$ specifies a configuration of the whole system. The energy of the system in the configuration specified by $\{s_i\}$ is defined to be

$$E_I\{s_i\} = -\sum_{\langle ij \rangle} \epsilon_{ij} s_i s_j - B \sum_{i=1}^{N} s_i$$

where the subscript I stands for Ising and the symbol $\langle ij \rangle$ denotes a nearest-neighbor pair of spins. There is no distinction between $\langle ij \rangle$ and $\langle ji \rangle$. Thus the sum over $\langle ij \rangle$ contains $\gamma N/2$ terms, where γ is the number of nearest neighbors of any given site. For example,

$$\gamma = \begin{cases} 4 \text{ (two-dimensional square lattice)} \\ 6 \text{ (three-dimensional simple cubic lattice)} \\ 8 \text{ (three-dimensional body-centered cubic lattice)} \end{cases}$$

The interaction energy ϵ_{ij} and the external magnetic field B are given constants. The geometry of the lattice enters the problem through γ and ϵ_{ij}. For simplicity we specialize the model to the case of isotropic interactions, so that all ϵ_{ij} are equal to a given number ϵ. Thus the energy will be taken as

$$E_I\{s_i\} = -\epsilon \sum_{\langle ij \rangle} s_i s_j - B \sum_{i=1}^{N} s_i \tag{16.1}$$

The case $\epsilon > 0$ corresponds to ferromagnetism and the case $\epsilon < 0$ to antiferromagnetism. We consider only the case $\epsilon > 0$. The partition function is

$$Q_I(B, T) = \sum_{s_1} \sum_{s_2} \cdots \sum_{s_N} e^{-\beta E_I\{s_i\}} \tag{16.2}$$

where each s_i ranges independently over the values ± 1. Hence there are 2^N terms in the summation. The thermodynamic functions are obtained in the usual manner from the Helmholtz free energy:

$$A_I(B, T) = -kT \log Q_I(B, T) \tag{16.3}$$

* E. Ising, *Z. Phys.*, **31**, 253 (1925).

Some of the interesting ones are

$$U_I(B, T) = -kT^2 \frac{\partial}{\partial T}\left(\frac{A_I}{kT}\right) \qquad \text{(internal energy)} \qquad (16.4)$$

$$C_I(B, T) = \frac{\partial U_I}{\partial I} \qquad \text{(heat capacity)} \qquad (16.5)$$

$$M_I(B, T) = -\frac{\partial}{\partial B}\left(\frac{A_I}{kT}\right) = \left\langle \sum_{i=1}^{N} s_i \right\rangle \qquad \text{(magnetization)} \qquad (16.6)$$

where $\langle \quad \rangle$ denotes ensemble average. The quantity $M_I(0, T)$ is called the spontaneous magnetization. If it is nonzero the system is said to be ferromagnetic.

Although a configuration of the system is specified by the N numbers $s_1, \ldots, s_N$, the energy value (16.1) is in general degenerate. There is another way of writing (16.1) that makes this manifest. In any given configuration of the lattice, let

$$N_+ = \text{total number of up spins}$$

$$N_- = \text{total number of down spins}$$

$$= N - N_+$$

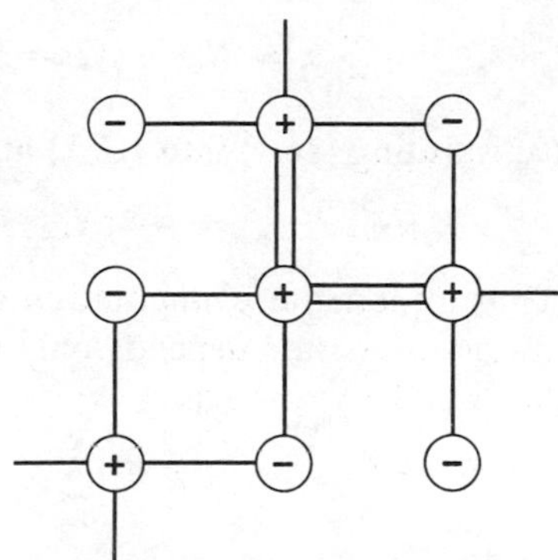

Fig. 16.1. Construction for the derivation of (16.7).

Each nearest-neighbor pair is one of the three types $(+\,+)$, $(-\,-)$, or $(+\,-)$, where $(+\,-)$ is not distinguished from $(-\,+)$. Let the respective number of such pairs be denoted by N_{++}, N_{--}, N_{+-}. These numbers are not independent of each other, nor of N_+, N_-. A relation among them may be found as follows. Choose a particular lattice site with spin up and draw a line connecting it to all its nearest neighbors. There should be γ lines drawn. Repeat this procedure for another site with spin up, and continue to do so until this is done for all sites with spin up. After the procedure is completed the total number of lines drawn is γN_+. A construction of the kind described is illustrated in Fig. 16.1 for a two-dimensional square lattice. The total number of lines drawn can also be counted by noting that between every $(+\,+)$ pair there are two lines, between every $(+\,-)$ pair there is one line, and between every $(-\,-)$ pair there is no line. Hence $\gamma N_+ = 2N_{++} + N_{+-}$. This relation remains valid if we interchange $+$ and $-$. Therefore we have the set of relations

$$\begin{aligned} \gamma N_+ &= 2N_{++} + N_{+-} \\ \gamma N_- &= 2N_{--} + N_{+-} \\ N_+ + N_- &= N \end{aligned} \qquad (16.7)$$

from which any three of the five numbers N_+, N_-, N_{++}, N_{--}, N_{+-} can be eliminated. If we choose to eliminate N_{+-}, N_{--}, N_-, we have

$$\begin{aligned} N_{+-} &= \gamma N_+ - 2N_{++} \\ N_- &= N - N_+ \\ N_{--} &= \frac{\gamma}{2} N + N_{++} - \gamma N_+ \end{aligned} \tag{16.8}$$

We further note that

$$\begin{aligned} \sum_{\langle ij \rangle} s_i s_j &= N_{++} + N_{--} - N_{+-} = 4N_{++} - 2\gamma N_+ + \frac{\gamma}{2} N \\ \sum_{i=1}^{N} s_i &= N_+ - N_- = 2N_+ - N \end{aligned} \tag{16.9}$$

Substituting (16.9) into (16.1) and using (16.8) we obtain

$$E_I(N_+, N_{++}) = -4\epsilon N_{++} + 2(\epsilon\gamma - B)N_+ - (\tfrac{1}{2}\gamma\epsilon - B)N \tag{16.10}$$

Thus although a configuration of the system depends on N numbers the energy of a state depends only on two numbers. The partition function can also be written as

$$e^{-\beta A_I(B,T)} = e^{N\beta(\frac{1}{2}\gamma\epsilon - B)} \sum_{N_+=0}^{N} e^{-2\beta(\epsilon\gamma - B)N_+} \sum_{N_{++}}' g(N_+, N_{++}) e^{4\beta\epsilon N_{++}} \tag{16.11}$$

where $g(N_+, N_{++})$ is the number of configurations that has a given set of values (N_+, N_{++}). The sum Σ' extends over all values of N_{++} consistent with the fact that there are N spins of which N_+ are up. Since $g(N_+, N_{++})$ is a complicated function, the form (16.11) is not a simplification over (16.2) for actual calculations.

16.2 EQUIVALENCE OF THE ISING MODEL TO OTHER MODELS

By a change of names the Ising model can be made to simulate other systems than a ferromagnet. Among these are a lattice gas and a binary alloy.

Lattice Gas

A lattice gas is a collection of atoms whose positions can take on only discrete values. These discrete values form a lattice of given geometry with γ nearest neighbors to each lattice site. Each lattice site can be occupied by at most one atom. Figure 16.2 illustrates a configuration of a two-dimensional lattice gas in which the atoms are represented by solid circles and the empty lattice sites by open circles. We neglect the kinetic energy of an atom and assume that only nearest neighbors interact, and the interaction energy for a pair of nearest neighbors is assumed to be a constant $-\epsilon_0$. Thus the potential energy of the system is equivalent to that of a gas in which the atoms are located only on lattice sites and interact through a two-body potential $v(|\mathbf{r}_i - \mathbf{r}_j|)$ with

$$v(r) = \begin{cases} \infty & (r = 0) \\ -\epsilon_0 & (r = \text{nearest-neighbor distance}) \\ 0 & (\text{otherwise}) \end{cases} \tag{16.12}$$

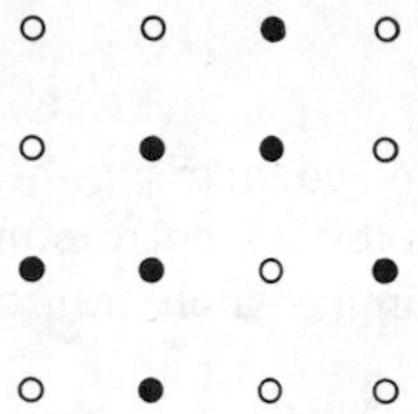

Let

$$\begin{aligned} N &= \text{total no. of lattice sites} \\ N_a &= \text{total no. of atoms} \\ N_{aa} &= \text{total no. of nearest-neighbor pairs of atoms} \end{aligned} \tag{16.13}$$

○ ○ ● ○
○ ● ● ○
● ● ○ ●
○ ● ○ ○

Fig. 16.2. A configuration of the lattice gas.

The total energy of the lattice gas is

$$E_G = -\epsilon_0 N_{aa} \tag{16.14}$$

and the partition function is

$$Q_G(N_a, T) = \frac{1}{N_a!} \sum{}^{a} e^{\beta\epsilon_0 N_{aa}} \tag{16.15}$$

where the sum Σ^a extends over all ways of distributing N_a distinguishable atoms over N lattice sites. If the volume of a unit cell of the lattice is chosen to be unity, then N is the volume of the system. The grand partition function is

$$\mathcal{Q}_G(z, N, T) = \sum_{N_a=0}^{\infty} z^{N_a} Q_G(N_a, T) \tag{16.16}$$

The equation of state is given, as usual, by

$$\begin{cases} \beta P_G = \dfrac{1}{N} \log \mathcal{Q}_G(z, N, T) \\ \dfrac{1}{v} = \dfrac{1}{N} z \dfrac{\partial}{\partial z} \log \mathcal{Q}_G(z, N, T) \end{cases} \tag{16.17}$$

To establish a correspondence between the lattice gas and the Ising model, let occupied sites correspond to spin up and empty sites to spin down. Then $N_a \leftrightarrow N_+$. In the Ising model a set of N numbers $\{s_1, \ldots, s_N\}$ uniquely defines a configuration. In the lattice gas an enumeration of the occupied sites determines not one, but $N_a!$ configurations. The difference arises from the fact that the atoms are supposed to be able to move from site to site. This difference, however, is obliterated by the adoption of "correct Boltzmann counting." Hence

$$Q_G(N_a, T) = \sum_{N_{++}}' g(N_+, N_{++})e^{\beta\epsilon_0 N_{++}} \tag{16.18}$$

where the function $g(N_+, N_{++})$ and the sum Σ' are identical with those appearing in (16.11). The grand partition function is

$$e^{\beta N P_G} = \mathcal{Q}_G(z, N, T) = \sum_{N_+=0}^{\infty} z^{N_+} \sum_{N_{++}}' g(N_+, N_{++})e^{\beta\epsilon_0 N_{++}} \tag{16.19}$$

A comparison between (16.19) and (16.11) yields the accompanying table of correspondence. Hence a solution of the Ising model can be immediately transcribed to be a solution of the lattice gas.

Ising Model		*Lattice Gas*
N_+	$\longleftrightarrow$	N_a
N_-	$\longleftrightarrow$	$N - N_a$
4ϵ	$\longleftrightarrow$	ϵ_0
$e^{2\beta(\epsilon\gamma - B)}$	$\longleftrightarrow$	z
$-\left(\frac{A_I}{N} + \frac{1}{2}\gamma\epsilon - B\right)$	$\longleftrightarrow$	P_G
$\frac{1}{2}\left(\frac{M_I}{N} + 1\right)$	$\longleftrightarrow$	$\frac{1}{v}$

The lattice gas does not directly correspond to any real system in nature. If we allow the lattice constant to approach zero, however, and then add to the resulting equation of state the pressure of an ideal gas, the model corresponds to a real gas of atoms interacting with one another through a zero-range potential. Thus it may be interesting to study the phase transition of a lattice gas.

The lattice gas has also been used as a model for the melting of a crystal lattice. When it is so used, however, the lattice constant must be kept finite. The kinetic energy of the atoms in the crystal lattice is appended in some *ad hoc* fashion. Such a model would only have a mathematical interest, because it is not clear that it describes melting.

Binary Alloy

Before introducing a model for the binary alloy, let us describe some of the salient features of an actual binary alloy, β-brass, which is a body-centered cubic lattice made up of Zn and Cu atoms. A unit cell of this lattice in its completely ordered state, which exists only at absolute zero,

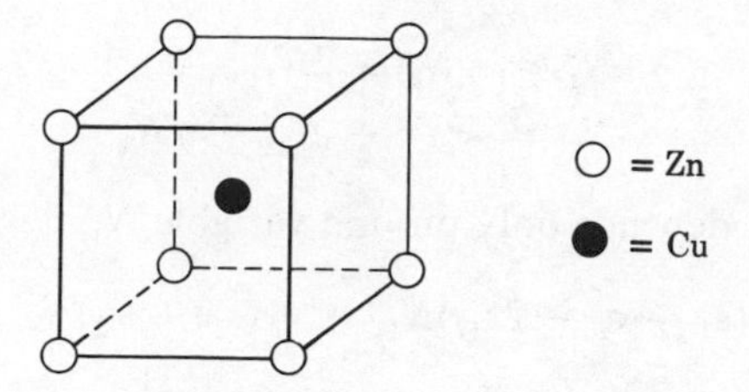

Fig. 16.3. Body-centered cubic lattice of β-brass.

is shown in Fig. 16.3, where an open circle indicates a Zn atom and a solid circle a Cu atom. As the temperature is increased some Zn atoms will exchange positions with Cu atoms, but the probability of finding a Zn atom in the "right" place is greater than $\frac{1}{2}$. Above the critical temperature of 742°K, however, the Zn and Cu atoms are thoroughly mixed, and the probability of finding a Zn atom in the "right" place becomes exactly $\frac{1}{2}$. This transition can be discovered experimentally through the Bragg reflection of x rays from the crystal. In the ordered state x ray reflection will reveal that there are two sets of atomic planes with spacing d, whereas in the state of disorder there is only one set of atomic planes with spacing $d/2$. It is observed experimentally that the specific heat c_P approaches infinity as the temperature approaches the critical temperature from both sides.

A model for a binary alloy follows. Let there be two kinds of atoms, called 1 and 2, of which there are N_1 and N_2 respectively. Let their positions be confined to the lattice sites of a given lattice, with γ nearest neighbors to each lattice site. At each lattice site there shall be one and only one atom. Thus the total number of sites is $N = N_1 + N_2$. There are three types of nearest-neighbor pair: (11), (22), and (12). The pair (12) is not distinguished from the pair (21). Let a configuration of the system be such that the number of pairs of each type present is respectively N_{11}, N_{22}, N_{12}. Neglecting the kinetic energy of the atoms and all but nearest-neighbor interactions, we take the energy of the system to be

$$E_A(N_{11}, N_{22}, N_{12}) = \epsilon_1 N_{11} + \epsilon_2 N_{22} + \epsilon_{12} N_{12} \tag{16.20}$$

where the subscript A stands for "alloy." Obviously E_A is in general degenerate. Moreover, the numbers N_{11}, N_{22}, N_{12} are not independent of one another. By analogy with (16.7) we have the relations

$$\begin{aligned} \gamma N_1 &= 2N_{11} + N_{12} \\ \gamma N_2 &= 2N_{22} + N_{12} \\ N_1 + N_2 &= N \end{aligned} \tag{16.21}$$

Thus

$$\begin{aligned} N_{12} &= \gamma N_1 - 2N_{11} \\ N_{22} &= \tfrac{1}{2}\gamma N + N_{11} - \gamma N_1 \end{aligned} \tag{16.22}$$

Hence the energy depends only on one variable N_{11}:

$$E_A(N_{11}) = (\epsilon_1 + \epsilon_2 - 2\epsilon_{12})N_{11} + [\gamma(\epsilon_{12} - \epsilon_2)N_1 + \tfrac{1}{2}\gamma\epsilon_2 N] \tag{16.23}$$

where the term in brackets is a constant. A correspondence between the binary alloy and the lattice gas may be established by identifying N_1 with N_a, which in turn is identified with N_+ of the Ising model. Comparison of (16.23) and the exponent in (16.19) immediately leads to the correspondence summarized in the accompanying table.

Lattice Gas		*Binary Alloy*
N_a	$\longleftrightarrow$	N_1
$N - N_a$	$\longleftrightarrow$	$N - N_1 = N_2$
$-\epsilon_0$	$\longleftrightarrow$	$\epsilon_1 + \epsilon_2 - 2\epsilon_{12}$
(Helmholtz free energy)	$\longleftrightarrow$	(Helmholtz free energy) $+ [\gamma(\epsilon_{12} - \epsilon_2)N_1 + \frac{1}{2}\gamma\epsilon_2 N]$

16.3 BRAGG-WILLIAMS APPROXIMATION

In the Ising model the energy of a configuration of the spin lattice depends not on the detailed distribution of spins over lattice sites but only on the two numbers N_+ and N_{++}, which express certain gross features of the spin distribution. The number N_+/N is said to be a measure of the "long-range order" in the lattice, and $N_{++}/(\gamma N/2)$ is said to be a measure of the "short-range order." The reason for this terminology is as follows. Let us imagine that the distribution of spin is random, except for the restriction that it possesses the given values of N_+ and N_{++}. If we know definitely that a given spin is up, then the number $N_{++}/(\gamma N/2)$ is the fraction of its nearest neighbors with spin up. This number, however, imposes less and less correlation as we consider the second-nearest neighbors, third-nearest neighbors, etc. It is therefore a measure of the local correlation of spins; hence the name short-range order. On the other hand, the number N_+/N requires no correlation between nearest neighbors. It does, however, require that in the entire lattice a fraction N_+/N of all the spins are up. Thus if the number N_+/N is known in the neighborhood of a given spin, we will know that no matter how far we go away from the given spin the order measured by it is the same. Hence the name long-range order.

We define the parameter of long-range order L and that of short-range order σ:

$$\frac{N_+}{N} \equiv \frac{1}{2}(L+1) \qquad (-1 \leq L \leq +1)$$
$$\frac{N_{++}}{\frac{1}{2}\gamma N} \equiv \frac{1}{2}(\sigma+1) \qquad (-1 \leq \sigma \leq +1) \tag{16.24}$$

From (16.9) we see that

$$\sum_{\langle ij \rangle} s_i s_j = \tfrac{1}{2}\gamma N(2\sigma - 2L + 1)$$
$$\sum_{i=1}^{N} s_i = NL \tag{16.25}$$

Thus the ensemble average of the long-range order is the magnetization per particle. The energy per spin is, by (16.1),

$$\frac{1}{N} E_I(L,\sigma) = -\tfrac{1}{2}\epsilon\gamma(2\sigma - 2L + 1) - BL \tag{16.26}$$

The Bragg-Williams approximation* is contained in the statement that "there is no short-range order apart from that which follows from long-range order." More precisely, the approximation consists of putting $N_{++}/(\frac{1}{2}\gamma N) \approx (N_+/N)^2$ or

$$\sigma \approx \tfrac{1}{2}(L+1)^2 - 1 \tag{16.27}$$

In this approximation the energy becomes

$$\frac{1}{N} E_I(L) \approx -\tfrac{1}{2}\epsilon\gamma L^2 - BL \tag{16.28}$$

This approximation clearly has heuristic value, but it is difficult to estimate the error involved.

With (16.28) the partition function (16.2) becomes

$$Q_I(B, T) = \sum_{\{s_i\}} e^{\beta N(\frac{1}{2}\epsilon\gamma L^2 + BL)} \tag{16.29}$$

The sum extends over all sets $\{s_i\}$, but the summand depends only on L. Hence we want to find the number of sets $\{s_i\}$ that share the same L. According to (16.24), L is determined by N_+. The number we seek is the number of ways to pick N_+ things out of N, namely $N!/N_+!\,(N - N_+)!$. Therefore

$$Q_I(B, T) = \sum_{L=-1}^{+1} \frac{N!}{[\frac{1}{2}N(1+L)]!\,[\frac{1}{2}N(1-L)]!} e^{\beta N(\frac{1}{2}\epsilon\gamma L^2 + BL)} \tag{16.30}$$

As $N \to \infty$ the logarithm of Q_I is equal to the logarithm of the largest

* W. L. Bragg and E. J. Williams, *Proc. Roy. Soc., A*, **145**, 699 (1934).

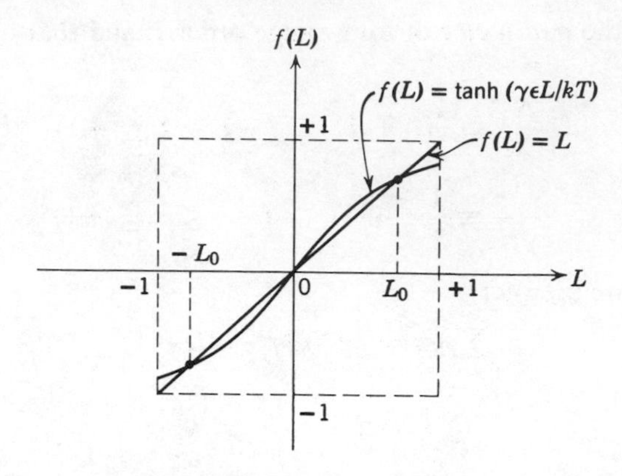

Fig. 16.4. Graphical solution of (16.35).

term in the summand. Using Sterling's approximation for $N!$ we find that

$$\frac{1}{N}\log Q_I(B,T) = \beta(\tfrac{1}{2}\epsilon\gamma\bar{L}^2 + B\bar{L}) - \frac{1+\bar{L}}{2}\log\frac{1+\bar{L}}{2} - \frac{1-\bar{L}}{2}\log\frac{1-\bar{L}}{2} \tag{16.31}$$

where $\bar{L}$ is the value of L that maximizes the summand of (16.30). We easily find that $\bar{L}$ is the root of the equation

$$\log\frac{1+\bar{L}}{1-\bar{L}} = 2\beta B + 2\beta\epsilon\gamma\bar{L} \tag{16.32}$$

which is equivalent to

$$\bar{L} = \tanh\left(\frac{B}{kT} + \frac{\gamma\epsilon\bar{L}}{kT}\right) \tag{16.33}$$

Thus (16.31) can also be rewritten

$$\frac{1}{N}A_I(B,T) = -\frac{kT}{N}\log Q_I(B,T) = \frac{\epsilon\gamma}{2}\bar{L}^2 + \frac{kT}{2}\log\frac{1-\bar{L}^2}{4} \tag{16.34}$$

We consider the case of no external magnetic field ($B = 0$). Then (16.33) becomes

$$\bar{L} = \tanh\left(\frac{\gamma\epsilon\bar{L}}{kT}\right) \tag{16.35}$$

which may be solved graphically as illustrated in Fig. 16.4. The main feature of the solution is that

$$\bar{L} = 0 \qquad \left(\frac{\gamma\epsilon}{kT} < 1\right)$$

$$\bar{L} = \begin{cases} L_0 \\ 0 \\ -L_0 \end{cases} \qquad \left(\frac{\gamma\epsilon}{kT} > 1\right)$$

In the second case the root $\bar{L} = 0$ must be rejected, because substituting it into (16.31) shows that it corresponds to a minimum instead of a maximum. If $\epsilon > 0$, there exists a critical temperature T_c, given by

$$kT_c = \gamma\epsilon \tag{16.36}$$

such that

$$\bar{L} = \begin{cases} 0 & (T > T_c) \\ \pm L_0 & (T < T_c) \end{cases} \tag{16.37}$$

where L_0 is the root of (16.35) that is greater than zero. Since $\bar{L}$ is the magnetization per particle, we immediately see that for $T < T_c$ the system is a ferromagnet, whereas for $T > T_c$ it has no magnetization. The temperature T_c is the Curie temperature of the system. The degeneracy $\bar{L} = \pm L_0$ arises from the fact that in the absence of an external magnetic field there is no intrinsic distinction between "up" and "down." This degeneracy has no effect on the free energy, which is an even function of $\bar{L}$.

In general L_0 must be computed numerically, but near $T = 0$ and $T = T_c$ an approximation can be easily worked out:

$$\begin{aligned} L_0 &\approx 1 - 2e^{-2T_c/T} & &\left(\frac{T_c}{T} \ll 1\right) \\ L_0 &\approx \sqrt{3\left(1 - \frac{T}{T_c}\right)} & &\left(0 < 1 - \frac{T}{T_c} \ll 1\right) \end{aligned} \tag{16.38}$$

A graph of L_0 is shown in Fig. 16.5.

The thermodynamic functions are summarized next:

$$\frac{1}{N} A_I(0, T) = \begin{cases} 0 & (T > T_c) \\ \dfrac{\gamma\epsilon}{2} L_0^2 + \dfrac{kT}{2} \log \dfrac{1 - L_0^2}{4} & (T < T_c) \end{cases} \tag{16.39}$$

$$\frac{1}{N} M_I(0, T) = \begin{cases} 0 & (T > T_c) \\ L_0 & (T < T_c) \end{cases} \tag{16.40}$$

$$\frac{1}{N} U_I(0, T) = \begin{cases} 0 & (T > T_c) \\ -\dfrac{\epsilon\gamma}{2} L_0^2 & (T < T_c) \end{cases} \tag{16.41}$$

$$\frac{1}{Nk} C_I(0, T) = \begin{cases} 0 & (T > T_c) \\ -\dfrac{\epsilon\gamma}{2} \dfrac{dL_0^2}{dT} & (T < T_c) \end{cases} \tag{16.42}$$

Using (16.38) we obtain

$$\frac{1}{Nk} C_I(0, T_c) = \tfrac{3}{2} \tag{16.43}$$

A graph of C_I is shown in Fig. 16.6. Above the critical temperature the specific heat vanishes. This is a consequence of the fact that both the long-range and the short-range order vanish in that region in the present approximation.

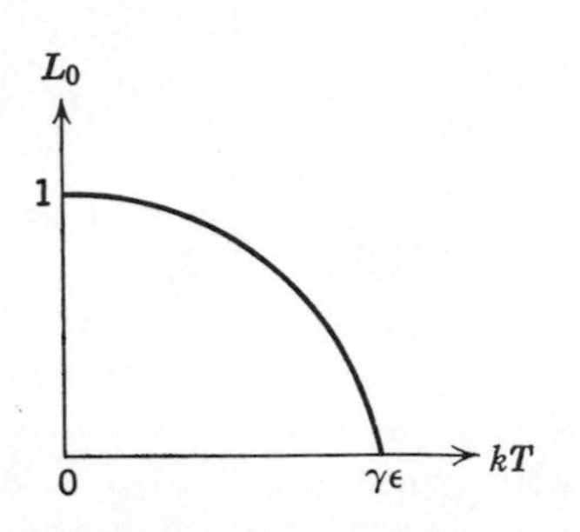

Fig. 16.5. Spontaneous magnetization in the Bragg-Williams approximation.

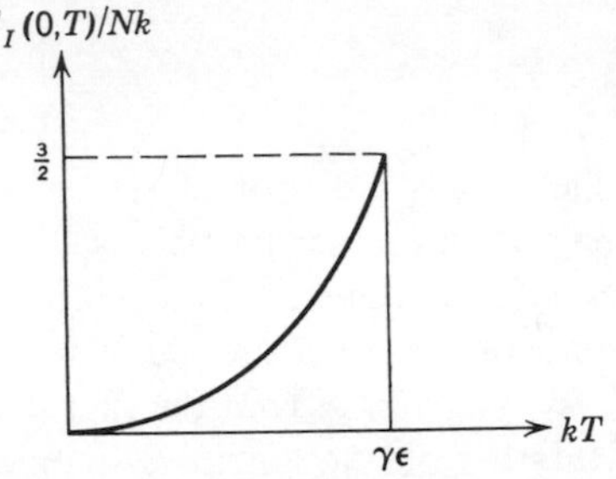

Fig. 16.6. Specific heat in the Bragg-Williams approximation.

We now turn to the lattice gas. The equation of state for the lattice gas in the Bragg-Williams approximation can be immediately obtained through the use of the table of correspondence in Sec. 16.2:

$$\begin{cases} P_G = B - \dfrac{\epsilon_0\gamma}{8}(1 + \bar{L}^2) - \dfrac{kT}{2}\log\left(\dfrac{1 - \bar{L}^2}{4}\right) \\ \dfrac{1}{v} = \dfrac{1}{2}(1 + \bar{L}) \end{cases} \tag{16.44}$$

where B is a free parameter, related to the fugacity z by the equation

$$B = \frac{\epsilon_0\gamma}{4} - \frac{kT}{2}\log z \tag{16.45}$$

and $\bar{L}$ is a function of B and T to be found by solving (16.33). To find P_G as a function of T and v we must eliminate B from (16.44). The explicit solution of the Ising model in the absence of external field ($B = 0$) corresponds only to a restricted region in the P–v diagram.

For $B = 0$ we have

$$\bar{L} = \begin{cases} 0 & (T > T_c) \\ \pm L_0 & (T < T_c) \end{cases} \tag{16.46}$$

where $T_c = \gamma\epsilon_0/4$ and L_0 is a function of temperature alone, being the nonzero root of the equation

$$L_0 = \tanh\left(\frac{T_c}{T}L_0\right) \tag{16.47}$$

If $L_0 \neq 0$ is a solution, so is $-L_0$. Thus fixing the temperature determines two values of v but only one value of P_G (since P_G is an even function of L_0). Without going into the details we present the results in the P–v diagram of Fig. 16.7. The solid curves indicate the points on the equation of state surface that correspond to $B = 0$. The two points marked T_1

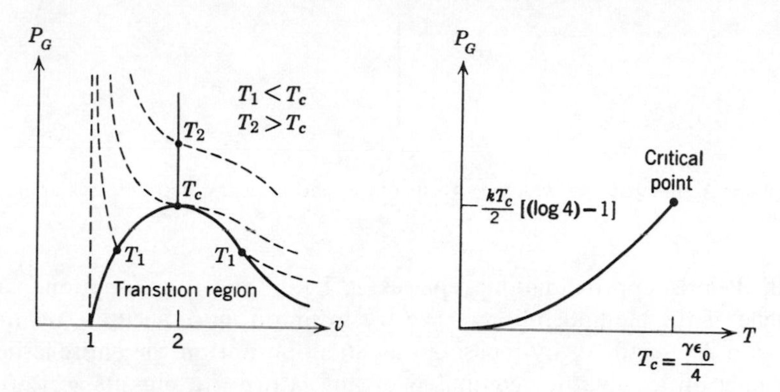

Fig. 16.7. Equations of state of the lattice gas in the Bragg-Williams approximation.

represent two points on the same isotherm with $T_1 < T_c$, whereas the point marked T_2 lies on an isotherm with $T_2 > T_c$. Thus with $B = 0$ it is only possible to obtain two points on each isotherm for $T < T_c$ and one point for $T > T_c$. To obtain a complete isotherm we must consider the case $B \neq 0$. We then obtain the isotherms represented by the dotted curves. In the Bragg-Williams approximation, the transition region is empty.* This indicates that the Bragg-Williams approximation is not satisfactory. We can see, however, that there is a first-order transition in the lattice gas and that T_c is the critical temperature of the transition region. It is to be noted that $v = 1$ is the smallest specific volume possible, because the volume of a unit cell of the lattice has been chosen to be unity. The profile of the transition region is shown in the P–T diagram of Fig. 16.7.

16.4 BETHE-PEIERLS APPROXIMATION

The Bethe-Peierls approximation is an improvement over the Bragg-Williams approximation, in that the former takes into account specific short-range order.

In the Bragg-Williams approximation the assumption $N_{++}/\frac{1}{2}\gamma N = (N_+/N)^2$ ignores the possibility of local correlation between spins. The

* Compare Fig. 8.8.

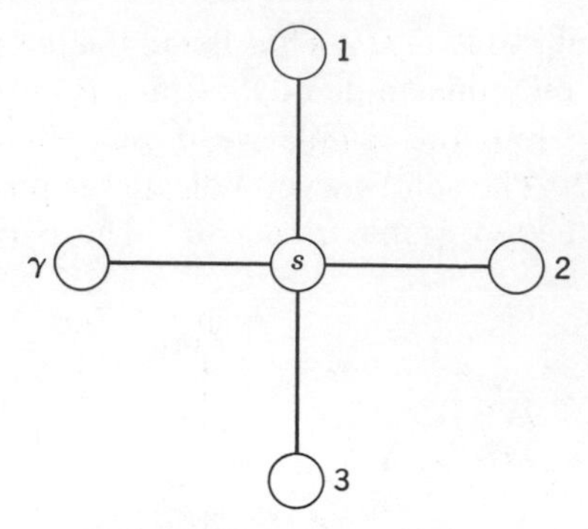

Fig. 16.8. A sublattice on which we focus our attention in the Bethe-Peierls approximation.

Bethe-Peierls approximation replaces it by a better assumption. An outline of the method follows. We try to find a more accurate relation between N_{++} and N_+ by focusing our attention not on the entire lattice, but only on a sublattice composed of any lattice site plus its γ nearest neighbors. We form the mental picture of this sublattice being "immersed" in the background provided by the rest of the lattice, just as a small-volume element in a liquid is "immersed" in the background provided by the rest of the liquid. It is assumed that the background influences the sublattice only through a single parameter that is similar to the fugacity in a liquid. The relation between N_{++} and N_+ will be obtained for the sublattice through heuristic arguments. It is then assumed that the same relation holds throughout the entire lattice.

We consider only the case $B = 0$. To begin, consider a sublattice consisting of any lattice site whose spin state is denoted by s, together with its γ nearest neighbors, as shown in Fig. 16.8. Let $P(s, n)$ be the probability that n of the nearest neighbors are in a spin-up state while the center site has the spin state s. If $s = +1$, then $P(s, n)$ refers to configurations of the sublattice in which there are n pairs $(+\,+)$ and $\gamma - n$ pairs $(+\,-)$. If $s = -1$, then $P(s, n)$ refers to configurations in which there are n pairs $(+\,-)$ and $n - \gamma$ pairs $(-\,-)$. For a given n there are $\binom{\gamma}{n}$ ways to decide which of the γ neighbors are the n spins in question. Thus we *assume* that

$$P(+1, n) = \frac{1}{q}\binom{\gamma}{n} e^{\beta\epsilon(2n-\gamma)} z^n \tag{16.48}$$

$$P(-1, n) = \frac{1}{q}\binom{\gamma}{n} e^{\beta\epsilon(\gamma-2n)} z^n \tag{16.49}$$

where q is a normalization factor and z is introduced to represent the effect of the background formed by the rest of the lattice. On account of the similarity between z and the fugacity, this method is also known as the *quasi-chemical method.* To determine q we require that

$$\sum_{n=0}^{\gamma} [P(+1, n) + P(-1, n)] = 1 \tag{16.50}$$

This leads to

$$\begin{aligned} q &= \sum_{n=0}^{\gamma} \binom{\gamma}{n} [(ze^{2\beta\epsilon})^n e^{-\beta\epsilon\gamma} + (ze^{-2\beta\epsilon})^n e^{\beta\epsilon\gamma}] \\ &= (e^{\beta\epsilon} + ze^{-\beta\epsilon})^\gamma + (ze^{\beta\epsilon} + e^{-\beta\epsilon})^\gamma \end{aligned} \tag{16.51}$$

It follows from the meaning given to $P(+1, n)$ that

$$\frac{1+L}{2} \equiv \frac{N_+}{N} = \sum_{n=0}^{\gamma} P(+1, n) = \frac{1}{q}(e^{\beta\epsilon} + ze^{-\beta\epsilon})^\gamma \tag{16.52}$$

$$\frac{1+\sigma}{2} \equiv \frac{N_{++}}{\frac{1}{2}\gamma N} = \frac{1}{\gamma}\sum_{n=0}^{\gamma} nP(+1, n) = \frac{z}{q} e^{\beta\epsilon}(e^{-\beta\epsilon} + ze^{\beta\epsilon})^{\gamma-1} \tag{16.53}$$

These equations express L and σ in terms of a single variable z. Since the energy of the Ising lattice depends on L and σ, we have an expression of the energy in terms of a single parameter z if we assume that (16.52) and (16.53) hold throughout the lattice. We may then use the resulting expression for the energy to calculate the partition function. This completes the statement of the Bethe-Peierls approximation.

It is not necessary to calculate the partition function, because there is a simpler procedure to obtain the magnetization. The interpretation that (16.48) and (16.49) are probabilities leads to the interpretation

$$\sum_{n=0}^{\gamma} P(+1, n) = \text{probability of finding an up spin at the center}$$

$$\frac{1}{\gamma}\sum_{n=0}^{\gamma} n[P(+1, n) + P(-1, n)] = \text{probability of finding an up spin among the neighbors}$$

Since these probabilities are not conditioned by the knowledge of anything else, they must be equal to each other for the interpretation to be consistent. Thus we require that

$$\sum_{n=0}^{\gamma} P(+1, n) = \frac{1}{\gamma}\sum_{n=0}^{\gamma} n[P(+1, n) + P(-1, n)] \tag{16.54}$$

This condition determines z. Using (16.48) and (16.49) we find that

$$\begin{aligned} (e^{-\beta\epsilon} + ze^{\beta\epsilon})^\gamma &= \frac{z}{\gamma}\frac{\partial}{\partial z}[(e^{-\beta\epsilon} + ze^{\beta\epsilon})^\gamma + (e^{\beta\epsilon} + ze^{-\beta\epsilon})^\gamma] \\ &= z[(e^{-\beta\epsilon} + ze^{\beta\epsilon})^{\gamma-1}e^{\beta\epsilon} + (e^{\beta\epsilon} + ze^{-\beta\epsilon})^{\gamma-1}e^{-\beta\epsilon}] \end{aligned}$$

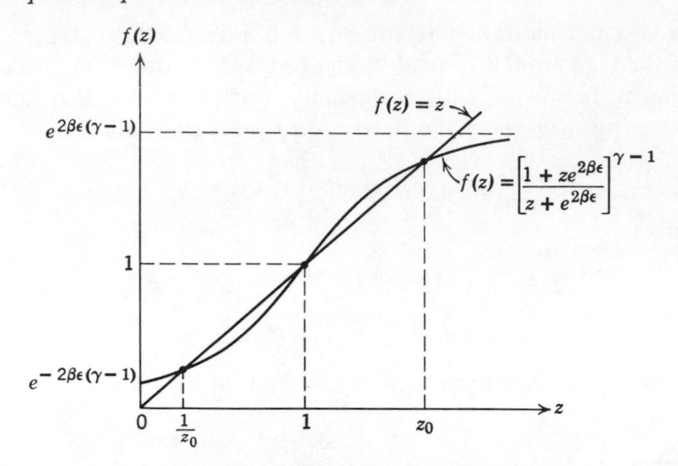

Fig. 16.9. Graphical solution of (16.55).

or

$$z = \left(\frac{1 + ze^{2\beta\epsilon}}{z + e^{2\beta\epsilon}}\right)^{\gamma-1} \tag{16.55}$$

After solving this equation for z, we can obtain $\bar{L}$ and $\bar{\sigma}$ from (16.52) and (16.53):

$$\bar{L} = \frac{z^x - 1}{z^x + 1}, \qquad x \equiv \frac{\gamma}{\gamma - 1} \tag{16.56}$$

$$\bar{\sigma} = \frac{2z^2}{(1 + ze^{-2\beta\epsilon})(1 + z^x)} - 1, \qquad x \equiv \frac{\gamma}{\gamma - 1} \tag{16.57}$$

The internal energy of the Ising lattice in the absence of a magnetic field is then given by

$$\frac{1}{N} U_I(0, T) = -\tfrac{1}{2}\epsilon\gamma(2\bar{\sigma} - 2\bar{L} + 1) \tag{16.58}$$

Thus it only remains to solve (16.55).

We note that

(*a*) $z = 1$ is always a solution of (16.55);

(*b*) if z is a solution of (16.55), then $1/z$ is also a solution of (16.55);

(*c*) interchanging z and $1/z$ interchanges $\bar{L}$ and $-\bar{L}$;

(*d*) $z = 1$ corresponds to $\bar{L} = 0$; $z = \infty$ corresponds to $\bar{L} = 1$.

The actual solution of (16.55) may be found graphically, as shown in Fig. 16.9. The slope of the right hand side of (16.55) at $z = 1$ is

$$c = \frac{(\gamma - 1)(e^{4\beta\epsilon} - 1)}{(1 + e^{2\beta\epsilon})^2} \tag{16.59}$$

Thus if $c < 1$ the only solution is $z = 1$. If $c > 1$, there are three solutions $z = 1, z_0, 1/z_0$, of which $z = 1$ is discarded by comparison with the solution in the Bragg-Williams approximation. The solution $1/z_0$ does not lead to anything new since it merely means interchanging up spin with down spin. It too will be ignored.

Let us define the critical temperature T_c by the equation

$$\frac{(\gamma - 1)e^{4\epsilon/kT_c}}{(1 + e^{2\epsilon/kT_c})^2} = 1 \tag{16.60}$$

which leads to the explicit expression

$$kT_c = \frac{2\epsilon}{\log\,[\gamma/(\gamma - 2)]} \tag{16.61}$$

For $T > T_c$ we have

$$\begin{aligned} z &= 1 \\ \bar{L} &= 0 \\ \bar{\sigma} &= \frac{1}{2(1 + e^{-2\beta\epsilon})} \end{aligned} \tag{16.62}$$

For $T < T_c$ we have

$$\begin{aligned} z &> 1 \\ \bar{L} &> 0 \end{aligned} \tag{16.63}$$

In this case we have spontaneous magnetization. Of the thermodynamic functions we only consider the specific heat, which can be shown to be

$$\frac{C_I(0, T)}{Nk} = \frac{1}{Nk}\frac{d}{dT}U_I(0, T) = -\frac{\epsilon\gamma}{Nk}\left(\frac{d\bar{\sigma}}{dT} - \frac{d\bar{L}}{dT}\right) \tag{16.64}$$

and which, in contradistinction to the Bragg-Williams result, does not

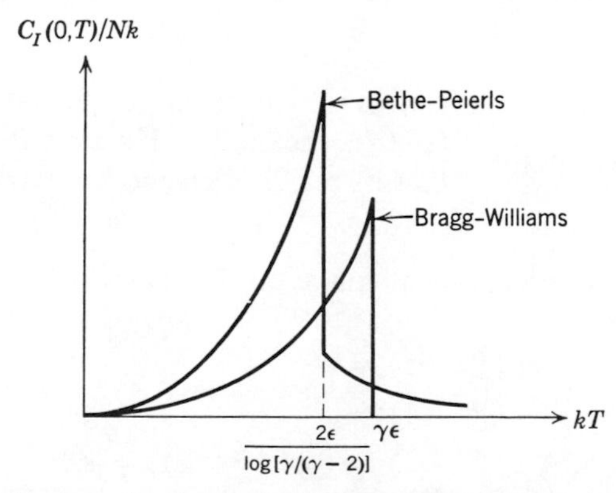

Fig. 16.10. Specific heat in the Bethe-Peierls approximation and in the Bragg-Williams approximation.

vanish for $T > T_c$:

$$\frac{C_I(0, T)}{Nk} = \frac{2\gamma\epsilon^2}{(kT)^2} \frac{e^{2\epsilon/kT}}{(1 + e^{2\epsilon/kT})^2} \qquad (T > T_c) \tag{16.65}$$

A more detailed calculation yields the graph shown in Fig. 16.10 for the specific heat. For comparison the Bragg-Williams result is also shown.

16.5 ONE-DIMENSIONAL ISING MODEL

The one-dimensional Ising model is a chain of N spins, each spin interacting only with its two nearest neighbors and with an external magnetic field. The energy for the configuration specified by $\{s_1, s_2, \ldots, s_N\}$ is

$$E_I = -\epsilon \sum_{k=1}^{N} s_k s_{k+1} - B \sum_{k=1}^{N} s_k \tag{16.66}$$

We impose the periodic boundary condition

$$s_{N+1} \equiv s_1 \tag{16.67}$$

making the topology of the chain that of a circle, as shown in Fig. 16.11. The partition function is

$$Q_I(B, T) = \sum_{s_1} \sum_{s_2} \cdots \sum_{s_N} \exp\left[\beta \sum_{k=1}^{N} (\epsilon s_k s_{k+1} + B s_k)\right] \tag{16.68}$$

where each s_k independently assumes the values ± 1.

The partition function can be expressed in terms of matrices.* Let us write

$$Q_I(B, T) = \sum_{s_1} \sum_{s_2} \cdots \sum_{s_N} \times \exp\left\{\beta \sum_{k=1}^{N} [\epsilon s_k s_{k+1} + \tfrac{1}{2} B(s_k + s_{k+1})]\right\} \tag{16.69}$$

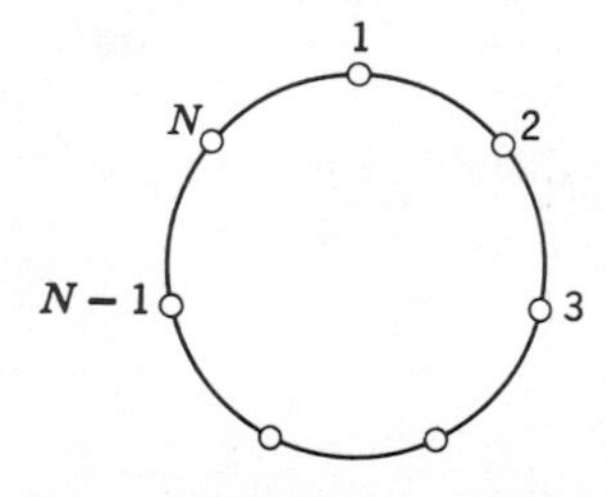

Fig. 16.11. Topology of the one-dimensional Ising lattice.

which is equivalent to (16.68) by virtue of (16.67). Let a 2×2 matrix P be so defined that its matrix elements are given by

$$\langle s \,|\, \mathsf{P} \,|\, s' \rangle = e^{\beta[\epsilon s s' + \frac{1}{2} B(s + s')]} \tag{16.70}$$

where s and s' may independently take on the values ± 1. A list of all the matrix elements is

$$\begin{aligned} \langle +1 \,|\, \mathsf{P} \,|\, +1 \rangle &= e^{\beta(\epsilon + B)} \\ \langle -1 \,|\, \mathsf{P} \,|\, -1 \rangle &= e^{\beta(\epsilon - B)} \\ \langle +1 \,|\, \mathsf{P} \,|\, -1 \rangle = \langle -1 \,|\, \mathsf{P} \,|\, +1 \rangle &= e^{-\beta\epsilon} \end{aligned} \tag{16.71}$$

* The matrix formulation of the Ising model in general is due to H. A. Kramers and G. H. Wannier, *Phys. Rev.*, **60**, 252 (1941).

Thus an explicit representation for P is

$$\mathsf{P} = \begin{bmatrix} e^{\beta(\epsilon+B)} & e^{-\beta\epsilon} \\ e^{-\beta\epsilon} & e^{\beta(\epsilon-B)} \end{bmatrix} \tag{16.72}$$

With these definitions we may rewrite (16.69) in the form

$$\begin{aligned} Q_I(B, T) &= \sum_{s_1}\sum_{s_2}\cdots\sum_{s_N} \langle s_1 | \mathsf{P} | s_2\rangle\langle s_2 | \mathsf{P} | s_3\rangle \cdots \langle s_N | \mathsf{P} | s_1\rangle \\ &= \sum_{s_1} \langle s_1 | \mathsf{P}^N | s_1\rangle = Tr\mathsf{P}^N = \lambda_+{}^N + \lambda_-{}^N \end{aligned} \tag{16.73}$$

where λ_+ and λ_- are the two eigenvalues of P, with $\lambda_+ \geq \lambda_-$. The fact that Q_I is the trace of the Nth power of a matrix is a consequence of the periodic boundary condition (16.67).

By an easy calculation we find that the two eigenvalues $\lambda_\pm$ are

$$\lambda_\pm = e^{\beta\epsilon}[\cosh(\beta B) \pm \sqrt{\cosh^2(\beta B) - 2e^{-2\beta\epsilon}\sinh(2\beta\epsilon)}\,] \tag{16.74}$$

Thus $\lambda_+ > \lambda_-$ for all B. As $N \to \infty$ only the larger one of the eigenvalues λ_+ is relevant, because

$$\frac{1}{N}\log Q_I(B, T) = \log\lambda_+ + \log\left[1 + \left(\frac{\lambda_-}{\lambda_+}\right)^N\right] \xrightarrow[N\to\infty]{} \log\lambda_+ \tag{16.75}$$

Thus the Helmholtz free energy per spin is given by

$$\frac{1}{N}A_I(B, T) = -\epsilon - kT\log[\cosh(\beta B) + \sqrt{\cosh^2(\beta B) - 2e^{-2\beta\epsilon}\sinh^2(2\beta\epsilon)}\,] \tag{16.76}$$

The magnetization per spin is

$$\frac{1}{N}M_I(B, T) = \frac{\sinh(\beta B)}{\sqrt{\cosh^2(\beta B) - 2e^{-2\beta\epsilon}\sinh^2(2\beta\epsilon)}} \tag{16.77}$$

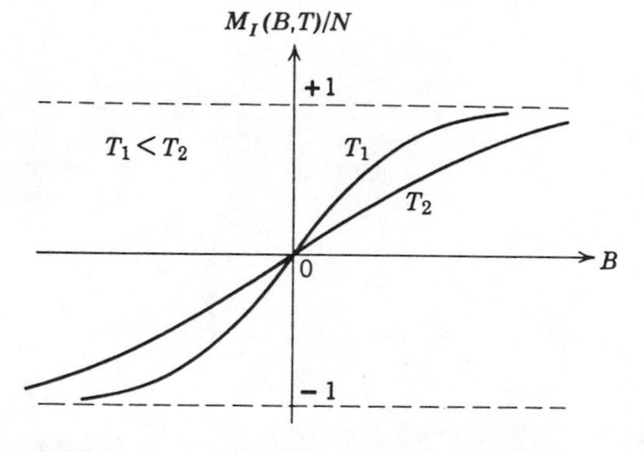

Fig. 16.12. Magnetization of the one-dimensional Ising model. There is no spontaneous magnetization.

Graphs of $N^{-1}M_I(B, T)$ for various temperatures are shown in Fig. 16.12.

For all $T > 0$,

$$\frac{1}{N} M_I(0, T) = 0 \tag{16.78}$$

Therefore the one-dimensional Ising model never exhibits ferromagnetism. The reason for this is that at any temperature the average configuration is determined by two opposite and competing tendencies: The tendency towards a complete alignment of spins to minimize the energy, and the tendency towards randomization to maximize the entropy. (The over-all tendency is to minimize the free energy $A = U - TS$.) For the one-dimensional model the tendency for alignment always loses out, because there are not enough nearest neighbors.

chapter 17

THE ONSAGER SOLUTION

17.1 FORMULATION OF TWO-DIMENSIONAL ISING MODEL

Matrix Formulation

We formulate the two-dimensional Ising model in terms of matrices as a preliminary step towards an exact solution of the model. Consider a square lattice of $N = n^2$ spins consisting of n rows and n columns, as shown in Fig. 17.1. Let us imagine the lattice to be enlarged by one row and one column with the requirement that the configuration of the $(n + 1)$th row and column be identical with that of the first row and column respectively. This boundary condition endows the lattice with the topology of a torus, as depicted in Fig. 17.2. Let μ_α $(\alpha = 1, \ldots, n)$ denote the collection of all the spin coordinates of the αth row:

$$\mu_\alpha \equiv \{s_1, s_2, \ldots, s_n\}_{\alpha\text{th row}} \tag{17.1}$$

The toroidal boundary condition implies the definition

$$\mu_{n+1} \equiv \mu_1 \tag{17.2}$$

A configuration of the entire lattice is then specified by $\{\mu_1, \ldots, \mu_n\}$. By assumption, the αth row interacts only with the $(\alpha - 1)$th and the $(\alpha + 1)$th row. Let $E(\mu_\alpha, \mu_{\alpha+1})$ be the interaction energy between the αth and the $(\alpha + 1)$th row. Let $E(\mu_\alpha)$ be the interaction energy of the spins within the

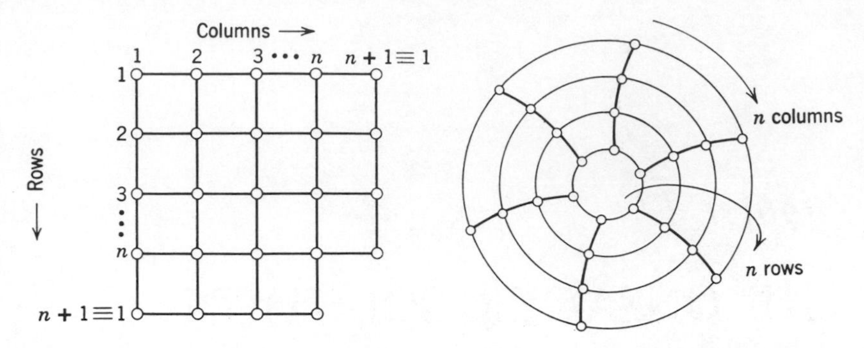

Fig. 17.1. Two-dimensional Ising lattice.

Fig. 17.2. Topology of the two-dimensional Ising lattice.

αth row plus their interaction energy with an external magnetic field. We can write

$$\begin{aligned} E(\mu, \mu') &= -\epsilon \sum_{k=1}^{n} s_k s_k' \\ E(\mu) &= -\epsilon \sum_{k=1}^{n} s_k s_{k+1} - B \sum_{k=1}^{n} s_k \end{aligned} \tag{17.3}$$

where μ and μ' respectively denote the collection of spin coordinates in two neighboring rows:

$$\begin{aligned} \mu &\equiv \{s_1, \ldots, s_n\} \\ \mu' &\equiv \{s_1', \ldots, s_n'\} \end{aligned} \tag{17.4}$$

The toroidal boundary condition implies that in each row

$$s_{n+1} \equiv s_1 \tag{17.5}$$

The total energy of the lattice for the configuration $\{\mu_1, \ldots, \mu_n\}$ is then given by

$$E_I\{\mu_1, \ldots, \mu_n\} = \sum_{\alpha=1}^{n} [E(\mu_\alpha, \mu_{\alpha+1}) + E(\mu_\alpha)] \tag{17.6}$$

The partition function is

$$Q_I(B, T) = \sum_{\mu_1} \cdots \sum_{\mu_n} \exp \{-\beta \sum_{\alpha=1}^{n} [E(\mu_\alpha, \mu_{\alpha+1}) + E(\mu_\alpha)]\} \tag{17.7}$$

Let a $2^n \times 2^n$ matrix* P be so defined that its matrix elements are

$$\langle \mu \mid \mathsf{P} \mid \mu' \rangle \equiv e^{-\beta[E(\mu,\mu')+E(\mu)]} \tag{17.8}$$

Then

$$\begin{aligned} Q_I(B, T) &= \sum_{\mu_1} \cdots \sum_{\mu_n} \langle \mu_1 \mid \mathsf{P} \mid \mu_2 \rangle \langle \mu_2 \mid \mathsf{P} \mid \mu_3 \rangle \cdots \langle \mu_n \mid \mathsf{P} \mid \mu_1 \rangle \\ &= \sum_{\mu_1} \langle \mu_1 \mid \mathsf{P}^n \mid \mu_1 \rangle = Tr \mathsf{P}^n \end{aligned} \tag{17.9}$$

* From now on all $2^n \times 2^n$ matrices are denoted by sans serif letters: P, Γ.

Since the trace of a matrix is independent of the representation of the matrix, the trace in (17.9)* may be evaluated by bringing P into its diagonal form:

$$\mathsf{P} = \begin{bmatrix} \lambda_1 & & & & \\ & \lambda_2 & & & \\ & & \cdot & & \\ & & & \cdot & \\ & & & & \cdot \\ & & & & & \lambda_{2^n} \end{bmatrix} \tag{17.10}$$

where $\lambda_1, \lambda_2, \ldots, \lambda_{2^n}$ are the 2^n eigenvalues of P. The matrix P^n is then also diagonal, with the diagonal matrix elements $(\lambda_1)^n, (\lambda_2)^n, \ldots, (\lambda_{2^n})^n$. Therefore

$$Q_I(B, T) = \sum_{\alpha=1}^{2^n} (\lambda_\alpha)^n \tag{17.11}$$

From the form of (17.8) we expect that the eigenvalues of P are in general of the order of e^n when n is large, since $E(\mu, \mu')$ and $E(\mu)$ are of the order of n. If $\lambda_{\max}$ is the largest eigenvalue of P, we expect that

$$\lim_{n\to\infty} \frac{1}{n} \log \lambda_{\max} = \text{finite number} \tag{17.12}$$

If this is true and if all the eigenvalues λ_α are positive, then

$$(\lambda_{\max})^n \le Q_I \le 2^n (\lambda_{\max})^n$$

or

$$\frac{1}{n} \log \lambda_{\max} \le \frac{1}{n^2} \log Q_I \le \frac{1}{n} \log \lambda_{\max} + \frac{1}{n} \log 2 \tag{17.13}$$

Therefore

$$\lim_{N\to\infty} \frac{1}{N} \log Q_I = \lim_{n\to\infty} \frac{1}{n} \log \lambda_{\max} \tag{17.14}$$

where $N = n^2$. It will turn out that (17.12) is true and that all the eigenvalues λ_α are positive. Thus it is sufficient to find the largest eigenvalue of P. The remaining parts of this section are devoted to the description of an explicit representation for P.

* The fact that Q_I is of the form of a trace is a consequence of (17.2) alone and does not require (17.5). In other words, to make Q_I a trace of some matrix it is only necessary to fold the two-dimensional lattice into a cylinder. The purpose of (17.5), which turns the cylinder into a torus, is to facilitate the actual diagonalization of P.

The Matrix P

From (17.8) and (17.3) we may obtain the matrix elements of P in the form

$$\langle s_1, \ldots, s_n \,|\, \mathsf{P} \,|\, s_1', \ldots, s_n' \rangle = \prod_{k=1}^{n} e^{\beta B s_k} e^{\beta \epsilon s_k s_{k+1}} e^{\beta \epsilon s_k s_k'} \tag{17.15}$$

Let us define three $2^n \times 2^n$ matrices V_1', V_2, and V_3 whose matrix elements are respectively given by

$$\langle s_1, \ldots, s_n \,|\, \mathsf{V}_1' \,|\, s_1', \ldots, s_n' \rangle \equiv \prod_{k=1}^{n} e^{\beta \epsilon s_k s_k'} \tag{17.16}$$

$$\langle s_1, \ldots, s_n \,|\, \mathsf{V}_2 \,|\, s_1', \ldots, s_n' \rangle \equiv \delta_{s_1 s_1'} \cdots \delta_{s_n s_n'} \prod_{k=1}^{n} e^{\beta \epsilon s_k s_{k+1}} \tag{17.17}$$

$$\langle s_1, \ldots, s_n \,|\, \mathsf{V}_3 \,|\, s_1', \ldots, s_n' \rangle \equiv \delta_{s_1 s_1'} \cdots \delta_{s_n s_n'} \prod_{k=1}^{n} e^{\beta B s_k} \tag{17.18}$$

where $\delta_{ss'}$ is the Kronecker symbol. Thus V_2 and V_3 are diagonal matrices in the present representation. It is easily verified that

$$\mathsf{P} = \mathsf{V}_3 \mathsf{V}_2 \mathsf{V}_1' \tag{17.19}$$

in the usual sense of matrix multiplication, namely

$$\langle s_1, \ldots, s_n \,|\, \mathsf{P} \,|\, s_1', \ldots, s_n' \rangle = \sum_{s_1'', \ldots, s_n''} \sum_{s_1''', \ldots, s_n'''} \langle s_1, \ldots, s_n \,|\, \mathsf{V}_3 \,|\, s_1'', \ldots, s_n'' \rangle \cdot$$
$$\langle s_1'', \ldots, s_n'' \,|\, \mathsf{V}_2 \,|\, s_1''', \ldots, s_n''' \rangle \langle s_1''', \ldots, s_n''' \,|\, \mathsf{V}_1' \,|\, s_1', \ldots, s_n' \rangle$$

Direct Product of Matrices

Before describing a convenient way to represent the matrices V_3, V_2, and V_1 we introduce the notion of a direct product of matrices. Let A and B be two $m \times m$ matrices whose matrix elements are respectively $\langle i \,|\, A \,|\, j \rangle$ and $\langle i \,|\, B \,|\, j \rangle$, where i and j independently take on the values $1, 2, \ldots, m$. Then the direct product $A \times B$ is the $m^2 \times m^2$ matrix whose matrix elements are

$$\langle ii' \,|\, A \times B \,|\, jj' \rangle \equiv \langle i \,|\, A \,|\, j \rangle \langle i' \,|\, B \,|\, j' \rangle \tag{17.20}$$

This definition can be immediately extended to define the direct product $A \times B \times \cdots \times C$ of any number of $m \times m$ matrices $A, B, \ldots, C$:

$$\langle ii' \cdots i'' \,|\, A \times B \times \cdots \times C \,|\, jj' \cdots j'' \rangle$$
$$\equiv \langle i \,|\, A \,|\, j \rangle \langle i' \,|\, B \,|\, j' \rangle \cdots \langle i'' \,|\, C \,|\, j'' \rangle \tag{17.21}$$

If AB denotes the product of the matrices A and B under ordinary matrix multiplication, then

$$(A \times B)(C \times D) = (AC) \times (BD) \tag{17.22}$$

To prove this, take matrix elements of the left-hand side:

$$
\begin{aligned}
\langle ii' \mid (A \times B)(C \times D) \mid jj' \rangle &= \sum_{kk'} \langle ii' \mid A \times B \mid kk' \rangle \langle kk' \mid C \times D \mid jj' \rangle \\
&= \sum_{k} \langle i \mid A \mid k \rangle \langle k \mid C \mid j \rangle \cdot \sum_{k'} \langle i' \mid B \mid k' \rangle \langle k' \mid C \mid j' \rangle \\
&= \langle i \mid AC \mid j \rangle \langle i' \mid BD \mid j' \rangle \\
&= \langle ii' \mid (AC) \times (BD) \mid jj' \rangle
\end{aligned}
$$

A generalization of (17.22) can be proved in the same way:

$$
(A \times B \times \cdots \times C)(D \times E \times \cdots \times F) = (AD) \times (BE) \times \cdots \times (CF) \tag{17.23}
$$

Spin Matrices

We now introduce some special matrices in terms of which V_1', V_2, and V_3 may be conveniently expressed. Let the three familiar 2×2 Pauli spin matrices be denoted by X, Y, and Z:

$$
X \equiv \begin{pmatrix} 0 & 1 \\ 1 & 0 \end{pmatrix}, \qquad Y \equiv \begin{pmatrix} 0 & -i \\ i & 0 \end{pmatrix}, \qquad Z \equiv \begin{pmatrix} 1 & 0 \\ 0 & -1 \end{pmatrix} \tag{17.24}
$$

The following properties are easily verified

$$
\begin{gathered}
X^2 = 1, \qquad Y^2 = 1, \qquad Z^2 = 1 \\
XY + YX = 0, \qquad YZ + ZY = 0, \qquad ZX + XZ = 0 \\
XY = iZ, \qquad YZ = iX, \qquad ZX = iY
\end{gathered} \tag{17.25}
$$

Let three sets of $2^n \times 2^n$ matrices X_α, Y_α, Z_α ($\alpha = 1, \ldots, n$) be defined as follows:*

$$
\begin{aligned}
\mathsf{X}_\alpha &\equiv 1 \times 1 \times \cdots \times X \times \cdots \times 1 \qquad (n \text{ factors}) \\
\mathsf{Y}_\alpha &\equiv 1 \times 1 \times \cdots \times Y \times \cdots \times 1 \qquad (n \text{ factors}) \\
\mathsf{Z}_\alpha &\equiv 1 \times 1 \times \cdots \times \underset{\substack{\uparrow \\ \alpha\text{th factor}}}{Z} \times \cdots \times 1 \qquad (n \text{ factors})
\end{aligned} \tag{17.26}
$$

For $\alpha \neq \beta$ we can easily verify that

$$
\begin{aligned}
[\mathsf{X}_\alpha, \mathsf{X}_\beta] = [\mathsf{Y}_\alpha, \mathsf{Y}_\beta] = [\mathsf{Z}_\alpha, \mathsf{Z}_\beta] = 0 \\
[\mathsf{X}_\alpha, \mathsf{Y}_\beta] = [\mathsf{X}_\alpha, \mathsf{Z}_\beta] = [\mathsf{Y}_\alpha, \mathsf{Z}_\beta] = 0
\end{aligned} \tag{17.27}
$$

* The matrices X_α, Y_α, Z_α are familiar in quantum mechanics. For example, for a system of n nonrelativistic electrons the spin matrices for the αth electron are precisely X_α, Y_α, and Z_α.

For any given α the $2^n \times 2^n$ matrices X_α, Y_α, Z_α formally satisfy all the relations (17.25).

The following identity holds for any matrix X whose square is the unit matrix

$$e^{\theta X} = \cosh\theta + X\sinh\theta \tag{17.28}$$

where θ is a number. The proof is as follows. Since $X^n = 1$ if n is even, and $X^n = X$ if n is odd,

$$e^{\theta X} = \sum_{n=0}^{\infty} \frac{\theta^n}{n!} X^n = \sum_{n\text{ even}} \frac{\theta^n}{n!} + X \sum_{n\text{ odd}} \frac{\theta^n}{n!} = \cosh\theta + X\sinh\theta$$

In particular (17.28) is satisfied separately by X, Y, Z and by X_α, Y_α, Z_α ($\alpha = 1, \ldots, n$).

The Matrices V_1', V_2, and V

By inspection of (17.16) it is clear that V_1' is a direct product of n 2×2 identical matrices:

$$\mathsf{V}_1' = \boldsymbol{a} \times \boldsymbol{a} \times \cdots \times \boldsymbol{a} \tag{17.29}$$

where

$$\langle s \mid \boldsymbol{a} \mid s' \rangle = e^{\beta\epsilon s s'} \tag{17.30}$$

Therefore

$$\boldsymbol{a} = \begin{bmatrix} e^{\beta\epsilon} & e^{-\beta\epsilon} \\ e^{-\beta\epsilon} & e^{\beta\epsilon} \end{bmatrix} = e^{\beta\epsilon} + e^{-\beta\epsilon} X \tag{17.31}$$

Using (17.28) we obtain

$$\boldsymbol{a} = \sqrt{2\sinh(2\beta\epsilon)}\, e^{\theta X} \tag{17.32}$$

where

$$\tanh\theta \equiv e^{-2\beta\epsilon} \tag{17.33}$$

Hence

$$\mathsf{V}_1' = [2\sinh(2\beta\epsilon)]^{n/2} e^{\theta X} \times e^{\theta X} \times \cdots \times e^{\theta X} \tag{17.34}$$

The following identity can be verified by a direct calculation of matrix elements:

$$e^{\theta X} \times e^{\theta X} \times \cdots \times e^{\theta X} = e^{\theta \mathsf{X}_1} e^{\theta \mathsf{X}_2} \cdots e^{\theta \mathsf{X}_n} = e^{\theta(\mathsf{X}_1 + \mathsf{X}_2 + \cdots + \mathsf{X}_n)} \tag{17.35}$$

Applying (17.35) to (17.34) we obtain

$$\mathsf{V}_1' = [2\sinh(2\beta\epsilon)]^{n/2} \mathsf{V}_1 \tag{17.36}$$

$$\mathsf{V}_1 = \prod_{\alpha=1}^{n} e^{\theta \mathsf{X}_\alpha}, \qquad \tanh\theta \equiv e^{-2\beta\epsilon} \tag{17.37}$$

A straightforward calculation of matrix elements shows that

$$\mathsf{V}_2 = \prod_{\alpha=1}^{n} e^{\beta\epsilon \mathsf{Z}_\alpha \mathsf{Z}_{\alpha+1}} \tag{17.38}$$

$$, \quad \mathsf{Z}_{n+1} \equiv \mathsf{Z}_1$$

$$\mathsf{V}_3 = \prod_{\alpha=1}^{n} e^{\beta B \mathsf{Z}_\alpha} \tag{17.39}$$

Therefore

$$\mathsf{P} = [2 \sinh (2\beta\epsilon)]^{n/2} \mathsf{V}_3 \mathsf{V}_2 \mathsf{V}_1 \tag{17.40}$$

For the case $B = 0$, $\mathsf{V}_3 = 1$. This completes the formulation of the two-dimensional Ising model.

17.2 MATHEMATICAL DIGRESSION

The following study of a general class of matrices is relevant to the solution of the two-dimensional Ising model in the absence of magnetic field ($B = 0$).

Let $2n$ matrices Γ_μ ($\mu = 1, \ldots, 2n$) be defined as a set of matrices satisfying the following anticommutation rule

$$\Gamma_\mu \Gamma_\nu + \Gamma_\nu \Gamma_\mu = 2\,\delta_{\mu\nu} \qquad \begin{matrix} (\mu = 1, \ldots, 2n) \\ (\nu = 1, \ldots, 2n) \end{matrix} \tag{17.41}$$

The following properties of $\{\Gamma_\mu\}$ are stated without proof.*

(*a*) The dimensionality of Γ_μ cannot be smaller than $2^n \times 2^n$.

(*b*) If $\{\Gamma_\mu\}$ and $\{\Gamma_\mu'\}$ are two sets of matrices satisfying (17.41), there exists a nonsingular matrix S such that $\Gamma_\mu = \mathsf{S}\Gamma_\mu'\mathsf{S}^{-1}$. The converse is obviously true.

(*c*) Any $2^n \times 2^n$ matrix is a linear combination of the unit matrix, the matrices Γ_μ (chosen to be $2^n \times 2^n$), and all the independent products $\Gamma_\mu\Gamma_\nu$, $\Gamma_\mu\Gamma_\nu\Gamma_\lambda$, ...

For $n = 1$, (17.41) defines two of the 2×2 Pauli spin matrices from which the third can be obtained as their product. It is obvious that any 2×2 matrix is a linear combination of the unit matrix and the Pauli spin matrices. For $n = 2$, (17.41) defines the four 4×4 Dirac matrices γ_μ.

A possible representation of $\{\Gamma_\mu\}$ by $2^n \times 2^n$ matrices is

$$\begin{array}{ll} \Gamma_1 = \mathsf{Z}_1 & \Gamma_2 = \mathsf{Y}_1 \\ \Gamma_3 = \mathsf{X}_1\mathsf{Z}_2 & \Gamma_4 = \mathsf{X}_1\mathsf{Y}_2 \\ \Gamma_5 = \mathsf{X}_1\mathsf{X}_2\mathsf{Z}_3 & \Gamma_6 = \mathsf{X}_1\mathsf{X}_2\mathsf{Y}_3 \\ \cdot & \cdot \\ \cdot & \cdot \\ \cdot & \cdot \end{array} \tag{17.42}$$

* These general properties are not necessary for future developments, since we work with an explicit representation. A general study of (17.41) is made by R. Brauer and H. Weyl, *Am. J. Math.*, **57**, 425 (1935).

That is,

$$\begin{aligned}\Gamma_{2\alpha-1} &= \mathsf{X}_1\mathsf{X}_2\cdots\mathsf{X}_{\alpha-1}\mathsf{Z}_\alpha \qquad (\alpha = 1, \ldots, n)\\ \Gamma_{2\alpha} &= \mathsf{X}_1\mathsf{X}_2\cdots\mathsf{X}_{\alpha-1}\mathsf{Y}_\alpha \qquad (\alpha = 1, \ldots, n)\end{aligned} \tag{17.43}$$

An equally satisfactory representation is obtained by interchanging the roles of X_α and Z_α ($\alpha = 1, \ldots, n$). It is also obvious that given any representation, such as (17.43), an equally satisfactory representation is obtained by an arbitrary permutation of the numbering of $\Gamma_1, \ldots, \Gamma_{2n}$.

It will presently be revealed that V_1 and V_2 are matrices that transform one set of $\{\Gamma_\mu\}$ into another equivalent set.

Let a definite set $\{\Gamma_\mu\}$ be given and let ω be the $2n \times 2n$ matrix describing a linear orthogonal transformation among the members of $\{\Gamma_\mu\}$:

$$\Gamma_\mu' = \sum_{\nu=1}^{2n} \omega_{\mu\nu}\Gamma_\nu \tag{17.44}$$

where $\omega_{\mu\nu}$ are complex numbers satisfying

$$\sum_{\mu=1}^{2n} \omega_{\mu\nu}\omega_{\mu\lambda} = \delta_{\nu\lambda} \tag{17.45}$$

This may be written in matrix form as

$$\omega^T\omega = 1 \tag{17.46}$$

where ω^T is the transpose of ω. If Γ_μ is regarded as a component of a vector in a $2n$-dimensional space, then ω induces a rotation in that space:

$$\begin{bmatrix}\Gamma_1'\\ \Gamma_2'\\ \cdot\\ \cdot\\ \cdot\\ \Gamma_{2n}'\end{bmatrix} = \begin{bmatrix}\omega_{11} & \omega_{12} & \cdots & \omega_{1,2n}\\ \omega_{21} & \omega_{22} & \cdots & \omega_{2,2n}\\ \cdot & & & \\ \cdot & & & \\ \cdot & & & \\ \omega_{2n,1} & \omega_{2n,2} & \cdots & \omega_{2n,2n}\end{bmatrix}\begin{bmatrix}\Gamma_1\\ \Gamma_2\\ \cdot\\ \cdot\\ \cdot\\ \Gamma_{2n}\end{bmatrix} \tag{17.47}$$

Substitution of (17.44) into (17.41) shows that the set $\{\Gamma_\mu'\}$ also satisfies (17.41), on account of (17.45). Therefore

$$\Gamma_\mu' = \mathsf{S}(\omega)\Gamma_\mu\mathsf{S}^{-1}(\omega) \tag{17.48}$$

where $\mathsf{S}(\omega)$ is a nonsingular $2^n \times 2^n$ matrix. The existence of $\mathsf{S}(\omega)$ will be demonstrated by explicit construction. Thus there is a correspondence

$$\omega \leftrightarrow \mathsf{S}(\omega) \tag{17.49}$$

which establishes $\mathsf{S}(\omega)$ as a $2^n \times 2^n$ matrix representation of a rotation in a $2n$-dimensional space. Combining (17.48) and (17.44) we have

$$\mathsf{S}(\omega)\Gamma_\mu\mathsf{S}^{-1}(\omega) = \sum_{\nu=1}^{2n} \omega_{\mu\nu}\Gamma_\nu \tag{17.50}$$

We call ω a *rotation* and $S(\omega)$ the *spin representative* of the rotation ω. It is obvious that if ω_1 and ω_2 are two rotations then $\omega_1\omega_2$ is also a rotation. Furthermore

$$S(\omega_1\omega_2) = S(\omega_1)S(\omega_2) \tag{17.51}$$

We now study some special rotations ω and their corresponding $S(\omega)$. Consider a rotation in a two-dimensional plane of the $2n$-dimensional space. A rotation in the plane $\mu\nu$ through the angle θ is defined by the transformation

$$\begin{cases} \Gamma_\lambda' = \Gamma_\lambda & (\lambda \neq \mu, \lambda \neq \nu) \\ \Gamma_\mu' = \Gamma_\mu \cos\theta - \Gamma_\nu \sin\theta & (\mu \neq \nu) \\ \Gamma_\nu' = \Gamma_\mu \sin\theta + \Gamma_\nu \cos\theta & (\mu \neq \nu) \end{cases} \tag{17.52}$$

where θ is a complex number. The rotation matrix, denoted by $\omega(\mu\nu \mid \theta)$, is explicitly given by

$$\omega(\mu\nu \mid \theta) = \begin{array}{c} \begin{matrix} & \mu\text{th column} & & \nu\text{th column} & \end{matrix} \\ \left[\begin{matrix} & \cdot & & \cdot & \\ & \cdot & & \cdot & \\ & \cdot & & \cdot & \\ \cdots & \cos\theta & \cdots & \sin\theta & \cdots \\ & \cdot & & \cdot & \\ & \cdot & & \cdot & \\ & \cdot & & \cdot & \\ \cdots & -\sin\theta & \cdots & \cos\theta & \cdots \\ & \cdot & & \cdot & \\ & \cdot & & \cdot & \\ & \cdot & & \cdot & \end{matrix}\right] \begin{matrix} \\ \\ \\ \mu\text{th row} \\ \\ \\ \\ \nu\text{th row} \\ \\ \\ \\ \end{matrix} \end{array} \tag{17.53}$$

where the matrix elements not displayed are unity along the diagonal and zero everywhere else. Now $\omega(\mu\nu \mid \theta)$ is called *the plane rotation in the plane* $\mu\nu$. It is easily verified that

$$\begin{aligned} \omega(\mu\nu \mid \theta) &= \omega(\nu\mu \mid -\theta) \\ \omega^T(\mu\nu \mid \theta)\omega(\mu\nu \mid \theta) &= 1 \end{aligned} \tag{17.54}$$

The properties of ω and $S(\omega)$ that are relevant to the solution of the Ising model are summarized in the following lemmas.*

Lemma 1. If $\omega(\mu\nu \mid \theta) \leftrightarrow S_{\mu\nu}(\theta)$, then

$$S_{\mu\nu}(\theta) = e^{-\frac{1}{2}\theta\Gamma_\mu\Gamma_\nu} \tag{17.55}$$

* It will be noted that the proofs of these lemmas make use only of the general property (17.41) and the special representation (17.42) of $\{\Gamma_\mu\}$.

Proof. Since $\Gamma_\mu\Gamma_\nu = -\Gamma_\nu\Gamma_\mu$ for $\mu \neq \nu$, $(\Gamma_\mu\Gamma_\nu)^2 = \Gamma_\mu\Gamma_\nu\Gamma_\mu\Gamma_\nu = -1$. An identity analogous to (17.28) is

$$e^{-\frac{1}{2}\theta\Gamma_\mu\Gamma_\nu} = \cos\frac{\theta}{2} - \Gamma_\mu\Gamma_\nu \sin\frac{\theta}{2}$$

Since $(\Gamma_\mu\Gamma_\nu)(\Gamma_\nu\Gamma_\mu) = (\Gamma_\nu\Gamma_\mu)(\Gamma_\mu\Gamma_\nu) = 1$, we have

$$e^{\frac{1}{2}\theta\Gamma_\mu\Gamma_\nu}e^{-\frac{1}{2}\theta\Gamma_\mu\Gamma_\nu} = e^{\frac{1}{2}\theta\Gamma_\mu\Gamma_\nu}e^{\frac{1}{2}\theta\Gamma_\nu\Gamma_\mu} = e^{\frac{1}{2}\theta(\Gamma_\mu\Gamma_\nu+\Gamma_\nu\Gamma_\mu)} = 1$$

Hence

$$\mathsf{S}_{\mu\nu}^{-1}(\theta) = e^{\frac{1}{2}\theta\Gamma_\mu\Gamma_\nu} \tag{17.56}$$

A straightforward calculation shows that

$$\begin{aligned}
\mathsf{S}_{\mu\nu}(\theta)\Gamma_\lambda\mathsf{S}_{\mu\nu}^{-1}(\theta) &= \Gamma_\lambda \qquad (\lambda \neq \mu, \lambda \neq \nu)\\
\mathsf{S}_{\mu\nu}(\theta)\Gamma_\mu\mathsf{S}_{\mu\nu}^{-1}(\theta) &= \Gamma_\mu\cos\theta + \Gamma_\nu\sin\theta\\
\mathsf{S}_{\mu\nu}(\theta)\Gamma_\nu\mathsf{S}_{\mu\nu}^{-1}(\theta) &= \Gamma_\mu\sin\theta - \Gamma_\nu\cos\theta
\end{aligned}$$

(QED)

Lemma 2. The eigenvalues of $\omega(\mu\nu \mid \theta)$ are 1 ($2n - 2$–fold degenerate), and $e^{\pm i\theta}$ (nondegenerate). The eigenvalues of $\mathsf{S}_{\mu\nu}(\theta)$ are $e^{\pm i\theta/2}$ (each 2^{n-1}-fold degenerate).

Proof. The first part is trivial. The second part can be proved by choosing a special representation for $\Gamma_\mu\Gamma_\nu$, since the eigenvalues of $\mathsf{S}_{\mu\nu}(\theta)$ are independent of the representation. As a representation for Γ_μ and Γ_ν we use (17.43) with X and Z interchanged. Since the numbering of the Γ_μ in (17.43) is not unique we are free to choose any two to be Γ_μ and Γ_ν. What we choose for the remaining $2n - 2$ matrices is irrelevant for the proof. We choose

$$\begin{aligned}
\Gamma_\mu &= \mathsf{Z}_1\mathsf{X}_2\\
\Gamma_\nu &= \mathsf{Z}_1\mathsf{Y}_2
\end{aligned}$$

Then

$$\Gamma_\mu\Gamma_\nu = \mathsf{X}_2\mathsf{Y}_2 = i\mathsf{Z}_2 = 1 \times \begin{pmatrix} i & 0 \\ 0 & -i \end{pmatrix} \times 1 \times \cdots \times 1$$

Therefore

$$\mathsf{S}_{\mu\nu}(\theta) = \cos\frac{\theta}{2} - \Gamma_\mu\Gamma_\nu\sin\frac{\theta}{2} = 1 \times \begin{pmatrix} e^{-i\theta/2} & 0 \\ 0 & e^{i\theta/2} \end{pmatrix} \times 1 \times \cdots \times 1$$

The matrix elements of $\mathsf{S}_{\mu\nu}(\theta)$ in this representation are

$$\langle s_1, \ldots, s_n \mid \mathsf{S}_{\mu\nu}(\theta) \mid s_1', \ldots, s_n' \rangle = e^{\frac{1}{2}i\theta s_2} \prod_{k=1}^{n} \delta_{s_k s_k'}$$

Thus $\mathsf{S}_{\mu\nu}(\theta)$ is diagonal. The diagonal elements are either $e^{i\theta/2}$ or $e^{-i\theta/2}$, each appearing the same number of times, i.e., 2^{n-1} times each. (QED)

Lemma 3. Let ω be a product of n commuting plane rotations:

$$\omega = \omega(\alpha\beta \mid \theta_1)\omega(\gamma\delta \mid \theta_2) \cdots \omega(\mu\nu \mid \theta_n) \tag{17.57}$$

where $\{\alpha, \beta, \ldots, \mu, \nu\}$ is a permutation of the set of integers $\{1, 2, \ldots, 2n-1, 2n\}$, and $\theta_1, \ldots, \theta_n$ are complex numbers. Then

(*a*) $\omega \leftrightarrow \mathsf{S}(\omega)$, with

$$\mathsf{S}(\omega) = e^{-\frac{1}{2}\theta_1\Gamma_\alpha\Gamma_\beta} e^{-\frac{1}{2}\theta_2\Gamma_\gamma\Gamma_\delta} \cdots e^{-\frac{1}{2}\theta_n\Gamma_\mu\Gamma_\nu} \tag{17.58}$$

(*b*) The $2n$ eigenvalues of ω are

$$e^{\pm i\theta_1}, e^{\pm i\theta_2}, \ldots, e^{\pm i\theta_n} \tag{17.59}$$

(*c*) The 2^n eigenvalues of $\mathsf{S}(\omega)$ are the values

$$e^{\frac{1}{2}i(\pm\theta_1 \pm \theta_2 \pm \cdots \pm \theta_n)} \tag{17.60}$$

with the signs $\pm$ chosen independently.

Proof. This lemma is an immediate consequence of lemmas 1 and 2 and the fact that $[\Gamma_\mu\Gamma_\nu, \Gamma_\alpha\Gamma_\beta] = 0$.

By this lemma, the eigenvalues of $\mathsf{S}(\omega)$ can be immediately obtained from those of ω, if the eigenvalues of ω are of the form (17.59).

The usefulness of these lemmas rests on the fact that $\mathsf{V}_2\mathsf{V}_1$ can be expressed in terms of $\mathsf{S}(\omega)$.

17.3 THE SOLUTION*

In the absence of an external magnetic field the formulas (17.14) and (17.40) lead to the following:

$$\lim_{N\to\infty} \frac{1}{N} \log Q_I(0, T) = \tfrac{1}{2} \log [2 \sinh (2\beta\epsilon)] + \lim_{n\to\infty} \frac{1}{n} \log \Lambda \tag{17.61}$$

where

$$\Lambda = \text{largest eigenvalue of } \mathsf{V} \tag{17.62}$$

and

$$\mathsf{V} = \mathsf{V}_1\mathsf{V}_2 \tag{17.63}$$

where V_1 is given by (17.37) and V_2 by (17.38). These formulas are valid if all eigenvalues of V are positive and if $\operatorname{Lim} n^{-1} \log \Lambda$ exists. Our main task is to diagonalize the matrix V. *Throughout the present section, all matrices are understood to be matrices in a definite representation.*

Expression of V in Terms of Spin Representatives

Using the representation (17.42), we note that

$$\Gamma_{2\alpha}\Gamma_{2\alpha-1} = \mathsf{Y}_\alpha\mathsf{Z}_\alpha = i\mathsf{X}_\alpha \qquad (\alpha = 1, \ldots, n) \tag{17.64}$$

* L. Onsager, *Phys. Rev.*, **65**, 117 (1944); B. Kaufmann, *Phys. Rev.*, **76**, 1232 (1949). The present account follows Kaufmann's treatment.

From (17.37) we immediately have

$$\mathsf{V}_1 = \prod_{\alpha=1}^{n} e^{\theta \mathsf{X}_\alpha} = \prod_{\alpha=1}^{n} e^{-i\theta\Gamma_{2\alpha}\Gamma_{2\alpha-1}} \tag{17.65}$$

Thus V_1 is a spin representative of a product of commuting plane rotations.

Again, from (17.42),

$$\begin{aligned} \Gamma_{2\alpha+1}\Gamma_{2\alpha} &= \mathsf{X}_\alpha \mathsf{Z}_{\alpha+1}\mathsf{Y}_\alpha = i\mathsf{Z}_\alpha\mathsf{Z}_{\alpha+1} \qquad (\alpha = 1, \ldots, n-1) \\ \Gamma_1\Gamma_{2n} &= \mathsf{Z}_1(\mathsf{X}_1 \cdots \mathsf{X}_{n-1})\mathsf{Y}_n = -i\mathsf{Z}_1\mathsf{Z}_n(\mathsf{X}_1 \cdots \mathsf{X}_n) \end{aligned} \tag{17.66}$$

By (17.38).

$$\mathsf{V}_2 = \left[\prod_{\alpha=1}^{n-1} e^{\beta\epsilon \mathsf{Z}_\alpha \mathsf{Z}_{\alpha+1}}\right] e^{\beta\epsilon\mathsf{Z}_n\mathsf{Z}_1}$$

The last factor commutes with the bracket. Therefore we can write

$$\mathsf{V}_2 = e^{\beta\epsilon\mathsf{Z}_n\mathsf{Z}_1}\left[\prod_{\alpha=1}^{n-1} e^{\beta\epsilon\mathsf{Z}_\alpha\mathsf{Z}_{\alpha+1}}\right] = e^{i\beta\epsilon\mathsf{U}\Gamma_1\Gamma_{2n}} \prod_{\alpha=1}^{n-1} e^{-i\beta\epsilon\Gamma_{2\alpha+1}\Gamma_{2\alpha}} \tag{17.67}$$

where

$$\mathsf{U} \equiv \mathsf{X}_1\mathsf{X}_2 \cdots \mathsf{X}_n \tag{17.68}$$

Were it not for the first factor in (17.67), V_2 would also be the spin representative of a product of commuting plane rotations. This factor owes its existence to the toroidal boundary condition imposed on the problem (i.e., the condition that $s_{n+1} \equiv s_1$ in every row of the lattice). At first sight this condition seems to be an unnecessary and artificial complication, but it actually simplifies our future task.

Substituting (17.67) and (17.65) into (17.63) we obtain

$$\mathsf{V} \equiv \mathsf{V}_2\mathsf{V}_1 = e^{i\phi\mathsf{U}\Gamma_1\Gamma_{2n}}\left[\prod_{\alpha=1}^{n-1} e^{-i\phi\Gamma_{2\alpha+1}\Gamma_{2\alpha}}\right]\left[\prod_{\lambda=1}^{n} e^{-i\theta\Gamma_{2\lambda}\Gamma_{2\lambda-1}}\right] \tag{17.69}$$

where

$$\phi = \beta\epsilon, \qquad \epsilon > 0, \qquad \theta \equiv \tanh^{-1} e^{-2\phi}$$

Some relevant properties of U are

(*a*) $\mathsf{U}^2 = 1, \qquad \mathsf{U}(1 + \mathsf{U}) = 1 + \mathsf{U}, \qquad \mathsf{U}(1 - \mathsf{U}) = -(1 - \mathsf{U})$ (17.70)

(*b*) $\mathsf{U} = i^n\Gamma_1\Gamma_2 \cdots \Gamma_{2n}$ (17.71)

(*c*) U commutes with a product of an even number of Γ_μ and anticommutes with a product of an odd number of Γ_μ.* A simple calculation shows that

$$\begin{aligned} e^{i\phi\Gamma_1\Gamma_{2n}\mathsf{U}} &= [\tfrac{1}{2}(1 + \mathsf{U}) + \tfrac{1}{2}(1 - \mathsf{U})][\cosh\phi + i\Gamma_1\Gamma_{2n}\mathsf{U}\sinh\phi] \\ &= \tfrac{1}{2}(1 + \mathsf{U})[\cosh\phi + i\Gamma_1\Gamma_{2n}\sinh\phi] \\ &\quad + \tfrac{1}{2}(1 - \mathsf{U})[\cosh\phi - i\Gamma_1\Gamma_{2n}\sinh\phi] \\ &= \tfrac{1}{2}(1 + \mathsf{U})e^{i\phi\Gamma_1\Gamma_{2n}} + \tfrac{1}{2}(1 - \mathsf{U})e^{-i\phi\Gamma_1\Gamma_{2n}} \end{aligned} \tag{17.72}$$

* This is an immediate consequence of (17.71) and (17.41). For $n = 2$, U is commonly denoted by $\gamma_5 \equiv \gamma_1\gamma_2\gamma_3\gamma_4$.

Substituting this result into (17.69) we obtain

$$\mathsf{V} = \tfrac{1}{2}(1 + \mathsf{U})\mathsf{V}^+ + \tfrac{1}{2}(1 - \mathsf{U})\mathsf{V}^- \tag{17.73}$$

where

$$\mathsf{V}^\pm \equiv e^{\pm i\phi\Gamma_1\Gamma_{2n}} \left[\prod_{\alpha=1}^{n-1} e^{-i\phi\Gamma_{2\alpha+1}\Gamma_{2\alpha}}\right]\left[\prod_{\lambda=1}^{n} e^{-i\theta\Gamma_{2\lambda}\Gamma_{2\lambda-1}}\right] \tag{17.74}$$

Thus both V^+ and V^- are spin representatives of rotations.

Representation in Which U is Diagonal

It is obvious that the three matrices U, V^+, and V^- commute with one another. Hence they can be simultaneously diagonalized. We first transform V into the representation in which U is diagonal (but in which $\mathsf{V}^\pm$ are not necessarily diagonal):

$$\mathsf{R}\mathsf{V}\mathsf{R}^{-1} \equiv \tilde{\mathsf{V}} = \tfrac{1}{2}(1 + \tilde{\mathsf{U}})\tilde{\mathsf{V}}^+ + \tfrac{1}{2}(1 - \tilde{\mathsf{U}})\tilde{\mathsf{V}}^- \tag{17.75}$$

$$\tilde{\mathsf{U}} \equiv \mathsf{R}\mathsf{U}\mathsf{R}^{-1} \tag{17.76}$$

$$\tilde{\mathsf{V}}^\pm \equiv \mathsf{R}\mathsf{V}^\pm\mathsf{R}^{-1} \tag{17.77}$$

Since $\mathsf{U}^2 = 1$, the eigenvalues of U are either $+1$ or -1. From (17.68) it is seen that U can also be written in the form $\mathsf{U} = X \times X \times \cdots \times X$. Therefore a diagonal form of U is $Z \times Z \times \cdots \times Z$, and the eigenvalues $+1$ and -1 occur with equal frequency. Other diagonal forms of U may be obtained by permuting the relative positions of the eigenvalues along the diagonal. We shall choose R in such a way that all the eigenvalues $+1$ are in one submatrix, and -1 in the other, so that the matrix $\tilde{\mathsf{U}}$ can be represented in the form

$$\tilde{\mathsf{U}} = \begin{pmatrix} I & 0 \\ 0 & -I \end{pmatrix} \tag{17.78}$$

where I is the $2^{n-1} \times 2^{n-1}$ unit matrix. Since $\tilde{\mathsf{V}}^\pm$ commute with $\tilde{\mathsf{U}}$, they must have the forms

$$\tilde{\mathsf{V}}^\pm = \begin{pmatrix} \mathfrak{A}^\pm & 0 \\ 0 & \mathfrak{B}^\pm \end{pmatrix} \tag{17.79}$$

where $\mathfrak{A}^\pm$ and $\mathfrak{B}^\pm$ are $2^{n-1} \times 2^{n-1}$ matrices and are not necessarily diagonal. It is now clear that $\tfrac{1}{2}(1 + \tilde{\mathsf{U}})$ annihilates the lower submatrix and $\tfrac{1}{2}(1 - \tilde{\mathsf{U}})$ annihilates the upper submatrix:

$$\tfrac{1}{2}(1 + \tilde{\mathsf{U}})\tilde{\mathsf{V}}^+ = \begin{pmatrix} \mathfrak{A}^+ & 0 \\ 0 & 0 \end{pmatrix} \tag{17.80}$$

$$\tfrac{1}{2}(1 - \tilde{\mathsf{U}})\tilde{\mathsf{V}}^- = \begin{pmatrix} 0 & 0 \\ 0 & \mathfrak{B}^- \end{pmatrix} \tag{17.81}$$

Therefore

$$\tilde{\mathsf{V}} = \begin{pmatrix} \mathfrak{A}^+ & 0 \\ 0 & \mathfrak{B}^- \end{pmatrix} \tag{17.82}$$

To diagonalize V, it is sufficient to diagonalize $\tilde{\mathsf{V}}$, which has the same set of eigenvalues as V. To diagonalize V it is sufficient to diagonalize (17.80) and (17.81) *separately and independently*, for each of them has only n nonzero eigenvalues. The combined set of their nonzero eigenvalues constitutes the set of eigenvalues of V.

To diagonalize (17.80) and (17.81), we first diagonalize $\tilde{\mathsf{V}}^+$ and $\tilde{\mathsf{V}}^-$ separately and independently, thereby obtaining twice too many eigenvalues for each. To obtain the eigenvalues of (17.80) and (17.81), we would then decide which eigenvalues so obtained are to be discarded. This last step will not be necessary, however, for we shall show that as $n \to \infty$ a knowledge of the eigenvalues of $\tilde{\mathsf{V}}^+$ and $\tilde{\mathsf{V}}^-$ suffices to determine the largest eigenvalue of V. The set of eigenvalues of $\tilde{\mathsf{V}}^\pm$, however, is respectively equal to the set of eigenvalues of $\mathsf{V}^\pm$. Therefore we shall diagonalize V^+ and V^- *separately and independently*.

Eigenvalues of V^+ and V^-

In order to find the eigenvalues of V^+ and V^- we first find the eigenvalues of the rotations, of which V^+ and V^- are spin representatives. These rotations shall be respectively denoted by Ω^+ and Ω^-, which are both $2n \times 2n$ matrices:

$$\mathsf{V}^\pm \leftrightarrow \Omega^\pm \tag{17.83}$$

From (17.74) we immediately have

$$\Omega^\pm = \omega(1, 2n \mid \mp 2i\phi)\left[\prod_{\alpha=1}^{n-1} \omega(2\alpha+1, 2\alpha \mid -2i\phi)\right]\left[\prod_{\lambda=1}^{n} \omega(2\lambda, 2\lambda - 1 \mid -2i\theta)\right] \tag{17.84}$$

where $\omega(\mu\nu \mid \alpha) = \omega(\nu\mu \mid -\alpha)$ is the plane rotation in the plane $\mu\nu$ through the angle α and is defined by (17.53). The eigenvalues of $\Omega^\pm$ are clearly the same as that of

$$\omega^\pm \equiv \Delta\Omega^\pm \Delta^{-1} \tag{17.85}$$

where Δ is the square root of the last factor in (17.84):

$$\Delta \equiv \sqrt{\prod_{\lambda=1}^{n} \omega(2\lambda, 2\lambda - 1 \mid -2i\theta)} = \prod_{\lambda=1}^{n} \omega(2\lambda, 2\lambda - 1 \mid -i\theta) \tag{17.86}$$

Thus

$$
\begin{aligned}
\omega^{\pm} &= \Delta\chi^{\pm}\Delta \\
\Delta &= \omega(12 \mid i\theta)\omega(34 \mid i\theta) \cdots \omega(2n-1, 2n \mid i\theta) \\
\chi^{\pm} &= \omega(1, 2n \mid \pm 2i\phi)[\omega(23 \mid 2i\phi)\omega(45 \mid 2i\phi) \cdots \omega(2n-2, 2n-1 \mid 2i\phi)]
\end{aligned} \tag{17.87}
$$

Explicitly,

$$
\Delta = \begin{bmatrix} \boxed{J} & \begin{matrix} 0 & 0 & \cdots \\ 0 & 0 & \cdots \end{matrix} & & \\ \begin{matrix} 0 & 0 \\ 0 & 0 \\ \cdot & \cdot \\ \cdot & \cdot \\ \cdot & \cdot \end{matrix} & \boxed{J} & & \\ & & \ddots & \\ & & & \boxed{J} \end{bmatrix}, \qquad \boxed{J} \equiv \begin{pmatrix} \cosh\theta & i\sinh\theta \\ -i\sinh\theta & \cosh\theta \end{pmatrix} \tag{17.88}
$$

$$
\chi^{\pm} = \begin{bmatrix} a & 0 \;\; 0 & \cdots & & \pm b \\ 0 & \boxed{K} & & & \\ 0 & & \boxed{K} & & \\ \cdot & & \ddots & & \cdot \\ \cdot & & & \ddots & \cdot \\ \cdot & & & & \cdot \\ & & & \boxed{K} & 0 \\ & & & & 0 \\ \mp b & & \cdots & 0 \;\; 0 & a \end{bmatrix}, \qquad \begin{aligned} \boxed{K} &\equiv \begin{pmatrix} \cosh 2\phi & i\sinh 2\phi \\ -i\sinh 2\phi & \cosh 2\phi \end{pmatrix} \\ a &\equiv \cosh 2\phi \\ b &\equiv i\sinh 2\phi \end{aligned} \tag{17.89}
$$

Performing the matrix multiplication $\Delta\chi^{\pm}\Delta$ in a straightforward way, we obtain

$$
\omega^{\pm} = \begin{bmatrix} A & B & 0 & 0 & \cdots & 0 & \mp B^* \\ B^* & A & B & 0 & & 0 & 0 \\ 0 & B^* & A & B & & & \cdot \\ \cdot & & & & & & \cdot \\ \cdot & & & & & & \cdot \\ \cdot & & & & & & \\ 0 & 0 & & & & & \\ \mp B & 0 & \cdots & & & B^* & A \end{bmatrix} \tag{17.90}
$$

where A and B are 2×2 matrices given by

$$A \equiv \begin{pmatrix} \cosh 2\phi \cosh 2\theta & -i \cosh 2\phi \sinh 2\theta \\ i \cosh 2\phi \sinh 2\theta & \cosh 2\phi \cosh 2\theta \end{pmatrix} \tag{17.91}$$

$$B \equiv \begin{pmatrix} -\frac{1}{2} \sinh 2\phi \sinh 2\theta & i \sinh 2\phi \sinh^2 \theta \\ -i \sinh 2\phi \cosh^2 \theta & -\frac{1}{2} \sinh 2\phi \sinh 2\theta \end{pmatrix} \tag{17.92}$$

and B^* is the hermitian conjugate of B.

To find the eigenvalues of $\omega^{\pm}$, try the following form for an eigenvector of $\omega^{\pm}$:

$$\psi = \begin{bmatrix} zu \\ z^2 u \\ \cdot \\ \cdot \\ \cdot \\ z^n u \end{bmatrix} \tag{17.93}$$

where z is a number and u is a two-component vector

$$u = \begin{pmatrix} u_1 \\ u_2 \end{pmatrix} \tag{17.94}$$

The requirement that

$$\omega^{\pm} \psi = \lambda \psi \tag{17.95}$$

leads to the following eigenvalue equations

$$\begin{aligned} (zA + z^2 B \mp z^n B^*)u &= z\lambda u \\ (z^2 A + z^3 B + z B^*)u &= z^2 \lambda u \\ (z^3 A + z^4 B + z^2 B^*)u &= z^3 \lambda u \\ &\cdot \\ &\cdot \\ &\cdot \\ (z^{n-1} A + z^n B + z^{n-2} B^*)u &= z^{n-1} \lambda u \\ (z^n A \mp z B + z^{n-1} B^*)u &= z^n \lambda u \end{aligned}$$

The second through the $(n - 1)$th equations are identical. Hence there are only three independent equations:

$$\begin{aligned} (A + zB \mp z^{n-1} B^*)u &= \lambda u \\ (A + zB + z^{-1} B^*)u &= \lambda u \\ (A \mp z^{1-n} B + z^{-1} B^*)u &= \lambda u \end{aligned} \tag{17.96}$$

These equations are solved by putting

$$z^n = \mp 1 \tag{17.97}$$

The three equations (17.96) then become the same one:

$$(A + zB + z^{-1}B^*)u = \lambda u \tag{17.98}$$

where the sign $\mp$ in (17.97) is associated with $\omega^\pm$. Thus, for ω^+ and for ω^-, there are n values of z:

$$z_k = e^{2i\pi k/n} \qquad (k = 0, 1, \ldots, 2n - 1) \tag{17.99}$$

where

$$\begin{aligned} k &= 1, 3, 5, \ldots, 2n - 1 \qquad (\text{for } \omega^+) \\ k &= 0, 2, 4, \ldots, 2n - 2 \qquad (\text{for } \omega^-) \end{aligned} \tag{17.100}$$

For each k, two eigenvalues λ_k are determined by the equation

$$(A + z_k B + z_k^{-1} B^*)u = \lambda_k u \tag{17.101}$$

and λ_k is to be associated with $\omega^\pm$ according to (17.100). This determines $2n$ eigenvalues each for $\omega^\pm$.

To find λ_k, note that according to (17.91) and (17.92)

$$\det|A| = 1, \qquad \det|B| = \det|B^*| = 0$$
$$\det|A + z_k B + z_k^{-1} B^*| = 1$$

Therefore the two values of λ_k must have the forms

$$\lambda_k = e^{\pm\gamma_k} \qquad (k = 0, 1, \ldots, 2n - 1) \tag{17.102}$$

The value of γ_k may be found from the equation

$$\tfrac{1}{2} Tr(A + z_k B + z_k^{-1} B^*) = \tfrac{1}{2}(e^{\gamma_k} + e^{-\gamma_k}) = \cosh \gamma_k \tag{17.103}$$

Evaluating the trace with the help of (17.91), (17.92), and (17.99) we obtain

$$\cosh \gamma_k = \cosh 2\phi \cosh 2\theta - \cos \frac{\pi k}{n} \sinh 2\phi \sinh 2\theta$$
$$(k = 0, 1, \ldots, 2n - 1) \tag{17.104}$$

If γ_k is a solution to (17.104) then $-\gamma_k$ is also a solution. But this possibility has already been taken into account in (17.102). Therefore we define γ_k to be the positive solution of (17.104).

It is easily verified that

$$\begin{aligned} &\gamma_k = \gamma_{2n-k} \\ &0 < \gamma_0 < \gamma_1 < \cdots < \gamma_n \end{aligned} \tag{17.105}$$

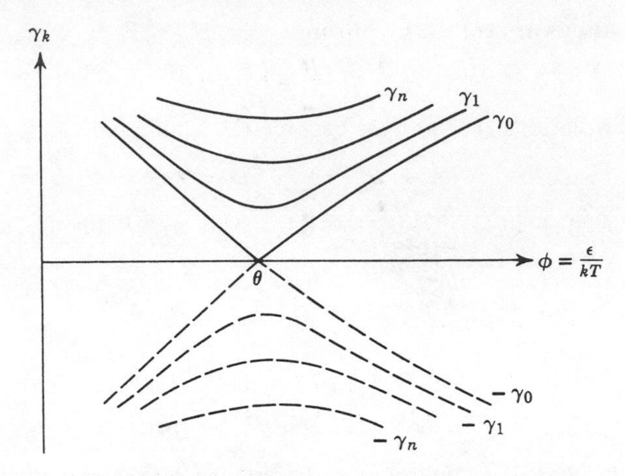

Fig. 17.3. The solutions of (17.104).

The first is obvious. The second can be seen by noting that $\partial\gamma_k/\partial k = (\pi/n)\sin(\pi k/n)/\sin\gamma_k$, which is positive for $k \leq n$. A plot of γ_k as a function of ϕ is shown in Fig. 17.3. As $n \to \infty$ these curves merge into a continuum.

The eigenvalues of $\Omega^{\pm}$ are the same as those of $\omega^{\pm}$ respectively. Therefore $\Omega^{\pm}$ are products of commuting plane rotations, although this fact is not obvious from (17.84).

The 2^n eigenvalues of $\mathsf{V}^{\pm}$ can now be written down immediately with the help of lemma 3 of the last section:

$$\text{eigenvalues of } \mathsf{V}^{-} \text{ are } e^{\frac{1}{2}(\pm\gamma_0\pm\gamma_2\pm\gamma_4\pm\cdots\pm\gamma_{2n-2})} \tag{17.106}$$

$$\text{eigenvalues of } \mathsf{V}^{+} \text{ are } e^{\frac{1}{2}(\pm\gamma_1\pm\gamma_3\pm\gamma_5\pm\cdots\pm\gamma_{2n-1})} \tag{17.107}$$

where all possible choices of the signs $\pm$ are to be made independently.

Eigenvalues of V

As we have explained, the set of eigenvalues of V consists of one-half the set of eigenvalues of V^{+} and one-half that of V^{-}. The eigenvalues of $\mathsf{V}^{\pm}$ are all positive and of order e^n. Therefore all eigenvalues of V are positive and of order e^n. This justifies the formula (17.61). To find explicitly the set of eigenvalues of V, it would be necessary to decide which half of the set of eigenvalues of V^{+} and V^{-} should be discarded. We are only interested, however, in the largest eigenvalue of V. For such a purpose it is not necessary to carry out this task.

Suppose the $2^n \times 2^n$ matrix F transforms (17.80) into diagonal form and the $2^n \times 2^n$ matrix G transforms (17.81) into diagonal form:

$$\mathsf{F}[\tfrac{1}{2}(1 + \tilde{\mathsf{U}})\tilde{\mathsf{V}}^+]\mathsf{F}^{-1} = \mathsf{V}_D^+ \tag{17.108}$$

$$\mathsf{G}[\tfrac{1}{2}(1 - \tilde{\mathsf{U}})\tilde{\mathsf{V}}^-]\mathsf{G}^{-1} = \mathsf{V}_D^- \tag{17.109}$$

where $\mathsf{V}_D^{\pm}$ are diagonal matrices with half the eigenvalues of (17.106) and (17.107) respectively appearing along the diagonal. It is possible to choose F and G in such a way that $\mathsf{F}\,\tilde{\mathsf{U}}\,\mathsf{F}^{-1}$ and $\mathsf{G}\,\tilde{\mathsf{U}}\,\mathsf{G}^{-1}$ remain diagonal matrices. Then F and G merely permute the eigenvalues of $\tilde{\mathsf{U}}$ along the diagonal. But the convention has been adopted that $\tilde{\mathsf{U}}$ has the form (17.78). Hence F and G either leave $\tilde{\mathsf{U}}$ unchanged or simply interchange the two submatrices 1 and -1 in (17.78). That is, F and G either commute or anticommute with $\tilde{\mathsf{U}}$. This means that

$$\mathsf{V}_D^+ = \tfrac{1}{2}(1 \pm \tilde{\mathsf{U}})\mathsf{F}\tilde{\mathsf{V}}^+\mathsf{F}^{-1} \tag{17.110}$$

$$\mathsf{V}_D^- = \tfrac{1}{2}(1 \pm \tilde{\mathsf{U}})\mathsf{G}\tilde{\mathsf{V}}^-\mathsf{G}^{-1} \tag{17.111}$$

where the signs $\pm$ can be definitely determined by an explicit calculation.* For our purpose this determination is not necessary.

We may write

$$\tfrac{1}{2}(1 \pm \tilde{\mathsf{U}}) = \tfrac{1}{2}(1 \pm \mathsf{Z}_1\mathsf{Z}_2 \cdots \mathsf{Z}_n) \tag{17.112}$$

$$\mathsf{F}\tilde{\mathsf{V}}^+\mathsf{F}^{-1} = \prod_{k=1}^{n} e^{\frac{1}{2}\gamma_{2k-1}\mathsf{Z}_{Pk}} \tag{17.113}$$

$$\mathsf{G}\tilde{\mathsf{V}}^-\mathsf{G}^{-1} = \prod_{k=1}^{n} e^{\frac{1}{2}\gamma_{2k-2}\mathsf{Z}_{Qk}} \tag{17.114}$$

where P and Q are two definite permutations (though so far unknown) of the integers $1, 2, \ldots, n$. The permutation P sends k into Pk and Q sends k into Qk. The forms (17.113) and (17.114) are arrived at by noting that $\mathsf{F}\tilde{\mathsf{V}}^+\mathsf{F}^{-1}$ and $\mathsf{G}\tilde{\mathsf{V}}^-\mathsf{G}^{-1}$ must respectively have the same eigenvalues as V^+ and V^-, except for possible different orderings of the eigenvalues. Since the eigenvalues of Z_k are ± 1, we have

$$\tfrac{1}{2}(1 \pm \tilde{\mathsf{U}}) = \begin{cases} 1 \text{ if an even number of } \mathsf{Z}_k \text{ are } \pm 1 \\ 0 \text{ if an odd number of } \mathsf{Z}_k \text{ are } \pm 1 \end{cases} \tag{17.115}$$

This condition is invariant under any permutation that sends $\{\mathsf{Z}_k\}$ into $\{\mathsf{Z}_{Pk}\}$. Therefore the eigenvalues of (17.110) consist of those eigenvalues (17.106) for which an even (odd) number of $-$ signs appears in the

* B. Kaufmann (*loc. cit.*) shows that the plus sign is to be chosen in both (17.110) and (17.111).

exponents, if the $+(-)$ sign is chosen in (17.110). From (17.106) and (17.105), we conclude that the

$$\text{largest eigenvalue of } \mathsf{V}_D^+ = e^{\frac{1}{2}(\pm\gamma_0+\gamma_2+\gamma_4+\cdots+\gamma_{2n-2})}$$

where the $\pm$ sign corresponds to the $\pm$ sign in (17.109). As $n \to \infty$, these two possibilities give the same result, for γ_0 is negligible compared to the entire exponent in the last equation. A similar conclusion can be reached for V_D^-. Therefore we conclude that as $n \to \infty$ the

$$\begin{aligned}\text{largest eigenvalue of } \mathsf{V}_D^+ &= e^{\frac{1}{2}(\gamma_0+\gamma_2+\gamma_4+\cdots+\gamma_{2n-2})}\\ \text{largest eigenvalue of } \mathsf{V}_D^- &= e^{\frac{1}{2}(\gamma_1+\gamma_3+\gamma_5+\cdots+\gamma_{2n-1})}\end{aligned} \tag{17.116}$$

The largest eigenvalue of V is the larger one of (17.116), which by (17.105) is that for V_D^-. Therefore the largest eigenvalue of V is

$$\Lambda = e^{\frac{1}{2}(\gamma_1+\gamma_3+\gamma_5+\cdots+\gamma_{2n-1})} \tag{17.117}$$

The Largest Eigenvalue of V

It is now necessary to evaluate explicitly the largest eigenvalue of V. Using (17.117), we obtain

$$\mathscr{L} \equiv \lim_{n\to\infty} \frac{1}{n} \log \Lambda = \lim_{n\to\infty} \frac{1}{2n}(\gamma_1 + \gamma_3 + \gamma_5 + \cdots + \gamma_{2n-1}) \tag{17.118}$$

Let

$$\begin{aligned}\gamma(\nu) &\equiv \gamma_{2k-1}\\ \nu &\equiv \frac{\pi}{n}(2k-1)\end{aligned} \tag{17.119}$$

As $n \to \infty$, ν becomes a continuous variable, and we have

$$\sum_{k=1}^{n} \gamma_{2k-1} \to \frac{n}{2\pi}\int_0^{2\pi} d\nu\, \gamma(\nu)$$

Therefore

$$\mathscr{L} = \frac{1}{4\pi}\int_0^{2\pi} d\nu\, \gamma(\nu) = \frac{1}{2\pi}\int_0^{\pi} d\nu\, \gamma(\nu) \tag{17.120}$$

where the last step results from (17.105), which states that $\gamma(\nu) = \gamma(2\pi - \nu)$. To express $\mathscr{L}$ in a more convenient form, we recall that $\gamma(\nu)$ is the positive solution of the equation

$$\cosh \gamma(\nu) = \cosh 2\phi \cosh 2\theta - \cos \nu \sinh 2\phi \sinh 2\theta \tag{17.121}$$

with

$$\begin{aligned}\phi &\equiv \beta\epsilon, \qquad \epsilon > 0\\ \theta &\equiv \tanh^{-1} e^{-2\phi}\end{aligned} \tag{17.122}$$

A straightforward calculation shows that

$$\begin{aligned} \sinh 2\theta &= \frac{1}{\sinh 2\phi} \\ \cosh 2\theta &= \coth 2\phi \end{aligned} \tag{17.123}$$

Hence (17.121) can also be written as

$$\cosh \gamma(\nu) = \cosh 2\phi \coth 2\phi - \cos \nu \tag{17.124}$$

We find the following identity helpful in the reduction of (17.120):

$$|z| = \frac{1}{\pi} \int_0^\pi dt \log (2 \cosh z - 2 \cos t) \tag{17.125}$$

With its help we see immediately that $\gamma(\nu)$ has the integral representation

$$\gamma(\nu) = \frac{1}{\pi} \int_0^\pi d\nu' \log (2 \cosh 2\phi \coth 2\phi - 2 \cos \nu - 2 \cos \nu') \tag{17.126}$$

Therefore

$$\mathscr{L} = \frac{1}{2\pi^2} \int_0^\pi d\nu \int_0^\pi d\nu' \log [2 \cosh 2\phi \coth 2\phi - 2(\cos \nu + \cos \nu')] \tag{17.127}$$

The double integral in (17.127) extends over the shaded square in the $\nu\nu'$ plane shown in Fig. 17.4. It is obvious that the integral remains unchanged if we let the region of integration be the rectangle shown in dotted lines, which corresponds to the range of integration

$$\begin{aligned} 0 \le \frac{\nu + \nu'}{2} \le \pi \\ 0 \le \nu - \nu' \le \pi \end{aligned} \tag{17.128}$$

Let

$$\begin{aligned} \delta_1 &\equiv \frac{\nu + \nu'}{2} \\ \delta_2 &\equiv \nu - \nu' \end{aligned} \tag{17.129}$$

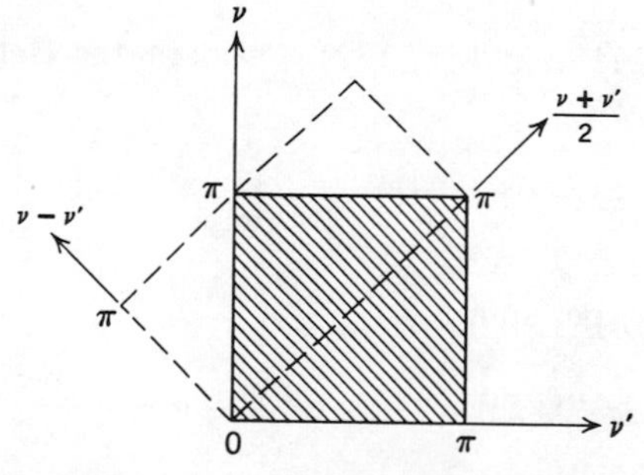

Fig. 17.4. Region of integration in (17.127).

Then

$$\mathscr{L} = \frac{1}{2\pi^2}\int_0^\pi d\delta_1 \int_0^\pi d\delta_2 \log\,(2\cosh 2\phi \coth 2\phi - 4\cos\delta_1 \cos\tfrac{1}{2}\delta_2)$$

$$= \frac{1}{\pi^2}\int_0^\pi d\delta_1 \int_0^{\pi/2} d\delta_2 \log\,(2\cosh 2\phi \coth 2\phi - 4\cos\delta_1 \cos\delta_2)$$

$$= \frac{1}{\pi^2}\int_0^\pi d\delta_1 \int_0^{\pi/2} d\delta_2 \log\,(2\cos\delta_2)$$

$$+ \frac{1}{\pi^2}\int_0^\pi d\delta_1 \int_0^{\pi/2} d\delta_2 \log\left(\frac{D}{\cos\delta_2} - 2\cos\delta_1\right)$$

$$= \frac{1}{\pi}\int_0^{\pi/2} d\delta_2 \log\,(2\cos\delta_2) + \frac{1}{\pi}\int_0^{\pi/2} d\delta_2 \cosh^{-1}\frac{D}{2\cos\delta_2}$$

where

$$D \equiv \cosh 2\phi \coth 2\phi \tag{17.130}$$

and where the identity (17.125) has been used once more. Since $\cosh^{-1} x = \log\,[x + \sqrt{x^2 - 1}]$, we can write

$$\mathscr{L} = \frac{1}{2\pi}\int_0^\pi d\delta \log\,[D(1 + \sqrt{1 - \kappa^2\cos^2\delta})]$$

where

$$\kappa \equiv \frac{2}{D} \tag{17.131}$$

It is clear that in the last integral $\cos^2\delta$ may be replaced by $\sin^2\delta$ without altering the value of the integral. Therefore

$$\mathscr{L} = \tfrac{1}{2}\log\left(\frac{2\cosh^2 2\beta\epsilon}{\sinh 2\beta\epsilon}\right) + \frac{1}{2\pi}\int_0^\pi d\phi \log\tfrac{1}{2}(1 + \sqrt{1 - \kappa^2\sin^2\phi}) \tag{17.132}$$

Thermodynamic Functions

From (17.6), (17.118), and (17.132) we obtain the Helmholtz free energy per spin $a_I(0, T)$:

$$\beta a_I(0, T) = -\log\,(2\cosh 2\beta\epsilon) - \frac{1}{2\pi}\int_0^\pi d\phi \log\tfrac{1}{2}(1 + \sqrt{1 - \kappa^2\sin^2\phi}) \tag{17.133}$$

The internal energy per spin is

$$u_I(0, T) = \frac{d}{d\beta}\,[\beta a_I(0, T)] = -2\epsilon\tanh 2\beta\epsilon + \frac{\kappa}{2\pi}\frac{d\kappa}{d\beta}\int_0^\pi d\phi\,\frac{\sin^2\phi}{\Delta(1 + \Delta)} \tag{17.134}$$

where $\Delta \equiv \sqrt{1 - \kappa^2 \sin^2 \phi}$. It is easily seen that

$$\int_0^\pi d\phi \frac{\sin^2 \phi}{\Delta(1 + \Delta)} = -\frac{\pi}{\kappa^2} + \frac{1}{\kappa^2}\int_0^\pi \frac{d\phi}{\Delta}$$

Therefore

$$u_I(0, T) = -2\epsilon \tanh 2\beta\epsilon + \frac{1}{2\kappa}\frac{d\kappa}{d\beta}\left[-1 + \frac{1}{\pi}\int_0^\pi \frac{d\phi}{\sqrt{1 - \kappa^2 \sin^2 \phi}}\right] \tag{17.135}$$

From (17.131) we obtain

$$\frac{1}{\kappa}\frac{d\kappa}{d\beta} = -2\epsilon \coth 2\beta\epsilon\,(2\tanh^2 2\beta\epsilon - 1) \tag{17.136}$$

$$-2\epsilon \tanh 2\beta\epsilon - \frac{1}{2\kappa}\frac{d\kappa}{d\beta} = -\epsilon \coth 2\beta\epsilon \tag{17.137}$$

Thus finally

$$u_I(0, T) = -\epsilon \coth 2\beta\epsilon\left[1 + \frac{2}{\pi}\kappa' K_1(\kappa)\right] \tag{17.138}$$

where $K_1(\kappa)$ is a tabulated function, the complete elliptic integral of the first kind:*

$$K_1(\kappa) \equiv \int_0^{\pi/2} \frac{d\phi}{\sqrt{1 - \kappa^2 \sin^2 \phi}} \tag{17.139}$$

and

$$\kappa \equiv \frac{2 \sinh 2\beta\epsilon}{\cosh^2 2\beta\epsilon} \tag{17.140}$$

$$\kappa' \equiv 2\tanh^2 2\beta\epsilon - 1 \tag{17.141}$$

$$\kappa^2 + \kappa'^2 = 1 \tag{17.142}$$

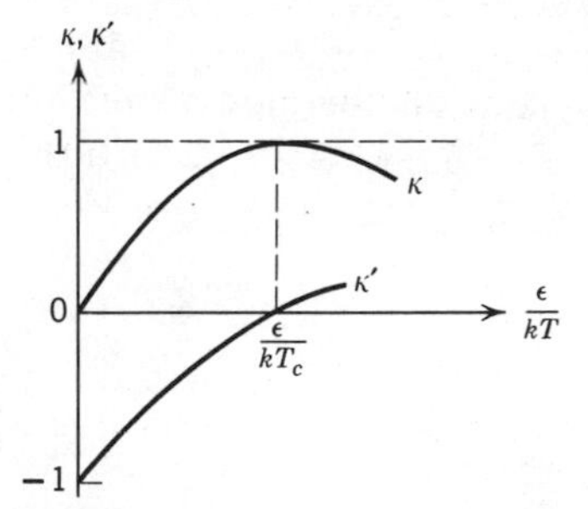

Fig. 17.5. The functions κ and κ'.

Graphs of κ and κ' are shown in Fig. 17.5.

The specific heat $c_I(0, T)$ is readily shown to be

$$\frac{1}{k}c_I(0, T) = \frac{2}{\pi}(\beta\epsilon \coth 2\beta\epsilon)^2\left\{2K_1(\kappa) - 2E_1(\kappa) - (1 - \kappa')\left[\frac{\pi}{2} + \kappa' K_1(\kappa)\right]\right\} \tag{17.143}$$

where $E_1(\kappa)$ is a tabulated function, the complete elliptic integral of the second kind:

$$E_1(\kappa) \equiv \int_0^{\pi/2} d\phi \sqrt{1 - \kappa^2 \sin^2 \phi} \tag{17.144}$$

* See, for example, H. Hancock, *Elliptic Integrals* (Dover, New York, 1958).

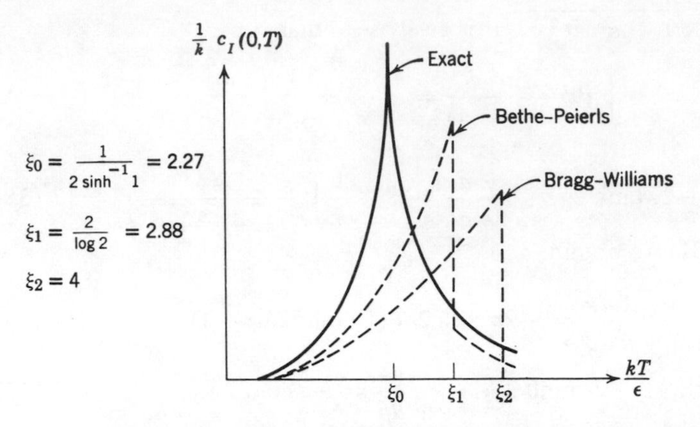

Fig. 17.6. Specific heat of the two-dimensional Ising model.

The elliptic integral $K_1(\kappa)$ has a singularity at $\kappa = 1$ (or $\kappa' = 0$), in which neighborhood

$$K_1(\kappa) \approx \log\frac{4}{\kappa'}, \qquad \frac{dK_1(\kappa)}{d\kappa} \approx \frac{\pi}{2} \qquad (17.145)$$
$$E_1(\kappa) \approx 1$$

Thus all thermodynamic functions have a singularity of some kind at $T = T_c$, where T_c is such that

$$2\tanh^2\frac{2\epsilon}{kT_c} = 1$$
$$\frac{\epsilon}{kT_c} = 0.440\,686\,8 \qquad (17.146)$$
$$kT_c = (2.269\,185)\epsilon$$

Other relations satisfied by T_c are

$$e^{-\epsilon/kT_c} = \sqrt{2} - 1$$
$$\cosh\frac{2\epsilon}{kT_c} = \sqrt{2} \qquad (17.147)$$
$$\sinh\frac{2\epsilon}{kT_c} = 1$$

Thus near $T = T_c$

$$\frac{1}{k}\,c_I(0, T) \approx \frac{2}{\pi}\left(\frac{2\epsilon}{kT_c}\right)^2\left[-\log\left|1 - \frac{T}{T_c}\right| + \log\left(\frac{kT_c}{2\epsilon}\right) - \left(1 + \frac{\pi}{4}\right)\right] \qquad (17.148)$$

It approaches infinity logarithmically as $|T - T_c| \to 0$. A graph of the specific heat is shown in Fig. 17.6, together with the results in the Bragg-Williams and the Bethe-Peierls approximations for comparison. It is seen from (17.138) and (17.145) that the internal energy is continuous at $T = T_c$. Thus the phase transition at $T = T_c$ involves no latent heat.

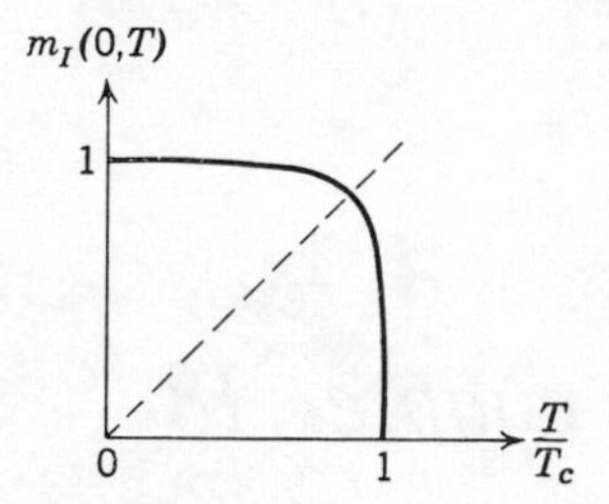

Fig. 17.7. Spontaneous magnetization of the two-dimensional Ising model. The curve is invariant under a reflection about the dotted line.

To justify calling the phenomenon at $T = T_c$ a phase transition, we must examine the long-range order, i.e., the spontaneous magnetization. This cannot be done within the calculations so far outlined, since we have set $B = 0$ from the beginning. To calculate the spontaneous magnetization we have to calculate the derivative of the free energy with respect to B at $B = 0$. This calculation, carried out by Yang,* is as complicated as the one we have presented. The result, however, is simple. Yang shows that the spontaneous magnetization per spin is

$$m_I(0, T) = \begin{cases} 0 & (T > T_c) \\ \dfrac{(1 + z^2)^{1/4}(1 - 6z^2 + z^4)^{1/8}}{\sqrt{1 - z^2}} & (T < T_c) \end{cases} \tag{17.149}$$

where the magnetic moment per spin has been taken to be unity and

$$z \equiv e^{-2\beta\epsilon} \tag{17.150}$$

The transition temperature T_c corresponds to the value $z_c = \sqrt{2} - 1$. A graph of the spontaneous magnetization is shown in Fig. 17.7.

* C. N. Yang, *Phys. Rev.*, **85**, 809 (1952).

chapter 18

LIQUID HELIUM

18.1 THE λ-TRANSITION

Everything about helium* is remarkable. Historically it was discovered in the sun before it was discovered on earth. Physically it has the unique characteristic of being a liquid at very low temperatures. Extrapolation of existing observations indicates that it would be a liquid even at absolute zero. The solid phase of helium comes into being only under an external pressure of at least 25 atm.

There are two isotopes of helium, He^3 and He^4. In the liquid phase of He^4 there is a phase transition that further divides the liquid into two phases called He I and He II. No such transition has been found in He^3.

The relation between the various phases of helium is summarized by the *P-T* diagram and the equation of state surface shown in Fig. 18.1. Along the vapor pressure curve, the λ-transition occurs at the temperature T_λ and the specific volume v_λ, given by

$$T_\lambda = 2.18°\text{K}$$
$$v_\lambda = 46.2\ \text{Å}^3/\text{atom}$$

The specific heat along the vapor curve becomes logarithmically infinite

* For general references, see F. London, *Superfluids*, Vol. II (John Wiley and Sons, New York, 1954); K. R. Atkins, *Liquid Helium* (Cambridge University Press, Cambridge, 1959).

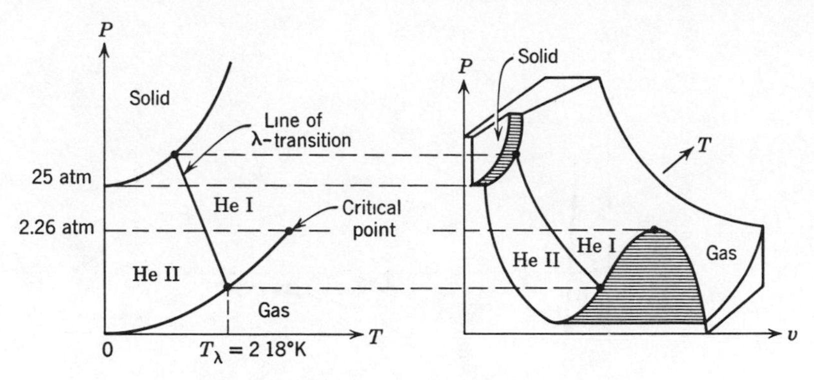

Fig. 18.1. The phases of He⁴ (not to scale).

as the point of λ-transition is approached from either side,* as shown in Fig. 18.2. There is no latent heat of transition. The shape of the specific-heat curve near T_λ gives rise to the name λ-transition. It is similar in appearance to those observed in ferromagnets near the Curie point and in binary alloys near the transition point of the order-disorder transition.

It is significant that He^4 exhibits the λ-transition and He^3 does not. Apart from a difference in atomic masses, the only difference between these two substances is that He^4 atoms are bosons, whereas He^3 atoms are fermions. Hence it is tempting to assume that the λ-transition is the Bose-Einstein condensation, modified, of course, by molecular interactions. In fact, an ideal Bose gas with the same mass and density as liquid He^4 would undergo the Bose-Einstein condensation at 3.14°K, which is of the same order of magnitude as T_λ.

The foregoing considerations are not independent of the fact that helium is a liquid at low temperature, because, if helium solidified before it reached the temperature T_λ, the wave functions of the individual atoms would not overlap, the symmetry of the total wave function would not have any important consequence, and the Bose-Einstein condensation would be completely irrelevant.

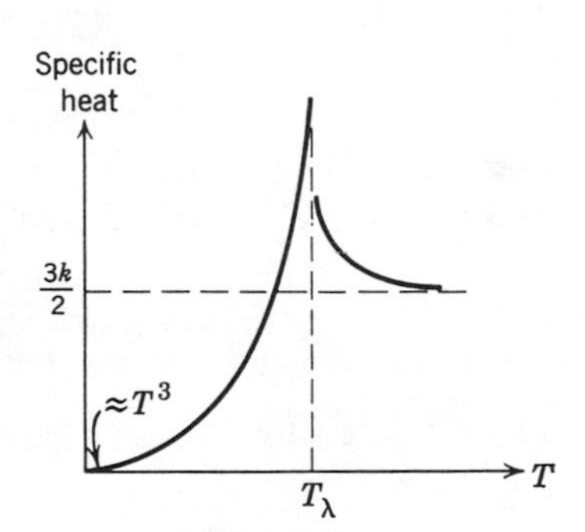

Fig. 18.2. Experimental specific heat of liquid He^4 along the vapor pressure curve.

* W. M. Fairbank, M. J. Buckingham, and C. F. Kellers, in *Low Temperature Physics and Chemistry*, edited by J. R. Dillinger (University of Wisconsin Press, Madison, 1958), p. 50.

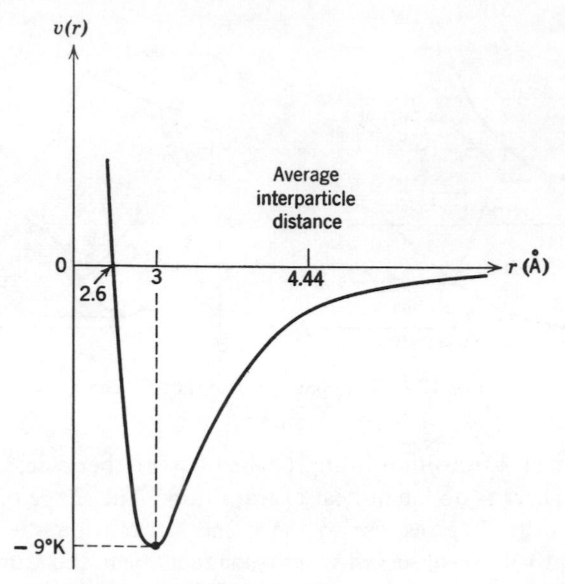

Fig. 18.3. Potential energy between two He atoms separated by distance r.

The qualitative reasons for the fluidity of helium are two. (*a*) The molecular interactions between He atoms are weak, as evidenced by the fact that He is a noble gas; (*b*) the mass of He is the smallest among the noble gases. These circumstances lead to a large zero-point motion of the He atoms, so that it becomes impossible to localize the atoms at well-defined lattice sites. To understand these reasons, we must first have a few facts.

The potential energy $v(r)$ between two He atoms separated by a distance r has been calculated by Slater and Kirkwood* on the basis of the electronic structure of the He atom. They find that

$$v(r) = (5.67 \times 10^6)e^{-21.5(r/\sigma)} - 1.08\left(\frac{\sigma}{r}\right)^6$$

where $\sigma = 4.64$ Å and $v(r)$ is in degrees K. A graph of $v(r)$ is shown in Fig. 18.3.

From the slope dP/dT of the experimental vapor pressure curve we can deduce the latent heat of vaporization of liquid helium. Extrapolation of experimental data shows that $dP/dT > 0$ at absolute zero. Hence liquid helium has a nonvanishing binding energy per atom at absolute zero. That is, the gound state of liquid helium is an N-body bound state that

* J. C. Slater and J. G. Kirkwood, *Phys. Rev.*, **37**, 682 (1931).

has a self-determined equilibrium density in the absence of external pressure.

Consider a collection of He atoms at absolute zero under no external pressure. The most probable configuration of the atoms is determined by the ground state wave function, which, according to the variational principle, must be such as to minimize the total energy of the system, with no external constraint imposed. Hence energy consideration alone determines the most probable configuration. We can then make the following qualitative argument. If a He atom is to have a well-defined location, it must be confined to within a distance Δx that is small compared to the range of the potential, say, $\Delta x \approx 0.5$ Å. By the uncertainty principle, we would then expect an uncertainty in energy (in units of Boltzmann's constant) of the order of

$$\Delta E \approx \frac{1}{2m}\left(\frac{h}{\Delta x}\right)^2 \approx 10°\text{K}$$

This is comparable to the depth of the potential well. Hence the localization is impossible. The fact that no other noble element can remain in liquid form down to very low temperatures is explained by their much greater masses. The fact that H_2, although lighter than He, solidifies at a finite temperature is explained by the strong molecular interactions between H_2 molecules. The argument we have given is independent of statistics and also explains why He^3 remains a liquid down to absolute zero.

Since the potential energy between the He^4 atoms is known, we can formally write down the partition function of liquid He^4. Unfortunately, it has so far defied explicit calculation. Hence the connection between the λ-transition and the Bose-Einstein condensation remains nothing more than a plausible conjecture.

We present an approximate calculation* to support the belief that the Bose-Einstein condensation occurs in liquid He^4. We make use of the formalism of quantized fields, as described in Appendix A, Sec. A.3. The average number of particles with momentum $\mathbf{p}$ is

$$\langle n_{\mathbf{p}} \rangle = \frac{Tr(e^{-\beta H} a_{\mathbf{p}}^{\dagger} a_{\mathbf{p}})}{Tr\, e^{-\beta H}}$$

where $a_{\mathbf{p}}$ is the annihilation operator for a free-particle state of momentum $\mathbf{p}$. Let N be the total number of particles in the system. The criterion for Bose-Einstein condition is as follows.

$$\lim_{N\to\infty} \frac{\langle n_0 \rangle}{N} > 0 \qquad \text{(Bose-Einstein condensation)}$$

$$\lim_{N\to\infty} \frac{\langle n_0 \rangle}{N} = 0 \qquad \text{(no Bose-Einstein condensation)}$$

* The calculation is that of O. Penrose and L. Onsager, *Phys. Rev.*, **104**, 576 (1956).

We consider a transitionally invariant system. To find a convenient formula for $\langle n_0 \rangle$, let the one-particle density matrix be defined as

$$\rho_1(\mathbf{x} - \mathbf{y}) \equiv N \sum_n \int d^3 r_2 \cdots d^3 r_N \Psi_n(\mathbf{y}, \mathbf{r}_2, \ldots, \mathbf{r}_N) e^{-\beta H} \Psi_n^*(\mathbf{x}, \mathbf{r}_2, \ldots, \mathbf{r}_N)$$

where $\{\Psi_n\}$ is a set of complete orthonormal wave functions. In terms of the quantized field operator $\psi(\mathbf{x})$, we can write

$$\rho_1(\mathbf{x} - \mathbf{y}) = \langle \psi^\dagger(\mathbf{x}) \psi(\mathbf{y}) \rangle = \frac{1}{V} \sum_{\mathbf{p},\mathbf{q}} e^{-i(\mathbf{p}\cdot\mathbf{x} - \mathbf{q}\cdot\mathbf{y})} \langle a_\mathbf{p}^\dagger a_\mathbf{q} \rangle$$

where the first form can be obtained through the use of (A.53), and the second form can be obtained through the use of (A.65). Let the total momentum operator be $\mathbf{P} = \Sigma \mathbf{p} a_\mathbf{p}^\dagger a_\mathbf{p}$. It is easily verified that

$$\langle [\mathbf{P}, a_\mathbf{p}^\dagger a_\mathbf{q}] \rangle = 0$$

because $[\mathbf{P}, H] = 0$ by assumption. Through a direct calculation we find that $[\mathbf{P}, a_\mathbf{p}^\dagger a_\mathbf{q}] = (\mathbf{p} - \mathbf{q}) a_\mathbf{p}^\dagger a_\mathbf{q}$. Therefore $(\mathbf{p} - \mathbf{q}) \langle a_\mathbf{p}^\dagger a_\mathbf{q} \rangle = 0$, and we can write

$$\langle a_\mathbf{p}^\dagger a_\mathbf{q} \rangle = \delta_{\mathbf{pq}} \langle n_\mathbf{p} \rangle$$

where $n_\mathbf{p} = a_\mathbf{p}^\dagger a_\mathbf{p}$. Hence

$$\rho_1(\mathbf{r}) = \frac{1}{V} \sum_\mathbf{p} e^{-i\mathbf{p}\cdot\mathbf{r}} \langle n_\mathbf{p} \rangle$$

As $r \to \infty$ all terms in the sum vanish except the term with $\mathbf{p} = 0$, because of the oscillations of the exponential factor. Therefore

$$\frac{\langle n_0 \rangle}{V} = \lim_{r \to \infty} \rho_1(\mathbf{r})$$

We now make an approximate calculation of $\langle n_0 \rangle$ for liquid He^4 at absolute zero. We have

$$\rho_1(\mathbf{r}) = N \int d^3 r_2 \cdots d^3 r_N \Psi_0(\mathbf{r}, \mathbf{r}_2, \ldots, \mathbf{r}_N) \Psi_0(0, \mathbf{r}_2, \ldots, \mathbf{r}_N)$$

where Ψ_0 is the ground state wave function. It is real, positive, and unique, as shown in the next section. Intuitively we expect Ψ_0 to be approximately constant except when two He atoms "touch." Then it is essentially zero. The main effect of the attractive part of the potential is to give the whole system a self-determined equilibrium density. It is plausible that this effect is not important for our considerations, as long as we assume the liquid to have the observed density of liquid He^4. We assume that

$$\Psi_0(\mathbf{r}_1, \mathbf{r}_2, \ldots, \mathbf{r}_N) = \frac{1}{\sqrt{Z_N}} F_N(\mathbf{r}_1, \ldots, \mathbf{r}_N)$$

where Z_N is a normalization constant, and

$$F_N(\mathbf{r}_1, \ldots, \mathbf{r}_N) = \begin{cases} 0 & (|\mathbf{r}_i - \mathbf{r}_j| \le a, \text{ any } i \ne j) \\ 1 & (\text{otherwise}) \end{cases}$$

$$a = 2.56 \text{ Å}$$

The constant $Z_N/N!$ is numerically equal to the configuration integral of a classical hard-sphere gas. We note that $F_N(\mathbf{r}, \mathbf{r}_2, \ldots, \mathbf{r}_N)F_N(0, \mathbf{r}_2, \ldots, \mathbf{r}_N) = F_{N+1}(\mathbf{r}, 0, \mathbf{r}_2, \ldots, \mathbf{r}_N)$. Hence

$$\rho_1(\mathbf{r}) = \frac{N}{Z_N}\int d^3r_2 \cdots d^3r_N F_{N+1}(\mathbf{r}, 0, \mathbf{r}_2, \ldots, \mathbf{r}_N)$$

which is proportional to the pair correlation function for $N + 1$ classical hard spheres (see Problem 10.3):

$$\rho_2(\mathbf{r}) \equiv \frac{N(N+1)}{Z_{N+1}}\int d^3r_2 \cdots d^3r_N F_{N+1}(\mathbf{r}, 0, \mathbf{r}_2, \ldots, \mathbf{r}_N)$$

which tends to $(N/V)^2$ as $r \to \infty$. We can now write

$$\rho_1(\mathbf{r}) = z\rho_2(\mathbf{r})$$

where

$$z = \frac{N!}{(N+1)!}\frac{Z_{N+1}}{Z_N}$$

As $r \to \infty$, we have $\text{Lim}\, \rho_1(\mathbf{r}) = z(N/V)^2$. Hence

$$\frac{\langle n_0 \rangle}{N} = \frac{zN}{V}$$

showing that Bose-Einstein condensation exists at absolute zero, if z is finite. Let

$$\lim_{N\to\infty} \frac{1}{N} \log \frac{Z_N}{N!} = f(v)$$

where $v \equiv V/N$. Then

$$\log z = f(v) - v\frac{\partial f(v)}{\partial v}$$

The atomic volume of liquid He^4 is $v = 46.2$ Å^3, which implies an average interatomic distance of 4.44 Å. Since this is somewhat larger than the effective hard-sphere radius of 1.28 Å, we may calculate $f(v)$ by making a virial expansion as discussed in Problem 8.9. The result is

$$f(v) = \log v + \frac{2\pi}{3}\frac{a^3}{v}$$

which leads to

$$\frac{\langle n_0 \rangle}{N} = e^{-[1+(4\pi/3)(a^3/v)]} = 0.08$$

18.2 TISZA'S TWO-FLUID MODEL

From now on we are concerned exclusively with the properties of liquid He^4 below the λ-point—the phase He II. Here liquid He^4 exhibits some interesting properties. Instead of presenting the experimental observations, we first present a model that describes them.

Tisza summarizes some of the experimental results by postulating that the phase He II is made up of two components called the *normal fluid* and the *superfluid*. In contradistinction, the phase He I is pure normal fluid. It is imagined that we may attribute characteristic mass densities ρ_n and ρ_s respectively to the normal fluid and the superfluid. If the liquid flows, we may attribute characteristic velocity fields $\mathbf{v}_n$ and $\mathbf{v}_s$ respectively to the normal and the superfluid. The mass density ρ and the velocity field $\mathbf{v}$ of He II are assumed to be given by

$$\rho = \rho_n + \rho_s$$

$$\rho\mathbf{v} = \rho_n\mathbf{v}_n + \rho_s\mathbf{v}_s$$

The normal fluid is supposed to behave like an ordinary classical fluid, whereas the superfluid has the unusual properties that

(*a*) its entropy is zero;

(*b*) it flows with no resistance through channels of extremely small diameters (10^{-2} cm or even smaller).

With no further assumptions many strange properties of He II can be qualitatively understood.

First, an extremely small opening in a tank of He II acts as a filter for the superfluid component because the latter can pass through the opening, leaving the normal-fluid component behind. Suppose two tanks of He II are connected by an extremely thin tube, and suppose some superfluid is made to flow from tank A to tank B by the establishment of a pressure differential. Then, since the superfluid has no entropy, the entropy per unit mass in tank A will increase whereas that of tank B will decrease. Therefore tank A will warm up while tank B cools. This is known as the *mechanocaloric effect*.

The inverse effect, that of the creation of a pressure differential by heating, is known as the *fountain effect*.

In the two fluid model a sound wave in He II must mean a sinusoidal oscillation of ρ_n and ρ_s in phase with each other, because only when they oscillate in phase can the total mass density vary sinusoidally. We can also imagine a new mode of oscillation, however, in which the normal fluid and the superfluid oscillate out of phase by 180°. This would not be a sound wave, for the total mass density would be constant throughout the liquid, but it would represent a sinusoidal variation of the entropy per unit mass, because the superfluid has no entropy. Therefore, we may hope to excite this new mode of oscillation by local heating of He II. The temperature gradient established would not propagate by diffusion, as in heat conduction, but would propagate like a wave, with a characteristic velocity. This phenomenon is called the *second sound*.

All these phenomena are observed in He II. Applying purely thermodynamic considerations, we can describe them quantitatively in terms of the two-fluid model.

In the two-fluid model the relative amount of normal fluid and superfluid present may be deduced from an experiment of Andronikashvili. A pile of discs spaced 0.2 mm apart are mounted on a shaft, and the assembly is made to rotate in He II, as shown in Fig. 18.4. The moment of inertia of the assembly is measured as a function of the temperature, when the He II is in equilibrium with its own vapor. By assuming that the superfluid remains completely unaffected by the rotating discs, whereas the normal fluid between the disks is dragged into rotation, we see that the moment of inertia must be proportional to ρ_n/ρ. The proportionality constant is determined by the fact that at the transition temperature $\rho_n/\rho = 1$. We obtain from the experimental result the empirical formula

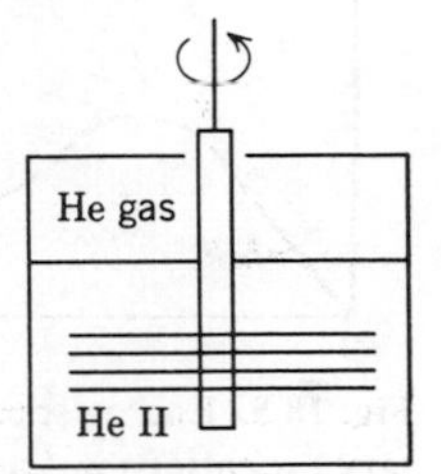

Fig. 18.4. Experiment of Andronikashvili.

$$\frac{\rho_n}{\rho} = \begin{cases} \left(\dfrac{T}{T_\lambda}\right)^{5.6} & (T < T_\lambda) \\ 1 & (T > T_\lambda) \end{cases}$$

This represents an empirical equation of state for He II along the vapor pressure curve.

The two-fluid model is of great value for the study of liquid He^4, because it is simple and it works. It does not, however, provide a complete understanding of liquid He. The reason is twofold:

(*a*) A complete hydrodynamic description of liquid He^4 is lacking.

(*b*) An explanation of the nature of the two fluids in molecular terms is lacking.

In the following we describe some theories that have partially removed these deficiencies.

18.3 THE THEORIES OF LANDAU AND FEYNMAN

The theories of Landau and Feynman attempt to lay a molecular foundation for the two-fluid model in the neighborhood of absolute zero.

The theory of Landau is based on the observation that the experimental specific heat of He II is proportional to T^3 as $T \to 0$. Such a behavior is characteristic of a gas of phonons. Accordingly Landau postulates that the quantum states of liquid He near the ground state can be described as a gas of noninteracting elementary excitations. The energy levels of such quantum states are

$$E_n = E_0 + \sum_{\mathbf{k}} \hbar\omega_{\mathbf{k}} n_{\mathbf{k}} \tag{18.1}$$

where $\hbar\omega_{\mathbf{k}}$ is the energy of an elementary excitation of wave number $\mathbf{k}$ and $\{n_{\mathbf{k}}\}$ is a set of occupation numbers. It is assumed that the elementary excitations are bosons, so that $n_{\mathbf{k}} = 0, 1, 2, \ldots$. Clearly, there is some restriction on the occupation numbers, just as in the Debye theory of the crystal lattice. This restriction is unknown; but it cannot be important for the low-temperature properties of the liquid, because only a few excitations would be present. Therefore we may take each $n_{\mathbf{k}}$ to be an independent number.

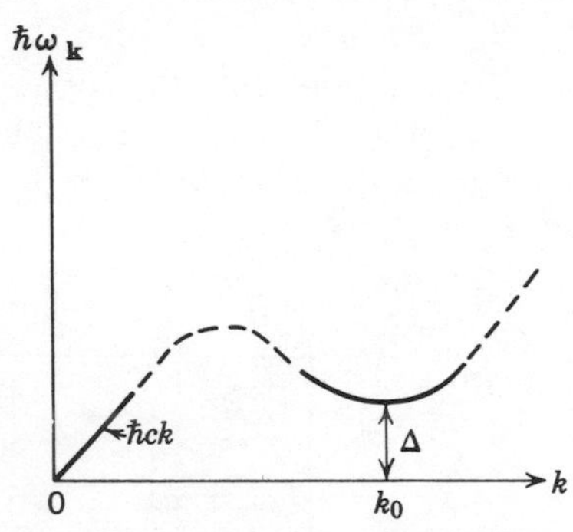

Fig. 18.5. Energy spectrum of elementary excitations in Landau's theory.

The energy $\hbar\omega_{\mathbf{k}}$ is to be determined by requiring that (18.1) leads to the correct experimental specific heat near absolute zero. From the fact that the specific heat vanishes like T^3 as $T \to 0$, it is immediately obvious that as $|\mathbf{k}| \to 0$, $\omega_{\mathbf{k}} \to \hbar c |\mathbf{k}|$, where c is the sound velocity of liquid He at absolute zero. For finite temperatures the specific heat contains an additional term which behaves like $\exp(-\Delta/kT)$, where Δ is a constant. Thus for larger values of $|\mathbf{k}|$, $\omega_{\mathbf{k}}$ should behave like an excitation with an "energy gap."* Landau postulates that

$$\hbar\omega = \begin{cases} \hbar ck & (k \ll k_0) \\ \Delta + \dfrac{\hbar^2(k-k_0)^2}{2\sigma} & (k \approx k_0) \end{cases} \tag{18.2}$$

where $k = |\mathbf{k}|$† and where the constants c, Δ, k_0, and σ are to be adjusted to fit the experimental specific heat. The function $\omega_{\mathbf{k}}$ is illustrated in Fig. 18.5. The linear part of $\omega_{\mathbf{k}}$ is called the phonon part, and the part near k_0 the roton part. The name "roton" was coined by Landau, who first thought that these were excitations of spin $\hbar$, distinct from the phonons. Actually, as we see later, the two parts of $\omega_{\mathbf{k}}$ seem to be different portions of a single continuous curve.

The specific heat at low temperatures can be trivially calculated from (18.2), for we have a system of noninteracting bosons whose total number is not conserved. The thermodynamic average of $n_{\mathbf{k}}$ is

$$\langle n_{\mathbf{k}} \rangle = \frac{1}{e^{\hbar\beta\omega_{\mathbf{k}}} - 1} \tag{18.3}$$

* See Problem 13.4.

† The use of the same symbol, k, for both the wave number and the Boltzmann constant should not, we hope, cause confusion.

where $\beta = 1/kT$. The internal energy of the liquid is

$$U = E_0 + \sum_{\mathbf{k}} \hbar\omega_{\mathbf{k}}\langle n_{\mathbf{k}}\rangle = E_0 + \frac{V}{2\pi^2}\int_0^\infty dk\, \frac{k^2\hbar\omega_{\mathbf{k}}}{e^{\beta\hbar\omega_{\mathbf{k}}} - 1} \tag{18.4}$$

where V is the volume occupied by the liquid. The specific heat at constant volume is

$$\frac{1}{Nk}\, C_V = \frac{1}{Nk}\frac{\partial U}{\partial T} \tag{18.5}$$

where N is the number of He^4 atoms in the liquid. For low temperatures, only the phonon part and the roton part contribute significantly to the integral in (18.4). The specific heat can therefore be written approximately as a sum of two terms, the phonon contribution and the roton contribution.

$$\frac{1}{Nk}\, C_{\text{phonon}} = \frac{2\pi^2 v(kT)^3}{15\hbar^3 c^3} \tag{18.6}$$

where $v = V/N$ and m is the mass of a He^4 atom. The roton part may be calculated by treating kT/Δ as a small quantity. We obtain

$$\frac{1}{Nk}\, C_{\text{roton}} \approx \frac{2\sqrt{\sigma}\, k_0{}^2 \Delta^2 v e^{-\Delta/kT}}{(2\pi)^{3/2}\hbar (kT)^{3/2}} \tag{18.7}$$

The total specific heat is the sum of (18.5) and (18.6). Fitting these to the experimental specific heat, Landau finds

$$\begin{aligned} c &= 226 \text{ m/sec} \\ \frac{\Delta}{k} &= 9°\text{K} \\ k_0 &= 2\text{Å}^{-1} \\ \frac{\sigma}{m} &= 0.3 \end{aligned} \tag{18.8}$$

The constant c, being the phonon velocity, should agree with the sound velocity in liquid He^4. Direct measurement of the latter, extrapolated to absolute zero, yields 239 ± 2 m/sec.

Feynman's theory* arrives at results similar to those of Landau's theory, but it throws more light on the nature of the elementary excitations. We first summarize the conclusions of Feynman's theory as follows.

* R. P. Feynman in *Progress in Low Temperature Physics*, Vol. I, edited by C. J. Gorter (North Holland Publishing Co., Amsterdam, 1955); *Phys. Rev.*, **94**, 262 (1954).

(*a*) The wave function of the state of liquid He^4 in which one elementary excitation is present is approximately given by

$$\Psi_{\mathbf{k}} = \text{const.} \sum_{j=1}^{N} e^{i\mathbf{k}\cdot\mathbf{r}_j}\Psi_o$$

where Ψ_0 is the ground state wave function and $\hbar\mathbf{k}$ is the momentum of the elementary excitation. As we shall explain, this wave function describes a density fluctuation in the liquid—a sound wave—for $k \to 0$. Hence the phonons are quantized sound waves. For finite k, $\Psi_{\mathbf{k}}$ describes internal motions of a more complicated type. In particular $\Psi_{\mathbf{k}}$ is still approximately valid when $k \approx k_0$, in the roton region.

(*b*) At very low energies, all excited states of liquid He^4 are accounted for by the wave function $\Psi_{\mathbf{k}}$. Any other type of excitation must be separated from the ground state by a finite energy gap. This conclusion is arrived at by a plausibility argument in which the Bose statistics plays a dominant role.

Let the Hamiltonian of liquid He^4 be

$$H = -\frac{\hbar^2}{2m}\sum_{j=1}^{N}\nabla_j^2 + \sum_{i<j} v_{ij} \tag{18.9}$$

where m is the mass of a He^4 atom and $v_{ij} = v(|\mathbf{r}_i - \mathbf{r}_j|)$ is the potential energy between the ith and the jth atoms. We refer to a set of values of the coordinates $\mathbf{r}_1, \ldots, \mathbf{r}_N$ as a configuration. The ground state wave function Ψ_0 satisfies

$$H\Psi_0 = E_0\Psi_0 \tag{18.10}$$

where E_0 is the lowest possible energy eigenvalue. Since He^4 atoms are bosons, Ψ_0 must be a symmetric formation of $\mathbf{r}_1, \ldots, \mathbf{r}_N$. It can be chosen to be real and positive. It is then unique. Qualitatively we may picture Ψ_0 as approximately constant for all configurations, except for those in which two atoms "touch." Then $\Psi_0 \approx 0$.

That Ψ_0 can be chosen to be real follows from the fact that H is a real operator. If Ψ is any eigenfunction of H, then Ψ^* is an eigenfunction belonging to the same eigenvalue. It follows that $\Psi_0 + \Psi_0^*$ and $i(\Psi_0 - \Psi_0^*)$ are eigenfunctions belonging to the same eigenvalue. Hence all eigenfunctions of H can be chosen to be real.

To show that Ψ_0 can be chosen to be positive,* we note that Ψ_0 minimizes the expression $(\Psi_0, H\Psi_0)$. Let $\Phi_0 \equiv |\Psi_0|$, which is a symmetric function.† It is easily shown that $(\Phi_0, H\Phi_0) = (\Psi_0, H\Psi_0)$. Therefore Φ_0 minimizes the expression $(\Phi_0, H\Phi_0)$. That is, if we replace Φ_0 by $\Phi_0 + \delta\Phi_0$, where $\delta\Phi_0$ is an arbitrarily small

* O. Penrose and L. Onsager, *op. cit.*

† Note that for Fermi statistics the proof fails at this point.

variation, then $\delta(\Phi_0, H\Phi_0) = 0$. This leads to the conclusion that Φ_0 satisfies (18.10). It follows that the first derivatives of Φ_0 must be continuous because the potential is finite. Hence Ψ_0 never changes sign, for if it did there would be a nodal surface on which not only Ψ_0 but also all the derivatives of Ψ_0 would vanish. Since (18.10) is a second-order partial differential equation, Ψ_0 would be identically zero. Therefore Ψ_0 never changes its sign, which we can take to be positive.

That Ψ_0 is unique follows from the fact that a ground state wave function never changes sign. If Ψ_1 and Ψ_2 are two normalized ground state wave functions, $\Psi_1 - \Psi_2$ is also a ground state wave function, which never changes sign. This means that one of them is always greater than the other. Hence Ψ_1 and Ψ_2 cannot be both normalized unless $\Psi_1 = \Psi_2$.

We now consider the low-energy excited states of the liquid. The wave function Ψ for an excited state must be orthogonal to Ψ_0 and must remain symmetric with respect to interchange of any pair of coordinates. Since Ψ_0 is positive for all configurations, Ψ must be positive for half the configurations and negative for the other half, in order that it can be orthogonal to Ψ_0. Let A denote the configuration at which Ψ assumes the largest positive value and B denote the configuration at which Ψ assumes the most negative value. How may A, B be chosen such that the energy of the state Ψ is as low as possible?

A configuration is given when the values of the coordinates $\mathbf{r}_1, \ldots, \mathbf{r}_N$ are all given. To change a given configuration, we change some or all of the values of the coordinates $\mathbf{r}_1, \ldots, \mathbf{r}_N$. The coordinates in A and B must be different, but they cannot differ only by a permutation, for the wave function is invariant under such a permutation. Apart from this, the coordinates for A and B must be as different as possible in order that the kinetic energy be small. Feynman argues that A and B should be qualitatively as follows: Divide the volume of the system into the parts 1 and 2. The configuration A is one in which there are more particles in 1 than in 2, and B is one in which there are more particles in 2 than in 1, roughly as illustrated in Fig. 18.6. For configuration A, Ψ is to have the positive maximum; for configuration B, Ψ is to have the negative minimum. Thus the state described by Ψ looks like a state in which there is a

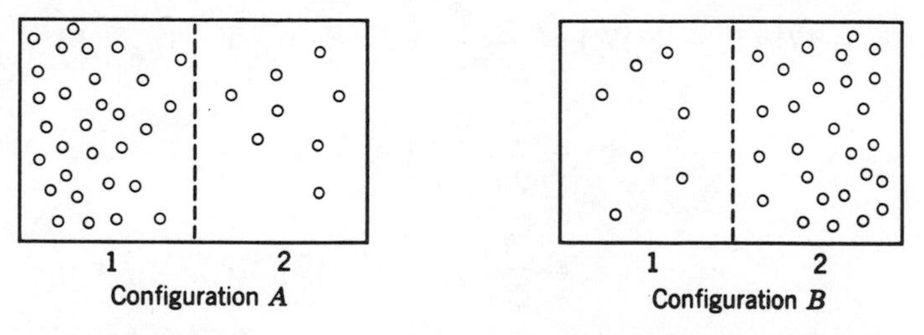

Fig. 18.6. Visualization of Feynman's wave function (18.8).

sound wave of very long wavelength. Accordingly Feynman postulates the form

$$\Psi = \sum_{j=1}^{N} f(\mathbf{r}_j)\Psi_0 \tag{18.11}$$

where $f(\mathbf{r})$ is a function to be determined by minimizing the energy of the state Ψ. It will later be explained why (18.11) may represent a sound wave. It is important to note that the argument fails if the wave function is antisymmetric instead of symmetric. Then a permutation of particles does not leave Ψ invariant; it changes its sign. The requirement that

$$\frac{(\Psi, H\Psi)}{(\Psi, \Psi)} \text{ is minimum}$$

determines $f(\mathbf{r}_j)$. Feynman has given various examples of possible forms of Ψ to support the contention that all forms other than (18.11) lead to an energy higher than the ground state energy by a finite amount. For these examples we refer to the reference previously given.*

Let

$$F \equiv \sum_{j=1}^{N} f(\mathbf{r}_j) \tag{18.12}$$

Then

$$\begin{aligned}(H - E_0)\Psi &= HF\Psi_0 - E_0F\Psi_0 \\ &= -\frac{\hbar^2}{2m}\sum_{j=1}^{N}[(\nabla_j^2 F)\Psi_0 + 2(\nabla_j F)\cdot(\nabla_j\Psi_0)] + [(H - E_0)\Psi_0]F\end{aligned}$$

The last term is zero by (18.10). Hence we can write

$$(H - E_0)\Psi = \frac{1}{\Psi_0}\left[-\frac{\hbar^2}{2m}\sum_{j=1}^{N}\nabla_j\cdot(\rho_N\nabla_j F)\right] \tag{18.13}$$

where

$$\rho_N \equiv \Psi_0^2(\mathbf{r}_1, \ldots, \mathbf{r}_N) \tag{18.14}$$

Next perform the following calculation:

$$\begin{aligned}\mathscr{E} \equiv (\Psi, (H - E_0)\Psi) &= -\frac{\hbar^2}{2m}\int d^{3N}r\, F^*\sum_{j=1}^{N}\nabla_j\cdot(\rho_N\nabla_j F) \\ &= \frac{\hbar^2}{2m}\sum_{j=1}^{N}\int d^{3N}r\,(\nabla_j F^*)\cdot(\nabla_j F)\rho_N \\ &= \frac{\hbar^2}{2m}\sum_{j=1}^{N}\int d^{3N}r[\nabla_j f^*(\mathbf{r}_j)\cdot\nabla_j f(\mathbf{r}_j)]\rho_N\end{aligned}$$

* Feynman, *op. cit.*

Since ρ_N is a symmetric function of $\mathbf{r}_1, \ldots, \mathbf{r}_N$,

$$\mathscr{E} = \frac{N\hbar^2}{2m}\int d^{3N}r[\nabla_1 f^*(\mathbf{r}_1)\cdot\nabla_1 f(\mathbf{r}_1)]\rho_N = \frac{N\hbar^2\rho_0}{2m}\int d^3r\,\nabla f^*(\mathbf{r})\cdot\nabla f(\mathbf{r}) \quad (18.15)$$

where

$$\rho_0 \equiv \int d^3r_2\cdots d^3r_N\,\rho_N \quad (18.16)$$

and is a constant for a sufficiently large system. Define

$$\begin{aligned}\mathscr{I} \equiv (\Psi,\Psi) &= \int d^{3N}r\,F^*F\rho_N\\ &= \sum_{i=1}^{N}\sum_{j=1}^{N}\int d^{3N}r\,f^*(\mathbf{r}_j)f(\mathbf{r}_i)\rho_N\\ &= \sum_{i=1}^{N}\int d^{3N}r\,|f(r_i)|^2\,\rho_N + \sum_{i\neq j}\int d^{3N}r\,f^*(\mathbf{r}_i)f(\mathbf{r}_j)\rho_N\\ &= N\rho_0\int d^3r\,|f(\mathbf{r})|^2 + N(N-1)\int d^3r_1\,d^3r_2 f^*(\mathbf{r}_1)f(\mathbf{r}_2)\rho_2(\mathbf{r}_1,\mathbf{r}_2)\end{aligned} \quad (18.17)$$

where

$$\rho_2(\mathbf{r}_1,\mathbf{r}_2) \equiv \int d^3r_3\cdots d^3r_N\rho_N \quad (18.18)$$

Let the pair correlation function $D(|\mathbf{r}_1 - \mathbf{r}_2|)$ be defined by

$$\begin{aligned}D(|\mathbf{r}_1 - \mathbf{r}_2|) &\equiv \frac{1}{\rho_0}\sum_{i=1}^{N}\sum_{j=1}^{N}\int d^{3N}r'\,\delta(\mathbf{r}_i' - \mathbf{r}_1)\delta(\mathbf{r}_j' - \mathbf{r}_2)\rho_N\\ &= N\delta(\mathbf{r}_1 - \mathbf{r}_2) + \frac{1}{\rho_0}N(N-1)\rho_2(\mathbf{r}_1,\mathbf{r}_2)\end{aligned} \quad (18.19)$$

which is the probability per unit volume of finding a particle at $\mathbf{r}_1$, if a particle is known to be at $\mathbf{r}_2$, in the ground state of the system. Then

$$\mathscr{I} = \rho_0\int d^3r_1\,d^3r_2 f^*(\mathbf{r}_1)f(\mathbf{r}_2)D(|\mathbf{r}_1 - \mathbf{r}_2|) \quad (18.20)$$

To determine $f(\mathbf{r})$ we minimize $\mathscr{E}$ subject to the condition $\mathscr{I} = 1$:

$$\delta\mathscr{E} - N\hbar\omega\,\delta\mathscr{I} = 0 \quad (18.21)$$

where ω is a Lagrange multiplier. Carrying out the variation we obtain for $f(\mathbf{r})$ the differentiation equation

$$-\frac{\hbar^2}{2m}\nabla^2 f(\mathbf{r}) - \hbar\omega\int d^3r\,D(r')f(\mathbf{r} + \mathbf{r}') = 0 \quad (18.22)$$

which is solved by requiring

$$f(\mathbf{r}) = \text{const. } e^{i\mathbf{k}\cdot\mathbf{r}}$$
$$\frac{\hbar^2k^2}{2m} - \hbar\omega \int d^3r\, D(r')e^{i\mathbf{k}\cdot\mathbf{r}'} = 0 \tag{18.23}$$

where $\mathbf{k}$ is an arbitrary nonzero vector. The corresponding value of ω is now denoted by $\omega_\mathbf{k}$:

$$\hbar\omega_\mathbf{k} = \frac{\hbar^2k^2}{2mS(k)} \tag{18.24}$$

where $S(k)$ is the Fourier transform of $D(r)$:

$$S(k) = \int d^3r\, e^{i\mathbf{k}\cdot\mathbf{r}} D(r) \tag{18.25}$$

Substituting (18.23) into (18.15), we see that the energy of the excited state is, for a given $\mathbf{k}$,

$$E_\mathbf{k} = E_0 + \hbar\omega_\mathbf{k} \tag{18.26}$$

The meaning of (18.23) and (18.26) is the following: For a given vector $\mathbf{k} \neq 0$, (18.23) minimizes the energy, and the energy is (18.26). It is meaningful to fix the value of $\mathbf{k}$ because the form (18.23) leads to an eigenstate of the total momentum of the system with the eigenvalue $\hbar\mathbf{k}$. To see this, note that (18.23) leads to the wave function

$$\Psi_\mathbf{k} = \text{const. } \sum_{j=1}^{N} e^{i\mathbf{k}\cdot\mathbf{r}_j}\Psi_0 \tag{18.27}$$

for an excited state. It is obvious that

$$\mathbf{P}_{\text{op}}\Psi_\mathbf{k} = \frac{\hbar}{i}\sum_{j=1}^{N}\nabla_j\Psi_\mathbf{k} = \hbar\mathbf{k}\Psi_\mathbf{k} \tag{18.28}$$

where $\mathbf{P}_{\text{op}}$ is the total momentum operator. The wave function $\Psi_\mathbf{k}$ is orthogonal to Ψ_0, for by (18.27) and the hermiticity of $\mathbf{P}_{\text{op}}$

$$\hbar\mathbf{k}(\Psi_\mathbf{k}, \Psi_0) = (\mathbf{P}_{\text{op}}\Psi_\mathbf{k}, \Psi_0) = (\Psi_\mathbf{k}, \mathbf{P}_{\text{op}}\Psi_0) = 0 \tag{18.29}$$

If $\mathbf{k} \neq 0$, then

$$(\Psi_\mathbf{k}, \Psi_0) = 0 \tag{18.30}$$

To find $\omega_\mathbf{k}$ we substitute into (18.24) the experimentally observed $S(k)$, which is plotted in Fig. 18.7. This leads to $\omega_\mathbf{k}$ as shown in Fig. 18.8. Quantitatively it differs considerably from the Landau curve which fits the specific heat, but the "roton" part comes out qualitatively right. The quantitative difference must be attributed to the restrictive form of the variational wave function (18.9), which is expected to be accurate only for

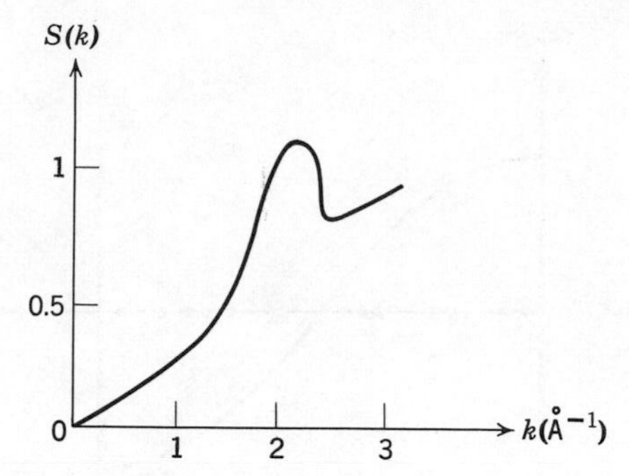

Fig. 18.7. Fourier transform of the pair correlation function of liquid He^4 (experimental).

very small k. The noteworthy feature of Feynman's theory is that there are no *ad hoc* adjustable parameters.*

For very small k the wave function represents a sound wave excitation of the system. That this is so follows from the property of Ψ_0. Suppose Ψ_0 is the ground state wave function of the ideal gas (i.e., a constant). Then (18.27) is the wave function for the state in which a single particle of momentum $\mathbf{k}$ is excited. Certainly this is no sound wave. How does $\Psi_{\mathbf{k}}$ look for the ideal gas? Since in this case Ψ_0 is a constant, it is sufficient to look at the factor $\sum_j \exp(i\mathbf{k} \cdot \mathbf{r}_j)$. It has maximum magnitude when every $\mathbf{r}_j$ is such that $\mathbf{k} \cdot \mathbf{r}_j = 2\pi n$ ($n = 0, 1, 2, \ldots$). Otherwise it is practically zero. Thus for isolated points in configuration space $\Psi_{\mathbf{k}}$ is very large. Everywhere else it is zero.

For liquid He^4, when Ψ_0 is of the form pictured earlier, the situation becomes completely different. To obtain an idea of what $\Psi_{\mathbf{k}}$ looks like

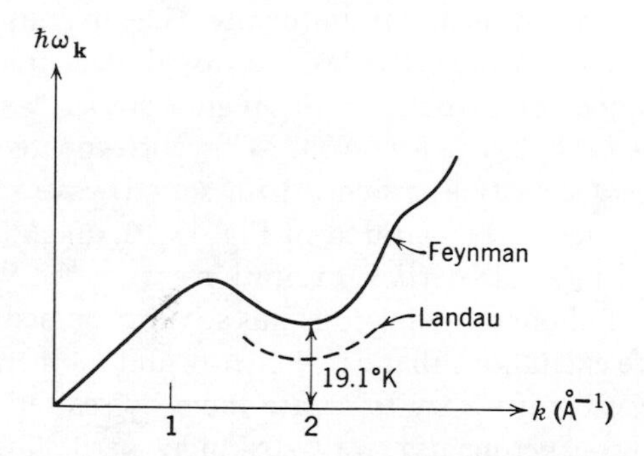

Fig. 18.8. Energy spectrum of elementary excitations in liquid He^4 in Feynman's theory.

* A better trial wave function that leads to more quantitatively satisfactory results has been devised by R. P. Feynman and M. Cohen, *Phys. Rev.*, **102,** 1189 (1956).

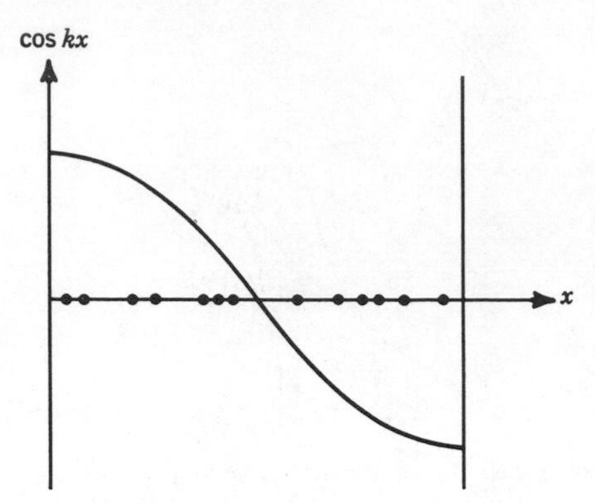

Fig. 18.9. To find $\Sigma_j \cos(kx_j)$, mark off the values of $x_1, \ldots, x_N$ on the abscissa and read off each term $\cos(kx_j)$ from this graph, which is drawn for $kr_0 \ll 1$, where $r_0 =$ average interparticle distance.

here, consider only one space dimension. The real part of $\Psi_{\mathbf{k}}$ is

$$\sum_{j=1}^{N} \cos(kx_j)\Psi_0(x_1, \ldots, x_N)$$

The behavior of this function is markedly different for $kr_0 \ll 1$ and for $kr_0 \gg 1$, where r_0 is the average interparticle distance. Plot $\cos(kx)$ as a function of x, and place N points on the abscissa representing the positions of the N particles. Then each term $\cos(kx_j)$ can be read off the graph. For $kr_0 \ll 1$, such a graph is shown in Fig. 18.9. The factor $\sum_j \cos(kx_j)$ is zero when the particles are uniformly distributed, for then kx_j tends to be random, but the factor Ψ_0 becomes zero when any two particles "touch." Thus there is an opposing tendency between the first factor and the second. The former tends to bunch the particles together, whereas the latter tends to spread them out uniformly. The magnitude of the wave function is largest when the particles are distributed in space in such a way that the number of particles in the region $\cos kx > 0$ exceeds those in the region $\cos kx < 0$, or vice versa. This corresponds to a sinusoidal spatial variation of density and hence to a sound wave. When $kr_0 \gg 1$, the situation becomes that illustrated in Fig. 18.10, and $\Psi_{\mathbf{k}}$ looks like that of the ideal gas (i.e., single-particle excitation).

The existence of phonons and rotons has so far been deduced indirectly; but since they are excitations that carry energy and momentum, they must be directly observable if we try to create them by scattering neutrons off liquid He^4.* The experiments† very strikingly verify the reality of the

* M. Cohen and R. P. Feynman, *Phys. Rev.*, **107**, 13 (1957).

† Palevsky, Otnes, and Larsson, *Phys. Rev.*, **112**, 11 (1958); Yarnell, Arnold, Bendt, and Kerr, *Phys. Rev.*, **113**, 1379 (1959). The measurements are made at about 1°K.

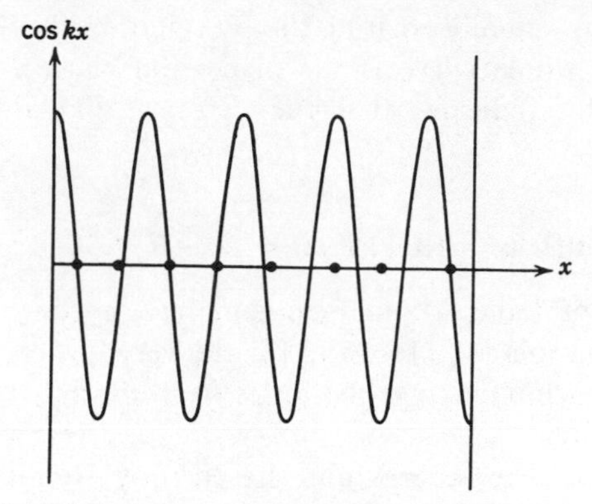

Fig. 18.10. This is the same graph as 18.9, but for $kr_0 \gg 1$.

phonon and the roton. In these experiments the energy spectrum $\omega_{\mathbf{k}}$ is directly observed. The results are shown in Fig. 18.11, together with the Landau curve for comparison. The experimental values of the constants of the Landau theory are

$$\begin{aligned} c &= (239 \pm 5)\ \text{m/sec} \\ \frac{\Delta}{k} &= (8.65 \pm 0.04)^\circ\text{K} \\ k_0 &= (1.92 \pm 0.01)\text{Å}^{-1} \\ \frac{\sigma}{m} &= 0.16 \pm 0.01 \end{aligned} \tag{18.31}$$

It is noted that the entire curve of $\omega_{\mathbf{k}}$, from the phonon part up through the roton part, including a maximum in between, has been measured

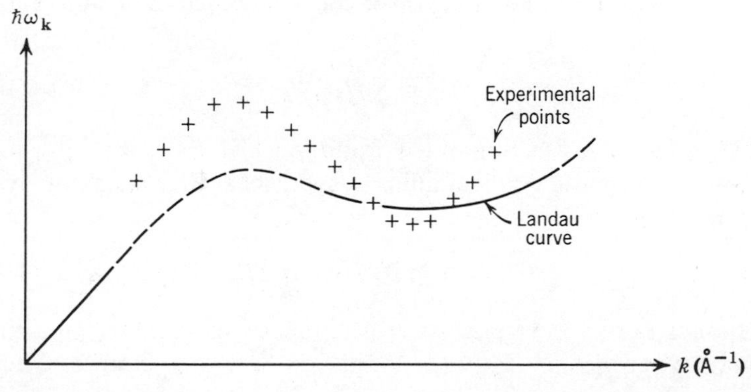

Fig. 18.11. Experimental-energy spectrum of elementary excitations in liquid He^4 from neutron scattering.

continuously by actually creating these excitations in the liquid. It would seem that the whole curve is the dispersion curve for a single type of excitation. This indicates that the "roton," like the phonon, has no intrinsic spin.*

18.4 EQUILIBRIUM PROPERTIES NEAR ABSOLUTE ZERO

The works of Landau and Feynman have made it plausible that the energy level formula (18.1) is valid for the very low excited states of liquid helium. All thermodynamic quantities near absolute zero can be deduced from this formula.

The Helmholtz free energy and the entropy are given by

$$A = E_0 - kT \sum_{\mathbf{k} \neq 0} \log [1 + \langle n_{\mathbf{k}} \rangle] \tag{18.32}$$

$$\frac{S}{k} = \frac{1}{kT} \sum_{\mathbf{k} \neq 0} \hbar\omega_{\mathbf{k}} \langle n_{\mathbf{k}} \rangle + \sum_{\mathbf{k} \neq 0} \log [1 + \langle n_{\mathbf{k}} \rangle] \tag{18.33}$$

where $\langle n_{\mathbf{k}} \rangle$ is given by (18.3). The internal energy and the specific heat have been calculated earlier. These formulas are valid only when

$$\frac{1}{N} \sum_{\mathbf{k} \neq 0} \langle n_{\mathbf{k}} \rangle \ll 1 \tag{18.34}$$

The entropy of the liquid is due to the elementary excitations. Thus we can interpret the dilute gas of elementary excitations to be the normal fluid, and the rest of the system—the background out of which the excitations arise—the superfluid. At absolute zero $\Sigma \langle n_{\mathbf{k}} \rangle = 0$, and the liquid is a pure superfluid. As the temperature increases, $\Sigma \langle n_{\mathbf{k}} \rangle$ increases, and the amount of normal-fluid component increases.

To present a more concrete picture of the two fluids let us consider the wave function of the liquid. Let $f(\mathbf{r})$ be the wave function of a wave packet and let

$$\Psi_f \equiv \sum_{j=1}^{N} f(\mathbf{r}_j) \Psi_0 \tag{18.35}$$

Then Ψ_f is the wave function of the liquid in which there is a wave packet of phonons. Suppose $g(\mathbf{r})$ is another wave packet that does not overlap with $f(\mathbf{r})$. Then

$$\Psi_{fg} \equiv \sum_{i=1}^{N} \sum_{j=1}^{N} f(\mathbf{r}_i) g(\mathbf{r}_j) \Psi_0 \tag{18.36}$$

corresponds to two wave packets of phonons in the liquids. In a similar manner we can define Ψ_{fgh}, etc. At a very low temperature, the wave

* T. D. Lee and F. Mohling, *Phys. Rev. Lett.*, **2**, 284 (1959), have suggested experiments to measure directly the spin of the "roton."

function of the liquid may be considered to be an *incoherent** superposition of such wave functions, representing 0, 1, 2, . . . wave packets of phonons, with coefficients depending on the temperature such that the average phonon occupation number is $\langle n_{\mathbf{k}} \rangle$. The ground state wave function Ψ_0, which is a factor common to (18.35), (18.36), and other similar phonon wave functions, may be looked on as the wave function of the superfluid. This interpretation is only valid in, and cannot be extended beyond, the region of very low temperatures, in which (18.34) holds. The fact that Ψ_0 is the same in (18.35) and (18.36) is an approximation which cannot be maintained if many phonons are excited.

If we are only concerned with properties of the liquid in absolute equilibrium, the two fluid model cannot be made more definite. For example, we do not know how to define ρ_n and ρ_s, because it is not clear what the "mass" of an elementary excitation should be. In fact, the main function of the two-fluid model is to describe nonequilibrium phenomena, such as the second sound. To derive the two-fluid model, we must consider the kinetic theory of He II.

18.5 MOTION OF THE SUPERFLUID

To discuss the kinetic theory† of He II it is important for us to make sure that no possible mode of motion of the liquid has so far escaped our attention. We suspect that some have.

The one-excitation wave function $\Psi_{\mathbf{k}}$ of (18.27) has the energy $E_0 + \hbar\omega_{\mathbf{k}}$ and the momentum $\mathbf{k}$. A state of the liquid characterized by the occupation numbers $\{n_{\mathbf{k}}\}$, with $\Sigma\, n_{\mathbf{k}} \ll N$, should have a total energy E_n and total momentum $\mathbf{P}_n$ given by‡

$$
\begin{aligned}
E_n &= E_0 + \sum_{\mathbf{k}\neq 0} \omega_{\mathbf{k}} n_{\mathbf{k}} \\
\mathbf{P}_n &= \sum_{\mathbf{k}\neq 0} \mathbf{k} n_{\mathbf{k}}
\end{aligned}
\tag{18.37}
$$

But the total momentum for such a state is due entirely to the elementary excitations—the normal fluid. It is intuitively obvious that there must exist states for which the total momentum is shared by both the normal and the superfluid. That this is the case will now be shown.

Let $\mathbf{r}_1, \ldots, \mathbf{r}_N$ be the positions of the N atoms in the liquid, and let

* Because we must represent the liquid by an ensemble. See Sec. 9.1.

† The discussion of the kinetic theory of He II will be developed along a line similar to that developed for the hard-sphere Bose gas by T. D. Lee and C. N. Yang, *Phys. Rev.*, **113**, 1406 (1959).

‡ From now on we use units in which $\hbar = 1$.

$\mathbf{p}_1, \ldots, \mathbf{p}_N$ be the momentum operators for the N atoms in the liquid, with the usual commutation rules. Consider the transformation

$$\begin{aligned} \mathbf{r}_j' &= \mathbf{r}_j \\ \mathbf{p}_j' &= \mathbf{p}_j + \mathbf{k}_s \end{aligned} \qquad (j = 1, \ldots, N) \tag{18.38}$$

where $\mathbf{k}_s$ is an arbitrary wave number vector. This transformation is canonical, because $[\mathbf{r}_i', \mathbf{p}_j'] = [\mathbf{r}_i, \mathbf{p}_j]$. In fact,

$$\begin{aligned} \mathbf{r}_j' &= e^{-iG}\mathbf{r}_j e^{iG} \\ \mathbf{p}_j' &= e^{-iG}\mathbf{p}_j e^{iG} \end{aligned} \qquad (j = 1, \ldots, N) \tag{18.39}$$

where G is the hermitian operator

$$G = \mathbf{k}_s \cdot \sum_{j=1}^{N} \mathbf{r}_j \tag{18.40}$$

Let H and $\mathbf{P}_{\text{op}}$ be respectively the Hamiltonian and the total momentum operator of the system:

$$\begin{aligned} H &= -\frac{1}{2m} \sum_{j=1}^{N} \nabla_j^2 + \sum_{i<j} v(|\mathbf{r}_i - \mathbf{r}_j|) \\ \mathbf{P}_{\text{op}} &= \frac{1}{i} \sum_{j=1}^{N} \nabla_j \end{aligned} \tag{18.41}$$

Then*

$$\begin{aligned} e^{-iG}He^{iG} &= H + \frac{1}{m}\mathbf{k}_s \cdot \mathbf{P}_{\text{op}} + \frac{Nk_s^2}{2m} \\ e^{-iG}\mathbf{P}_{\text{op}}e^{iG} &= \mathbf{P}_{\text{op}} + N\mathbf{k}_s \end{aligned} \tag{18.42}$$

Therefore, if Ψ is a simultaneous eigenfunction of H and $\mathbf{P}_{\text{op}}$, so is $e^{-iG}\Psi$. In addition, if (18.37) are eigenvalues of H and $\mathbf{P}_{\text{op}}$, so are the following:

$$E_n(\mathbf{k}_s) = E_0 + \sum_{\mathbf{q} \neq 0} \left(\omega_{\mathbf{q}} + \frac{1}{m} \mathbf{k}_s \cdot \mathbf{q} \right) n_{\mathbf{q}} + \frac{Nk_s^2}{2m} \tag{18.43}$$

$$\mathbf{P}_n(\mathbf{k}_s) = \sum_{\mathbf{q} \neq 0} \mathbf{q} n_{\mathbf{q}} + N\mathbf{k}_s \tag{18.44}$$

The vector $\mathbf{k}_s$ is a new quantum number labeling a set of eigenvalues which includes (18.37) as a special case, for which $\mathbf{k}_s = 0$.

For a physical interpretation of the formulas (18.43) and (18.44), we note that the transformation generated by G is a Gallilean transformation. If Ψ_0 is the ground state wave function for a liquid at rest, then $e^{-iG}\Psi_0$ is

**Proof.* $e^{-iG}He^{iG} = H - i[G, H] - \frac{1}{2!}[G, [G, H]] - \frac{i}{3!}[G, [G, [G, H]]] + \cdots$.
The series terminates after the third term.

the same wave function seen in a uniformly moving coordinate system. We have

$$\mathbf{P}_{\text{op}}\, e^{-iG}\Psi_0 = e^{-iG}(e^{iG}\mathbf{P}_{\text{op}}\, e^{-iG})\Psi_0 = e^{-iG}(\mathbf{P}_{\text{op}} + N\mathbf{k}_s)\Psi_0$$
$$= N\mathbf{k}_s\, e^{-iG}\Psi_0$$

Therefore $N\mathbf{k}_s$ is the momentum of the superfluid. Now consider the wave function $e^{-iG}\Psi_{\mathbf{q}}$ where $\Psi_{\mathbf{q}}$ is defined by (18.27). We have

$$e^{-iG}\Psi_{\mathbf{q}} = e^{-iG}\sum_{j=1}^{N} e^{i\mathbf{q}\cdot\mathbf{r}_j}\Psi_0 = \sum_{j=1}^{N} e^{i\mathbf{q}\cdot\mathbf{r}_j}(e^{-iG}\Psi_0)$$

Furthermore,

$$\mathbf{P}_{\text{op}}\, e^{-iG}\Psi_{\mathbf{q}} = (\mathbf{q} + N\mathbf{k}_s)e^{-iG}\Psi_{\mathbf{q}}$$

Therefore the vector $\mathbf{q}$ in (18.43) and (18.44) is the wave number vector of an excitation with respect to the moving superfluid. The momentum of the excitation is $\mathbf{q}$ in a frame moving with the superfluid; but it is $\mathbf{q} + N\mathbf{k}_s$ in a frame at rest in the laboratory.

In the calculation of the last section we did not take the states with $\mathbf{k}_s \neq 0$ into account. Have we then committed an error? The answer is no, because in absolute equilibrium $\langle \mathbf{k}_s \rangle = 0$, if the liquid is at rest. To prove this we prove a more general result, namely, in absolute equilibrium

$$\langle \mathbf{k}_s \rangle = \frac{\mathbf{P}}{N} \tag{18.45}$$

where $\mathbf{P}$ is a given parameter, the total momentum of the liquid.

Let us calculate the partition function of the liquid with the energy levels (18.43), subject to the condition that

$$\mathbf{P} = \sum_{\mathbf{q}\neq 0} \mathbf{q} n_{\mathbf{q}} + n\mathbf{k}_s \tag{18.46}$$

where $\mathbf{P}$ is a given vector. The partition function will be a function of $\mathbf{P}$ and also of N, v, and T, but we leave the latter variables understood. Thus

$$Q(\mathbf{P}) = e^{-\beta E_0} \sum_{\mathbf{k}_s, \{n\}}{}' \exp\left\{-\beta\left[\frac{Nk_s^{\,2}}{2m} + \sum_{\mathbf{q}\neq 0}\left(\omega_{\mathbf{q}} + \frac{\mathbf{q}\cdot\mathbf{k}_s}{m}\right)n_{\mathbf{q}}\right]\right\} \tag{18.47}$$

where the sum Σ' extends over all sets $\{n_{\mathbf{q}}\}$ and all values of $\mathbf{k}_s$ that satisfy (18.46). Define the following generating function:

$$\mathscr{G}(\mathbf{w}) \equiv \sum_{\mathbf{P}} e^{\beta \mathbf{w}\cdot\mathbf{P}} Q(\mathbf{P}) \tag{18.48}$$

where the sum over $\mathbf{P}$ extends over all vectors of the form

$$\mathbf{P} = \frac{2\pi \mathbf{n}}{L} \tag{18.49}$$

where $L = V^{1/3}$ and $\mathbf{n}$ is a vector whose components are $0, \pm 1, \pm 2, \ldots$. If $t_j = e^{(2\pi\beta/L)w_j}$ $(j = 1, 2, 3)$, then

$$Q(\mathbf{P}) = \frac{1}{(2\pi i)^3} \oint dt_1 \oint dt_2 \oint dt_3 \frac{\mathscr{G}(\mathbf{w})}{t_1^{n_1+1} t_2^{n_2+1} t_3^{n_3+1}}$$

where $n_j \equiv (L/2\pi)P_j$ and the contours of integrations are circles about $t_j = 0$. We can also write

$$Q(\mathbf{P}) = \frac{1}{V}\left(\frac{\beta}{i}\right)^3 \int_0^{2\pi} dw_1 \int_0^{2\pi} dw_2 \int_0^{2\pi} dw_3 e^{-\beta \mathbf{P}\cdot\mathbf{w}} \mathscr{G}(\mathbf{w}) \tag{18.50}$$

It is straightforward to verify that

$$\mathscr{G}(\mathbf{w}) = \frac{V}{(2\pi)^3} e^{-\beta E_0 + \frac{1}{2}N\beta m w^2} \int d^3K e^{-N[\beta(k^2/2m) + g(\mathbf{K})]} \tag{18.51}$$

where

$$g(\mathbf{K}) = \frac{v}{(2\pi)^3} \int d^3q \log\left(1 - e^{-\beta(\omega_\mathbf{q} + \mathbf{K}\cdot\mathbf{q})}\right) \tag{18.52}$$

As $N \to \infty$, we can evaluate $\mathscr{G}(\mathbf{w})$ by the method of saddle point integration* and find that

$$\frac{1}{N} \log \mathscr{G}(\mathbf{w}) = -\frac{\beta E_0}{N} + \frac{1}{2}\beta m w^2 - \frac{v}{(2\pi)^3} \int d^3q \log\left(1 - e^{-\beta\omega_\mathbf{q}}\right) \tag{18.53}$$

Let Q_0 denote the partition function for the energy levels (18.37). By substituting (18.53) into (18.50) we obtain, as $N \to \infty$,

$$\frac{1}{N} \log Q(\mathbf{P}) = \frac{1}{N} \log Q_0 - \frac{\beta}{2m}\left(\frac{\mathbf{P}}{N}\right)^2 \tag{18.54}$$

Therefore $\langle n_\mathbf{q} \rangle$ is still given by (18.3). Taking the ensemble average of (18.46), and noting that $\Sigma\, \mathbf{q}\langle n_\mathbf{q}\rangle = 0$, we obtain

$$\mathbf{P} = N\langle \mathbf{k}_s \rangle$$

which is (18.45). This result shows that in absolute equilibrium, in a coordinate system moving with the superfluid, the total momentum of the gas of excitations is zero. In other words, *there is no relative motion between the center of masses of the normal fluid and the superfluid.*

Suppose an initial situation violating (18.45) is created. Then the system is not in equilibrium, and after a certain length of time (18.45) will be established. Statistical mechanics does not tell us, however, how long it will take for equilibrium to be established. In the absence of a general

* See Sec. 10.1.

kinetic theory we can only guess the answer. We argue that the establishment of (18.45) will take a macroscopically long time. In fact, to the extent that the energy levels (18.43) are valid the absolute equilibrium can never be established. The reason is as follows. To establish the relation (18.45) momentum must be transferred from the gas of excitations to somewhere else. According to (18.43), however, the excitations are stable, and hence there is no mechanism for momentum transfer. In a more accurate treatment, the elementary excitations should interact with one another and have finite lifetimes. Only in such an accurate treatment can we discuss the approach to absolute equilibrium.

We assume that it is a good approximation to take the lifetime of an elementary excitation to be infinite. The kinetic theory which results from this assumption is analogous to the zero-order approximation in the classical kinetic theory of gases and leads to nonviscous hydrodynamics. Specifically, the assumption is that $\mathbf{P}$ and $\mathbf{k}_s$ are *independent* variables. Such a treatment does not correspond to the situation of absolute equilibrium; it corresponds instead to situations of *quasi-equilibrium*.

By this assumption, the liquid is endowed with a new degree of freedom, namely, the relative motion between the normal fluid and the superfluid. This new degree of freedom is the essence of the transport phenomena in He II known collectively as superfluidity.

18.6 KINETIC THEORY NEAR ABSOLUTE ZERO

Formulation of the Problem

To dispel any delusions, we should point out from the beginning that no general kinetic theory exists in the way that a general statistical mechanics exists. The present discussion is heuristic and is modeled after the familiar classical kinetic theory of gases. These considerations are based on the assumption that there exists a length d such that the following conditions are satisfied:

(*a*) d is large compared to the thermal wavelength and the average interatomic distance but small by macroscopic standards, so that the properties of a sample of liquid of volume d^3 are essentially those of an infinite system. Thus in a volume d^3 there should be enough atoms for us to apply statistics to them, but at the same time the volume should be so small that we may regard it as a "point" from the macroscopic point of view. It is clear that for this condition to be satisfied the temperature cannot be arbitrarily small.

(*b*) The liquid is such that all thermodynamic functions vary negligibly over the distance d. Then in any volume d^3 the system may be regarded as in equilibrium, with a given temperature, pressure, density, and total average momentum.

Let ϵ, s, $\mathbf{p}$ denote respectively the internal-energy density, entropy density, and momentum density of the liquid. We postulate definite expressions for the following currents, which, in the approximation in which all dissipative effects are neglected, are conserved:

$$\begin{aligned} &\mathbf{J}_\rho &&\text{(mass current)} \\ &\mathbf{J}_\epsilon &&\text{(energy current)} \\ &\mathbf{J}_s &&\text{(entropy current)} \\ &T_{ij} &&\text{(momentum current)} \end{aligned} \tag{18.55}$$

where T_{ij} is the ith component of the "current of the jth component of momentum." The laws that govern the variation in space and time of the local thermodynamic functions are the conservation laws

$$\frac{\partial \rho}{\partial t} + \nabla \cdot \mathbf{J}_\rho = 0 \qquad \text{(conservation of mass)} \tag{18.56}$$

$$\frac{\partial \epsilon}{\partial t} + \nabla \cdot \mathbf{J}_\epsilon = 0 \qquad \text{(conservation of energy)} \tag{18.57}$$

$$\frac{\partial s}{\partial t} + \nabla \cdot \mathbf{J}_s = 0 \qquad \text{(conservation of entropy)} \tag{18.58}$$

$$\frac{\partial p_i}{\partial t} + \sum_{j=1}^{3} \frac{\partial T_{ij}}{\partial x_j} = 0 \qquad \text{(conservation of momentum)} \tag{18.59}$$

When the explicit forms for the currents are substituted into these equations, they become the equations of hydrodynamics in the lowest approximation.

Quasi-Equilibrium Thermodynamics

According to the scheme just formulated we should first investigate the equilibrium properties of an infinite liquid with uniform temperature T, specific volume v, and total momentum $\mathbf{P}$. The term "equilibrium," however, includes both absolute equilibrium and quasi-equilibrium, in the sense explained at the end of Sec. 18.5. Therefore we admit a new independent thermodynamic parameter, the vector $\mathbf{k}_s$ appearing in (18.44).

The quasi-equilibrium thermodynamics can be derived from the partition function

$$Q(\mathbf{P}, \mathbf{k}_s) = \exp\left(-\beta E_0 - N\beta \frac{k_s^2}{2m}\right) \sum_{\{n\}}{}' \exp\left[-\beta \sum_{\mathbf{q}} \left(\omega_{\mathbf{q}} + \frac{\mathbf{k}_s \cdot \mathbf{q}}{m}\right) n_{\mathbf{q}}\right] \tag{18.60}$$

where the sum Σ' is subject to the condition

$$\sum_{\mathbf{q}\neq 0} \mathbf{q} n_{\mathbf{q}} = \mathbf{P} - N\mathbf{k}_s \tag{18.61}$$

and the dependence of the partition function on T and v is understood. We can again express the partition function in a form analogous to (18.50), namely

$$Q(\mathbf{P}, \mathbf{k}_s) = \frac{1}{V}\left(\frac{\beta}{i}\right)^3 e^{-\beta E_0 - N\beta(k_s^2/2m)} \int_0^{2\pi} dw_1 \int_0^{2\pi} dw_2 \int_0^{2\pi} dw_3 e^{-\beta(\mathbf{P}-N\mathbf{k}_s)\cdot\mathbf{w}}\, \mathscr{G}(\mathbf{w}, \mathbf{k}_s) \tag{18.62}$$

where

$$\mathscr{G}(\mathbf{w}, \mathbf{k}_s) \equiv \sum_{\{n\}} \exp\left\{-\beta \sum_{\mathbf{q}} \left[\omega_{\mathbf{q}} + \left(\frac{\mathbf{k}_s}{m} - \mathbf{w}\right)\cdot \mathbf{q}\right] n_{\mathbf{q}}\right\} \tag{18.63}$$

or

$$\frac{1}{V}\log \mathscr{G}(\mathbf{w}, \mathbf{k}_s) = \int \frac{d^3q}{(2\pi)^3} \log\left(1 - e^{-\beta(\omega_{\mathbf{q}} - \mathbf{u}\cdot\mathbf{q})}\right) \tag{18.64}$$

where $\mathbf{u} = \mathbf{w} - (\mathbf{k}_s/m)$. Substituting this into (18.62) and using the method of saddle point integration, we find that as $N \to \infty$, $V \to \infty$, with fixed $\rho = mN/V$,

$$\frac{1}{V}\log Q(\mathbf{P}, \mathbf{k}_s) = -\frac{\beta E_0}{V} - \beta\mathscr{E} - \int \frac{d^3q}{(2\pi)^3} \log\left(1 - e^{-\beta(\omega_{\mathbf{q}} - \mathbf{u}\cdot\mathbf{q})}\right) \tag{18.65}$$

where

$$\mathscr{E} \equiv \tfrac{1}{2}\rho\left(\frac{\mathbf{k}_s}{m}\right)^2 + \left(\frac{\mathbf{P}}{V} - \frac{\rho\mathbf{k}_s}{m}\right)\cdot\frac{\mathbf{k}_s}{m} \tag{18.66}$$

and where $\mathbf{u}$ is determined by the saddle point condition

$$\frac{\mathbf{P}}{V} - \frac{\rho\mathbf{k}_s}{m} = \int \frac{d^3q}{(2\pi)^3} \frac{\mathbf{q}}{e^{\beta(\omega_{\mathbf{q}} - \mathbf{u}\cdot\mathbf{q})} - 1} \tag{18.67}$$

If we set $\mathbf{k}_s = \mathbf{P}/N$, then $\mathbf{u} = 0$, and (18.65) reduces to (18.54), the case of absolute equilibrium. The average occupation number of elementary excitations is easily found to be

$$\langle n_{\mathbf{q}}\rangle = \frac{1}{e^{\beta(\omega_{\mathbf{q}} - \mathbf{u}\cdot\mathbf{q})} - 1} \tag{18.68}$$

The thermodynamic functions follow straightforwardly from (18.65). The pressure is

$$P = P_0 + kT\int \frac{d^3q}{(2\pi)^3} \log\left(1 + \langle n_{\mathbf{q}}\rangle\right) \tag{18.69}$$

where P_0 is the pressure of the liquid at absolute zero. The Helmholtz free energy per unit volume is

$$a = \epsilon_0 + \mathscr{E} - (P - P_0) \tag{18.70}$$

where $\epsilon_0 = E_0/V$ is the ground state energy per unit volume. The Gibbs potential per unit volume is

$$g = \epsilon_0 + \mathscr{E} + P_0 \tag{18.71}$$

The internal energy per unit volume is

$$\epsilon = \epsilon_0 + \mathscr{E} + \int \frac{d^3q}{(2\pi)^3}\, \omega_{\mathbf{q}} \langle n_{\mathbf{q}} \rangle \tag{18.72}$$

The entropy per unit volume (divided by Boltzmann's constant) is

$$\frac{s}{k} = \frac{\epsilon - a}{kT} = \int \frac{d^3q}{(2\pi)^3} \langle n_{\mathbf{q}} \rangle \left[\frac{\omega_{\mathbf{q}}}{kT} + \frac{\log(1 + \langle n_{\mathbf{q}} \rangle)}{\langle n_{\mathbf{q}} \rangle} \right] \tag{18.73}$$

To give the vector $\mathbf{u}$ a physical interpretation, note the identity

$$\mathbf{u} = \frac{\int d^3q \langle n_{\mathbf{q}} \rangle \nabla_{\mathbf{q}} \omega_{\mathbf{q}}}{\int d^3q \langle n_{\mathbf{q}} \rangle} \tag{18.74}$$

The proof is as follows. By (18.68)

$$\nabla_{\mathbf{q}} \langle n_{\mathbf{q}} \rangle = -\beta \langle n_{\mathbf{q}} \rangle (1 + \langle n_{\mathbf{q}} \rangle)(\nabla_{\mathbf{q}} \omega_{\mathbf{q}} - \mathbf{u})$$

Multiplying both sides by $\langle n_{\mathbf{q}} \rangle^j$ $(j = 0, 1, 2, \ldots)$ we obtain

$$\nabla_{\mathbf{q}} \langle n_{\mathbf{p}} \rangle^{j+1} = -\beta (j+1) \langle n_{\mathbf{q}} \rangle^{j+1} (1 + \langle n_{\mathbf{q}} \rangle)(\nabla_{\mathbf{q}} \omega_{\mathbf{q}} - \mathbf{u})$$

Integrating over all $\mathbf{q}$ yields

$$0 = \int d^3q \langle n_{\mathbf{q}} \rangle^{j+1} (1 + \langle n_q \rangle)(\nabla_{\mathbf{q}} \omega_{\mathbf{q}} - \mathbf{u}) \qquad (j = 0, 1, 2, \ldots)$$

It follows that if $f(x)$ is a function that can be expanded in powers of x, then

$$\int d^3q \langle n_{\mathbf{q}} \rangle (\nabla_{\mathbf{q}} \omega_{\mathbf{q}} - \mathbf{u}) f(\langle n_{\mathbf{q}} \rangle) = 0$$

Putting $f(x) = 1$, we obtain (18.74).

Now $\omega_{\mathbf{q}}$ is the energy of an excitation whose momentum with respect to the superfluid is $\mathbf{q}$. Therefore $\nabla_{\mathbf{q}} \omega_{\mathbf{q}}$ is the group velocity of the excitation with respect to the superfluid. By (18.74), $\mathbf{u}$ is the average group velocity of an excitation with respect to the superfluid.

At sufficiently low temperatures only the phonon part of $\omega_{\mathbf{q}}$ need be taken into account. All the thermodynamic functions can then be evaluated explicitly. As an example we calculate $\mathbf{u}$ from (18.67). The integral on the right-hand side of (18.67) must be proportional to $\mathbf{u}$, because that is the only fixed vector in the integrand. Hence

$$\int d^3q \frac{\mathbf{q}}{e^{\beta(cq - \mathbf{u}\cdot\mathbf{q})} - 1} = \frac{\mathbf{u}}{u^2} \int d^3\mathbf{q} \frac{\mathbf{q}\cdot\mathbf{u}}{e^{\beta(cq - \mathbf{u}\cdot\mathbf{q})} - 1}$$

Let θ be the angle between $\mathbf{u}$ and $\mathbf{q}$, $x = -\cos\theta$, and $y = \beta cq[1 + (u/c)x]$. Then

$$\frac{\mathbf{u}}{u^2}\int d^3q\,\frac{\mathbf{q}\cdot\mathbf{u}}{e^{\beta(cq-\mathbf{u}\cdot\mathbf{q})}-1} = -\frac{\mathbf{u}}{u(\beta c)^4}\int_{-1}^{+1}dx\,\frac{x}{[1+(u/c)x]^4}\int_0^\infty dy\,\frac{y^3}{e^y-1}$$
$$= \frac{16\pi^5}{45\beta^4c^5}\frac{\mathbf{u}}{[1-(u/c)^2]^3}$$

Therefore $\mathbf{u}$ is determined by

$$\frac{\mathbf{P}}{V} - \frac{\rho\mathbf{k}_s}{m} = \frac{2\pi^2}{45}\frac{(kT)^4}{c^5}\frac{\mathbf{u}}{[1-(u/c)^2]^3} \tag{18.75}$$

Similarly we find that

$$P = P_0 + \frac{\pi^2}{90}\frac{(kT)^4}{c^3}\frac{1}{[1-(u/c)^2]^2} \tag{18.76}$$

$$\epsilon = \epsilon_0 + \mathscr{E} + \frac{\pi^2}{30}\frac{(kT)^4}{c^3}\frac{1+\frac{1}{3}(u/c)^2}{[1-(u/c)^2]^3} \tag{18.77}$$

$$\frac{s}{k} = \frac{2\pi^2}{45}\left(\frac{kT}{c}\right)^3\frac{1+\frac{1}{4}(u/c)^2}{[1-(u/c)^2]^3} \tag{18.78}$$

Thus in absolute equilibrium, when $\mathscr{E} = 0$ and $\mathbf{u} = 0$, we have

$$P - P_0 = \tfrac{1}{3}(\epsilon - \epsilon_0) = \tfrac{4}{9}Ts$$

A Two-Fluid Model

We now define, *purely for mathematical convenience*, the mass densities ρ_n, ρ_s and the velocities $\mathbf{v}_n$, $\mathbf{v}_s$ of the normal fluid and the superfluid:

$$\mathbf{v}_s \equiv \frac{\mathbf{k}_s}{m} \tag{18.79}$$

$$\mathbf{v}_n = \mathbf{v}_s + \frac{\int d^3q\,\langle n_\mathbf{q}\rangle\nabla_\mathbf{q}\omega_\mathbf{q}}{\int d^3q\,\langle n_\mathbf{q}\rangle} = \mathbf{v}_n + \mathbf{u} \tag{18.80}$$

$$\rho_s \equiv \rho - \rho_n \tag{18.81}$$

$$\rho_n\mathbf{u} \equiv \int\frac{d^3q}{(2\pi)^3}\,\mathbf{q}\langle n_\mathbf{q}\rangle \tag{18.82}$$

In the definition for ρ_n we note that the integral $\int d^3q\,\mathbf{q}\langle n_\mathbf{q}\rangle$ must be proportional to $\mathbf{u}$, because there is no other vector in the integrand. Thus

$$\rho_n = \frac{1}{u^2}\int\frac{d^3q}{(2\pi)^3}\,\mathbf{u}\cdot\mathbf{q}\langle n_\mathbf{q}\rangle \tag{18.83}$$

With these definitions the mass density and the momentum density of the liquid can be expressed respectively as

$$\rho = \rho_n + \rho_s \tag{18.84}$$

$$\mathbf{p} \equiv \frac{\mathbf{P}}{V} = \rho_n \mathbf{v}_n + \rho_s \mathbf{v}_s \tag{18.85}$$

The quantity $\mathscr{E}$ defined by (18.66) becomes

$$\mathscr{E} = \tfrac{1}{2}\rho_s v_s^2 + \tfrac{1}{2}\rho_n v_n^2 - \tfrac{1}{2}\rho_n |\mathbf{v}_s - \mathbf{v}_n|^2 \tag{18.86}$$

At temperatures low enough so that only the phonon part of $\omega_{\mathbf{q}}$ need be considered, we have

$$\rho_n = \frac{2\pi^2}{45c^5} \frac{(kT)^4}{[1 - (u/c)^2]^3} \tag{18.87}$$

As a numerical example, $\rho_n/\rho \approx 10^{-6}$ at 0.5°K.

The definitions (18.79)–(18.82) correspond to the quantities introduced by Tisza in the phenomenological two-fluid model. Here we supply them on the basis of the molecular theory of Landau and Feynman. The following points, however, cannot be overemphasized:

(*a*) The definitions (18.79)–(18.82) are not unique.

(*b*) The definitions (18.79)–(18.82) are not necessary but merely convenient.

(*c*) The picture of He II as composed of two fluids cannot be taken too literally, because it is impossible to associate each fluid with definite groups of He atoms.

Equations of Hydrodynamics

The definitions (18.79)-(18.82) suggest a two-fluid physical picture of He II. On the basis of this picture, it seems reasonable to postulate the following:

$$\mathbf{J}_\epsilon = \rho_n \mathbf{v}_n + \rho_s \mathbf{v}_s \tag{18.88}$$

$$\mathbf{J}_s = s\mathbf{v}_n \tag{18.89}$$

$$T_{ij} = (\mathbf{v}_s)_i(\rho_s \mathbf{v}_s)_j + (\mathbf{v}_n)_i(\rho_n \mathbf{v}_n)_j + \delta_{ij} P \tag{18.90}$$

These currents have a feature not shared by the corresponding currents in an ordinary liquid; namely, there can be a flow of momentum and of entropy without a flow of mass. This feature clearly owes its origin to the independent degree of freedom $\mathbf{k}_s$. The conservation laws for these currents lead to the hydrodynamic equations

$$\frac{\partial \rho}{\partial t} + \nabla \cdot (\rho_n \mathbf{v}_n + \rho_s \mathbf{v}_s) = 0 \tag{18.91}$$

$$\frac{\partial s}{\partial t} + \nabla \cdot (s\mathbf{v}_n) = 0 \tag{18.92}$$

$$\frac{\partial}{\partial t}(\rho_n \mathbf{v}_n + \rho_s \mathbf{v}_s) + \sum_{i=1}^{3} \frac{\partial}{\partial x_i} [(\mathbf{v}_s)_i \rho_s \mathbf{v}_s + (\mathbf{v}_n)_i \rho_n \mathbf{v}_n] = -\nabla P \tag{18.93}$$

The equation corresponding to the conservation of energy remains to be discussed. With the present model it is not possible for us to arrive at an unambiguous energy current $\mathbf{J}_\epsilon$, because we do not know how $\omega_\mathbf{q}$, which contains the velocity of sound, depends on ρ_n and ρ_s. Consequently we cannot deduce which part of the energy density (18.72) is due to the superfluid and which part to the normal fluid.* Accordingly we adopt an alternative procedure through the following argument.†

Let the kinetic energy density of the relative motion between the normal fluid and the superfluid be defined as

$$\mathscr{E}' = \tfrac{1}{2}\rho_n \, |\mathbf{v}_n - \mathbf{v}_s|^2$$

Consider a flat slab of liquid of unit area and of thickness dx. Suppose that during the time dt the kinetic energy of relative motion within the slab decreases by the amount $d\mathscr{E}'\, dx$. Then there must be a net outflow of normal fluid from the cylinder. Consequently there is a heat current flowing from the slab, which we postulate to be

$$\mathbf{h} = Ts(\mathbf{v}_n - \mathbf{v}_s)$$

such that the amount of heat lost by the slab during the time dt is $\mathbf{h} \cdot \mathbf{n}\, dt$, where $\mathbf{n}$ is the normal vector to a flat face of the slab. Since there are no dissipative effects in our model, this heat flow must be reversible; and a reversible heat flow is equivalent to that produced by a Carnot engine. Therefore, if the two faces of the slab differ in temperature by dT, the outflow of the amount of heat $\mathbf{q} \cdot \mathbf{n}\, dt$ requires the amount of work $(\mathbf{q} \cdot \mathbf{n}\, dt)(dT/T)$ to be supplied. We assume that this work is supplied by $d\mathscr{E}'\, dx$. Hence

$$-d\mathscr{E}'\, dx = \mathbf{h} \cdot \mathbf{n} \frac{dT}{T}$$

* Lee and Yang, *Phys. Rev.*, **113,** 1406 (1959), were able to postulate $\mathbf{J}_\epsilon$ of the dilute hard-sphere Bose gas, for which $\omega_\mathbf{q}$ is precisely known. London (*op. cit.*, pp. 135–137) derives $\mathbf{J}_\epsilon$, but not from the molecular point of view.

† London, *op. cit.*, p. 78.

or
$$\frac{\partial \mathscr{E}'}{\partial t} + \frac{1}{T}\mathbf{h} \cdot \nabla T = 0$$

Using the explicit forms assumed for $\mathscr{E}'$ and $\mathbf{h}$, we obtain the desired equation

$$(\mathbf{v}_n - \mathbf{v}_s) \cdot \left\{\frac{\partial}{\partial t}\left[\rho_n(\mathbf{v}_n - \mathbf{v}_s)\right] + s\nabla T\right\} = 0 \tag{18.94}$$

Together with (18.91)–(18.93) this makes up the equations of hydrodynamics for He II.

In absolute equilibrium both $\mathbf{v}_n$ and $\mathbf{v}_s$ vanish, and ρ_n, ρ_s not only are uniform in space and constant in time but are also definite functions of T and ρ. The equations of hydrodynamics (18.91)–(18.94) certainly cannot claim to be valid except for situations only slightly different from that of absolute equilibrium. Accordingly we treat $\mathbf{v}_n$, $\mathbf{v}_s$, the space and time derivatives of ρ_n, ρ_s, and all thermodynamic functions as first-order small quantities. Neglecting quantities of the second-order smallness, we obtain from (18.91)–(18.94) the following linear equations of hydrodynamics:

$$\frac{\partial \rho}{\partial t} + \rho_n \nabla \cdot \mathbf{v}_n + \rho_s \nabla \cdot \mathbf{v}_s = 0 \tag{18.95}$$

$$\frac{\partial s}{\partial t} + s\nabla \cdot \mathbf{v}_n = 0 \tag{18.96}$$

$$\rho_n \frac{\partial \mathbf{v}_n}{\partial t} + \rho_s \frac{\partial \mathbf{v}_s}{\partial t} = -\nabla P \tag{18.97}$$

$$\rho_n\left(\frac{\partial \mathbf{v}_n}{\partial t} - \frac{\partial \mathbf{v}_s}{\partial t}\right) = -s\nabla T \tag{18.98}$$

Strictly speaking, we should add to the right-hand side of (18.98) an arbitrary vector perpendicular to $\mathbf{v}_n - \mathbf{v}_s$. We arbitrarily take this vector to be zero.

The fountain effect is an immediate consequence of (18.97) and (18.98). By subtracting one from the other we obtain

$$\rho \frac{\partial \mathbf{v}_s}{\partial t} + \nabla P - s\,\nabla T = 0$$

In a steady flow of the superfluid we have $\partial \mathbf{v}_s/\partial t = 0$, hence $\nabla P = s\,\nabla T$. This means that if two points on a streamline of $\mathbf{v}_s$ differ in temperature by ΔT, they must differ in pressure by

$$\Delta P = s\,\Delta T$$

18.7 SUPERFLUIDITY

First and Second Sound

We can eliminate $\mathbf{v}_n$ and $\mathbf{v}_s$ from (18.95)–(18.98) by subtracting or adding the appropriate equations, after taking their time derivatives or their gradients as may be necessary. This reduces the four equations to two, which in linearized forms read

$$\frac{\partial^2 \rho}{\partial t^2} - \nabla^2 P = 0$$
$$\frac{\partial^2 \rho}{\partial t^2} - \frac{\rho}{s}\frac{\partial^2 s}{\partial t^2} + \frac{\rho_s}{\rho_n} s \nabla^2 T = 0 \tag{18.99}$$

Although there are four quantities ρ, s, P, T in the equations (18.99), only two are to be regarded as independent. We choose these to be T and ρ. Then in the linear approximation we can write*

$$\frac{\partial^2 s}{\partial t^2} = \left(\frac{\partial s}{\partial T}\right)_\rho \frac{\partial^2 T}{\partial t^2} + \left(\frac{\partial s}{\partial \rho}\right)_T \frac{\partial^2 \rho}{\partial t^2}$$
$$\nabla^2 P = \left(\frac{\partial P}{\partial T}\right)_\rho \nabla^2 T + \left(\frac{\partial P}{\partial \rho}\right)_T \nabla^2 \rho$$

where the thermodynamic derivatives are to be calculated in absolute equilibrium. Substituting these into (18.99) we obtain a set of coupled equations for ρ and T:

$$\nabla^2 \rho - \frac{1}{c_1^2}\frac{\partial^2 \rho}{\partial t^2} + \gamma_1 \nabla^2 T = 0$$
$$\nabla^2 T - \frac{1}{c_2^2}\frac{\partial^2 T}{\partial t^2} + \gamma_2 \frac{\partial^2 \rho}{\partial t^2} = 0 \tag{18.100}$$

where

$$c_1 = \frac{1}{\sqrt{(\partial P/\partial \rho)_T}}$$
$$c_2 = \sqrt{\frac{s^2 \rho_s}{\rho \rho_n (\partial s/\partial T)_\rho}} \tag{18.101}$$
$$\gamma_1 = \frac{(\partial P/\partial T)_\rho}{(\partial P/\partial \rho)_T} = -\left(\frac{\partial \rho}{\partial T}\right)_P$$
$$\gamma_2 = \frac{\rho_n}{s\rho_s}\left[1 - \frac{\rho}{s}\left(\frac{\partial s}{\partial \rho}\right)_T\right]$$

* Note that, if P were a function of ρ alone, the first of (18.99) would be the usual sound wave equation. But P is a function of both ρ and T, because the usual adiabatic condition is lacking. (Compare Sec. 5.3).

At absolute zero $\gamma_1 = 0$, because it is proportional to the coefficient of thermal expansion, which vanishes by the third law of thermodynamics. Furthermore, $\gamma_2 = 0$ because $\rho_n/s \propto T$, as we can see from (18.78) and (18.87). By definition

$$c_1 \xrightarrow[T\to 0]{} c \tag{18.102}$$

which is an experimental parameter in our model. From (18.78) and (18.87) we obtain trivially

$$c_2 \xrightarrow[T\to 0]{} \frac{c}{\sqrt{3}} \tag{18.103}$$

Thus at absolute zero the two equations (18.100) are decoupled. The first describes a density wave of velocity c, whereas the second describes a thermal wave of velocity $c/\sqrt{3}$. These are called respectively first and second sound.

At finite temperatures (18.100) remain coupled. To solve them we seek solutions of the form

$$\begin{aligned} \rho &= \rho_0 + \rho_1 e^{i(\omega t - \mathbf{k}\cdot\mathbf{r})} \\ T &= T_0 + T_1 e^{i(\omega t - \mathbf{k}\cdot\mathbf{r})} \end{aligned} \tag{18.104}$$

When these are substituted into (18.100) we obtain, for given k, two equations for the two unknowns ω and ρ_1/T_1. The ratio ω/k is the velocity of these waves. We easily find that

$$\left(\frac{\omega}{k}\right)^2 = \tfrac{1}{2}\{[c_1^2 + c_2^2 + (c_1c_2)^2\gamma_1\gamma_2] \pm \sqrt{[(c_1+c_2)^2 + (c_1c_2)^2\gamma_1\gamma_2][(c_1-c_2)^2 + (c_1c_2)^2\gamma_1\gamma_2]}\} \tag{18.105}$$

This gives the two sound velocities. At finite temperatures the numbers $c_1, c_2, \gamma_1, \gamma_2$ cannot be calculated from our model; but they may be taken from experiments.

Superfluid Flow

The hydrodynamic flow of He II is not completely determined by the equations (18.95)–(18.98), because the boundary conditions for $\mathbf{v}_n$ and $\mathbf{v}_s$ have not been given. Hence we cannot use these equations to discuss such problems as the flow of He II past a wall, or the behavior of He II in a rotating container. In the following we merely mention some ideas that have been advanced in these problems. A complete theory is at present lacking.

Landau* suggests that near absolute zero He II can flow past a wall

* L. D. Landau, *J. Phys. U.S.S.R.*, **5**, 71 (1940).

with no friction whatsoever, provided the relative velocity is less than the velocity of sound c. To understand this we need only consider what happens if an external object with a velocity $\mathbf{v}_e$ is dragged through a stationary liquid. Since the only excitations of the liquid are phonons, the only ways energy and momentum can be imparted to the liquid are by (*a*) excitation of new phonons and (*b*) scattering of existing phonons. Suppose an amount of energy ΔE is transferred through excitation of a number of phonons, specified by the occupation numbers $\{n_{\mathbf{k}}\}$

$$\Delta E = \sum_{\mathbf{k}\neq 0} c\,|\mathbf{k}|\,n_{\mathbf{k}}$$

The momentum imparted to the system is necessarily

$$\Delta \mathbf{P} = \sum_{\mathbf{k}\neq 0} \mathbf{k} n_{\mathbf{k}}$$

Therefore

$$|\Delta \mathbf{P}| \leq \sum_{\mathbf{k}\neq 0} |\mathbf{k}|\,n_{\mathbf{k}}$$

or

$$c\,|\Delta \mathbf{P}| \leq \Delta E$$

On the other hand, if the external object loses the amount of energy ΔE and momentum $\Delta \mathbf{P}$, we must have

$$\Delta E = \mathbf{v}_e \cdot \Delta \mathbf{P}$$

This is impossible unless $|\mathbf{v}_e| > c$. At low temperatures the transfer of energy and momentum through scattering of existing phonons can be neglected because as $T \to 0$ the number of phonons becomes zero. It is noted that the argument depends on the linearity of the phonon energy spectrum and does not apply to an ideal gas of bosons.

The foregoing argument is valid at absolute zero, but it breaks down at a higher temperature when there are many phonons present. The external object may now transfer energy and momentum by scattering the phonons. Experiments on the flow of He II past a wall at 1°K indicate that the critical velocity is orders of magnitude smaller than c.

London* advances the proposition that we should append to the equations of hydrodynamics the *equation of motion*

$$\nabla \times \mathbf{v}_s = 0$$

With this additional equation of motion we would, for example, be able to reach definite conclusions concerning the flow of He II in a rotating vessel. This proposal is based on the assumption that the superfluid may be identified with the ground state wave function Ψ_0 of (18.27), which is a factor common to all wave functions of the liquid when very few phonons

* London, *op. cit.*, p. 142.

are present. London assumes that a flow of the superfluid component is induced by a slow change of the macroscopic boundary conditions, so that Ψ_0 changes adiabatically. The flow pattern resulting from the adiabatic change of a single quantum state should be irrotational. Hence $\nabla \times \mathbf{v}_s = 0$.

London's suggestion appears to be highly plausible below 10^{-20} °K, when the thermal wavelength is of the same order as the size of the container of the liquid. The wave function will then have a strong correlation over macroscopic distances and will "feel" as a single body any change in the macroscopic boundary conditions. But such low temperatures are beyond the reach of experiments. Experiments performed at 1°K have not borne out London's proposal.

Onsager and Feynmann made the proposal that in addition to the elementary excitations characterized by the energy spectrum ω_k of (18.2) there is another type of excitation with the following properties: The state of the liquid corresponding to the presence of one such excitation has an energy separated from the ground state energy by a finite gap. If Ψ is the wave function of such a state, the velocity distribution $(\hbar/2mi)(\Psi^*\nabla\Psi - \Psi\nabla\Psi^*)$ is similar to that of a vortex line in classical hydrodynamics. Such an excitation may be called a quantum vortex line. It is so far the most interesting idea advanced towards an understanding of superfluid flow.*

* For a discussion of the quantum vortex line see W. F. Vinen in *Progress in Low Temperature Physics*, Vol. III, edited by C. J. Gorter (North Holland Publishing Co., Amsterdam, 1961). Another interesting idea, that of phenomenologically appending to the hydrodynamic equations a slip boundary condition, in analogy with that we have discussed in Sec. 5.5, is advanced by C. C. Lin, *Phys. Rev. Lett.*, **2**, 245 (1959).

chapter 19

HARD-SPHERE BOSE GAS

19.1 STATEMENT OF THE PROBLEM

A hard-sphere Bose gas is a collection of N spinless bosons, each of mass m, contained in a box of volume V, with the boundary condition that the wave function of the system shall vanish whenever any pair of particles are separated by a distance less than the hard-sphere diameter a. For definiteness we require in addition that the wave function obey periodic boundary conditions with respect to the volume V, which is taken to be a cube whose edges are of length $L = V^{1/3}$.

The motivation for studying the hard-sphere Bose gas is the desire to understand the behavior of liquid He^4 from an atomic point of view. In the last chapter we presented the theories of Landau and Feynman that establish an atomic basis on which the low-temperature behavior of liquid He^4 can be understood. The theories of Landau and Feynman are not, however, rigorous. To lend support to their results it is desirable to study a simpler model of an interacting system of boson that can be treated with some mathematical rigor. The hard-sphere Bose gas is such a model.

We formulate the problem of the hard-sphere Bose gas with the help of the method of pseudopotentials, which is discussed in Secs. 13.2 and 13.3. The Hamiltonian of the system is taken to be

$$H = -\frac{\hbar^2}{2m}\sum_{j=1}^{N}\nabla_j^{\,2} + \frac{4\pi a\hbar^2}{m}\sum_{i<j}\delta(\mathbf{r}_i - \mathbf{r}_j)\frac{\partial}{\partial r_{ij}}r_{ij} \qquad (19.1)$$

where $r_{ij} \equiv |\mathbf{r}_i - \mathbf{r}_j|$. The validity of this Hamiltonian as an approximation to that of a hard-sphere gas has been discussed in Sec. 13.3. We need only remind ourselves that (19.1) is arrived at by neglecting terms that are proportional to a^n with $n \geq 3$. If we wish to calculate the energy levels of the system for finite values of N and V, then (19.1) is valid only up to the order a^2. We seek, however, the energy levels in the limit

$$\begin{gathered} N \to \infty \\ V \to \infty \\ \frac{V}{N} = v \end{gathered} \tag{19.2}$$

where v is a given finite number. Here the result of applying ordinary perturbation theory shows that an expansion of the energy in powers of a does not exist. [See equation (19.31).] Hence the limit of validity of (19.1) is unclear; it is discussed later, after we have performed some calculations.

The object of the present investigation is to find the eigenvalues of (19.1), under the assumption that the gas is dilute, i.e.,

$$\frac{a^3}{v} \ll 1 \tag{19.3}$$

Having found the eigenvalues we may use them to evaluate the partition function of the system and to discuss its macroscopic equilibrium and transport properties.*

19.2 PERTURBATION THEORY

The Hamiltonian

In this section we are interested in solving, by perturbation theory, the Schrödinger equation†

$$(K + \Omega)\Psi = E\Psi \tag{19.4}$$

where

$$\begin{aligned} K &= -\frac{1}{2m}\sum_{j=1}^{N}\nabla_j^2 \\ \Omega &= \sum_{i<j} v_{ij} \\ v_{ij} &= \frac{4\pi a}{m}\,\delta(\mathbf{r}_i - \mathbf{r}_j)\frac{\partial}{\partial r_{ij}}\,r_{ij} \end{aligned} \tag{19.5}$$

* The general references are K. Huang and C. N. Yang, *Phys. Rev.*, **105**, 767 (1957); T. D. Lee, K. Huang, and C. N. Yang, *Phys. Rev.*, **106**, 1135 (1957).

† We choose units such that $\hbar = 1$.

in which Ω is to be considered the perturbation. Both N and V are very large but finite numbers.

It is convenient to rewrite (19.4) in the language of quantized fields.* Let $\psi(r)$, with $\psi^\dagger(r)$ its hermitian conjugate, be the field operator of the boson field we are dealing with, and let it be Fourier-analyzed:

$$\psi(\mathbf{r}) = \sum_{\mathbf{k}} a_{\mathbf{k}} \frac{e^{i\mathbf{k}\cdot\mathbf{r}}}{\sqrt{V}} \tag{19.6}$$

where, in order to satisfy the periodic boundary conditions imposed, we must take

$$\mathbf{k} = \frac{2\pi\mathbf{n}}{L} \tag{19.7}$$

where $\mathbf{n}$ is a vector whose components are independently $0, \pm 1, \pm 2, \ldots$. The annihilation operator $a_{\mathbf{k}}$ and creation operator $a_{\mathbf{k}}{}^\dagger$ satisfy the commutation rules

$$[a_{\mathbf{k}}, a_{\mathbf{k}'}{}^\dagger] = \delta_{\mathbf{k}\mathbf{k}'} \tag{19.8}$$

The Hamiltonian can then be put into the form

$$H = K + \Omega$$

where

$$K = -\frac{1}{2m}\int d^3r\, \psi^\dagger(\mathbf{r})\nabla^2\psi(\mathbf{r}) = \frac{1}{2m}\sum_{\mathbf{k}} k^2 a_{\mathbf{k}}{}^\dagger a_{\mathbf{k}} \tag{19.9}$$

$$\Omega = \frac{2\pi a}{m}\int d^3r_1\, d^3r_2 \psi^\dagger(\mathbf{r}_1)\psi^\dagger(\mathbf{r}_2)\,\delta(\mathbf{r}_1 - \mathbf{r}_2)\,\frac{\partial}{\partial r_{12}}\,[r_{12}\psi(\mathbf{r}_1)\psi(\mathbf{r}_2)]$$

The expression of Ω in terms of $a_{\mathbf{k}}$ and $a_{\mathbf{k}}{}^\dagger$ requires a little discussion, on account of the presence of the operator $(\partial/\partial r)r$. This operator has the property that

$$\frac{\partial}{\partial r}[rf(r)]_{r=0} = f(0) \qquad (\text{if } f \text{ is regular at } r = 0)$$

whereas

$$\frac{\partial}{\partial r}[rf(r)]_{r=0} = 0 \qquad \left(\text{if } f \xrightarrow[r\to 0]{} \frac{1}{r}\right)$$

The operator $(\partial/\partial r)r$ weeds out, so to speak, any singularity of the $1/r$ type in the function to which it is applied.†

If we evaluate the matrix elements of v_{ij} as given by (19.5) with respect to single-particle plane wave states, then for each single matrix element the operator $(\partial/\partial r)r$ is effectively unity, because a plane wave is regular

* See Appendix A, Sec. A.3.

† See Problem 13.1.

at the origin. On the other hand, if a matrix element is to be taken with an initial state that is a superposition of plane waves, the operator $(\partial/\partial r)r$ may not be put equal to unity, because a sum of plane waves may produce a function containing a $1/r$ singularity. This means that when the initial state is written as a sum of plane waves, the operator $(\partial/\partial r)r$ may not be applied to the sum term by term. The termwise application, which amounts to putting $(\partial/\partial r)r = 1$, is correct only if the sum is uniformly convergent. Accordingly in substituting (19.6) into the expression for Ω given by (19.9) we must write

$$\Omega = \frac{2\pi a}{mV} \sum_{\alpha,\beta} a_\alpha^\dagger a_\beta^\dagger \frac{\partial}{\partial r}\left[r \sum_{\mu,\nu} e^{i(\mathbf{k}_\mu - \mathbf{k}_\nu)\cdot \mathbf{r}/2}\, \delta(\mathbf{k}_\alpha + \mathbf{k}_\beta - \mathbf{k}_\mu - \mathbf{k}_\nu) a_\mu a_\nu \right]_{r=0}$$

where $a_\alpha \equiv a_{\mathbf{k}_\alpha}$, and the δ-function is a Kronecker symbol expressing the conservation of momentum in each elementary interaction. Thus

$$\Omega = \frac{2\pi a}{mV} \sum_{\mathbf{p},\mathbf{q}} a_\mathbf{p}^\dagger a_\mathbf{q}^\dagger \frac{\partial}{\partial r}\left[r \sum_{\mathbf{k}} e^{i\mathbf{k}\cdot\mathbf{r}} a_{\mathbf{p}+\mathbf{k}} a_{\mathbf{q}-\mathbf{k}} \right]_{r=0} \tag{19.10}$$

This expression seems complicated but is actually simple to use, as we shall see. The total Hamiltonian H is now given by

$$H = \frac{1}{2m} \sum k^2 a_\mathbf{k}^\dagger a_\mathbf{k} + \frac{2\pi a}{mV} \sum_{\mathbf{p},\mathbf{q}} a_\mathbf{p}^\dagger a_\mathbf{q}^\dagger \frac{\partial}{\partial r}\left[r \sum_{\mathbf{k}} e^{i\mathbf{k}\cdot\mathbf{r}} a_{\mathbf{p}+\mathbf{k}} a_{\mathbf{q}-\mathbf{k}} \right]_{r=0} \tag{19.11}$$

First-Order Calculation

The unperturbed problem is defined by the Hamiltonian K and the perturbation Hamiltonian is Ω. An unperturbed state is specified by a set of occupation numbers $\{n_0, n_1, \ldots\}$ of the single-particle plane wave states. We denote such an unperturbed state by

$$| n \rangle \equiv | n_0, n_1, \ldots \rangle \tag{19.12}$$

whose associated unperturbed eigenvalue is

$$E_n^{(0)} = \frac{1}{2m} \sum_{\mathbf{k}} k^2 n_\mathbf{k} \tag{19.13}$$

The occupation numbers satisfy the condition

$$\sum_{\mathbf{k}} n_\mathbf{k} = N \tag{19.14}$$

To the first order in the perturbation the energy levels are given by

$$E_n = E_n^{(0)} + E_n^{(1)}$$

where

$$E_n^{(1)} = \langle n | \Omega | n \rangle \tag{19.15}$$

Substituting Ω from (19.10) we see that for this matrix element the operator $(\partial/\partial r)r$ may be set equal to unity. We then obtain

$$E_n^{(1)} = \frac{2\pi a}{mV}\left\langle n \left| \sum_{\mathbf{p},\mathbf{q},\mathbf{k}} a_\mathbf{p}{}^\dagger a_\mathbf{q}{}^\dagger a_{\mathbf{p}+\mathbf{k}} a_{\mathbf{q}-\mathbf{k}} \right| n \right\rangle \tag{19.16}$$

In the sum in (19.16) only the following types of term have diagonal matrix elements:

$$\mathbf{p} = \mathbf{q}, \quad \mathbf{k} = 0: \qquad \sum_\mathbf{p} a_\mathbf{p}{}^\dagger a_\mathbf{p}{}^\dagger a_\mathbf{p} a_\mathbf{p} = \sum_\mathbf{p} n_\mathbf{p}{}^2 - N$$

$$\mathbf{p} \neq \mathbf{q}, \quad \mathbf{k} = 0: \qquad \sum_{\mathbf{p}\neq\mathbf{q}} a_\mathbf{p}{}^\dagger a_\mathbf{q}{}^\dagger a_\mathbf{p} a_\mathbf{q} = N^2 - \sum_\mathbf{p} n_\mathbf{p}{}^2$$

$$\mathbf{p} \neq \mathbf{q}, \quad \mathbf{k} = \mathbf{q} - \mathbf{p}: \qquad \text{the same}$$

The sum of these gives*

$$E_n^{(1)} = \frac{4\pi a}{mV}\left(N^2 - \tfrac{1}{2}N - \tfrac{1}{2}\sum_\mathbf{p} n_\mathbf{p}{}^2\right) \tag{19.17}$$

The ground state energy to this approximation, denoted by $E_0^{(1)}$, is that for which all $n_\mathbf{k} = 0$ except n_0, which equals N:

$$\frac{E_0^{(1)}}{N} = \frac{2\pi a}{mv} \tag{19.18}$$

The physical interpretation of these results has been given in Sec. 13.5.

Second-Order Calculation

The energy of the ground state is now calculated to the second order. The main purpose of doing this is to illustrate the role of the operator $(\partial/\partial r)r$ in the pseudopotential. The second-order correction to the ground state energy is given by the usual perturbation formula

$$E_0^{(2)} = \langle 0 \mid \Omega \mid \Psi^{(1)} \rangle \tag{19.19}$$

where $\mid 0 \rangle$ is the *unperturbed ground state* and $\mid \Psi^{(1)} \rangle$ is the first-order correction to the ground state wave function:

$$\mid \Psi^{(1)} \rangle = \sum_{n\neq 0} \frac{\langle n \mid \Omega \mid 0 \rangle}{-E_n^{(0)}} \mid n \rangle \tag{19.20}$$

The operator Ω in (19.19) may not be applied term by term to the sum in (19.20), because the latter is not uniformly convergent, as we shall see.

To calculate $\mid \Psi^{(1)} \rangle$ note that, because of momentum conservation, a state $\mid n \rangle$ that appears in the sum in (19.20) is one in which all particles

* In agreement with the configuration space calculation of equation (A.36) in Appendix A.

except two have momentum zero, and those two particles have momenta $\mathbf{k}$, $-\mathbf{k}$, respectively. We denote such a state by $|\,\mathbf{k}\rangle$, which is properly normalized:

$$|\,\mathbf{k}\rangle = \frac{1}{\sqrt{N(N-1)}}\, a_{\mathbf{k}}{}^{\dagger} a_{-\mathbf{k}}{}^{\dagger} a_0 a_0 \,|\, 0\rangle \tag{19.21}$$

Consider the matrix element

$$\langle \mathbf{k} \,|\, \Omega \,|\, 0\rangle = \frac{1}{\sqrt{N(N-1)}}\, \langle 0 \,|\, a_0{}^{\dagger} a_0{}^{\dagger} a_{-\mathbf{k}} a_{\mathbf{k}} \Omega \,|\, 0\rangle$$

The operation $(\partial/\partial r)r$ in Ω may clearly be set equal to unity. The only term of Ω [see (19.10)] that contributes to the matrix element is $(4\pi a/mV)\, a_{\mathbf{k}}{}^{\dagger} a_{-\mathbf{k}}{}^{\dagger} a_0 a_0$, in which a factor of 2 is included because $\mathbf{p} = \mathbf{k}$, $\mathbf{q} = -\mathbf{k}$, or $\mathbf{p} = -\mathbf{k}$, $\mathbf{q} = \mathbf{k}$. We thus obtain

$$\langle \mathbf{k} \,|\, \Omega \,|\, 0\rangle = \frac{4\pi a}{mV}\sqrt{N(N-1)} \tag{19.22}$$

Substituting (19.22) and (19.21) into (19.20), we obtain

$$|\,\Psi^{(1)}\rangle = -\frac{4\pi a}{mV}\sum_{\mathbf{k}>0}\frac{1}{2k^2}\, a_{\mathbf{k}}{}^{\dagger} a_{-\mathbf{k}}{}^{\dagger} a_0 a_0 \,|\, 0\rangle \tag{19.23}$$

where the summation extends over half of the momentum space excluding the origin. This condition is denoted by $\mathbf{k} > 0$. Substituting this into (19.19) we obtain

$$\begin{aligned} E_0^{(2)} &= \frac{2\pi a}{mV}\langle 0 \,|\, \Omega \sum_{\mathbf{k}>0}\frac{1}{k^2}\, a_{\mathbf{k}}{}^{\dagger} a_{-\mathbf{k}}{}^{\dagger} a_0 a_0 \,|\, 0\rangle \\ &= \frac{1}{2m}\left(\frac{4\pi a}{V}\right)^2 \lim_{r\to 0}\frac{\partial}{\partial r}\left[r\sum_{\mathbf{k}\neq 0}\frac{e^{i\mathbf{k}\cdot\mathbf{r}}}{k^2}\langle 0 \,|\, a_0{}^{\dagger} a_0{}^{\dagger} a_{\mathbf{k}} a_{-\mathbf{k}} a_{\mathbf{k}}{}^{\dagger} a_{-\mathbf{k}}{}^{\dagger} a_0 a_0 \,|\, 0\rangle\right] \\ &= \frac{1}{2m}\left(\frac{4\pi a}{V}\right)^2 N(N-1)\frac{\partial}{\partial r}\,[rF(r)]_{r=0} \end{aligned} \tag{19.24}$$

where*

$$F(r) \equiv \sum_{\mathbf{k}\neq 0}\frac{e^{i\mathbf{k}\cdot\mathbf{r}}}{k^2} \xrightarrow[r\to 0]{} \frac{V}{4\pi}\left(\frac{1}{r} - \frac{2.37}{L}\right) \tag{19.25}$$

Since $F(r)$ has a $1/r$ singularity, we see that had we ignored the operation $(\partial/\partial r)r$ in (19.24) we would have obtained a divergent result. Actually we have

$$E_0^{(2)} = \frac{1}{2m}\left(\frac{4\pi a}{V}\right)^2 N(N-1)\frac{V}{4\pi}\frac{2.37}{L} = 2.37\,\frac{a}{L}\,E_0^{(1)} \tag{19.26}$$

* For the details of the verification of (19.25) see Huang and Yang, *loc. cit.*

In the limit (19.2) we have

$$\frac{E_0^{(2)}}{E_0^{(1)}} \to 0 \tag{19.27}$$

The second-order correction can therefore be ignored.

Convergence of the Perturbation Series

The third-order correction to the energy is given by the formula

$$E_n^{(3)} = -(\Psi_n^{(1)}, \Psi_n^{(1)})E_n^{(1)} + \left(\Psi_n^{(0)}, \Omega \sum_{m \neq n} \frac{(\Psi_m^{(0)}, \Omega\Psi_n^{(1)})}{E_n^{(0)} - E_m^{(0)}} \Psi_m^{(0)}\right)$$

which, for the ground state, yields*

$$E_n^{(3)} = \left[(2.37)^2 + \frac{x}{\pi^2}(2N - 5)\right]\left(\frac{a}{L}\right)^2 E_0^{(1)} \tag{19.28}$$

where

$$x \equiv \sum_{l,m,n}{}' \frac{1}{(l^2 + m^2 + n^2)^2} \tag{19.29}$$

The summation Σ' includes all integer values of l, m, n from $-\infty$ to $+\infty$, excepting the value $l = m = n = 0$. We see that as $N \to \infty$

$$\frac{E_n^{(3)}}{N} = \frac{1}{2m}\frac{8x}{\pi}\frac{1}{NL^2}\left(\frac{aN}{L}\right)^3\left[1 + O\left(\frac{1}{N}\right)\right] \tag{19.30}$$

which diverges in the limit (19.2). Therefore the perturbation series in powers of a certainly does not converge unless $a = 0$.

It is possible to show that the nth-order correction to the energy per particle diverges for $n > 2$ in the limit (19.2), the most strongly divergent term being proportional to $(NL^2)^{-1}(aN/L)^n$. That is,

$$\frac{E_0^{(n)}}{N} = \text{const.}\,\frac{1}{NL^2}\left(\frac{aN}{L}\right)^n\left[1 + O\left(\frac{1}{N}\right)\right] \tag{19.31}$$

The dimensionless parameter for the expansion in powers of a appears to be aN/L. If, in each order, we keep only the most strongly divergent term, we obtain a series of the form

$$\frac{E_0}{N} = \frac{2\pi a}{mv} + \frac{1}{NL^2}\left[A\left(\frac{aN}{L}\right)^3 + B\left(\frac{aN}{L}\right)^4 + \cdots\right] + \text{“remainder”} \tag{19.32}$$

where A, B, . . . are numerical constants independent of a, N, and L. We now conjecture that the sum of the series displayed in (19.32) and the term

* Huang and Yang, *loc. cit.*

"remainder" separately approach finite numbers in the desired limit. This amounts to the definition of a new perturbation scheme, in which a new unperturbed Hamiltonian is so chosen that its lowest energy is (19.32) with omission of the "remainder". This conjecture will be verified.

Without detailed calculations it is possible to see what might be the result if our conjecture turns out to be correct. Since the series in (19.32) is a power series in aN/L, its sum must be of the form

$$\frac{1}{NL^2} f\left(\frac{aN}{L}\right)$$

The conjecture is that this quantity approaches a finite number when $N \to \infty$, $V \to \infty$, with $V/N = v$. Let us try the assumption

$$f\left(\frac{aN}{L}\right) = \left(\frac{aN}{L}\right)^x$$

where x is some number. If the conjecture is to be correct we must have $x = \frac{5}{2}$, for only then can $f(aN/L)/NL^2$ approach a finite limit. This leads to the result

$$\frac{E_0}{N} = \frac{2\pi a}{mv}\left[1 + C\sqrt{\frac{a^3}{v}}\right] + \text{"remainder"} \tag{19.33}$$

where C is a numerical constant. That this is actually the case will be shown. It can be shown further that "remainder" is small if $\sqrt{a^3/v} \ll 1$.

We might raise the question of whether, in trying to calculate the series in (19.32), we should include higher-order corrections to the pseudopotential (19.10). Since (19.32) is a power series involving all powers of a, it may seem at first sight that to be consistent we need the exact pseudopotentials. This is actually not so. In every term of the series appearing in (19.32), N is raised to the same power as a. It is easily seen that any additional contributions coming from the higher-order pseudopotentials (S-wave effective range, P-wave, etc.) give a power series in a. In every term of this power series N is raised to a *smaller* power than a. Thus these higher-order pseudopotentials can contribute only to what is designated as "remainder" in (19.32). Therefore we can continue to use (19.11) as the Hamiltonian.

19.3 A NEW PERTURBATION METHOD

We describe a new perturbation method by which, in the lowest-order approximation, we obtain the ground state energy in the form (19.33). As explained before, the basis of the new perturbation method is the conjecture that the series in (19.32) is a finite number in the limit (19.2) although every term of the series diverges.

The new method is guided by the observation that the series in (19.32) contains the most divergent terms in every order of a. That is, for a given order in a, N occurs with the maximum power possible. Accordingly we examine the matrix elements of the Hamiltonian for their dependence on N. Only those involving N with the maximum possible power are retained; the others are discarded. In this manner we obtain an effective Hamiltonian that will be the unperturbed Hamiltonian of the new method. This effective Hamiltonian turns out to be trivially diagonalizable, yielding for the ground state energy an expression of the form (19.33).

We first study the off-diagonal matrix elements of the Hamiltonian, which are the off-diagonal matrix elements of the operator Ω given by (19.10). It has nonvanishing matrix elements only between states $|\, n\rangle$ and $|\, m\rangle$ that are of the form

$$\begin{aligned} |\, m\rangle &= |\, \ldots, n_{\mathbf{p}}, \ldots, n_{\mathbf{q}}, \ldots, n_{\mathbf{p}+\mathbf{k}}, \ldots, n_{\mathbf{q}-\mathbf{k}}, \ldots\rangle \\ |\, n\rangle &= |\, \ldots, n_{\mathbf{p}} - 1, \ldots, n_{\mathbf{q}} - 1, \ldots, n_{\mathbf{p}+\mathbf{k}} + 1, \ldots, n_{\mathbf{q}-\mathbf{k}} + 1, \ldots\rangle \end{aligned}$$

where the occupation numbers not displayed are the same for both states and where $\mathbf{p}, \mathbf{q}, \mathbf{k}$ are arbitrary vectors except for the requirement that $|\, n\rangle \neq |\, m\rangle$. Thus the only nonvanishing matrix elements of Ω are those that describe the collision of two particles, with momentum conservation. A typical one is

$$\langle n\,|\Omega|\,m\rangle = \frac{2\pi a}{mV}\langle n\,|\, a^\dagger_{\mathbf{p}+\mathbf{k}} a^\dagger_{\mathbf{q}-\mathbf{k}} a_{\mathbf{p}} a_{\mathbf{q}}\,|\, m\rangle = \frac{2\pi a}{mV}\sqrt{(n_{\mathbf{p}+\mathbf{k}}+1)(n_{\mathbf{q}-\mathbf{k}}+1)n_{\mathbf{p}}n_{\mathbf{q}}} \tag{19.34}$$

As a shorthand, we write

$$\langle n\,|\Omega|\,m\rangle = \begin{array}{ccc} \mathbf{p}+\mathbf{k} & & \mathbf{p} \\ & \times & \\ \mathbf{q}-\mathbf{k} & & \mathbf{q} \end{array} \tag{19.35}$$

We now distinguish three classes of off-diagonal matrix elements:

$$\begin{array}{lll}
\text{(I)} & \begin{array}{ccc} \mathbf{k} & & 0 \\ & \times & \\ -\mathbf{k} & & 0, \end{array} \quad \begin{array}{ccc} 0 & & \mathbf{k} \\ & \times & \\ 0 & & -\mathbf{k} \end{array} & (\mathbf{k} \neq 0) \\[2ex]
\text{(II)} & \begin{array}{ccc} \mathbf{k} & & 0 \\ & \times & \\ \mathbf{q}-\mathbf{k} & & \mathbf{q}, \end{array} \quad \begin{array}{ccc} 0 & & \mathbf{k} \\ & \times & \\ \mathbf{q} & & \mathbf{q}-\mathbf{k} \end{array} & (\mathbf{k} \neq 0, \mathbf{q} \neq 0, \mathbf{q}-\mathbf{k} \neq 0) \\[2ex]
\text{(III)} & \begin{array}{ccc} \mathbf{p}+\mathbf{k} & & \mathbf{p} \\ & \times & \\ \mathbf{q}-\mathbf{k} & & \mathbf{q}, \end{array} & (\mathbf{p} \neq 0, \mathbf{q} \neq 0, \mathbf{p}+\mathbf{k} \neq 0, \mathbf{p}+\mathbf{k} \neq \mathbf{q}, \mathbf{q}-\mathbf{k} \neq 0)
\end{array} \tag{19.36}$$

In short, the classification is made according to whether two, one, or none of the momenta involved in the matrix element is zero. It is clear that any off-diagonal matrix element of the Hamiltonian belongs to one of these classes. The usefulness of such a classification will now be explained.

For finite values of N and V, no matter how large, a state of the perturbed system continuously approaches a unique state of the free-particle system as the interaction is gradually "turned off." There is a one-to-one correspondence between free and perturbed states. The ground state is the state that corresponds to the free-particle ground state, denoted by $|0\rangle$, in which all particles occupy the level $\mathbf{k} = 0$. The perturbed state vector $|\Psi\rangle$ can be written as a linear superposition of $|0\rangle$ and such unperturbed states $|n\rangle$ as are connected to $|0\rangle$ directly or indirectly through nonvanishing off-diagonal matrix elements of the Hamiltonian. These states $|n\rangle$ differ from $|0\rangle$ by virtue of the fact that a number of particles have been taken out of the level $\mathbf{k} = 0$ and delivered into other levels. It is clear that for finite N there will be states $|n\rangle$ in which the level $\mathbf{k} = 0$ is empty, or that some level other than $\mathbf{k} = 0$ becomes occupied with a finite fraction of the N particles, or both. To be able to proceed, we make the simplifying assumption that in the linear decomposition of $|\Psi\rangle$ only the states $|n\rangle$ whose n_0 is a finite fraction of N need be considered, and no other occupation number is a finite fraction of N. Such states $|n\rangle$ span a subspace S of the Hilbert space of the system. The three classes of matrix elements (19.36), taken between states in S, have different orders of magnitude. It is easy to see that those of class I are proportional to N, those of class II are proportional to $\sqrt{N}$, and those of class III are independent of N. To obtain the result to the lowest order in the new perturbation method *we neglect all off-diagonal matrix elements except those of class* I.

The matrix elements of class I describe binary collisions in which either the initial or the final particles both have zero momentum. Therefore the expansion of $|\Psi\rangle$ contains only such states $|n\rangle$ as are obtainable from $|0\rangle$ through repeated collisions of this kind. The most general state $|n\rangle$ of the subspace S is specified by the following occupation numbers:

Level	Occupation Number
0	$N - \sum_{k \neq 0} l_{\mathbf{k}}$
$\mathbf{k}$	$l_{\mathbf{k}}$
$-\mathbf{k}$	$l_{\mathbf{k}}$

(19.37)

where $\mathbf{k}$ ranges through all possible momenta except zero. As each $l_{\mathbf{k}}$ independently ranges through the values

$$l_{\mathbf{k}} = 0, 1, 2, \ldots \tag{19.38}$$

the entire subspace S is generated. It is to be noted that by assumption we regard any $l_{\mathbf{k}}$ as negligible compared to N. Thus the range of values (19.38) must be imagined to terminate at some point much less than N. As $N \to \infty$ the terminal point approaches infinity in such a way that

$$\frac{l_{\mathbf{k}}}{N} \xrightarrow[N\to\infty]{} 0 \qquad (\mathbf{k} \neq 0) \tag{19.39}$$

This is the fundamental assumption of the new perturbation method.

Let $|\ldots, l_{\mathbf{k}}, \ldots\rangle$ be a typical state described by (19.37). By (19.34) we have

$$\begin{aligned}\langle \ldots, l_{\mathbf{k}} + 1, \ldots |\,\Omega\,| \ldots, l_{\mathbf{k}}, \ldots\rangle &= \frac{2\pi a}{mV}\Big(N - \sum_{\mathbf{p}\neq 0} l_{\mathbf{p}}\Big)(l_{\mathbf{k}} + 1)\\ &\approx \frac{2\pi a}{mv}(l_{\mathbf{k}} + 1)\end{aligned} \tag{19.40}$$

$$\begin{aligned}\langle \ldots, l_{\mathbf{k}}, \ldots |\,\Omega\,| \ldots, l_{\mathbf{k}} + 1, \ldots\rangle &= \frac{2\pi a}{mV}\Big(N - \sum_{\mathbf{p}\neq 0} l_{\mathbf{p}} + 1\Big)(l_{\mathbf{k}} + 1)\\ &\approx \frac{2\pi a}{mv}(l_{\mathbf{k}} + 1)\end{aligned} \tag{19.41}$$

where $\mathbf{k} \neq 0$ and where we have assumed that

$$\frac{1}{N}\sum_{\mathbf{p}\neq 0} l_{\mathbf{p}} \ll 1 \tag{19.42}$$

That this is so is demonstrated later. [See equation (19.70).] The formulas (19.40) and (19.41) contain all the nonvanishing off-diagonal matrix elements of the Hamiltonian that are to be retained for the lowest-order calculation.

To complete the definition of the new perturbation method, we give the diagonal matrix elements of the Hamiltonian to be used. From (19.17) the diagonal matrix element for any states whose occupation numbers are $\{n_{\mathbf{k}}\}$ is

$$\frac{4\pi a N}{mv}\left[1 - \frac{1}{2}\sum_{\mathbf{k}}\left(\frac{n_{\mathbf{k}}}{N}\right)^2\right]$$

The second term may be rewritten as

$$\begin{aligned}\sum_{\mathbf{k}} n_{\mathbf{k}}^2 = n_0^2 + \sum_{\mathbf{k}\neq 0} n_{\mathbf{k}}^2 &= \Big(N - \sum_{\mathbf{k}\neq 0} n_{\mathbf{k}}\Big)^2 + \sum_{\mathbf{k}\neq 0} n_{\mathbf{k}}^2\\ &= N^2 - 2N\sum_{\mathbf{k}\neq 0} n_{\mathbf{k}} + \Big(\sum_{\mathbf{k}\neq 0} n_{\mathbf{k}}\Big)^2 + \sum_{\mathbf{k}\neq 0} n_{\mathbf{k}}^2\end{aligned} \tag{19.43}$$

Thus for any state of the subspace S we have

$$\langle \ldots, l_{\mathbf{k}}, \ldots |\Omega| \ldots, l_{\mathbf{k}}, \ldots \rangle$$
$$= \frac{1}{2m}\left\{\frac{4\pi aN}{v} + \frac{8\pi a}{v}\sum_{\mathbf{k}\neq 0} l_{\mathbf{k}} - \frac{4\pi a}{v}\frac{1}{N}\left[\left(\sum_{\mathbf{k}\neq 0} l_{\mathbf{k}}\right)^2 + \sum_{\mathbf{k}\neq 0} l_{\mathbf{k}}^2\right]\right\} \quad (19.44)$$

In accordance with (19.42), we neglect the last term as compared to the second term. The second term, although much smaller than the first, should not be neglected. If we did so, there would be no problem left. Therefore we take

$$\langle \ldots, l_{\mathbf{k}}, \ldots |\Omega| \ldots, l_{\mathbf{k}}, \ldots \rangle = \frac{1}{2m}\left(\frac{4\pi aN}{v} + \frac{8\pi a}{v}\sum_{\mathbf{k}\neq 0} l_{\mathbf{k}}\right) \quad (19.45)$$

The formulas (19.40), (19.41), and (19.45) completely define the perturbation scheme. It is noted that in these formulas the relevant states are labeled by the quantum numbers $\{l_{\mathbf{k}}\}$ with $\mathbf{k} \neq 0$. Consider the effective Hamiltonian H_{eff}, defined by

$$2mH_{\text{eff}} \equiv \frac{4\pi aN}{v} + \sum_{\mathbf{k}\neq 0}{}' \left[\left(k^2 + \frac{8\pi a}{v}\right) a_{\mathbf{k}}^{\dagger} a_{\mathbf{k}} + \frac{4\pi a}{v}\left(a_{\mathbf{k}}^{\dagger} a_{-\mathbf{k}}^{\dagger} + a_{\mathbf{k}} a_{-\mathbf{k}}\right)\right] \quad (19.46)$$

It is obvious that with respect to the states in the subspace S this operator has the same matrix elements given by (19.40), (19.41), and (19.45). The lowest eigenvalue of H_{eff} therefore gives the ground state energy to the lowest order in the new perturbation method. This is the unperturbed Hamiltonian in the new perturbation method.

In this discussion so far, the effect of the differential operator $(\partial/\partial r)r$ in the pseudopotential has not been taken into account. This can now be incorporated by appending the rule that in the lowest eigenvalue of (19.46) the term proportional to a^2 is to be subtracted [by virtue of (19.27)]. Alternatively we may understand by the sum Σ' in (19.46) the operation

$$\sum_{k\neq 0}{}' f_{\mathbf{k}} \equiv \lim_{r\to 0} \frac{\partial}{\partial r}\left(r \sum_{\mathbf{k}\neq 0} f_{\mathbf{k}}\right) \quad (19.47)$$

19.4 THE GROUND STATE AND LOW EXCITED STATES

Energy Spectrum

The effective Hamiltonian (19.46) can be immediately diagonalized. Only the ground state and states immediately above the ground state are significant, however, for the approximations leading to (19.46) become increasingly worse for higher excited states. How far from the ground state we can still use (19.46) is discussed later.

To diagonalize (19.46) we introduce a linear transformation first used by Bogolubov:*

$$\begin{aligned} a_{\mathbf{k}} &= \frac{1}{\sqrt{1-\alpha_{\mathbf{k}}^2}}(b_{\mathbf{k}} - \alpha_{\mathbf{k}} b_{-\mathbf{k}}^{\dagger}) \\ a_{\mathbf{k}}^{\dagger} &= \frac{1}{\sqrt{1-\alpha_{\mathbf{k}}^2}}(b_{\mathbf{k}}^{\dagger} - \alpha_{\mathbf{k}} b_{-\mathbf{k}}) \end{aligned} \tag{19.48}$$

or

$$\begin{aligned} b_{\mathbf{k}} &= \frac{1}{\sqrt{1-\alpha_{\mathbf{k}}^2}}(a_{\mathbf{k}} + \alpha_{\mathbf{k}} a_{-\mathbf{k}}^{\dagger}) \\ b_{\mathbf{k}}^{\dagger} &= \frac{1}{\sqrt{1-\alpha_{\mathbf{k}}^2}}(a_{\mathbf{k}}^{\dagger} + \alpha_{\mathbf{k}} a_{-\mathbf{k}}) \end{aligned} \tag{19.49}$$

where $\alpha_{\mathbf{k}}$ is assumed to be a real number less than one. It is clear that $b_{\mathbf{k}}$ and $b_{\mathbf{k}}^{\dagger}$ satisfy the same commutation rules as $a_{\mathbf{k}}$ and $a_{\mathbf{k}}^{\dagger}$, namely

$$\begin{aligned} [b_{\mathbf{k}}, b_{\mathbf{k}'}] &= [b_{\mathbf{k}}^{\dagger}, b_{\mathbf{k}'}^{\dagger}] = 0 \\ [b_{\mathbf{k}}, b_{\mathbf{k}'}^{\dagger}] &= \delta_{\mathbf{k}\mathbf{k}'} \end{aligned} \tag{19.50}$$

Therefore $b_{\mathbf{k}}$ and $b_{\mathbf{k}}^{\dagger}$ can be interpreted respectively as annihilation and creation operators, just as $a_{\mathbf{k}}$ and $a_{\mathbf{k}}^{\dagger}$. If we substitute (19.48) into the effective Hamiltonian (19.46), we find that H_{eff} is diagonalized by choosing

$$\alpha_{\mathbf{k}} = 1 + x^2 - x\sqrt{x^2+2}, \qquad x^2 \equiv \frac{k^2}{8\pi a/v} \tag{19.51}$$

Clearly $\alpha_{\mathbf{k}}$ is real and is less than one, as we earlier assumed.

After we have performed the subtraction procedure required by (19.47) the Hamiltonian in diagonal form reads

$$\begin{aligned} 2mH_{\text{eff}} = \frac{4\pi a N}{v} - \frac{1}{2}\sum_{\mathbf{k}\neq 0}\left[\frac{8\pi a}{v} + k^2 - k\sqrt{k^2 + \frac{16\pi a}{v}} - \frac{1}{2}\left(\frac{8\pi a}{v}\right)^2\frac{1}{k^2}\right] \\ + \sum_{\mathbf{k}\neq 0} k\sqrt{k^2 + \frac{16\pi a}{v}}\, b_{\mathbf{k}}^{\dagger} b_{\mathbf{k}} \end{aligned} \tag{19.52}$$

where the term $-\frac{1}{2}(8\pi a/v)^2 k^{-2}$ is the subtraction. It cancels the a^2 term in the expansion of $k\sqrt{k^2 + (16\pi a/v)}$. If we had not made this subtraction, the sum over $\mathbf{k}$ would be divergent. It is now of course finite and can be converted into an integral as $V \to \infty$:

$$-\frac{V}{(2\pi)^2}\left(\frac{8\pi a}{v}\right)^{5/2}\int_0^\infty dx\, x^2\left(1 + x^2 - x\sqrt{x^2+2} - \frac{1}{2x^2}\right) = \frac{4\pi a N}{v}\frac{128}{15}\sqrt{\frac{a^3}{\pi v}} \tag{19.53}$$

* N. N. Bogolubov, *J. Phys. U.S.S.R.*, II, 23 (1947).

The Hamiltonian now reads

$$H_{\text{eff}} = \frac{2\pi aN}{mv}\left(1 + \frac{128}{15}\sqrt{\frac{a^3}{\pi v}}\right) + \sum_{\mathbf{k}\neq 0} \frac{k}{2m}\sqrt{k^2 + \frac{16\pi a}{v}}\, b_{\mathbf{k}}^\dagger b_{\mathbf{k}} \quad (19.54)$$

It follows from the commutation rules (19.50) that the eigenvalues of $b_{\mathbf{k}}^\dagger b_{\mathbf{k}}$ are 0, 1, 2, ... Therefore the lowest eigenvalue of H_{eff} is

$$E_0 = \frac{2\pi aN}{mv}\left(1 + \frac{128}{15}\sqrt{\frac{a^3}{\pi v}}\right) \quad (19.55)$$

This verifies (19.33), with C explicitly given.

As we mentioned in the beginning of this section, the effective Hamiltonian also correctly describes the excited states immediately above the ground state, to the same order of approximation. As we can see from (19.54), the excited states are characterized by the occupation numbers $b_{\mathbf{k}}^\dagger b_{\mathbf{k}} = 0, 1, 2, \ldots$ of noninteracting excitations whose energies are given by $(k/2m)\sqrt{k^2 + 16\pi a/v}$. For very small k, these energies become $(k/2m)\sqrt{16\pi a/v}$. These excitations are therefore phonons. The velocity of sound for very long wave lengths ($k \to 0$) is

$$c = \sqrt{\frac{4\pi a}{m^2 v}} \quad (19.56)$$

We may check the consistency of this interpretation by calculating the sound velocity independently from the compressibility of the system at absolute zero:

$$c = \frac{1}{\sqrt{\rho\kappa_s}} = \sqrt{\frac{v}{m}\frac{\partial P_0}{\partial v}} \quad (19.57)$$

where P_0 is the pressure at absolute zero:

$$P_0 = -\frac{\partial}{\partial v}\frac{E_0}{N} \quad (19.58)$$

Using (19.55) we find that

$$c = \sqrt{\frac{4\pi a}{m^2 v}}\left(1 + 16\sqrt{\frac{a^3}{\pi v}}\right) \quad (19.59)$$

We see that the first term agrees with (19.56). The next term is of the order $16\sqrt{a^3/v}$ and is beyond the accuracy of the calculation from which (19.56) is obtained.

Wave Functions

We now calculate the wave functions of the system, both for the ground state and for states with phonons.

The eigenstates of (19.54) may be labeled by phonon occupation numbers $\sigma_{\mathbf{k}}$, which can independently assume the values 0, 1, 2, ... We denote an eigenstate by $|\ldots, \sigma_{\mathbf{k}}, \ldots\rangle$. It has the properties that

$$\begin{aligned} b_{\mathbf{k}}^{\dagger} b_{\mathbf{k}} |\ldots, \sigma_{\mathbf{k}}, \ldots\rangle &= \sigma_{\mathbf{k}} |\ldots, \sigma_{\mathbf{k}}, \ldots\rangle \\ b_{\mathbf{k}} |\ldots, \sigma_{\mathbf{k}}, \ldots\rangle &= \sqrt{\sigma_{\mathbf{k}}}\, |\ldots, \sigma_{\mathbf{k}} - 1, \ldots\rangle \\ b_{\mathbf{k}}^{\dagger} |\ldots, \sigma_{\mathbf{k}}, \ldots\rangle &= \sqrt{\sigma_{\mathbf{k}} + 1}\, |\ldots, \sigma_{\mathbf{k}} + 1, \ldots\rangle \end{aligned} \tag{19.60}$$

The ground state is denoted by

$$|\Psi_0\rangle \equiv |0, 0, 0, \ldots\rangle$$

It is defined by

$$b_{\mathbf{k}} |\Psi_0\rangle = 0 \tag{19.61}$$

Let $|n_1 m_1; n_2 m_2; \ldots\rangle$ be the unperturbed state in which there are

$$\left.\begin{aligned} &n_{\mathbf{k}} \text{ particles of momentum } \mathbf{k} \\ &m_{\mathbf{k}} \text{ particles of momentum } -\mathbf{k} \end{aligned}\right. \qquad (\mathbf{k} > 0)$$

Then we may expand $|\Psi_0\rangle$ as follows:

$$|\Psi_0\rangle = \sum_{\substack{n_1=0\\ m_1=0}}^{\infty} \sum_{\substack{n_2=0\\ m_2=0}}^{\infty} \cdots (C_{n_1 m_1} C_{n_2 m_2} \cdots) |n_1 m_1; n_2 m_2; \ldots\rangle \tag{19.62}$$

Only the amplitudes for $\mathbf{k}$, $-\mathbf{k}$ are coupled together because the transformation (19.49) has this property. Substituting (19.61) into (19.62) and using (19.49) we obtain the following equation for C_{nm}:

$$\sum_{n=0}^{\infty} \sum_{m=0}^{\infty} [C_{nm}\sqrt{n}\, |n - 1, m\rangle + \alpha C_{nm}\sqrt{m + 1}\, |n, m + 1\rangle] = 0$$

where

$$\begin{aligned} |n, m\rangle &\equiv |n_1 m_1; n_2 m_2; \ldots\rangle \\ \alpha &\equiv \alpha_{\mathbf{k}} \end{aligned}$$

By changing the indices of summation, we can write

$$\sum_{n=0}^{\infty} \sum_{m=0}^{\infty} [C_{n+1,m}\sqrt{n + 1} + \alpha C_{n,m-1}\sqrt{m}]\, |n, m\rangle = 0$$

which implies

$$C_{n+1,m}\sqrt{n + 1} + \alpha C_{n,m-1}\sqrt{m} = 0$$

From this we can deduce that

$$C_{nm} = 0 \qquad (n \neq m)$$

for this is obviously true for $m = 0$, and the general case can be proved by induction. Hence it is sufficient to consider C_{mm}, which satisfies the equation

$$C_{mm} + \alpha C_{m-1,\, m-1} = 0 \tag{19.63}$$

The solution is

$$C_{mm} = (-\alpha)^m C_{00} \tag{19.64}$$

where C_{00} is to be determined by normalizing the total wave function. Therefore the unperturbed states that appear in the expansion of $|\Psi_0\rangle$ are states in which pairs of particles **k**, $-$**k**, are excited. We denote such a state by $|l_1, l_2, \ldots\rangle$, in which there are $l_{\mathbf{k}}$ particles with momentum **k**, and the same number of particles with momentum $-$**k**. Thus

$$|\Psi_0\rangle = Z \sum_{l_1=0}^{\infty} \sum_{l_2=0}^{\infty} \cdots [(-\alpha_1)^{l_1}(-\alpha_2)^{l_2} \cdots] \, |l_1, l_2, \ldots\rangle \tag{19.65}$$

where $\alpha_{\mathbf{k}}$ is defined by (19.51) and one factor $\alpha_{\mathbf{k}}$ appears for each $\mathbf{k} > 0$. The normalization constant Z can be shown to be

$$Z = \prod_{k>0} \sqrt{1 - \alpha_{\mathbf{k}}^2} = \exp\left[-\tfrac{4}{9}N(3\pi - 8)\sqrt{\frac{a^3}{\pi v}}\right] \tag{19.66}$$

In (19.65) the term with all $l_{\mathbf{k}} = 0$ corresponds to the unperturbed ground state $|0\rangle$. Hence Z is none other than $\langle 0 | \Psi_0\rangle$, the probability amplitude of finding the unperturbed ground state in the perturbed ground state. According to (19.66) these two states become orthogonal to each other as $N \to \infty$.

The wave functions for excited states can be easily calculated. For example, for the state with one phonon of momentum **k**, the normalized wave function is defined by

$$|\Psi_{\mathbf{k}}\rangle = b_{\mathbf{k}}^{\dagger} \, |\Psi_0\rangle \tag{19.67}$$

By a straightforward calculation we obtain

$$|\Psi_{\mathbf{k}}\rangle = \sqrt{1 - \alpha_{\mathbf{k}}^2} \, a_{\mathbf{k}}^{\dagger} \, |\Psi_0\rangle \tag{19.68}$$

Thus the one-phonon state is a superposition of unperturbed states in which there are any number of particles **p**, $-$**p** for all **p**, *plus* an additional particle of momentum **k**.

The average number of particles that have momentum **k** in the perturbed ground state is

$$\langle n_{\mathbf{k}}\rangle = \langle \Psi_0 | \, a_{\mathbf{k}}^{\dagger} a_{\mathbf{k}} \, | \Psi_0\rangle = \frac{\alpha_{\mathbf{k}}^2}{1 - \alpha_{\mathbf{k}}^2} \qquad (\mathbf{k} \neq 0) \tag{19.69}$$

Therefore the total number of excited particles in the perturbed ground state is

$$\sum_{\mathbf{k}\neq 0} \langle n_{\mathbf{k}}\rangle = \sum_{\mathbf{k}\neq 0} \frac{\alpha_{\mathbf{k}}^2}{1 - \alpha_{\mathbf{k}}^2} = \frac{8}{3}\sqrt{\frac{a^3}{\pi v}}\, N \tag{19.70}$$

The number of particles of zero momentum in the perturbed ground state is

$$\langle n_0 \rangle = N\left[1 - \frac{8}{3}\sqrt{\frac{a^3}{\pi v}}\right] \tag{19.71}$$

Thus only a small fraction $\sqrt{a^3/v}$ of all the particles are excited. This justifies the neglect in (19.44) of the term $\left(\sum_{\mathbf{k}\neq 0} n_\mathbf{k}\right)^2 + \sum_{\mathbf{k}\neq 0} n_\mathbf{k}^2$ as compared to n_0^2.

It is instructive to calculate the wave functions in configuration space. In each term of the sum in (19.65), the number $n \equiv \sum_{\mathbf{k}>0} l_\mathbf{k}$ is half the total number of particles with nonzero momentum. Thus we must have $N \geq 2n$. We rewrite (19.65) as follows:

$$| \Psi_0 \rangle = Z \sum_{n=0}^{N/2} \sum_{\substack{l_1, l_2, \ldots \\ \Sigma l_\mathbf{k} = n}} [(-\alpha_1)^{l_1}(-\alpha_2)^{l_2} \cdots] \mid l_1, l_2, \ldots \rangle \tag{19.72}$$

We imagine $N \to \infty$ at the end of the calculation. The normalized configuration space wave function for an unperturbed state specified by the occupation numbers $\{n_0, n_1, \ldots\}$ is

$$\langle \mathbf{r}_1, \ldots, \mathbf{r}_N \mid n_0, n_1, \ldots \rangle \equiv \frac{1}{V^{N/2}} \frac{1}{\sqrt{N!\, \Pi(n_\mathbf{k}!)}} \sum_P P e^{i(\mathbf{p}_1 \cdot \mathbf{r}_1 + \cdots + \mathbf{p}_N \cdot \mathbf{r}_N)} \tag{19.73}$$

where among the N momenta $\mathbf{p}_1, \ldots, \mathbf{p}_N$, n_0 are 0, n_1 are $\mathbf{k}_1$, etc. The symbol P denotes a permutation of $\mathbf{r}_1, \ldots, \mathbf{r}_N$. Hence

$$\langle \mathbf{r}_1, \ldots, \mathbf{r}_N \mid l_1, l_2, \ldots \rangle = \frac{1}{V^{N/2}} \frac{1}{\sqrt{N!\,(N-2n)!}} \frac{1}{\Pi(l_\mathbf{k}!)} \sum_P P e^{i[\mathbf{p}_1 \cdot (\mathbf{r}_1 - \mathbf{r}_2) + \cdots + \mathbf{p}_n \cdot (\mathbf{r}_3 - \mathbf{r}_4)]} \tag{19.74}$$

where among the n vectors $\mathbf{p}_1, \ldots, \mathbf{p}_n$, l_1 are $\mathbf{k}_1$, l_2 are $\mathbf{k}_2$, etc.

Now consider the sum appearing in (19.74):

$$| \chi_n \rangle \equiv \sum_{\substack{l_1, l_2, \ldots \\ \Sigma l_k = n}} [(-\alpha_1)^{l_1}(-\alpha_2)^{l_2} \cdots] \mid l_1, l_2, \ldots \rangle$$

This may be rewritten as

$$\begin{aligned} | \chi_n \rangle &= \sum_{\mathbf{k}_1 \leqslant \mathbf{k}_2 \leqslant \cdots} [(-\alpha_1)(-\alpha_2)(-\alpha_3) \cdots] \mid l_1, l_2, \ldots \rangle \\ &= \frac{\Pi(l_\mathbf{k}!)}{n!} \sum_{\mathbf{k}_1 > 0} \sum_{\mathbf{k}_2 > 0} \cdots [(-\alpha_1)(-\alpha_2)(-\alpha_3) \cdots] \mid l_1, l_2, \ldots \rangle \end{aligned} \tag{19.75}$$

where, in the first line, $\mathbf{k}_1 \leq \mathbf{k}_2 \leq \cdots$ denotes any ordering of the momenta. In the second line each momentum independently ranges through half of momentum space excluding $\mathbf{k} = 0$ (as denoted by $\mathbf{k} > 0$). The configuration space representation of (19.75) is obtained by substituting (19.74) for $|\, l_1, l_2, \ldots\rangle$:

$$\chi_n(\mathbf{r}_1, \ldots, \mathbf{r}_N) = \frac{1}{V^{N/2}} \frac{1}{n!\sqrt{N!\,(N-2n)!}} \sum_P P \times \left[\sum_{k_1>0} (-\alpha_1) e^{i\mathbf{k}_1 \cdot (\mathbf{r}_1 - \mathbf{r}_2)} \cdots \sum_{k_n>0} (-\alpha_n) e^{i\mathbf{k}_n \cdot (\mathbf{r}_3 - \mathbf{r}_4)}\right]$$

Let us define

$$f(ij) \equiv f(\mathbf{r}_i - \mathbf{r}_j)$$

$$f(\mathbf{r}) = -\frac{2}{N} \sum_{\mathbf{k}>0} \alpha_{\mathbf{k}} e^{i\mathbf{k}\cdot\mathbf{r}} = -\frac{v}{8\pi^3} \int d^3k\, \alpha_{\mathbf{k}} e^{i\mathbf{k}\cdot\mathbf{r}} \tag{19.76}$$

Since $\alpha_{\mathbf{k}}$ depends only on $|\mathbf{k}|$, it follows that $f(\mathbf{r})$ depends only on $|\mathbf{r}|$. We can then write

$$\chi_n(\mathbf{r}_1, \ldots, \mathbf{r}_N) = \frac{1}{V^{N/2}} \frac{1}{n!\sqrt{N!\,(N-2n)!}} \left(\frac{N}{2}\right)^n \sum_P P[f(12) \cdots f(34)]$$

in which there are n factors $f(ij)$.

The number of distinct ways of choosing the arguments of $f(ij)$ from the N coordinates $\mathbf{r}_1, \ldots, \mathbf{r}_N$ can be found by filling the n boxes in the following with N balls. The boxes are identical, each holding two balls, and the order of the two balls is irrelevant.

1 2 ... $n-1$ n — $N - 2n$ balls left over

It is evident that there are

$$\frac{N!}{(N-2n)!\, n!\, 2^n}$$

distinct ways to fill the boxes, whereas there are $N!$ permutations of $\mathbf{r}_1, \ldots, \mathbf{r}_N$. Hence

$$\chi_n(\mathbf{r}_1, \ldots, \mathbf{r}_N) = \frac{1}{V^{N/2}} \sqrt{\frac{(N-2n)!}{N!}}\, N^n \sum [f(12) \cdots f(34)] \tag{19.77}$$

where the sum is extended over all distinct ways of filling the "boxes." Therefore

$$\begin{aligned}\Psi_0(\mathbf{r}_1, \ldots, \mathbf{r}_N) &= Z \sum_{n=0}^{N/2} \chi_n(\mathbf{r}_1, \ldots, \mathbf{r}_N) \\ &= \frac{Z}{V^{N/2}} \{1 + [f(12) + f(34) + \cdots] \\ &\qquad + [f(12)f(34) + f(12)f(56) + \cdots] \\ &\qquad + [f(12)f(34)f(56) + \cdots] + \cdots\}\end{aligned} \tag{19.78}$$

It differs from the function

$$\Psi_0'(\mathbf{r}_1, \ldots, \mathbf{r}_N) \equiv \frac{Z}{V^{N/2}} \prod_{i<j} [1 + f(ij)] \tag{19.79}$$

only in that (19.79) contains extra terms of the type

$$f(12)f(13), f(12)f(13)f(34), \ldots$$

in which the same particle appears in more than one "box." These terms may be shown to belong to a higher order in the calculation than we have considered. Therefore we may take (19.79) to be the wave function to the order $\sqrt{a^3/v}$. The function $f(\mathbf{r})$ defined in (19.76) has the following asymptotic behavior:

$$\begin{aligned} f(\mathbf{r}) &\approx -\frac{a}{r} && (r \ll r_0) \\ f(\mathbf{r}) &\approx -32\sqrt{\frac{a^3}{\pi v}}\left(\frac{r_0}{r}\right)^4 && (r \gg r_0) \end{aligned} \tag{19.80}$$

where

$$r_0 = \sqrt{\frac{v}{8\pi a}} \tag{19.81}$$

Thus although (19.79) vanishes at $r_{ij} = a$ (19.78) does so only approximately, but its value at $r_{ij} = a$ is of a higher order than $\sqrt{a^3/v}$.

The one-phonon wave function (19.67) is easily shown to have the following form in configuration space:

$$\Psi_{\mathbf{k}}(\mathbf{r}_1, \ldots, \mathbf{r}_N) = \text{const.} \sum_{j=1}^{N} e^{i\mathbf{k}\cdot\mathbf{r}_j} \Psi_0(\mathbf{r}_1, \ldots, \mathbf{r}_N) \tag{19.82}$$

This is of the same form as the phonon wave function that Feynman obtains for liquid He^4, on the basis of plausibility arguments.

Finally we can calculate the pair correlation function $D(r)$, which is defined by

$$D(|\mathbf{r}_1 - \mathbf{r}_2|) = v^2 \int d^{3N}r' \sum_{i\neq j} \delta(\mathbf{r}_1 - \mathbf{r}_i')\delta(\mathbf{r}_2 - \mathbf{r}_j')\,|\Psi_0(\mathbf{r}_1', \ldots, \mathbf{r}_N')|^2 \tag{19.83}$$

It is normalized to unity as $|\mathbf{r}_1 - \mathbf{r}_2| \to \infty$ and represents the relative probability of finding two particles at a distance $|\mathbf{r}_1 - \mathbf{r}_2|$ apart in the ground state of the system (i.e., at the absolute zero of temperature). For actual calculation it is convenient to represent $D(r)$ in the equivalent form

$$D(|\mathbf{r}_1 - \mathbf{r}_2|) = v^2 \langle \Psi_0 \,|\, \psi^+(\mathbf{r}_1)\psi^+(\mathbf{r}_2)\psi(\mathbf{r}_1)\psi(\mathbf{r}_2) \,|\, \Psi_0 \rangle \tag{19.84}$$

where $\psi(\mathbf{r})$ is the quantized field operator (19.6). This can be calculated with the help of (19.65). We only give the result graphically, as shown in

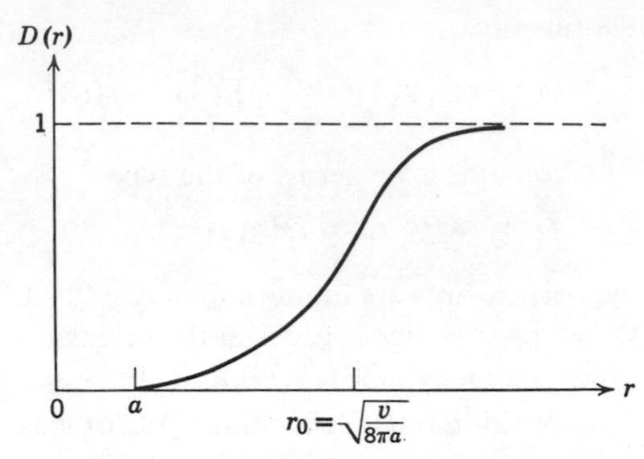

Fig. 19.1. Pair correlation function of hard-sphere Bose gas.

Fig. 19.1. It is interesting to note that the function $D(r)$ changes from an r^{-2} behavior at small distances to an r^{-4} behavior at large distances. The approximate distance at which the change-over takes place is r_0, which may be termed a "correlation length," because beyond this distance the two particles rapidly become uncorrelated in position.

19.5 HIGHER EXCITED STATES

The results in the previous section concerning the excited states have been deduced with the help of the effective Hamiltonian (19.46). We recall that in deriving (19.46) we had in mind solely the calculation of the ground state energy. Its validity for a calculation of the excited states must be separately investigated. Such an investigation is the main purpose of this section. We find that the effective Hamiltonian is correct only for excited states very close to the ground state, i.e., states with few phonons. When many phonons are excited, both the energy of the excited state and the sound velocity are no longer correctly given by (19.46).

We recall that a one-phonon state **k** arises from an unperturbed state in which all particles have momentum 0 except one, which has momentum **k**. Since (19.46) was based on the assumption that the level with momentum 0 is occupied by almost all N particles, we see that it cannot correctly describe a perturbed state in which the total number of phonons is a finite fraction of N.

In the derivation of (19.46) we have assumed that it is consistent to limit ourselves to only a subspace of the Hilbert space of the system. That "allowed" subspace is spanned by unperturbed states in which

(*a*) the ground level is occupied by essentially all N particles;

(*b*) the combined occupation of all other levels is vanishingly small compared to N.

We retained only the matrix elements of the original Hamiltonian (19.11) with respect to the "allowed" subspace, and we arrived at the effective Hamiltonian (19.46).

To generalize the method it is evident that we need only generalize the "allowed" subspace. The "allowed" subspace now is spanned by the unperturbed states for which

(*a*) the ground level is occupied by a finite fraction of all the particles;

(*b*) no other single level is so occupied;

(*c*) the *combined* occupation of all other levels is a finite fraction of all the particles.

The assumption is again made that any state in the "allowed" subspace remains in the "allowed" subspace despite any number of repeated interactions. The new "allowed" subspace, although larger than the old one, still maintains the privileged position of the ground level—a fact that makes the calculation simple. The ground level is privileged because it is regarded as an inexhaustible "sea" of particles (like the vacuum of the Dirac positron theory).

Let us choose as the unperturbed state any state in our "allowed" subspace. We specify this state, called $|f\rangle$, by the following occupation numbers:

Level	Occupation Number	
0	ξN	$(0 < \xi \leq 1)$
$\mathbf{k}$	$f_{\mathbf{k}}$	$(\mathbf{k} \neq 0)$

(19.85)

We require that

$$\sum_{\mathbf{k}\neq 0} f_{\mathbf{k}} = (1 - \xi)N$$
$$\frac{f_{\mathbf{k}}}{N} \underset{N\to\infty}{\longrightarrow} 0 \qquad (\mathbf{k} \neq 0) \tag{19.86}$$

When $\xi = 1$, we are back to the case discussed in the last section.* The largest off-diagonal matrix elements of the Hamiltonian, just as before, are those exciting two particles from the ground level to the levels $\mathbf{k}$, $-\mathbf{k}$ respectively. A state connected through such matrix elements to the state (19.85) either directly or indirectly has the following occupation numbers:

Level	Occupation Number
0	$\xi N - \sum_{\mathbf{k}\neq 0} l_{\mathbf{k}}$
$\mathbf{k}$	$f_{\mathbf{k}} + l_{\mathbf{k}}$
$-\mathbf{k}$	$f_{-\mathbf{k}} + l_{\mathbf{k}}$

(19.87)

* Note that the parameter ξ is the same as that introduced in Sec. 12.4.

This state we call $|f; l_1, l_2, \ldots\rangle$ or simply $|l_1, l_2, \ldots\rangle$. It is specified by a set of integers $\{l_{\mathbf{k}}\}$, such that $l_{\mathbf{k}}$ is the number of particles in the level $\mathbf{k}$ over and above $f_{\mathbf{k}}$ and $l_{-\mathbf{k}} = l_{\mathbf{k}}$. When each $l_{\mathbf{k}}$ independently ranges over the values*

$$l_{\mathbf{k}} = -f_{\mathbf{k}}, -f_{\mathbf{k}} + 1, \ldots, 0, 1, 2, \ldots \tag{19.88}$$

we generate all the states (19.87).

We now write down all the relevant matrix elements of the Hamiltonian. As before, we may ignore the operator $(\partial/\partial r)r$ in the Hamiltonian and adopt a subtraction procedure in its stead, as explained in the last section.

The relevant diagonal matrix elements are

$$\begin{aligned}
&\langle l_1, l_2, \ldots | H | l_1, l_2, \ldots\rangle \\
&= \frac{1}{2m}\sum_{\mathbf{k}} k^2(f_{\mathbf{k}} + l_{\mathbf{k}}) + \frac{4\pi a}{mV}\left\{N^2 - \frac{N}{2} - \frac{1}{2}\left[\left(\xi N - \sum_{k\neq 0} l_{\mathbf{k}}\right)^2 + \sum_{\mathbf{k}\neq 0}(f_{\mathbf{k}} + l_{\mathbf{k}})^2\right]\right\} \\
&= \frac{1}{2m}\left\{\sum_{\mathbf{k}} k^2 f_{\mathbf{k}} + \frac{4\pi a N}{v}\left[(2 - \xi^2) - \underbrace{\frac{2}{N^2}\sum_{\mathbf{k}\neq 0} f_{\mathbf{k}}^2}\right]\right. \\
&\qquad \left. + \sum_{\mathbf{k}\neq 0}\left[k^2 + \frac{8\pi a\xi}{v} - \underbrace{\frac{8\pi a}{Nv} f_{\mathbf{k}}}\right] l_{\mathbf{k}} - \underbrace{\frac{4\pi a}{v}\left[1 + \frac{1}{N}\left(\sum_{\mathbf{k}\neq 0} l_{\mathbf{k}}\right)^2 + \frac{1}{N}\sum_{\mathbf{k}\neq 0} l_{\mathbf{k}}^2\right]}\right\}
\end{aligned}$$

where terms underlined with a wavy line are to be neglected. Hence we take

$$\langle l_1, l_2, \ldots | H | l_1, l_2, \ldots\rangle = \frac{1}{2m}\left[\sum_{\mathbf{k}} k^2 f_{\mathbf{k}} + \frac{4\pi a N}{v}(2 - \xi^2) + \sum_{\mathbf{k}\neq 0}\left(k^2 + \frac{8\pi a\xi}{v}\right) l_{\mathbf{k}}\right] \tag{19.89}$$

The relevant off-diagonal matrix elements are the matrix elements of the operator

$$\frac{2\pi a}{mV}\sum_{\mathbf{k}\neq 0}(a_{\mathbf{k}}^\dagger a_{-\mathbf{k}}^\dagger a_0 a_0 + a_{\mathbf{k}} a_{-\mathbf{k}} a_0^\dagger a_0^\dagger)$$

They are

$$\begin{aligned}
\langle \ldots, l_{\mathbf{k}} + 1, \ldots | H | \ldots, l_{\mathbf{k}}, \ldots\rangle &= \frac{2\pi a\xi}{mV}(f_{\mathbf{k}} + l_{\mathbf{k}} + 1) \\
\langle \ldots, l_{\mathbf{k}}, \ldots | H | \ldots, l_{\mathbf{k}} + 1, \ldots\rangle &= \frac{2\pi a\xi}{mV}(f_{\mathbf{k}} + l_{\mathbf{k}})
\end{aligned} \tag{19.90}$$

* We assume, without loss of generality, that $f_{\mathbf{k}} \leq f_{-\mathbf{k}}$.

We may represent (19.89) and (19.90) by matrices in which a row or a column refers to a state $|\, l_1, l_2, \ldots \rangle$. By (19.88) $l_{\mathbf{k}}$ is the diagonal matrix

$$l_{\mathbf{k}} = \begin{bmatrix} -f_{\mathbf{k}} & & & & & & & \\ & -f_{\mathbf{k}} + 1 & & & & & & \\ & & \cdot & & & & & \\ & & & \cdot & & & & \\ & & & & 0 & & & \\ & & & & & 1 & & \\ & & & & & & 2 & \\ & & & & & & & \cdot \\ & & & & & & & & \cdot \\ & & & & & & & & & \cdot \end{bmatrix} = -f_{\mathbf{k}} + \begin{bmatrix} 0 & & & & & \\ & 1 & & & & \\ & & 2 & & & \\ & & & \cdot & & \\ & & & & \cdot & \\ & & & & & \cdot \end{bmatrix} \tag{19.91}$$

The matrix that has the diagonal elements (19.89) and off-diagonal elements (18.90) is given by

$$\frac{2\pi a N}{mv}[1 + (1 - \xi)^2] + \frac{1}{2m}\sum_{\mathbf{k}\neq 0}\left\{\left[\left(k^2 + \frac{8\pi a\xi}{v}\right)\begin{pmatrix} 0 & & & & & \\ & 1 & & & & \\ & & 2 & & & \\ & & & \cdot & & \\ & & & & \cdot & \\ & & & & & \cdot \end{pmatrix}\right]\right.$$

$$\left. + \frac{4\pi a\xi}{v}\begin{bmatrix} 0 & 1 & 0 & 0 & \cdots \\ 1 & 0 & 2 & 0 & \\ 0 & 2 & 0 & 3 & \\ & \cdot & \cdot & \cdot & \\ & & \cdot & \cdot & \cdot \\ & & & \cdot & \cdot & \cdot \end{bmatrix}\right\} \tag{19.92}$$

In the matrix in (19.92) the row or column corresponding to the unperturbed state $|f\rangle$ is the one with all $l_{\mathbf{k}} = 0$, i.e., the $f_{\mathbf{k}}$th row or column. Therefore we are interested only in the $f_{\mathbf{k}}$th eigenvalue of this matrix. To find this eigenvalue we note that it is the same as the $f_{\mathbf{k}}$th eigenvalue of the operator

$$H_{\text{eff}} = \frac{1}{2m}\left\{\frac{4\pi a N}{v}[1 + (1 - \xi)^2] + \sum_{\mathbf{k}\neq 0}\left[\left(k^2 + \frac{8\pi a\xi}{v}\right)a_{\mathbf{k}}^{\dagger}a_{\mathbf{k}} + \frac{4\pi a\xi}{v}(a_{\mathbf{k}}^{\dagger}a_{-\mathbf{k}}^{\dagger} + a_{\mathbf{k}}a_{-\mathbf{k}})\right]\right\} \tag{19.93}$$

This is the new effective Hamiltonian. We can diagonalize (19.93) immediately, since it differs only trivially from (19.46). The eigenvalues are

$$\frac{1}{2m}\left\{\frac{4\pi aN}{v}\left[1+(1-\xi)^2+\frac{128}{15}\sqrt{\frac{a^3\xi^5}{\pi v}}\right]+\sum_{\mathbf{k}\neq 0}k\sqrt{k^2+\frac{16\pi a\xi}{v}}\,\sigma_{\mathbf{k}}\right\} \tag{19.94}$$

where $\{\sigma_{\mathbf{k}}\}$ is a set of independent integers 0, 1, 2, ... The eigenvalue we are interested in is that for which

$$\sigma_{\mathbf{k}} = f_{\mathbf{k}} \tag{19.95}$$

Other eigenvalues are to be discarded.

We have thus shown that the unperturbed state $|f\rangle$ leads to a perturbed state that contains $(1-\xi)N$ phonons, with $f_{\mathbf{k}}$ phonons having momentum $\mathbf{k}$. The perturbed energy is

$$2mE\{f_{\mathbf{k}}\} = \frac{4\pi aN}{v}\left[1+(1-\xi)^2+\frac{128}{15}\sqrt{\frac{a^3\xi^5}{\pi v}}\right]+\sum_{\mathbf{k}\neq 0}k\sqrt{k^2+\frac{16\pi a\xi}{v}}\,f_{\mathbf{k}} \tag{19.96}$$

where

$$\xi \equiv 1-\frac{1}{N}\sum_{\mathbf{k}\neq 0}f_{\mathbf{k}} \tag{19.97}$$

The energy depends on the phonon occupation numbers $\{f_{\mathbf{k}}\}$ in a nonlinear way. When $\xi = 1$, (19.97) reduces to the previous result (19.54), but $\xi = 1$ also implies that

$$\frac{1}{N}\sum_{\mathbf{k}\neq 0}f_{\mathbf{k}} = 0$$

Hence the result (19.54) is valid only if the total number of phonons excited is vanishingly small compared to N.

The velocity of sound is now given by

$$c = \sqrt{\frac{4\pi a\xi}{m^2 v}} \tag{19.98}$$

which depends on the number of phonons present. If so many phonons are excited that $\xi = 0$, then $c = 0$. In this case the energy (19.96) reduces to the free-particle energy plus a constant correction term:

$$E(\xi = 0) = \sum_{\mathbf{k}}\frac{k^2}{2m}f_{\mathbf{k}} + \left(\frac{4\pi a}{mv}\right)N \tag{19.99}$$

The excitations are therefore no longer phonons but single particles. For $\xi = 0$ the ground level is completely depleted. Thus we might question the applicability of the method for $\xi = 0$. But the result (19.99) agrees with the first-order calculation (19.18). Therefore we may conclude *a posteriori* that the method does not break down at $\xi = 0$.

We have calculated the energy of certain states of the perturbed system to the lowest order in the parameter $\sqrt{a^3/v}$. All the states that have been included in the calculation have the common property that, when the interaction is switched off, they approach free-particle states whose occupation numbers $\{n_{\mathbf{k}}\}$ have the following properties:

(*a*) $n_0 = \xi N$, where ξ is a fixed number between 0 and 1

(*b*) $\dfrac{n_{\mathbf{k}}}{N} \xrightarrow[N\to\infty]{} 0 \qquad (\mathbf{k} \neq 0)$

(*c*) $\sum_{\mathbf{k}\neq 0} n_{\mathbf{k}} = (1 - \xi)N$

It is clear that these states do not comprise all possible states of the system. In particular, wc have not included any state in which a level other than $\mathbf{k} = 0$ is occupied by a finite fraction of all the particles. Such states may be divided into two classes: (*a*) Those for which $n_{\mathbf{k}} = \xi N$ for some given $\mathbf{k}$, whereas no other single occupation number is a finite fraction of N: (*b*) those for which at least two different occupation numbers are finite fractions of N. The first class is essentially included in our calculation, because they may be obtained from a state we considered through a Galilean transformation in the manner discussed in Sec. 18.5. The second class must be separately investigated. They may be shown to have very high energies. For the purpose of statistical mechanics, therefore, it is plausible that the states of the second class are of no importance.

19.6 CRITICAL DISCUSSION

The calculations in the last two sections are subject to some obvious limitations of validity. For the energy levels and the wave functions we must require that

$$\sqrt{\frac{a^3}{v}} \ll 1 \qquad (19.100)$$

$$ka \ll 1$$

where k is the wave number of any phonon. The first condition is required by the fact that $\sqrt{a^3/v}$ is the expansion parameter for the ground state energy, whereas the second condition is required for the validity of the S-wave pseudopotential.

To see that $\sqrt{a^3/v}$ is the relevant parameter in the calculation, higher-order corrections [i.e., the "remainder" of (19.32)] to the ground state energy must be investigated. This may be done by making the transformation (19.48) in the original Hamiltonian (19.11), and treating the left-over nondiagonal terms by standard perturbation theory. If we write down the correction to the energy (19.55) in the usual form of a sum over intermediate states, then by dimensional analysis we conclude that it is proportional to

$$N \frac{4\pi a}{v} \frac{a^3}{v}$$

This would indicate that we have an expansion in powers of $\sqrt{a^3/v}$, except that the proportionality constant turns out to be infinite. Thus the correct calculation of the next order requires some caution. Wu* has made the calculation, which gives the answer

$$E_0 = \frac{2\pi a N}{mv}\left[1 + \frac{128}{15}\sqrt{\frac{a^3}{\pi v}} + 8\left(\frac{4\pi}{3} - \sqrt{3}\right)\frac{a^3}{v}\log\frac{a^3}{v} + \text{const.}\left(\frac{a^3}{v}\right)\right] \tag{19.101}$$

It shows that the correction is small if $\sqrt{a^3/v} \ll 1$ but that we do not have an expansion in powers of $\sqrt{a^3/v}$.

We point out two questions that may legitimately be raised but which have not been answered in a rigorous way.

The first has to do with boundary conditions. We have imposed periodic boundary conditions on the wave functions, and this boundary condition has played an essential role in the practicality of the calculation. Although the pseudopotentials are independent of the asymptotic boundary conditions, our whole perturbation scheme depends on the periodic boundary conditions. It is because of the periodic boundary conditions that momentum is conserved in the elementary interactions between particles. This results in reducing the number of matrix elements we have, and it allows us to classify them into the various classes of different orders of magnitude. If we had imposed the boundary condition, for example, that the wave function vanishes on the surface of a large box, our perturbation scheme would not have worked. The question, then, is whether our results are sensitive to boundary conditions. As long as we do not have a method of calculation for general boundary conditions, we do not have a rigorous answer. An answer based on physical arguments, however, may be indicated.

We may argue that our results are independent of boundary conditions. Suppose the ground state wave function is required to vanish at the surface of the normalization box. Since the system has a finite "correlation length" as defined in (19.81), we expect the density of particles to be substantially constant in the interior of the volume and drop off to zero

* T. T. Wu, *Phys. Rev.*, **115**, 1390 (1959).

at the surface of the box within the distance of one correlation length. For small a, the correlation length is very large. It is nevertheless a finite quantity, whereas the box is infinitely large. The results for this boundary condition therefore must be the same as those we have obtained. This argument also indicates why a perturbation treatment may be difficult if we use the last-mentioned boundary condition. In that case, the unperturbed wave function gives a particle density of the form $[\sin(\pi x/L)]^N$, which is very far from being uniform.

The second question concerns the fundamental assumption of the calculation, namely that it is legitimate to limit our considerations to a subspace S of the Hilbert space. Our calculations have shown the assumption to be consistent with the results, but we have not proved its validity. In order to investigate its validity the assumption must be stated in a form that is mathematically precise. Even this has not been achieved, because of the difficulty in defining the limit of the Hilbert space as $N \to \infty$ (if such a limit exists). This question therefore remains unanswered.

19.7 MACROSCOPIC PROPERTIES*

The thermodynamic properties of the hard-sphere Bose gas can be worked out by evaluating the partition function with the help of the energy levels (19.96). Since the low-lying excited states of the system can be described in terms of phonons, it is clear that the thermodynamic properties near absolute zero are qualitatively the same as those of liquid He^4. Furthermore, a heuristic discussion of the kinetic theory of the hard-sphere Bose gas can be made along the lines described in Chapter 18. The only difference is that here we can calculate everything, including the two sound velocities at finite temperatures. We do not go into the details of these calculations.

The energy levels (19.96) enable us to calculate the partition function at all temperatures, but the results are significant only in the region in which

$$\sqrt{\frac{a^3}{v}} \ll 1$$

$$\frac{a}{\lambda} \ll 1$$

where λ is the thermal wavelength. This region includes a portion of the transition line of the Bose-Einstein condensation of the ideal Bose gas.

* For a detailed discussion of the topics mentioned in this section, see K. Huang in J. de Boer and G. E. Uhlenbeck, *Studies in Statistical Mechanics*, Vol. 2 (North-Holland Publishing Co., Amsterdam, 1963).

Thus we may hope to discuss the Bose-Einstein condensation of the hard-sphere Bose gas in the present approximation. The result is not qualitatively different from that discussed in Sec. 13.5.

Finally we note that in the hard-sphere Bose gas there can only be two thermodynamic phases, namely those analogous to the gas phase and the condensed phase of the ideal Bose gas. This situation is qualitatively different from that of liquid He^4, which (ignoring the solid phase) has three phases: gas, He I, and He II. The difference is due to the fact that there is no attractive interaction in the hard-sphere Bose gas. Hence the system does not form a bound state that can be identified as a liquid. We obtain three phases, with the same qualitative relationship between the phases as in He, if we add an attractive interaction to the hard-sphere interaction.

D

APPENDICES

appendix A

N-BODY SYSTEM OF IDENTICAL PARTICLES

A.1 THE TWO KINDS OF STATISTICS

An N-body system of identical particles is characterized by a Hamiltonian operator H that is invariant under the interchange of all the coordinates of any two particles. Any wave function for the system can be written as a linear superposition of eigenfunctions Ψ_n of the Hamiltonian:

$$H\Psi_n(q_1, \ldots, q_N) = E_n\Psi_n(q_1, \ldots, q_N) \tag{A.1}$$

where q_i denotes the collection of all the coordinates of the ith particle, including the position coordinates, and the spin and other internal coordinates, if any. To study the general symmetry property of any wave function, it is sufficient to study the general symmetry property of Ψ_n.

Let $\mathscr{P}$ be an operator that, when applied to Ψ_n, interchanges the positions of q_i and q_j:

$$\mathscr{P}\Psi_n(\ldots, q_i, \ldots, q_j, \ldots) = \Psi_n(\ldots, q_j, \ldots, q_i, \ldots) \tag{A.2}$$

By definition we have

$$\mathscr{P}^{-1}H\mathscr{P} = H \tag{A.3}$$

Therefore

$$H(\mathscr{P}\Psi_n) = E_n(\mathscr{P}\Psi_n) \tag{A.4}$$

i.e., if Ψ_n is an eigenfunction of H belonging to the eigenvalue E_n, then $\mathscr{P}\Psi_n$ is also an eigenfunction of H belonging to the same eigenvalue.

A possible property of Ψ_n is that $\mathscr{P}\Psi_n$ is proportional to Ψ_n. If this is so, then $\mathscr{P}^2\Psi_n = \Psi_n$. Hence the proportionality constant is either $+1$ or -1, and Ψ_n is either symmetric or antisymmetric under the interchange of two coordinates:

$$\mathscr{P}\Psi_n = \pm\Psi_n \tag{A.5}$$

Suppose that for a given system (A.5) is fulfilled. The question naturally arises whether we can have some Ψ_n symmetric while others are antisymmetric. The answer is as follows. There is nothing to forbid such a situation; but then the symmetric wave functions $\{\Psi_n^{(+)}\}$ and the antisymmetric ones $\{\Psi_n^{(-)}\}$ form two classes of wave functions that do not mix, in the sense of the "superselection rule"

$$(\Psi_m^{(+)}, \mathcal{O}\Psi_n^{(-)}) = 0 \tag{A.6}$$

Proof. $(\Psi_m^{(+)}, \mathcal{O}\Psi_n^{(-)}) = (\Psi_m^{(+)}, \mathscr{P}^{-1}\mathcal{O}\mathscr{P}\Psi_n^{(-)}) = (\mathscr{P}\Psi_m^{(+)}, \mathcal{O}\mathscr{P}\Psi_n^{(-)})$

$$= -(\Psi_m^{(+)}, \mathcal{O}\Psi_n^{(-)}) \qquad \text{(QED)}$$

where $\mathcal{O}$ is any operator invariant under a permutation of particles. Since only such operators will ever be considered (otherwise the particles are not identical), the sets $\{\Psi_n^{(+)}\}$ and $\{\Psi_n^{(-)}\}$ form two disjoint systems of wave functions that do not influence each other. Hence it is sufficient to consider a system for which all Ψ_n are either $\Psi_n^{(+)}$ or $\Psi_n^{(-)}$.

According to (A.5), two wave functions differing only by an interchange of two coordinates correspond to one and the same state of the system. This has a direct bearing on the correct counting of states for a given energy. Hence (A.5) is said to define the statistics of the system. The plus sign refers to Bose statistics and the minus sign refers to Fermi statistics.

The property (A.5) is not the only one consistent with (A.4). In general $\mathscr{P}\Psi_n$ may be a linear combination of eigenfunctions $\Psi_1, \ldots, \Psi_g$, which must all have the eigenvalue E_n. If this is the case, the eigenvalue E_n has an intrinsic degeneracy that cannot be removed by the introduction of interactions among the particles, because any perturbation $\mathcal{O}$ added to the Hamiltonian must have the property $\mathscr{P}\mathcal{O}\mathscr{P}^{-1} = \mathcal{O}$. Nothing in what we have said so far rules out such intrinsic degeneracies. Nevertheless nature apparently abhors degeneracies, because it is an experimental fact that (A.5) is the only possibility so far observed. If future experiments discover a type of particle that does not obey (A.5), new types of symmetry in addition to those for bosons and fermions must be explored.

When the principle of Lorentz covariance is imposed on physical systems, it is possible to show,* in the formalism of relativistic quantum field theory, that particles of integer spin must obey Bose statistics and

* W. Pauli, *Phys.Rev.*, **58,** 716 (1940).

The permutation P is even or odd according to whether it is equivalent to an even or an odd number of successive interchanges. It is obvious that a permutation of $\alpha_1, \ldots, \alpha_N$ changes (A.14) by at most a sign and does not lead to a new independent wave function. For fermions, (A.14) is equivalent to the definition

$$\Phi_\alpha(1, \ldots, N) = \frac{1}{\sqrt{N!}} \begin{vmatrix} u_{\alpha_1}(1) & u_{\alpha_2}(1) & \cdots & u_{\alpha_N}(1) \\ \cdot & & & \cdot \\ \cdot & & & \cdot \\ \cdot & & & \cdot \\ u_{\alpha_1}(N) & u_{\alpha_2}(N) & \cdots & u_{\alpha_N}(N) \end{vmatrix} \tag{A.16}$$

It is obvious from this form that for fermions Φ_α vanishes unless $\alpha_1, \ldots, \alpha_N$ are distinct quantum numbers.

From the orthogonality of the functions $u_\alpha(\mathbf{r})$ it easily follows that

$$(\Phi_\alpha, \Phi_\beta) = \int d^{3N}r\, \Phi_\alpha^*(1, \ldots, N)\Phi_\beta(1, \ldots, N) = 0$$
$$(\text{if } \{\alpha_1, \ldots, \alpha_N\} \text{ and } \{\beta_1, \ldots, \beta_N\} \text{ are not the same})$$

Therefore (A.14) defines an orthogonal set of wave functions. We now calculate the norm of Φ_α:

$$(\Phi_\alpha, \Phi_\alpha) = \int d^{3N}r \frac{1}{N!} \sum_P \sum_Q \delta_P \delta_Q [u^*_{P\alpha_1}(1) u_{Q\alpha_1}(1)] \cdots [u^*_{P\alpha_N}(N) u_{Q\alpha_N}(N)]$$

For fermions we must have $P = Q$, because the α_i are all distinct. Therefore

$$(\Phi_\alpha, \Phi_\alpha) = 1 \qquad \text{(fermions)} \tag{A.17}$$

For bosons the α_i are not necessarily all distinct. Suppose that among $\alpha_1, \ldots, \alpha_N$ there are n_α having the value α. Then

$$(\Phi_\alpha, \Phi_\alpha) = \prod_\alpha (n_\alpha!) \qquad \text{(bosons)} \tag{A.18}$$

which is not necessarily unity.

The integer n_α is called the occupation number of the single-particle level α. Obviously we have the conditions

$$\begin{aligned} &\sum_\alpha n_\alpha = N \\ &n_\alpha = 0, 1, \ldots, N \qquad \text{(boson)} \\ &n_\alpha = 0, 1 \qquad \text{(fermion)} \end{aligned} \tag{A.19}$$

Instead of $\{\alpha_1, \ldots, \alpha_N\}$ we can equally well label the wave function by the occupation numbers $\{n_0, n_1, \ldots\}$. Thus we also introduce the notation Φ_n, defined by

$$\Phi_n \equiv \Phi_\alpha \tag{A.20}$$

where n stands for the set $\{n_0, n_1, \ldots\}$.

If we wish, we may use in place of Φ_α the wave function

$$\Phi'_\alpha \equiv \frac{\Phi_\alpha}{\sqrt{\prod_\alpha (n_\alpha !)}} \tag{A.21}$$

which is normalized to unity for both bosons and fermions. It is, however, neither necessary nor convenient to do this. The reason is as follows. Suppose we are to calculate the trace of an operator. We may write

$$Tr\mathcal{O} = \sum_{\{\alpha\}} (\Phi_\alpha', \mathcal{O}\Phi_\alpha') \tag{A.22}$$

where the sum extends over all distinct sets $\{\alpha_1, \ldots, \alpha_N\}$. A convenient way to calculate this sum is to sum over each α_i independently and to take into account the fact that a permutation of the α_i must not be counted as a new term in the sum. That is, we write

$$Tr\mathcal{O} = \sum_{\alpha_1, \ldots, \alpha_N} \frac{\prod_\alpha (n_\alpha !)}{N!} (\Phi_\alpha', \mathcal{O}\Phi_\alpha')$$

By (A.21) we have

$$Tr\mathcal{O} = \frac{1}{N!} \sum_{\alpha_1, \ldots, \alpha_N} (\Phi_\alpha, \mathcal{O}\Phi_\alpha) \tag{A.23}$$

Thus it is actually more convenient to use the wave functions Φ_α as defined in (A.14).

Free-Particle Wave Functions

A useful choice of $u_\alpha(\mathbf{r})$ is the single-particle wave function for a free particle of momentum $\mathbf{p}$. The quantum number α is now explicitly $\mathbf{p}$, and we have

$$u_\mathbf{p}(\mathbf{r}) \equiv \frac{1}{\sqrt{V}} e^{i\mathbf{p}\cdot\mathbf{r}/\hbar} \tag{A.24}$$

The allowed values of $\mathbf{p}$ are determined by the periodic boundary conditions

$$u_\mathbf{p}(\mathbf{r} + \mathbf{n}L) = u_\mathbf{p}(\mathbf{r}) \tag{A.25}$$

where $\mathbf{n}$ and L are defined in (A.11). This implies that the allowed values of $\mathbf{p}$ are

$$\mathbf{p} = \frac{2\pi\hbar\mathbf{n}}{L} \tag{A.26}$$

These values form a cubic lattice in momentum space with the lattice constant $2\pi\hbar/L$, which approaches 0 as $V \to \infty$. In this limit a volume element of size d^3p in momentum space contains $(V/h^3)\, d^3p$ lattice points. Thus as $V \to \infty$ a sum over $\mathbf{p}$ may be replaced by an integration over $\mathbf{p}$ in the following manner:

$$\sum_{\mathbf{p}} \to \frac{V}{h^3} \int d^3p \tag{A.27}$$

The functions defined by (A.24) obviously form an orthogonal set. The completeness of this set follows from the fact that any function can be Fourier-analyzed.

The N-body wave functions built up from $u_{\mathbf{p}}(\mathbf{r})$ according to (A.14) are the N-body free-particle wave functions. They are denoted by $\Phi_p(1, \ldots, N)$ and are eigenfunctions of the kinetic-energy operator K:

$$K\Phi_p(1, \ldots, N) = \frac{1}{2m}(p_1{}^2 + \cdots + p_N{}^2)\Phi_p(1, \ldots, N) \tag{A.28}$$

where $\mathbf{p}_1, \ldots, \mathbf{p}_N$ are the momenta of the N single-particle wave functions contained in Φ_p.

Example of Calculation: System of Bosons

We calculate $(\Phi_\alpha, \Omega\Phi_\alpha)$ for a system of bosons:

$$(\Phi_\alpha, \Omega\Phi_\alpha) = \int d^{3N}r\, \Phi_\alpha{}^* \sum_{i<j} v_{ij}\Phi_\alpha = \tfrac{1}{2}N(N-1)\int d^{3N}r\, \Phi_\alpha{}^* v_{12}\Phi_\alpha$$

where the second equality is obtained by renaming the integration variables $\mathbf{r}_1, \ldots, \mathbf{r}_N$ in an appropriate fashion in each term of the sum $\Sigma\, v_{ij}$. Using (A.14) we have

$$\begin{aligned}
(\Phi_\alpha, \Omega\Phi_\alpha) \\
&= \frac{N(N-1)}{N!\,2} \sum_P \sum_Q \int d^{3N}r [u^*_{P\alpha_1}(1) \cdots u^*_{P\alpha_N}(N)] v_{12} [u_{Q\alpha_1}(1) \cdots u_{Q\alpha_N}(N)] \\
&= \frac{N(N-1)}{N!\,2} \sum_P \sum_Q \langle P\alpha_1, P\alpha_2 \,|\, v \,|\, Q\alpha_1, Q\alpha_2\rangle (\delta_{P\alpha_3, Q\alpha_3} \cdots \delta_{P\alpha_N, Q\alpha_N})
\end{aligned} \tag{A.29}$$

where

$$\langle \alpha, \beta \,|\, v \,|\, \gamma, \lambda\rangle \equiv \int d^3r_1\, d^3r_2\, u_\alpha{}^*(1) u_\beta{}^*(2) v_{12} u_\gamma(1) u_\lambda(2) \tag{A.30}$$

In (A.29) only two terms in the sum Σ_Q are nonzero, namely the terms satisfying the conditions (*a*) or (*b*):

$$\begin{aligned}
&(a)\quad Q\alpha_1 = P\alpha_1, \quad Q\alpha_2 = P\alpha_2, \quad Q\alpha_j = P\alpha_j \quad (j = 3, \ldots, N) \\
&(b)\quad Q\alpha_1 = P\alpha_2, \quad Q\alpha_2 = P\alpha_1, \quad Q\alpha_j = P\alpha_j \quad (j = 3, \ldots, N)
\end{aligned} \tag{A.31}$$

Hence

$$(\Phi_\alpha, \Omega\Phi_\alpha) = \frac{N(N-1)}{N!\,2}$$

$$\times \sum_P (\langle P\alpha_1, P\alpha_2 \mid v \mid P\alpha_1, P\alpha_2\rangle + \langle P\alpha_1, P\alpha_2 \mid v \mid P\alpha_2, P\alpha_1\rangle) \qquad \text{(A.32)}$$

As P ranges through all the $N!$ permutations of the set $\{\alpha_1, \ldots, \alpha_N\}$, the pair $\{P\alpha_1, P\alpha_2\}$ takes on all possible pairs of values $\{\alpha, \beta\}$ chosen from the set $\{\alpha_1, \ldots, \alpha_N\}$. Suppose the occupation numbers for the single-particle states α, β are respectively n_α, n_β. Then the number of ways in which the pair $\{\alpha, \beta\}$ can be chosen from the set $\{\alpha_1, \ldots, \alpha_N\}$ is

$$f_{\alpha\beta} = \begin{cases} n_\alpha n_\beta & (\alpha \neq \beta) \\ \frac{1}{2}n_\alpha(n_\alpha - 1) & (\alpha = \beta) \end{cases}$$

or

$$f_{\alpha\beta} = (1 - \delta_{\alpha\beta})n_\alpha n_\beta + \tfrac{1}{2}\delta_{\alpha\beta}n_\alpha(n_\alpha - 1)$$

Furthermore, there are $(N-2)!$ permutations that affect only the quantum numbers $\{\alpha_3, \ldots, \alpha_N\}$ and thus leave $\{\alpha_1, \alpha_2\}$ unchanged. Changing the label of Φ_α to occupation numbers we obtain

$$(\Phi_n, \Omega\Phi_n) = \tfrac{1}{2}N(N-1)\frac{(N-2)!}{N!}\sum_{\alpha,\beta} f_{\alpha\beta}(\langle\alpha, \beta \mid v \mid \alpha, \beta\rangle + \langle\alpha, \beta \mid v \mid \beta, \alpha\rangle)$$

or

$$(\Phi_n, \Omega\Phi_n) = \tfrac{1}{2}\sum_{\alpha,\beta}[(1 - \delta_{\alpha\beta})n_\alpha n_\beta + \tfrac{1}{2}\delta_{\alpha\beta}n_\alpha(n_\alpha - 1)]$$

$$\times (\langle\alpha, \beta \mid v \mid \alpha, \beta\rangle + \langle\alpha, \beta \mid v \mid \beta, \alpha\rangle) \qquad \text{(A.33)}$$

This result can be derived without all the tedious counting if we use the method of quantized fields.

For the free-particle wave functions (A.24), and for $v_{12} = \delta(\mathbf{r}_1 - \mathbf{r}_2)$, (A.30) reduces to

$$\langle \mathbf{p}_1, \mathbf{p}_2 \mid \delta \mid \mathbf{p}_1', \mathbf{p}_2'\rangle = \frac{1}{V} \qquad \text{(A.34)}$$

Therefore for free-particle wave functions we have

$$\left(\Phi_n, \sum_{i<j}\delta(\mathbf{r}_i - \mathbf{r}_j)\Phi_n\right) = \frac{1}{V}\left[\sum_{\mathbf{p}\neq\mathbf{k}} n_\mathbf{p} n_\mathbf{k} + \tfrac{1}{2}\sum_\mathbf{p} n_\mathbf{p}(n_\mathbf{p} - 1)\right]$$

Since

$$\sum_{\mathbf{p}\neq\mathbf{k}} n_\mathbf{p} n_\mathbf{k} = \sum_\mathbf{p} n_\mathbf{p}\sum_\mathbf{k} n_\mathbf{k} - \sum_\mathbf{p} n_\mathbf{p}^2 = N^2 - \sum_\mathbf{p} n_\mathbf{p}^2 \qquad \text{(A.35)}$$

we have

$$\left(\Phi_n, \sum_{i<j}\delta(\mathbf{r}_i - \mathbf{r}_j)\Phi_n\right) = \frac{1}{V}\left(N^2 - \tfrac{1}{2}N - \tfrac{1}{2}\sum_\mathbf{p} n_\mathbf{p}^2\right) \qquad \text{(A.36)}$$

Example of Calculation: System of Fermions

We calculate $(\Phi_\alpha, \Omega\Phi_\alpha)$ for a system of fermions. The formula (A.29) remains valid if we insert the factor $\delta_P\delta_Q$ before the summand in (A.29). The conditions in (A.31) remain pertinent. Noting that $\delta_P\delta_Q = 1$ under the condition (*a*) of (A.30), and that $\delta_P\delta_Q = -1$ under the condition (*b*) of (A.30), we obtain in place of (A.32) the formula

$$(\Phi_\alpha, \Omega\Phi_\alpha) = \frac{N(N-1)}{N!\,2} \sum_P (\langle P\alpha_1, P\alpha_2 | v | P\alpha_1, P\alpha_2\rangle - \langle P\alpha_1, P\alpha_2 | v | P\alpha_2, P\alpha_1\rangle) \tag{A.37}$$

Noting that $n_\alpha = 0, 1$ we obtain in place of (A.33) the formula

$$(\Phi_n, \Omega\Phi_n) = \tfrac{1}{2} \sum_{\alpha,\beta} n_\alpha n_\beta (\langle \alpha, \beta | v | \alpha, \beta\rangle - \langle \alpha, \beta | v | \beta, \alpha\rangle) \tag{A.38}$$

To illustrate how the N-body wave functions may be generalized to include the spin coordinates of the particles, let us consider fermions of spin $\hbar/2$. In addition to the position coordinate $\mathbf{r}$, each particle now has a spin coordinate σ which can take on only the values ± 1. The free-particle wave function $u_{\mathbf{p}s}(\mathbf{r}, \sigma)$ of a particle is now labeled by the momentum $\mathbf{p}$ and the spin quantum number s, which can assume only the values ± 1. When $s = +1$ the particle is said to be in a state of up spin, and when $s = -1$, in down spin. Explicitly we have

$$u_{\mathbf{p}s}(\mathbf{r}, \sigma) = \frac{1}{\sqrt{V}} e^{i\mathbf{p}\cdot\mathbf{r}/\hbar}\, \delta(s, \sigma) \tag{A.39}$$

where

$$\delta(s, \sigma) = \begin{cases} 1 & (s = \sigma) \\ 0 & (s \neq \sigma) \end{cases} \tag{A.40}$$

Sometimes we write out the two values of $u_{\mathbf{p}s}(\mathbf{r}, \sigma)$ for $\sigma = \pm 1$, as follows

$$\begin{pmatrix} u_{\mathbf{p},+1}(\mathbf{r}, +1) \\ u_{\mathbf{p},+1}(\mathbf{r}, -1) \end{pmatrix} = \frac{1}{\sqrt{V}} e^{i\mathbf{p}\cdot\mathbf{r}/\hbar} \begin{pmatrix} 1 \\ 0 \end{pmatrix}$$

$$\begin{pmatrix} u_{\mathbf{p},-1}(\mathbf{r}, +1) \\ u_{\mathbf{p},-1}(\mathbf{r}, -1) \end{pmatrix} = \frac{1}{\sqrt{V}} e^{i\mathbf{p}\cdot\mathbf{r}/\hbar} \begin{pmatrix} 0 \\ 1 \end{pmatrix}$$

We do not use such a representation here.

If we let α stand for the collection of quantum numbers $\{\mathbf{p}_\alpha, s_\alpha\}$ and let $u_\alpha(\mathbf{r}_1, \sigma_1)$ be abbreviated by $u_\alpha(1)$, then (A.14) defines a complete orthonormal set of wave functions for the N-fermion system and (A.38) continues to be valid. Let $v_{12} = \delta(\mathbf{r}_1 - \mathbf{r}_2)$, which is independent of spin coordinates. Then

$$\langle \alpha, \beta \mid \delta \mid \alpha, \beta \rangle = \frac{1}{V}$$
$$\langle \alpha, \beta \mid \delta \mid \beta, \alpha \rangle = \frac{1}{V} \delta(s_\alpha, s_\beta)$$

Let $n_{\mathbf{p}s}$ denote the occupation number of the single-particle state with momentum $\mathbf{p}$ and spin quantum number s. Then (A.38) becomes

$$\left(\Phi_n, \sum_{i<j} \delta(\mathbf{r}_i - \mathbf{r}_j)\Phi_n\right) = \frac{1}{2V} \sum_{s,s'} \sum_{\mathbf{p},\mathbf{k}} n_{\mathbf{p}s} n_{\mathbf{k}s'} [1 - \delta(s, s')]$$
$$= \frac{1}{2V}\left(N^2 - \sum_s \sum_{\mathbf{p},\mathbf{k}} n_{\mathbf{p}s} n_{\mathbf{k}s}\right)$$

Let

$$N_+ \equiv \sum_{\mathbf{p}} n_{\mathbf{p},+1}$$
$$N_- \equiv \sum_{\mathbf{p}} n_{\mathbf{p},-1} = N - N_+ \tag{A.41}$$

Then

$$\left(\Phi_n, \sum_{i<j} \delta(\mathbf{r}_i - \mathbf{r}_j)\Phi_n\right) = \frac{N_+ N_-}{V} \tag{A.42}$$

A.3 METHOD OF QUANTIZED FIELDS

A system of N particles is equivalent to a quantized field. This equivalence is often used to great advantage in the calculation of the energy levels and the partition function of an N-particle system.

A quantized field is a system characterized by field operators $\psi(\mathbf{r})$ that are defined for all values of the coordinate $\mathbf{r}$ and operate on a Hilbert space. A vector in this Hilbert space corresponds to a state of the quantized field. It is our purpose to show that a quantized field can be so defined that its Hilbert space contains the Hilbert space of a given N-particle system. For simplicity we consider the N particles to be either all identical spinless bosons or all identical spinless fermions.

First we define the quantized fields that corresponds to bosons and fermions. The field operators of the two cases are defined by the following commutation rules.

Bosons	Fermions
$[\psi(\mathbf{r}), \psi^\dagger(\mathbf{r}')] = \delta(\mathbf{r} - \mathbf{r}')$	$\{\psi(\mathbf{r}), \psi^\dagger(\mathbf{r}')\} = \delta(\mathbf{r} - \mathbf{r}')$
$[\psi(\mathbf{r}), \psi(\mathbf{r}')] = 0$	$\{\psi(\mathbf{r}), \psi(\mathbf{r}')\} = 0$
$[\psi^\dagger(\mathbf{r}), \psi^\dagger(\mathbf{r}')] = 0$	$\{\psi^\dagger(\mathbf{r}), \psi^\dagger(\mathbf{r}')\} = 0$

(A.43)

where $\psi^\dagger$ is the hermitian conjugate of ψ and $[A, B] \equiv AB - BA$, $\{A, B\} \equiv AB + BA$.

The definition of the quantized field is completed by defining two hermitian operators—the Hamiltonian operator H and the number operator N_{op}. The Hamiltonian operator is

$$\begin{aligned} H &\equiv K + \Omega \\ K &= -\frac{\hbar^2}{2m} \int d^3r\, \psi^\dagger(\mathbf{r}) \nabla^2 \psi(\mathbf{r}) \\ \Omega &= \frac{1}{2} \int d^3r_1\, d^3r_2\, \psi^\dagger(\mathbf{r}_1)\psi^\dagger(\mathbf{r}_2) v_{12} \psi(\mathbf{r}_2)\psi(\mathbf{r}_1) \end{aligned} \tag{A.44}$$

where $v_{12} \equiv v(\mathbf{r}_1, \mathbf{r}_2)$. The number operator is

$$N_{\text{op}} \equiv \int d^3r\, \psi^\dagger(\mathbf{r})\psi(\mathbf{r}) \tag{A.45}$$

These definitions hold for both bosons and fermions. We can easily verify that

$$[H, N_{\text{op}}] = 0 \tag{A.46}$$

Therefore H and N_{op} can be simultaneously diagonalized. We show that a simultaneous eigenstate of H and N_{op} is an energy eigenstate of a system of a definite number of particles.

Let a complete orthonormal basis of the Hilbert space be so chosen that any vector $| \Phi_n \rangle$ of the basis is a simultaneous eigenstate of H and N_{op}. Let a particular member of the basis be denoted by $| \Psi_{EN} \rangle$, with the properties that

$$\begin{aligned} \langle \Psi_{EN} | \Psi_{EN} \rangle &= 1 \\ H | \Psi_{EN} \rangle &= E | \Psi_{EN} \rangle \\ N_{\text{op}} | \Psi_{EN} \rangle &= N | \Psi_{EN} \rangle \end{aligned} \tag{A.47}$$

The state $| 0 \rangle \equiv | \Psi_{00} \rangle$, called the vacuum state, is assumed to be unique. Its properties are:

$$\begin{aligned} \langle 0 | 0 \rangle &= 1 \\ H | 0 \rangle &= 0 \\ N_{\text{op}} | 0 \rangle &= 0 \end{aligned} \tag{A.48}$$

From (A.45) and (A.43) it is easily verified that

$$\begin{aligned} [\psi(\mathbf{r}), N_{\text{op}}] &= \psi(\mathbf{r}) \\ [\psi^\dagger(\mathbf{r}), N_{\text{op}}] &= -\psi^\dagger(\mathbf{r}) \end{aligned} \tag{A.49}$$

Hence

$$\begin{aligned} N_{\text{op}}\psi(\mathbf{r}) | \Psi_{EN} \rangle &= (N - 1)\psi(\mathbf{r}) | \Psi_{EN} \rangle \\ N_{\text{op}}\psi^\dagger(\mathbf{r}) | \Psi_{EN} \rangle &= (N + 1)\psi^\dagger(\mathbf{r}) | \Psi_{EN} \rangle \end{aligned} \tag{A.50}$$

Thus $\psi(\mathbf{r})$ decreases N by 1, and $\psi^\dagger(\mathbf{r})$ increases N by 1. By repeated application of $\psi^\dagger(\mathbf{r})$ to $|\,0\rangle$, we prove that the eigenvalues of $N_{\rm op}$ are

$$N = 0, 1, 2, \ldots \tag{A.51}$$

Since $\psi(\mathbf{r})$ decreases N by 1, and the state with $N = 0$ is assumed to be unique, we have the identity

$$\langle \Phi_n \,|\, \psi(1)\psi(2)\cdots\psi(N) \,|\, \Psi_{EN}\rangle = 0 \qquad \text{unless } |\,\Phi_n\rangle \equiv |\,0\rangle \tag{A.52}$$

where $\psi(j) \equiv \psi(\mathbf{r}_j)$.

Let a function of the N position coordinates $\mathbf{r}_1, \ldots, \mathbf{r}_N$ be defined by

$$\Psi_{EN}(1, \ldots, N) \equiv \frac{1}{\sqrt{N!}} \langle 0 \,|\, \psi(1)\cdots\psi(N) \,|\, \Psi_{EN}\rangle \tag{A.53}$$

By (A.43) this function is symmetric (antisymmetric) with respect to the exchange of any two coordinates for bosons (fermions). The norm of $\Psi_{EN}(1, \ldots, N)$ is unity, i.e.,

$$\int d^{3N}r\, \Psi_{EN}{}^*(1, \ldots, N)\Psi_{EN}(1, \ldots, N) = 1 \tag{A.54}$$

Proof. Let

$$\begin{aligned} I &\equiv \int d^{3N}r\, \Psi_{EN}{}^*(1, \ldots, N)\Psi_{EN}(1, \ldots, N) \\ &= \frac{1}{N!}\int d^{3N}r \langle \Psi_{EN} \,|\, \psi^\dagger(N)\cdots\psi^\dagger(1) \,|\, 0\rangle\langle 0 \,|\, \psi(1)\cdots\psi(N) \,|\, \Psi\rangle \end{aligned}$$

By (A.52) we can write

$$\begin{aligned} I &= \frac{1}{N!}\int d^{3N}r \sum_n \langle \Psi_{EN} \,|\, \psi^\dagger(N)\cdots\psi^\dagger(1) \,|\, \Phi_n\rangle\langle \Phi_n \,|\, \psi(1)\cdots\psi(N) \,|\, \Psi_{EN}\rangle \\ &= \frac{1}{N!}\int d^{3N}r \langle \Psi_{EN} \,|\, [\psi^\dagger(N)\cdots\psi^\dagger(1)][\psi(1)\cdots\psi(N)] \,|\, \Psi_{EN}\rangle \end{aligned}$$

Now carry out the integration over $\mathbf{r}_1$. The relevant factor is

$$\int d^3r_1\, \psi^\dagger(1)\psi(1) = N_{\rm op}$$

Next carry out the integration over $\mathbf{r}_2$. The relevant factor is

$$\int d^3r_2\, \psi^\dagger(2)N_{\rm op}\psi(2) = N_{\rm op}(N_{\rm op} - 1)$$

By induction we can show that

$$I = \frac{1}{N!}\langle \Psi_{EN} \,|\, N_{\rm op}(N_{\rm op} - 1)(N_{\rm op} - 2)\cdots 1 \,|\, \Psi_{EN}\rangle = 1 \quad \text{(QED)}$$

The connection between the quantized field and an N-body system is furnished by the following theorem.

THEOREM

$$\left(-\frac{\hbar^2}{2m}\sum_{j=1}^{N}\nabla_j^{\,2}+\sum_{i<j}v_{ij}\right)\Psi_{EN}(1,\ldots,N)=E\Psi_{EN}(1,\ldots,N) \quad \text{(A.55)}$$

Proof. By (A.47) and (A.53)

$$\frac{1}{\sqrt{N!}}\langle 0\mid[\psi(1)\cdots\psi(N)]H\mid\Psi_{EN}\rangle=E\Psi_{EN}(1,\ldots,N) \quad \text{(A.56)}$$

Since $H\mid 0\rangle=0$, and H is hermitian, we also have $\langle 0\mid H=0$. Hence the left-hand side of (A.56) has the form of a commutator:

$$\begin{aligned}
J &\equiv \frac{1}{\sqrt{N!}}\langle 0\mid[\psi(1)\cdots\psi(N)]H\mid\Psi_{EN}\rangle\\
&=\frac{1}{\sqrt{N!}}\langle 0\mid[\psi(1)\cdots\psi(N),H]\mid\Psi_{EN}\rangle\\
&=\frac{1}{\sqrt{N!}}\sum_{j=1}^{N}\langle 0\mid\psi(1)\cdots[\psi(j),H]\cdots\psi(N)\mid\Psi_{EN}\rangle \qquad \text{(A.57)}
\end{aligned}$$

where the last step is obtained through repeated use of the identity

$$[AB,C]=[A,C]B+A[B,C]$$

We explicitly calculate $[\psi(j),H]$. From (A.44) we have

$$[\psi(i),H]=[\psi(j),K]+[\psi(j),\Omega]$$

For Bosons

$$\begin{aligned}
[\psi(j),K] &= -\frac{\hbar^2}{2m}\int d^3r[\psi(j),\psi^\dagger(\mathbf{r})\nabla^2\psi(\mathbf{r})]\\
&= -\frac{\hbar^2}{2m}\int d^3r[\psi(j),\psi^\dagger(\mathbf{r})]\nabla^2\psi(\mathbf{r})\\
&= -\frac{\hbar^2}{2m}\nabla_j^{\,2}\psi(j)
\end{aligned}$$

$$\begin{aligned}
[\psi(j),\Omega] &= \tfrac{1}{2}\int d^3r_1\,d^3r_2[\psi(j),\psi^\dagger(1)\psi^\dagger(2)]v_{12}\psi(2)\psi(1)\\
&= \tfrac{1}{2}\int d^3r_1\,d^3r_2\,\{[\psi(j),\psi^\dagger(1)]\psi^\dagger(2)+\psi^\dagger(1)[\psi(j),\psi^\dagger(2)]\}v_{12}\psi(2)\psi(1)\\
&= \left[\int d^3r\,\psi^\dagger(\mathbf{r})v(\mathbf{r},\mathbf{r}_j)\psi(\mathbf{r})\right]\psi(j)
\end{aligned}$$

For Fermions

$$[\psi(j), K] = -\frac{\hbar^2}{2m}\int d^3r[\psi(j), \psi^\dagger(\mathbf{r})\nabla^2\psi(\mathbf{r})]$$
$$= -\frac{\hbar^2}{2m}\int d^3r\{\psi(j), \psi^\dagger(\mathbf{r})\}\nabla^2\psi(\mathbf{r})$$
$$= -\frac{\hbar^2}{2m}\nabla_j^{\,2}\psi(j)$$

$$[\psi(j), \Omega] = \tfrac{1}{2}\int d^3r_1\, d^3r_2[\psi(j), \psi^\dagger(1)\psi^\dagger(2)]v_{12}\psi(2)\psi(1)$$
$$= \tfrac{1}{2}\int d^3r_1\, d^3r_2[\{\psi(j), \psi^\dagger(1)\}\psi^\dagger(2) - \psi^\dagger(1)\{\psi(j), \psi^\dagger(2)\}]v_{12}\psi(2)\psi(1)$$
$$= \left[\int d^3r\ \psi^\dagger(\mathbf{r})v(\mathbf{r}, \mathbf{r}_j)\psi(\mathbf{r})\right]\psi(j)$$

Hence for both bosons and fermions we have

$$[\psi(j), H] = \left[-\frac{\hbar^2}{2m}\nabla_j^{\,2} + X(j)\right]\psi(j) \tag{A.58}$$

where

$$X(j) = \int d^3r\ \psi^\dagger(\mathbf{r})v(\mathbf{r}, \mathbf{r}_j)\psi(\mathbf{r}) \tag{A.59}$$

The following properties of $X(j)$ are trivial:

$$[\psi(i), X(j)] = v_{ij}\psi(i)\psi(j) \tag{A.60}$$

$$X(j)\,|\,0\rangle = 0, \qquad \langle 0\,|\,X(j) = 0 \tag{A.61}$$

Substitution of (A.59) into (A.57) yields

$$J = -\frac{\hbar^2}{2m}\sum_{j=1}^{N}\nabla_j^{\,2}\Psi_{EN}(1, \ldots, N)$$
$$+\frac{1}{\sqrt{N!}}\sum_{j=1}^{N}\langle 0\,|\,\psi(1)\cdots\psi(j-1)X(j)\psi(j)\cdots\psi(N)\,|\,\Psi_{EN}\rangle \tag{A.62}$$

We now commute $X(j)$ all the way to the left with the help of (A.60):

$$[\psi(1)\cdots\psi(j-1)X(j)\psi(j)\cdots\psi(N)]$$
$$= [\psi(1)\cdots\psi(j-2)X(j)\psi(j-1)\cdots\psi(N)] + v_{j-1,j}[\psi(1)\cdots\psi(N)]$$
$$= [\psi(1)\cdots\psi(j-3)X(j)\psi(j-2)\cdots\psi(N)] + (v_{j-2,j} + v_{j-1,j})[\psi(1)\cdots\psi(N)]$$
$$= \cdots$$
$$= \left[X(j) + \sum_{i=1}^{j-1}v_{ij}\right][\psi(1)\cdots\psi(N)] \tag{A.63}$$

Substituting this into (A.62) and using (A.61) we obtain

$$J = \left[-\frac{\hbar^2}{2m} \sum_{j=1}^{N} \nabla_j^2 + \sum_{i<j} v_{ij} \right] \Psi_{EN}(1, \ldots, N) \qquad \text{(QED)}$$

For convenience in actual calculations, let us introduce a complete orthonormal set of single-particle wave functions $\{u_\alpha(\mathbf{r})\}$ with

$$\int d^3r\, u_\alpha^*(\mathbf{r}) u_\beta(\mathbf{r}) = \delta_{\alpha\beta} \tag{A.64}$$

Then we may expand the field operators $\psi(\mathbf{r})$ and $\psi^\dagger(\mathbf{r})$ in the following manner:

$$\begin{aligned} \psi(\mathbf{r}) &= \sum_\alpha a_\alpha u_\alpha(\mathbf{r}) \\ \psi^\dagger(\mathbf{r}) &= \sum_\alpha a_\alpha^\dagger u_\alpha^*(\mathbf{r}) \end{aligned} \tag{A.65}$$

where, in accordance with (A.43), a_α and $a_\alpha^\dagger$ are operators satisfying the following commutation rules:

Bosons	Fermions
$[a_\alpha, a_\beta^\dagger] = \delta_{\alpha\beta}$	$\{a_\alpha, a_\beta^\dagger\} = \delta_{\alpha\beta}$
$[a_\alpha, a_\beta] = 0$	$\{a_\alpha, a_\beta\} = 0$
$[a_\alpha^\dagger, a_\beta^\dagger] = 0$	$\{a_\alpha^\dagger, a_\beta^\dagger\} = 0$

(A.66)

It easily follows that the eigenvalues of $a_\alpha^\dagger a_\alpha$ are

$$n_\alpha \equiv a_\alpha^\dagger a_\alpha = \begin{cases} 0, 1, 2, \ldots & \text{(bosons)} \\ 0, 1 & \text{(fermions)} \end{cases} \tag{A.67}$$

In terms of a_α and $a_\alpha^\dagger$ we have

$$N_{\text{op}} = \sum_\alpha a_\alpha^\dagger a_\alpha \tag{A.68}$$

$$H = \frac{\hbar^2}{2m} \sum_{\alpha,\beta} \langle \alpha | -\nabla^2 | \beta \rangle a_\alpha^\dagger a_\beta + \tfrac{1}{2} \sum_{\alpha,\beta,\gamma,\lambda} \langle \alpha\beta | v | \gamma\lambda \rangle (a_\alpha a_\beta)^\dagger (a_\gamma a_\lambda) \tag{A.69}$$

where

$$\begin{aligned} \langle \alpha | -\nabla^2 | \beta \rangle &= -\int d^3r\, u_\alpha^* \nabla^2 u_\beta \\ \langle \alpha\beta | v | \gamma\lambda \rangle &= \int d^3r_1\, d^3r_2\, u_\alpha^*(1) u_\beta^*(2) v_{12} u_\gamma(1) u_\lambda(2) \end{aligned} \tag{A.70}$$

Let a set of integers $\{n_0, n_1, \ldots\}$ be given, such that each n_α is a possible value of $a_\alpha^\dagger a_\alpha$ as given by the rule (A.67). Define the state $| n \rangle$ by

$$| n \rangle \equiv | n_0, n_1, \ldots \rangle \equiv C_n [(a_0^\dagger)^{n_0} (a_1^\dagger)^{n_1} \cdots] | 0 \rangle \tag{A.71}$$

where C_n is a normalization constant so chosen that $\langle n \mid n \rangle = 1$:

$$C_n = \frac{1}{\sqrt{\prod_\alpha (n_\alpha !)}} \tag{A.72}$$

It can be verified that

For Bosons

$$\begin{aligned} a_\alpha \mid \ldots, n_\alpha, \ldots \rangle &= \sqrt{n_\alpha} \mid \ldots, n_\alpha - 1, \ldots \rangle \\ a_\alpha^\dagger \mid \ldots, n_\alpha, \ldots \rangle &= \sqrt{n_\alpha + 1} \mid \ldots, n_\alpha + 1, \ldots \rangle \end{aligned} \tag{A.73}$$

*For Fermions**

$$\begin{aligned} a_\alpha \mid \ldots, n_\alpha, \ldots \rangle &= \xi_\alpha \sqrt{n_\alpha} \mid \ldots, n_\alpha - 1, \ldots \rangle \\ a_\alpha^\dagger \mid \ldots, n_\alpha, \ldots \rangle &= \xi_\alpha \sqrt{n_\alpha + 1} \mid \ldots, n_\alpha + 1, \ldots \rangle \end{aligned} \tag{A.74}$$

where $\xi_\alpha = \pm 1$ according to whether $\sum_{\beta<\alpha} n_\beta$ is an even or odd integer. The operator a_α is called the annihilation operator for the single-particle state α, and the operator $a_\alpha^\dagger$ is called the creation operator for the single-particle state α.

For both bosons and fermions we have

$$a_\alpha^\dagger a_\alpha \mid n \rangle = n_\alpha \mid n \rangle \tag{A.75}$$

Therefore

$$N_{\text{op}} \mid n \rangle = \left(\sum_\alpha n_\alpha \right) \mid n \rangle \tag{A.76}$$

By the use of (A.73) and (A.74) any matrix element $\langle n \mid H \mid n' \rangle$ can be obtained trivially from (A.69).

By (A.55) and (A.54) the complete set of wave functions Φ_n defined in (A.14) can also be represented in the form

$$\frac{1}{\sqrt{\prod_\alpha (n_\alpha !)}} \Phi_n(1, \ldots, n) = \frac{1}{\sqrt{N!}} \langle 0 \mid \psi(1) \ldots \psi(N) \mid n \rangle \tag{A.77}$$

Therefore

$$\left(\Phi_{n}, \left[-\frac{\hbar^2}{2m} \sum_{j=1}^{N} \nabla_j^2 + \sum_{i<j} v_{ij} \right] \Phi_{n'} \right) = \langle n \mid H \mid n' \rangle \tag{A.78}$$

In particular, the results (A.33) and (A.38) can be trivially obtained through the use of this relation.

* Note that for fermions $\mid n \rangle \equiv 0$ if any $n_\alpha > 1$.

appendix B

THE PSEUDOPOTENTIAL

We consider the two-body problem for which the Schrödinger equation in the center-of-mass system is

$$\frac{\hbar^2}{2\mu}(\nabla^2 + k^2)\psi(\mathbf{r}) = v(r)\psi(\mathbf{r}) \tag{B.1}$$

where $v(r)$ is a finite-ranged central potential with no bound state. We wish to find the Schrödinger equation that contains a potential which is nonvanishing only at $\mathbf{r} = 0$ and that has the same eigenvalues k as (B.1).

By assumption $v(r) \to 0$ sufficiently rapidly as $r \to \infty$, so that

$$\psi(\mathbf{r}) \xrightarrow[r\to\infty]{} \psi_\infty(\mathbf{r})$$

where $\psi_\infty(\mathbf{r})$, the asymptotic wave function, is a solution of the equation

$$(\nabla^2 + k^2)\psi_\infty(\mathbf{r}) = 0 \qquad (r > 0)$$

Thus we can write, for $r > 0$,

$$\psi_\infty(\mathbf{r}) = \sum_{l=0}^{\infty} \sum_{m=-l}^{+l} Y_{lm}(\theta, \phi)\psi_{lm}(kr) \tag{B.2}$$

where $Y_{lm}(\theta, \phi)$ is a normalized spherical harmonic and

$$\psi_{lm}(kr) = A_{lm}[j_l(kr) - n_l(kr)\tan\eta_l] \tag{B.3}$$

The constants A_{lm} depend on the boundary conditions at large r, and the functions $j_l(x)$, $n_l(x)$ are the usual spherical Bessel functions.* The number $\tan \eta_l$, the scattering phase shift of the lth partial wave, is a function of k. It depends on the form of the potential $v(r)$ and is assumed to be known. If $v(r)$ is the hard-sphere potential of diameter a, then

$$\tan \eta_l = \frac{j_l(ka)}{n_l(ka)} \qquad \text{(hard-sphere potential)} \tag{B.4}$$

Obviously the number k appearing in (B.2) is the same as the number k appearing in (B.1). To find the possible values of k under a given boundary condition for (B.1), we proceed as follows. We define $\psi_\infty(\mathbf{r})$ for *all* $\mathbf{r}$ by (B.2), find the equation that $\psi_\infty(\mathbf{r})$ satisfies *everywhere*, and solve it under the same boundary condition as for (B.1).

To find the equation that $\psi_\infty(\mathbf{r})$ satisfies everywhere, let us calculate the quantity $(\nabla^2 + k^2)(Y_{lm}\psi_{lm})$. It is clear that $(\nabla^2 + k^2)(Y_{lm}\psi_{lm}) = 0$ everywhere except at $\mathbf{r} = 0$. Thus it is sufficient to consider the behavior of ψ_{lm} near $\mathbf{r} = 0$. Using the well-known asymptotic formulas

$$\begin{aligned} j_l(x) &\xrightarrow[x\to 0]{} \frac{x^l}{(2l+1)!!} \\ n_l(x) &\xrightarrow[x\to 0]{} -\frac{(2l-1)!!}{x^{l+1}} \end{aligned} \tag{B.5}$$

where $(2l + 1)!! \equiv 1 \cdot 3 \cdot 5 \cdots (2l + 1)$, we find that

$$\psi_{lm}(kr) \xrightarrow[r\to 0]{} r^l B_{lm}\left\{1 + (2l+1)[(2l-1)!!]^2 \frac{\tan \eta_l}{(kr)^{2l+1}}\right\} \tag{B.6}$$

where

$$B_{lm} = \frac{A_{lm}k^l}{(2l+1)!!} \tag{B.7}$$

We may express B_{lm} in terms of ψ_{lm} as follows:

$$B_{lm} = \frac{1}{(2l+1)!}\left[\left(\frac{d}{dr}\right)^{2l+1}(r^{l+1}\psi_{lm})\right]_{r=0} \tag{B.8}$$

Thus A_{lm}, which depends on the boundary condition at large r, need not be explicitly mentioned.

We notice immediately that

$$(\nabla^2 + k^2)[Y_{lm}j_l(kr)] = 0 \qquad \text{(everywhere)} \tag{B.9}$$

* See, e.g., L. I. Schiff, *Quantum Mechanics*, 2nd ed. (McGraw-Hill, New York, 1955), p. 77.

Thus it is sufficient to calculate the quantity

$$Y_{lm}F_l(r) \equiv (\nabla^2 + k^2)[Y_{lm}n_l(kr)] \tag{B.10}$$

for

$$(\nabla^2 + k^2)\psi_\infty(\mathbf{r}) = -\sum_{l=0}^{\infty}\sum_{m=-l}^{+l} A_{lm} \tan\eta_l Y_{lm}(\theta, \phi)F_l(r) \tag{B.11}$$

It is easily shown that

$$F_l(r) = \left[\frac{1}{r^2}\frac{d}{dr}r^2\frac{d}{dr} + k^2 - \frac{l(l+1)}{r^2}\right]n_l(kr) \tag{B.12}$$

which is zero except at $r = 0$. Multiplying $F_l(r)$ by r^l and integrating the result over a small sphere of infinitesimal radius ϵ about the origin, we find that as $\epsilon \to 0$

$$\begin{aligned}\int d^3r\, r^l F_l(r) &= \int d^3r\, r^l\left[\frac{1}{r^2}\frac{d}{dr}r^2\frac{d}{dr}n_l - \frac{l(l+1)}{r^2}n_l\right] \\ &= \int d^3r\, r^l\left[\nabla^2 n_l - \frac{l(l+1)}{r^2}n_l\right] = \int d\mathbf{S}\cdot(r^l\nabla n_l) \\ &= (4\pi\epsilon^2)\epsilon^l\frac{(2l-1)!!\,(l+1)}{k^{l+1}\epsilon^{l+2}} = \frac{4\pi(2l-1)!!\,(l+1)}{k^{l+1}}\end{aligned} \tag{B.13}$$

Hence we conclude that

$$F_l(r) = \frac{(2l-1)!!\,(l+1)}{k^{l+1}}\frac{\delta(r)}{r^{l+2}} \tag{B.14}$$

Therefore

$$(\nabla^2 + k^2)\psi_\infty(\mathbf{r}) = -\frac{4\pi}{k\cot\eta_0}\delta(\mathbf{r})\frac{\partial}{\partial r}(r\psi_\infty) + \sum_{l=1}^{\infty}\sum_{m=-l}^{+l} f_l(k)Y_{lm}\frac{\delta(r)}{r^{l+2}} \tag{B.15}$$

where

$$f_l(k) \equiv -\frac{1}{k^{2l+1}\cot\eta_l}\cdot\frac{(2l-1)!!\,(l+1)}{l!\,2^l}\left\{\left(\frac{d}{dr}\right)^{2l+1}\left[r^{l+1}\int d\Omega\, Y_{lm}\psi_\infty(\mathbf{r})\right]\right\}_{r=0} \tag{B.16}$$

in which $\int d\Omega$ denotes the integration over all the angles of $\mathbf{r}$ in a spherical coordinate system. The right-hand side of (B.15) contains the exact pseudopotential.

In general a solution to (B.15) is not unique. It becomes unique, however, if we specify that $\psi_\infty(\mathbf{r})$ must *everywhere* reduce to a free-particle wave function when all phase shifts are put equal to 0. This condition is that $v(r)$ has no bound state.

appendix C

THE THEOREMS OF YANG AND LEE

C.1 TWO LEMMAS

In this appendix the system considered is the same as that in Chapter 15, as is the notation, unless otherwise explained.

Let $Q_N(V)$ be the classical partition function for the system. The volume V is assumed to be a union of cubes. Accordingly, the volume V can be covered with a number γ of elementary cubes of the same size. Within each elementary cube we construct a smaller cube, called a *cell*, of which each face is a distance $r_0/2$ from the nearest face of the elementary cube that contains it. Thus the volume V contains γ cells, each separated by a distance r_0 from its nearest neighbors. Two particles lying in different cells do not interact with each other. The construction is schematically represented by Fig. C.1. The space between the cells shall be called the corridor. Let the volume of each cell be V_0 and let the volume of the corridor be V_c. Then

$$V_c = V - \gamma V_0 \tag{C.1}$$

Clearly*

$$V_c < c\gamma r_0 V_0^{2/3} \tag{C.2}$$

where c is a numerical constant, so that as $V_0 \to \infty$, $V_c/V \to 0$ and $V \to \gamma V_0$. We assume that N, V, and V_0 are always such that all the N particles can be placed in the cells with no two particles "touching."

* See the argument preceding (15.25)

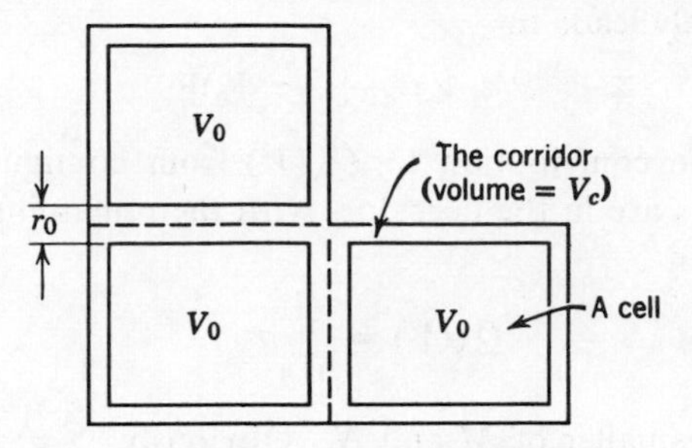

Fig. C.1. Cells in the volume V.

We define the comparison partition function $\tilde{Q}_N(V, V_0)$ to be the partition function of the system under the restriction that *no particle shall be in the corridor.*

If the grand partition function of the system in the volume V is $\mathscr{Q}(z, V)$, the corresponding comparison grand partition function is

$$\sum_{N=0}^{\infty} z^N \tilde{Q}_N(V, V_0) = [\mathscr{Q}(z, V_0)]^\gamma \tag{C.3}$$

We prove the following lemmas.

Lemma 1

$$[\mathscr{Q}(z, V_0)]^\gamma \leq \mathscr{Q}(z, V) \leq [\mathscr{Q}(z, V_0)]^\gamma e^{z\sigma' M}$$

where σ' is a finite constant independent of z, V, and V_0; and M is the maximum number of particles in the corridor:

$$M = \frac{V_c}{\frac{4}{3}\pi a^3} < \text{const.} \frac{r_0}{a^3} \gamma V_0^{2/3} \tag{C.4}$$

Lemma 2

$$\tilde{Q}_N(V, V_0) \leq Q_N(V) \leq \tilde{Q}_N(V, V_0) e^{\sigma M}$$

where σ is a constant.

Proof of Lemmas. By definition

$$Q_N(V) = \frac{1}{N!\,\lambda^{3N}} \int_V d^{3N}r\, e^{-\beta\Omega(1,\ldots,N)} \tag{C.5}$$

$$\tilde{Q}_N(V, V_0) = \frac{1}{N!\,\lambda^{3N}} \int_{\tilde{V}} d^{3N}r\, e^{-\beta\Omega(1,\ldots,N)} \tag{C.6}$$

where the subscript $\tilde{V}$ indicates that each coordinate $\mathbf{r}_i$ is integrated over the interior volume of the γ cells. The integral in (C.6) is obtainable from that in (C.5) when we omit a certain region of integration. Since the integrand of (C.5) is non-negative, we have

$$Q_N(V) \geq \tilde{Q}_N(V, V_0) \tag{C.7}$$

which immediately leads to

$$\mathscr{Q}(z, V) \geq [\mathscr{Q}(z, V_0)]^{\gamma} \tag{C.8}$$

Let $q_{N,l}$ be the contribution to $Q_N(V)$ from configurations in which exactly l particles are in the corridor, with the remaining $N - l$ particles in the cells. Then

$$Q_N(V) = \sum_{l=0}^{M'} q_{N,l} \tag{C.9}$$

where M' is the smaller of M and N. Obviously $\sum_{N=l}^{\infty} z^N q_{N,l}$ is the contribution to $\mathscr{Q}(z, V)$ from configurations in which exactly l particles are in the corridor. Hence

$$\mathscr{Q}(z, V) = \sum_{l=0}^{M} \sum_{N=l}^{\infty} z^N q_{N,l} \tag{C.10}$$

Let y denote collectively the l coordinates that are restricted to lie in the corridor, and let x denote the remaining $N - l$ coordinates restricted to lie within the cells. The potential energy can, in an obvious fashion, be decomposed into three terms:

$$\Omega = \Omega_{xx} + \Omega_{yy} + \Omega_{xy}$$

Hence

$$q_{N,l} = \frac{1}{l!\,(N-l)!\,\lambda^{3N}} \int dx\, dy\, e^{-\beta\Omega_{xx}}\, e^{-\beta(\Omega_{yy}+\Omega_{xy})} \tag{C.11}$$

Since each particle can interact with at most $(r_0/a)^3$ others, and each interaction contributes to the integrand of (C.11) at most the factor $e^{\beta\epsilon}$, we obtain an upper bound for $q_{N,l}$ by replacing $\exp[-\beta(\Omega_{xx} + \Omega_{xy})]$ by $\exp[l\beta\epsilon(r_0/a)^3]$. Thus

$$q_{N,l} \leq \left[\frac{1}{(N-l)!\,\lambda^{3N-3l}} \int dx\, e^{-\beta\Omega_{xx}}\right] \frac{V_c^{\,l}\, e^{l\beta\epsilon(r_0/a)^3}}{l!\,\lambda^{3l}}$$

The first factor is none other than $\tilde{Q}_{N-l}(V, V_0)$. Hence

$$q_{N,l} \leq \frac{(\sigma' M)^l}{l!} \tilde{Q}_{N-l}(V, V_0) \tag{C.12}$$

where σ is a finite number:

$$\sigma' = 4\left(\frac{a}{\lambda}\right)^3 e^{\beta\epsilon(r_0/a)^3} \tag{C.13}$$

From (C.12) we obtain

$$\sum_{N=l}^{\infty} z^N q_{N,l} \leq \frac{(z\sigma' M)^l}{l!} \sum_{N=0}^{\infty} z^N \tilde{Q}_N(V, V_0) = \frac{(z\sigma' M)^l}{l!} [\mathscr{Q}(z, V_0)]^{\gamma}$$

Substituting this into (C.10), we obtain (noting that $z \geq 0$)

$$\mathcal{Q}(z, V) \leq [\mathcal{Q}(z, V_0)]^{\gamma} \sum_{l=0}^{M} \frac{(z\sigma' M)^l}{l!} \leq [\mathcal{Q}(z, V_0)]^{\gamma} e^{z\sigma' M} \tag{C.14}$$

Together with (C.18), this proves Lemma 1.

If we increase N by one in $\tilde{Q}_{N-1}$, the attractive interactions of the added particle can only increase $\tilde{Q}_{N-1}$. The total volume available to the particle is greater than a numerical multiple of Na^3. Thus it may be easily seen that

$$\tilde{Q}_N \geq \text{const.} \left(\frac{a}{\lambda}\right)^3 \tilde{Q}_{N-1}$$

Hence

$$\tilde{Q}_{N-l} \leq \left[\text{const.} \left(\frac{a}{\lambda}\right)^3\right]^l \tilde{Q}_N$$

Substituting this into (C.12), we find that

$$q_{N,l} \leq \frac{(\sigma M)^l}{l!} \tilde{Q}_N(V, V_0) \tag{C.15}$$

Substituting this into (C.9) we obtain

$$Q_N(V) \leq \tilde{Q}_N(V, V_0) \sum_{l=0}^{M} \frac{(\sigma M)^l}{l!} \leq \tilde{Q}_N(V, V_0) e^{\sigma M} \tag{C.16}$$

which, in conjunction with (C.7), leads to lemma 2.

C.2 THEOREM 1 OF YANG AND LEE

THEOREM 1. $\lim_{V \to \infty} [V^{-1} \log \mathcal{Q}(z, V)]$ exists for all $z > 0$. This limit is independent of the shape of the volume V and is a continuous, non-decreasing function of z.

It is assumed that as $V \to \infty$ the surface area of V increases no faster than $V^{2/3}$.

Proof. Suppose V is a union of cubes. Construct γ cells of volume V_0 within V in the manner described in the last section. We can obtain equally valid constructions by further constructing cells within cells, and cells within cells within cells, etc., as shown schematically in Fig. C.2. If we carry out this subdivision n times, we obtain a sequence of cells of smaller and smaller sizes. Let the sequence of cell volumes be denoted in the order of increasing size by $W_0, W_1, W_2, \ldots$. Let W_0 be so large that

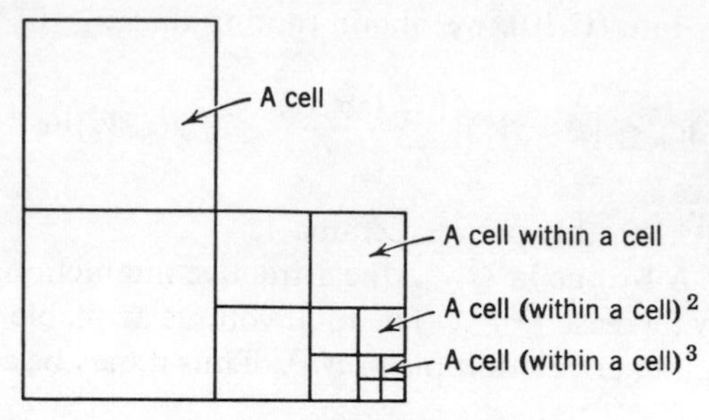

Fig. C.2. Cells within cells.

the volume of the corridor is always negligible compared to V. Then we may write

$$
\begin{aligned}
W_1 &= 8W_0 \\
W_2 &= 8^2 W_0 \\
&\cdot \\
&\cdot \\
&\cdot \\
W_n &= 8^n W_0 \equiv V \\
V &= \gamma V_0
\end{aligned}
\tag{C.17}
$$

For fixed W_0, the limit $V \to \infty$ may be approached by letting $n \to \infty$. Let

$$
F_V \equiv \frac{1}{V} \log \mathcal{Q}(z, V)
$$

$$
F_k \equiv \frac{1}{W_k} \log \mathcal{Q}(z, W_k) \qquad (k = 0, 1, \ldots, n) \tag{C.18}
$$

Then we have, by lemma 1,

$$
F_k \le F_{k+1} \le F_k + \frac{z\sigma M_k}{W_{k+1}} \qquad (k = 0, 1, \ldots, n) \tag{C.19}
$$

$$
F_n \le F_V \le F_n + \frac{z\sigma M}{V} \tag{C.20}
$$

where M is given by (C.4) and

$$
M_k < \frac{3r_0}{4\pi a^3} \, 8W_k^{2/3} \tag{C.21}
$$

Thus (C.19) and (C.20) can be respectively rewritten as

$$
|F_{k+1} - F_k| < \frac{c}{2^k W_0^{1/3}} \qquad (k = 0, 1, \ldots, n) \tag{C.22}
$$

$$
|F_V - F_n| < \frac{c'}{2^n W_0^{1/3}} \tag{C.23}
$$

where c and c' are finite constants. From (C.22) it can be easily shown that

$$|F_{k+p} - F_k| < \frac{c''}{2^k W_0^{1/3}} \tag{C.24}$$

where c'' is a finite constant. Hence by the Cauchy criterion of convergence* the sequence $F_0, F_1, \ldots$ converges to a limit for any fixed value of W_0. That is, $\lim_{n\to\infty} F_n$ exists. By (C.23),

$$\lim_{V\to\infty} F_V = \lim_{n\to\infty} F_n \tag{C.25}$$

Therefore $\lim_{V\to\infty} F_V$ exists. If V is not a union of cubes, we can place V between two volumes V_1, V_2, each of which is a union of cubes. By lemma 1, $\lim |F_{V1} - F_{V2}| = 0$ as $V \to \infty$. Hence $\lim F_V = \lim F_{V1}$, which exists and is independent of the shape of V.

The limit (C.25) is an increasing function of z because it holds for every value of V [see (15.8)]. That it is a continuous function of z follows from the fact that its derivative is bounded for every value of V:

$$z\frac{\partial}{\partial z}\left[\frac{1}{V}\log \mathcal{Q}(z, V)\right] \leq \frac{N_m}{V} \tag{C.26}$$

which follows from (15.9) and (15.11). This completes the proof of the theorem.

C.3 THEOREM 2 OF YANG AND LEE

THEOREM 2. Let R be a region in the complex z plane that contains a segment of the positive real axis and contains no root of the equation $\mathcal{Q}(z, V) = 0$ for any V. Then for all z in R the quantity $V^{-1}\log \mathcal{Q}(z, V)$ converges *uniformly* to a limit as $V \to \infty$. This limit is an analytic function of z, for all z in R.

Proof. Within the region R draw a circle D of finite radius, with center lying on the positive real axis. We first prove the theorem in D.

Let $F_V(z) = V^{-1}\log \mathcal{Q}(z, V)$. By (15.8),

$$F_V(z) = \frac{1}{V}\log\,[1 + zQ_1(V) + \cdots + z^{N_m}Q_{N_m}(V)]$$

We note the following properties:

(*a*) For any finite V, $F_V(z)$ is an analytic function in D, because $\mathcal{Q}(z, V)$ has no root in D.

* See E. T. Whittaker and G. N. Watson, *A Course of Modern Analysis* (Cambridge University Press, Cambridge, 1948), p. 13.

(*b*) By theorem 1, $\lim_{V\to\infty} F_V(z)$ exists for z lying along a continuous path in D.

(*c*) For any V, and for any z in D, $|F_V(z)|$ is bounded by a number independent of V and z. This can be seen as follows: Since $Q_N(V) \geq 0$,

$$|F_V(z)| \leq \frac{1}{V}\log\left[1 + |z|\, Q_1(V) + \cdots + |z|^{N_m} Q_{N_m}(V)\right]$$

In D we have $|z| < \sigma$, where σ is a real number. Hence

$$|F_V(z)| \leq \frac{1}{V}\log\left[1 + \sigma Q_1(V) + \cdots + \sigma^{N_m} Q_{N_m}(V)\right]$$

The right-hand side is bounded because it is bounded for any finite V and because by theorem 1 it approaches a limit as $V \to \infty$.

The desired theorem holds in the circle D, by virtue of Vitali's convergence theorem.*

Within the region R construct a circle D' whose center lies in D. Since we have proved the theorem for D, the properties (*a*), (*b*), and (*c*) now hold in D'. Thus the theorem holds in D'. By repeating this process, we extend the theorem to the entire region R. (QED)

* *Vitali's Convergence Theorem.* Let $f_n(z)$ be a sequence of functions, each regular in a region D; let $|f_n(z)| \leq M$ for every n and z in D; and let $f_n(z)$ tend to a limit, as $n \to \infty$, at a set of points having a limit point inside D. Then $f_n(z)$ tends uniformly to a limit in any region bounded by a contour interior to D, the limit being, therefore, an analytic function of z.

For a proof see E. C. Titchmarsh, *The Theory of Functions*, 2nd ed. (Oxford University Press, Oxford, 1939), p. 168.

INDEX